Lecture N
Mathemati

Edited by A. Dold and B. Eckmann

Subseries: USSR
Adviser: L. D. Faddeev, Leningrad

1043

Linear and Complex Analysis Problem Book

199 Research Problems

Edited by V. P. Havin, S. V. Hruščëv and N. K. Nikol'skii

Springer-Verlag
Berlin Heidelberg New York Tokyo 1984

Editors

Victor P. Havin
Leningrad State University
Stary Peterhof, 198904 Leningrad, USSR

Sergei V. Hruščëv
Nikolai K. Nikol'skii
Leningrad Branch of the V.A. Steklov Mathematical Institute
Fontanka 27, 191011 Leningrad, USSR

Scientific Secretary to the Editorial Board
V.I. Vasyunin

AMS Subject Classifications (1980): 30, 31, 32, 41, 42, 43, 46, 47, 60, 81

ISBN 3-540-12869-7 Springer-Verlag Berlin Heidelberg New York Tokyo
ISBN 0-387-12869-7 Springer-Verlag New York Heidelberg Berlin Tokyo

Library of Congress Cataloging in Publication Data. Main entry under title: Linear and complex analysis problem book. (Lecture notes in mathematics; 1043) 1. Mathematical analysis–Problems, exercises, etc. I. Khavin, Viktor Petrovich. II. Krushchev, S. V. III. Nikol'skii, N. K. (Nikolai Kapitonovich) IV. Series: Lecture notes in mathematics (Springer-Verlag; 1043) QA3.L28 no. 1043 [QA301] 510s [515'.076] 83-20344
ISBN 0-387-12869-7 (U.S.)

Printing and binding: Beltz Offsetdruck, Hemsbach/Bergstr.
2146/3140-543210

CONTENTS

PROBLEMS

SOLUTIONS

LIST OF PARTICIPANTS

ACKNOWLEDGEMENTS

This book was created by a very large body of mathematicians. We were in touch with more than 200 colleagues, and approximately twenty (mostly members of our seminar) assisted us in preparing this volume. Many of our correspondents will find their problems in the pages of the book, and - regardless of whether they supplied us with a mathematical text or with a criticism of our intentions - WE ARE GRATEFUL TO ALL WHOSE PARTICIPATION CONTRIBUTED TO THIS BOOK.

The goodwill and enthusiasm of many colleagues were crucial for our work. Pushing their own investigations aside they generously rendered us invaluable help - invaluable both in its amount and its skill. This help ranged from writing commentary to organizing the material, from critical analysis of problems to linguistic consultations, to preparing of a huge mass of references and - last but not least - to the technical scissors-and-glue toil (the proof-reading, removing misprints, compiling indexes etc.), duties which, we daresay, are rarely allotted to mathematicians of comparable qualifications.

We list below in deep gratitude and respect our "informal editorial board".

WRITING COMMENTARY

The following colleagues put at our disposal valuable and sometimes very detailed information, used extensively in our commentary:

V.M.Adamyan (Адамян) A.B.Aleksandrov (Александров)
L.A.Aizenberg (Айзенберг) D.Z.Arov (Аров)

V.S.Azarin (Азарин)

Ch.Berg

B.Bollobás

L. de Branges

P.G.Casazza

D.N.Clark

A.A.Gol'dberg (Гольдберг)

A.Ya.Gordon (Гордон)

E.A.Gorin (Горин)

L.I.Hedberg

P.P.Kargayev (Каргаев)

R.P.Kaufman

S.V.Kisliakov (Кисляков)

M.G.Krein (Крейн)

M.M.Malamud (Маламуд)

V.G.Maz'ya (Мазья)

I.V.Ostrovskii (Островский)

V.V.Peller (Пеллер)

D.E.Sarason

F.A.Shamoyan (Шамоян)

B.M.Solomyak (Соломяк)

V.A.Tkachenko (Ткаченко)

I.E.Verbitskii (Вербицкий)

A.M.Vershik (Вершик)

A.L.Vol'berg (Вольберг)

H.Wallin

V.P.Zaharyuta (Захарюта)

S.V.Znamenskii (Знаменский)

ADVICE AND MATHEMATICAL CONSULTATIONS

Helpful advice and consultations of the following colleagues have been used on many occasions:

B.I.Batikyan (Батикян)

M.S.Birman (Бирман)

A.G.Chernyavskii (Чернявский)

E.M.Dyn'kin (Дынькин)

Yu.B.Farforovskaya (Фарфоровская)

A.B.Gulisashvili (Гулисашвили)

G.M.Henkin (Хенкин)

S.V.Kisliakov (Кисляков)

P.Koosis

Lee Lorch

A.S.Markus (Маркус)

V.P.Palamodov (Паламодов)

V.V.Peller (Пеллер)

V.A.Tolokonnikov (Толоконников)

J.Zemánek

LINGUISTIC AID

We thank J.Brennan, E.M.Dyn'kin (Дынькин), A.B.Gulisashvili (Гулисашвили), B.Jöricke, S.V.Kisliakov (Кисляков), N.G.Makarov

(Макаров), V.V.Peller (Пеллер), B.M.Solomyak (Соломяк) and A.L.Vol'-berg (Вольберг)who helped us to translate about 100 problems from Russian into English. We had also to write commentary and introductions in English, and A.B.Gulisashvili, S.V.Kisliakov and V.V.Peller participated in solving many linguistic problems. P.Koosis, Lee Lorch, S.C.Power and J.A.Siddiqi checked some parts of the text, and Lee Lorch helped to translate the preface from Russian English into English.

REFERENCES CONTROL AND INDEXES:

These were prepared by L.N.Dovbysh (Довбыш) and V.V.Peller (Пеллер).

PROOF-READING AND CORRECTION OF MISPRINTS

This task was allotted to L.N.Dovbysh (Довбыш), A.B.Gulisashvili (Гулисашвили), S.V.Kisliakov (Кисляков), V.V.Peller (Пеллер).

EDITORS

PREFACE

This volume offers a collection of problems concerning analytic functions (mainly of one complex variable), linear function spaces and linear operators.

The most exciting challenge to a mathematician is usually not what he understands, but what still eludes him. This book reports what eluded a rather large group of analysts in 1983 whose interests have a large overlap with those of our Seminar*). Consequently, therefore, the materials contained herein are chosen for some sort of mild homogeneity, and are not at all encyclopaedic. Thus, this volume differs markedly from some well-known publications which aim at universality. We confine ourselves to the (not very wide) area of Analysis in which we work, and try - within this framework - to make our collection as representative as possible. However, we confess to obeying the Bradford law (the exponential increase of difficulties in obtaining complete information). One of our purposes is to publish these problems promptly, before they lose the flavour of topicality or are solved by their proposers or other colleagues.

This Problem Book evolved from the earlier version published as volume 81 of "Zapiski Nauchnyh Seminarov LOMI" in 1978 (by the way, much of the work arising from the above mentioned Seminar is regularly published in this journal). It is now twice the size, reflecting the current interests of a far wider circle of mathematicians. For

---

*) i.e., the Seminar on Spectral Theory and Complex Analysis consisting principally of mathematicians working in the Leningrad Branch of the V.A.Steklov Mathematical Institute (LOMI) and in Leningrad University.

five years now the field of interests of the "invisible community" of analysts we belong to has enlarged and these interests have drifted towards a more intense mixing of Spectral Theory with Function Theory. And the volume as a whole is a rather accurate reflection of this process (see especially Chapters 4-7 below).

We are pleased that almost a half of the problems recorded in the first edition, 50 of 99, have been solved, partly or completely. This book contains a l l the problems of 1978 (we call them "old" problems). They are sometimes accompanied with commentary reporting what progress towards their solution has come to our attention. Moreover, those "old" problems which have been c o m p l e t e l y solved are assembled under the title "SOLUTIONS" at the end of the book (including information as to how and by whom they have been solved).

When we decided to prepare this new edition we solicited the cooperation of many colleagues throughout the world. Some two hundred responded with ample and helpful materials, doubling the number of collaborators of the first edition. Their contributions ranged from carefully composed articles (not always short) to brief remarks. This flow it was our task to organize and to compress into the confines of a single volume. To effectuate this we saw no alternative to making extensive revisions (more exactly, abbreviations) in the texts supplied. We hope that we have succeeded in preserving the essential features of all contributions and have done no injustice to any.

At first sight the problems may appear very heterogeneous. But they display a certain intrinsic unity, and their approximate classification (i.e. division into chapters) did not give us much trouble. We say "approximate" because every real manifestation of life resists systematization. Some problems did not fit into our initial outline and so some very interesting ones are collected under the title "Miscellaneous Problems" as Chapter 13. We took the liberty to provide almost all chapters with introductions. In these introductions we try to help the reader to grasp quickly the main point of the chapter, to record additional bibliography, and sometimes also to explain our point of view on the subject or to make historical comments.

Chapters are divided into sections. They total 199 (in 1978 there were 99). We treat the words "section" and "problem" as synonymous for the purposes of classification (though a section may contain more than one problem). "Problem 1.25" means the 25-th section of the first chapter; "Problem 1.26 old" means that Problem 1.26 is reproduced from the first edition and has not been completely solved (as far as we know); "Problem S.27" means the 27-th section of "SOLUTIONS". Problems accompanied by commentary are designated in the table of contents by the letter "c". Some notation (used sometimes without further explanations) is indicated at the end of the book. A subject index and an author index are provided.

EDITORS

CHAPTER 1

ANALYSIS IN FUNCTIONAL SPACES

Views on the place of Banach Functional Analysis in Analysis as a whole have undergone many changes during its relatively short history. Early successes had produced the extreme (and incorrect) opinion that this branch would eventually absorb all (or at least almost all) of Analysis and that every concrete analytic problem could be solved just by inventing an appropriate abstract Banach framework. Later, when the fundamental ideas of the theory of Banach Spaces became commonplace (as normally happens with all really important ideas) another extreme view emerged. According to it, the mathematical activity of Banach space theorists was doomed to mere technical details and insignificant variants. The "golden age" of the theories of Linear Topological Spaces and Distributions in the fifties and early sixties also contributed to shunting the Banach Space theory aside.

But developments of the last two decades have shown that the second extremism was also unjustified. The most significant results in Linear Analysis obtained in this period (and in particular those connected with "Hard Analysis") are undoubtedly of a Banach-theoretic nature. Avoiding another commonplace, i.e. a mere statement of mutual benefits brought by the interplay of "Concrete" (or "Classical", or "Hard") and "Abstract" branches of Analysis, we review briefly some important features of the present situation reflected in this Chapter.

One of them is the remarkably increased interest in c o n c r e t e   f u n c t i o n   s p a c e s , and not only in traditional ones, like $L^p(\mu)$ or $C(K)$, but in many others (various spaces of smooth and analytic functions, of Fourier and power series, BMO etc.). Some delicate invariants (discovered relatively recently) of a Banach space X , namely, properties of special classes of operators (X-valued or defined on X), play a prominent role today. Classes of p-absolutely summing operators and other analogous classes may serve as examples. We indicate in this connection the Problem of A.Pełczyński whose solution required ingenious analytical tools and culminated in interesting new results of a classical character (see S.1). In the present Chapter it is replaced (in a sense) by Problem 1.1. It fits into the same circle of ideas but deals with the space U of uniformly convergent Fourier series (instead of the disc-algebra). The "abstract" theory of p-summing operators is presented in Problems 1.2 and 1.3.

Another important feature of modern Banach-theoretic investigations is the special attention paid to the q u a n t i t a t i v e   r e f i n e m e n t s of qualitative results concerning concrete spaces. These refinements are often based on estimates of certain quantities associated with finite dimensional subspaces of a space (the so-called "local theory" of Banach spaces). This tendency is well illustrated by Problem 1.4 (now almost completely solved although it arises in new forms in the context of other spaces, e.g. $C^n(\mathbb{T})$ or U).

It is quite usual nowadays for "concrete" function-theoretic problems to appear in connection with general ideas of "abstract" Linear Analysis. So, for instance, general problems of Banach Geometry read in the context of a concrete function space become fascinating questions on "individual" functions: for example, problems on the description of extreme points of balls like in Problem 1.6 or on complemented subspaces like in Problem 1.5, the isomorphic classification of spaces resulting in a deeper comprehension of properties of functions forming the space under consideration (e.g. Problems 1.10 and 1.11 related to non-normed spaces). The same can be said for the problems of finding a basis or investigating the approximation property in concrete (Banach and non-Banach) spaces (Problems 1.7, 1.8 and 1.12) and of describing dual spaces (Problems 1.13, 1.14 and S.2). Even more traditional kinds of spaces (say measurable function spaces) still supply interesting unsolved problems (see e.g. Problem 1.9).

1.1. SOME QUESTIONS ON THE STRUCTURE OF THE SPACE OF UNIFORMLY CONVERGENT FOURIER SERIES

$A = A(\mathbb{D})$ denotes the disc algebra and

$$U = U(\mathbb{T}) = \{f \in A : f \quad \text{is uniform limit of} \quad f * D_n\},$$

$$D_n = \sum_{0 \leq k \leq n} e^{ik\theta}.$$

The norm on U is given by $\|f\|_U = \sup_n \|f * D_n\|_\infty$.
Various analogies between the spaces A and U are known now. It was shown by D.Oberlin [3] that measure-zero compact subsets of $\mathbb{T}$ are peak-interpolation sets for U, an improvement of the Rudin-Carleson theorem. Using related techniques, I obtained the following

PROPOSITION 1: <u>Let K be a compact subset of $\mathbb{T}$ and $\varepsilon > 0$. Then there exists $f \in U$ satisfying</u>

(i) $\|f\|_U \leq C$

(ii) $|f| < \varepsilon$ on K

(iii) $\|1 - f\|_2 < C(\log \frac{1}{\varepsilon})|K|^{1/2}$,

<u>where C is a constant.</u>

Fixing a finite sequence $g_1, \dots, g_r$ in U, $\|g_s\|_U \leq M$ $(1 \leq s \leq r)$, condition (i) can be strenghtened by requiring in addition

$$\|f \cdot g_s\|_U \leq C(r, M) \qquad (1 \leq s \leq r).$$

Here are some corollaries of Prop.1 for the Banach space theory of U (see[2]).

PROPOSITION 2: <u>The dual space U^* of U is weakly complete. In fact, bounded sequences in U have either a w^*-complemented ℓ^1-subsequence or a weakly convergent subsequence.</u>

2. <u>Reflexive subspaces of U^* are isomorphic to subspaces of L^1. In particular, they are of cotype 2.</u>

Recall that a normed space X has cotype $q \geq 2$ provided following inequality holds for all finite sequences $\{x_i\}$ in X :

$$\int \|\sum \varepsilon_i x_i\| d\varepsilon \geqslant \delta (\sum \|x_i\|^q)^{1/q},$$

where δ is a fixed constant, and (ε_i) is the usual Rademacher sequence.

No results seem to be known as far as the finite dimensional properties of U , U^* are concerned. In particular, the following problems can be posed.

PROBLEM 1: <u>Does</u> U^* <u>have any cotype</u> $q<\infty$? <u>Is</u> U <u>of cotype</u> 2?

PROBLEM 2: <u>Assume</u> E <u>a</u> λ <u>-complemented subspace of</u> U , <u>of dimension</u> n . <u>Is it true that</u> E <u>contains</u> ℓ^∞_m<u>-subspaces for</u> $m \sim n$? <u>How well can</u> E <u>be embedded as a complemented subspace of</u> ℓ^∞ ?

These questions are solved for the disc algebra A (see [1]). Their solution for the space U probably requires different techniques.

REFERENCES

1. B o u r g a i n J. New Banach space properties of the disc algebra and H^∞, to appear in Acta Math.
2. B o u r g a i n J. Quelques propriétés linéaires topologiques de l'espace des séries de Fourier uniformément convergentes. - C.R.A.S. Paris, 1982, 295, Sér.1, 623-625.
3. O b e r l i n D.M. A Rudin-Carleson theorem for uniformly convergent Taylor series. - Michigan Math. J., 1980, 27, N 3, 309-314.

J.BOURGAIN

Vrije Universiteit Brussel
Dept, Mathematics
Pleinlaan 2
1050 Brussel
Belgium

1.2. COMPACTNESS OF ABSOLUTELY SUMMING OPERATORS

Every concrete absolutely summing surjection with an infinite dimensional range allows to prove the classical Grothendieck theorem on absolutely summing operators from ℓ^1 into ℓ^2. On the other hand, given a Banach space X so that every absolutely summing operator from every ultrapower of X to ℓ^2 is compact, one can consider new local characteristics of operators on X (e.g. in spirit of Problem $3'$ below).

Our knowledge of what concerns the existence of non-compact absolutely summing operators with a given domain space is however less than satisfactory.

PROBLEM 1. Let X be a Banach space. Are the following conditions equivalent:

(a) there is an absolutely summing non-compact operator from X into a Hilbert space ℓ^2;

(b) there is an absolutely summing surjection from X onto ℓ^2?

Observe that if one replaces in (a) and (b) "absolutely summing" by "2-absolutely summing" then the "new (a)" is equivalent to the "new(b)" and is equivalent to the fact that X contains an isomorph of ℓ^1 (cf.[1],[2]).

An obvious example of a space satisfying (a) and (b) is any $\mathcal{L}^1$-space of infinite dimension (by the Grothendieck theorem). Another example is the disc algebra A. A well-known example of a 1-absolutely summing surjection from A onto ℓ^2 is the so called "Paley projection"

$$f \longmapsto \left(\frac{1}{2\pi}\int_0^{2\pi} e^{-i2^n t} f(t)\,dt\right)_{1\leqslant n\leqslant\infty}$$

cf.[3].

PROBLEM 2. Are the following conditions equivalent:

(a) every absolutely summing operator from X to ℓ^2 factors through a Hilbert-Schmidt operator;

(b) X^{**} is isomorphic to a quotient of a $C(K)$-space?

It is a well-known consequence of the classical result of Grotendieck that (b) implies (a).

Observe that if ℓ^2 in (a) is replaced by ℓ^1 then the modified property (a) is equivalent to (b), cf. J.Bourgain and A.Pełczyński (in preparation).

Every $\mathcal{L}^\infty$-space satisfies (a). The $\mathcal{L}^\infty$ -space constructed by Bourgain [4] which does not contain C_0 is not isomorphic to any quotient of a $C(K)$-space. So „X^{**}" in (b) can not be replaced by „X".

Let $L(\ell^2)$ (respectively $K(\ell^2)$) stand for the spaces of all bounded operators (respectively all compact operators) from ℓ^2 into itself.

PROBLEM 3. Is every absolutely summing operator from $L(\ell^2)$ into a Hilbert space compact?

Obviously every absolutely summing operator from $K(\ell^2)$ into ℓ^2 is compact because the dual of $K(\ell^2)$ is separable. However Problem 3 has a local counterpart for $K(\ell^2)$.

PROBLEM 3'. Does there exist a "modulus of capacity" $\varepsilon \mapsto N(\varepsilon)$ such that if $\pi_1(u: K(\ell^2) \to \ell^2) \leq 1$ then the ε-capacity of $u(B_{K(\ell^2)})$ does not exceed $N(\varepsilon)$ (here $B_{K(\ell^2)} = \{T \in K(\ell^2): \|T\| \leq 1\}$)?

The positive answer to PROBLEM 3 will follow if one could establish the following structural property of $L(\ell^2)$:

let X be a subspace of $L(\ell^2)$ isomorphic to ℓ^1; then there exists a subspace Y of $L(\ell^2)$ isomorphic to a $C(K)$-space such that $Y \cap X$ is infinite-dimensional.

Our last problem concerns spaces of smooth functions.

PROBLEM 4. Is every absolutely summing operator from $C^k(\mathbb{T}^n)$ into a Hilbert space compact?

We do not know whether there exists a "Paley phenomenon" for $C^1(\mathbb{T}^n)$, i.e. whether there is a absolutely summing surjection from $C^1(\mathbb{T}^n)$ onto ℓ^2 (cf. comments to PROBLEM 1). It seems to be unlikely that there exists an i n v a r i a n t absolutely summing surjection, as in the case of the disc algebra A.

The author would like to thank Prof. S.V.Kisliakov for a valuable discussion.

REFERENCES

1. O v s e p i a n R.I., P e ł c z y ń s k i A. On the existence of a fundamental total and bounded biorthogonal sequence in every separable Banach space, and related constructions of uniformly bounded orthonormal systems in L^2. - Studia Math., 1975, 54, 149-159.
2. W e i s L. On strictly singular and strictly cosingular operators. - ibid., 285-290.
3. P e ł c z y ń s k i A. Banach spaces of analytic functions and absolutely summing operators. Regional conference series in mathematics, N 30. AMS, Providence, 1977.
4. B o u r g a i n J., D e l b a e n F. A class of special $\mathcal{L}_\infty$-spaces. - Acta Math., 1980, 145, N 3-4, 155-176.

A. PEŁCZYŃSKI

Institute of Mathematics
Polish Academy of Sciences
Śniadeckich 8,
00-950 Warsaw, Poland

1.3. WHEN IS $\Pi_2(X,\ell^2)=L(X,\ell^2)$?
old

Let X and Y be two infinite dimensional Banach spaces and let $L(X,Y)$ denote the space of all continuous linear operators from X to Y. An operator $T\in L(X,Y)$ is said to be p- a b s o l u t e l y s u m m i n g if there exists a positive constant C such that

$$\sum_{k=1}^{n} \|Tx_k\|^p \leq C^p\cdot \sup\{\sum_{k=1}^{n} |\langle x_k, x'\rangle|^p : x'\in X^*, \|x'\|\leq 1\}$$

for each n in $\mathbb{N}$ and $x_1, x_2, \ldots, x_n\in X$. The set of all p-absolutely summing operators from X to Y is denoted by $\Pi_p(X,Y)$. The conditions for $\Pi_p(X,Y)$ to coincide with $\Pi_r(X,Y)$ or with $L(X,Y)$ have been the subject of a great number of publications (see [1]-[4]). The results obtained are not only of their own interest but also are widely used in problems connected with the isomorphic classification of Banach spaces.

It is easy to see that $\Pi_p(X,Y)\subset\Pi_r(X,Y)$ for $p<r$. The Dvoretzky theorem on almost Euclidean sections of convex bodies shows that the equality $\Pi_p(X,Y)=L(X,Y)$ has the highest chance to hold if Y is isomorphic to a Hilbert space, i.e. that $\Pi_p(X,\ell^2)=L(X,\ell^2)$ provided $\Pi_p(X,Y)=L(X,Y)$ at least for one infinite dimensional space Y. Besides, it is well-known that $\Pi_p(X,\ell^2)=\Pi_2(X,\ell^2)$ for $p\geq 2$. Thus the investigation of the problem whether $\Pi_p(X,Y)$ coincides with $L(X,Y)$ leads immediately to the question of conditions ensuring the equality

$$\Pi_2(X,\ell^2)=L(X,\ell^2)\ . \qquad (1)$$

A space X satisfying (1) will be called 2-t r i v i a l (cf.[5]). Obviously a space X and its dual X^* are 2-trivial (or not) simultaneously.

A GENERAL QUESTION we want to raise is <u>to find out conditions</u> (in particular the conditions of geometrical nature) <u>under which a Banach space</u> X <u>is</u> (or is not) <u>2-trivial</u>.

It is known [6] that (1) is impossible for the space X not containing ℓ_n^1 uniformly (for example if X is uniformly convex).

On the other hand it is easy to verify that this condition is not sufficient for the 2-triviality. Indeed, the sequential Lorentz space $\Lambda(C)$ not only contains ℓ^1_n uniformly but is even saturated by subspaces isomorphic to ℓ^1 (to wit every infinite dimensional subspace of $\Lambda(C)$ contains a subspace isomorphic to ℓ^1). Nevertheless $\Lambda(C)$ fails to be 2-trivial and moreover it is a space of type $(\mathcal{E})$ (see the definition below).

It can be proved that X is not 2-trivial provided X satisfies the following condition: there exist two sequences $\{A_n\}_{n\geq 1}$ and $\{B_n\}_{n\geq 1}$ of operators such that

$$A_n : \ell^2_n \to X, \quad B_n : X \to \ell^2_n, \quad B_n \circ A_n = id_{\ell^2_n}$$

and $\underline{\lim}_n \, n^{-1/2} |A_n| \cdot |B_n| = 0$. A space satisfying these conditions is said to be of type ($\mathcal{E}$).

It has been essentially proved in [6] that a space not containing ℓ^1_n uniformly is the space of type $(\mathcal{E})$. However the condition of being of type $(\mathcal{E})$ is also not necessary for the non-2-triviality. As S.V.Kisljakov has pointed out, the reflexive "non sufficiently Euclidean" space built in [7] fails to be of type $(\mathcal{E})$ and simultaneously it can be proved that this space fails to be 2-trivial.

What has been said above indicates that the class of all 2-trivial spaces cannot be too large. The following conjecture looks therefore rather plausible.

CONJECTURE 1. <u>No infinite-dimensional reflexive Banach space is 2-trivial.</u>

An equivalent statement: there exists no infinite-dimensional reflexive Banach space X, such that each operator from $L(\ell^2, \ell^2)$, which can be factored through X , is a Hilbert-Schmidt operator. We note that a positive solution to CONJECTURE 1 would obviously imply the solution (in the class of reflexive Banach spaces) to the GROTHENDICK PROBLEM on the coincidence of the spaces of nuclear and compact operators.

The following QUESTIONS arise naturally.

1. <u>Under what conditions does 2-triviality of a space</u> X <u>imply the equality</u> $\Pi_2(X, \ell^1) = L(X, \ell^1)$?

2. <u>Which of the spaces of analytic or smooth functions are of type</u> $(\mathcal{E})$?

3. <u>Is it true that in any space</u> X <u>of type</u> $(\mathcal{E})$ <u>there exists a sequence of subspaces</u> $\{X_n\}$ $(\dim X_n = n)$ <u>with one of the following two properties</u>: (a) $\sup \lambda(X_n, X) < \infty$, $d(X_n, \ell^2_n) = O(\sqrt{n})$ <u>or</u> (b) $\sup_n d(X, \ell^2_n) < \infty$, $\lambda(X_n, X) = O(\sqrt{n})$?
(Here $\lambda(X_n, X)$ is the relative projection constant).

The assumption that a 2-trivial space has an unconditional basis apparently rather drastically diminishes the class of such spaces. For example, each reflexive Banach space with an unconditional basis is not 2-trivial [8]. On the other hand, as it is shown in [9], the space $(\sum C_o)_{\ell^1}$ also fails to be 2-trivial (more precisely it is of type $(\mathcal{E})$). These results give some ground to the following

CONJECTURE 2. <u>If a 2-trivial infinite dimensional Banach space</u> X <u>has an unconditional basis, then</u> X <u>is isomorphic to either</u> C_o <u>or</u> ℓ^1 <u>or</u> $C_o \oplus \ell^1$.

To illustrate conjecture 2 we mention a result which follows from Theorem 1 in [8]: If X has an unconditional basis, if Y is not isomorphic to a Hilbert space, and if $\Pi_2(X,Y) = L(X,Y)$ then X is isomorphic to C_o .

REFERENCES

1. Lindenstrauss J., Pełczyński A. Absolutely summing operators in $\mathcal{L}_p$-spaces and their applications. - Studia Math., 1968, 29, 275-326.
2. Kwapień S. On a theorem of L.Schwartz and its applications to absolutely summing operators. - ibid., 1970, 38, 193-201.
3. Dubinsky E., Pełczyński A., Rosenthal H. On Banach spaces X for which $\Pi_2(\mathcal{L}_\infty, X) = B(\mathcal{L}_\infty, X)$ - ibid., 1972, 44, 617-648.
4. Maurey B. Théorèmes de factorisation pour les opérateurs linéaires à valeurs dans les espaces L^p. - Astérisque, 1974, 11, 1-163.
5. Morrell J.S., Retherford J.R. p-trivial Banach spaces. - Studia Math., 1972, 43, 1-25.
6. Davis W.J., Johnson W.B. Compact nonnuclear operators. - Studia Math., 1974, 51, 81-85.

7. J o h n s o n W.B. A reflexive Banach space which is not sufficiently Euclidean. - ibid., 1976, 55, 201-205.

8. К о м а р ч е в И.А. О 2-абсолютно суммирующих операторах в банаховых решетках. - Вестник ЛГУ, сер.матем., мех., астрон., 1980, № 19, 97-98.

9. F i g i e l T., L i n d e n s t r a u s s J., M i l m a n V. The dimension of almost spherical sections of convex bodies. - Acta Math., 1977, 133, 53-94.

I.A.KOMARCHEV
(И.А.КОМАРЧЕВ)
B.M.MAKAROV
(Б.М.МАКАРОВ)

СССР, 198904, Петродворец,
Библиотечная пл.2,
математико-механический
факультет Ленинградского
Университета

1.4. old FINITE DIMENSIONAL OPERATORS ON SPACES OF ANALYTIC FUNCTIONS

Let A be the Banach space of all functions continuous in $\operatorname{clos}\mathbb{D}$ and analytic in $\mathbb{D}$, equipped with the supremum norm and let H^1 be the Hardy space. We consider A as a subspace of $C(\mathbb{T})$ and H^1 as a subspace of $L^1(m)$. We would like to know the relation between finite dimensional subspaces and finite dimensional operators in A and those in $C(\mathbb{T})$. This question is of importance in the theory of the Banach space A. We feel also that such connection, when expressed in precise terms, can lead to some new isomorphic invariants of Banach spaces. Let us start with the following

PROBLEM 1. <u>Let X be an n-dimensional subspace of A. Does every projection $P: A \xrightarrow{\text{onto}} X$ extend to a projection $\tilde{P}: C(\mathbb{T}) \xrightarrow{\text{onto}} X$ with $\|\tilde{P}\| \le C(1+\log n)\|P\|$?</u>

<u>May be we have only</u>

$$\inf\{\|P\| : P: A \xrightarrow{\text{onto}} X\} \geqslant C(1+\log n)^{-1} \inf\{\|P\| : P: C(\mathbb{T}) \longrightarrow X\}.$$

This problem is obviously a special case of the following

PROBLEM 2. <u>Let Y be a Banach space and let $T: A \longrightarrow Y$ be an operator of rank n. Does there exist an extension $\tilde{T}: C(\mathbb{T}) \longrightarrow Y$ with $\|\tilde{T}\| \le C(1+\log n)\|T\|$.</u>

Another particular case of this problem is also of interest:

PROBLEM 3. <u>Let X be an n-dimensional subspace of $L^1(m)/H^1$. Does there exists a map $T: X \longrightarrow L^1(m)$ with $\|T\| \le C(1+\log n)$ such that $\pi \circ T = id_X$, where π is the canonical quotient map from $L^1(m)$ onto $L^1(m)/H^1$?</u>

It seems that the estimates of the above type can be useful in proving the non-isomorphism of spaces of analytic functions of different numbers of variables. There are also some problems of this type connected with Schauder bases. Let us recall that a system $(f_n)_{n\geqslant 1}$ of elements of a Banach space Y is called a *Schauder basis* if for every y, $y \in Y$, there exists a unique sequence of scalars $(a_n)_{n\geqslant 1}$ such that the series $\sum_{n=1}^{\infty} a_n f_n$ converges to y in the norm of Y. If it is so, then there exists a constant K

such that for every N $\|\sum_{n=1}^{N} a_n f_n\| \leq K \|y\|$ The best such constant is called a basis constant of the basis (f_n). S.V.Bočkariov [1] has proved that the disc algebra A has a Schauder basis. On the other hand it was proved in [2] that A does not have a Schauder basis with constant 1. So the question arises.

PROBLEM 4. <u>Does there exist a constant</u> q, $q>1$, <u>such that every basis for the disc algebra</u> A <u>has the basis constant</u> $>q$? <u>It was proved by P.Enflo</u> [3] <u>that there exists a Banach space which has the property described in the Problem 4.</u>

Our last problem is connected with the space of polynomials. Let W_n^p denote the linear span of $1, e^{it}, e^{2it}, \dots, e^{nit}$ considered in the $L^p(m)$ norm. It is known [4] that the norm of the best projection from A onto W_n^∞ and from H^1 onto W_n^1 is of the order $\log n$. If X and Y are two n-dimensional Banach spaces then we define the B a n a c h - M a z u r d i s t a n c e between X and Y by

$$d(X,Y) = \inf\{\|T\| : \|T^{-1}\| : T : X \xrightarrow{1-1} Y\}.$$

PROBLEM 5. (a) <u>Let</u> X <u>be an</u> $(n+1)$ -<u>dimensional subspace of</u> H^1. <u>Is it true that for every projection</u> P <u>from</u> H^1 <u>onto</u> X <u>we have</u> $\|P\| \geq C \log n \cdot d(X, W_n)^{-1}$?

(b) <u>Let</u> X <u>be an</u> $(n+1)$ -<u>dimensional subspace of</u> H^1. <u>Is it true that for every projection</u> P <u>from</u> H^1 <u>onto</u> X <u>we have</u>

$$\|P\| \geq C \log n \cdot d(X, W_n^1)^{-1} ?$$

A positive solution to Problem 5(b) immediately yields that H^1 and $H^1(\mathbb{T}\times\mathbb{T}, m\times m)$ are non-isomorphic Banach spaces.

REMARK. In the above problems C means an absolute constant.

REFERENCES

1. Б о ч к а р е в С.В. Существование базиса в пространстве функций, аналитических в круге, и некоторые свойства системы Франклина. - Матем.сб., 1974, 95, № 1, 3-18.

2. W o j t a s z c z y k P. On projections in spaces of bounded analytic functions with applications. - Stud.Math., 1979, 65, N 2, 147-173.
3. E n f l o P. The Banach space with basis constant > 1. - Arch. för Mat.,1973, 11, 103-107.
4. Z y g m u n d A. Trigonometric series, v.1, Cambridge Univ. Press, 1959.

P.WOJTASZCZYK

Institute of Math.
Polish Academy of Sciences
Śniadeckich 8,
00950 Warsaw, Poland

* * *

COMMENTARY

The AFFIRMATIVE answer to PROBLEM 2 (and therefore to PROBLEMS 1 and 3) has been obtained by J.Bourgain (cf. the references mentioned in the Commentary to S.1).

PROBLEM 4 seems to be open.

PROBLEM 5 has a NEGATIVE solution. Namely, there exist a sequence $\{V_n\}$ of subspaces of C_A (of H^1) and a sequence $\{P_n\}$ of projections $P_n : C_A \xrightarrow{onto} V_n$ (resp. $P_n : H^1 \xrightarrow{onto} V_n$) with

$$\sup_n \|P_n\| < \infty, \quad \sup_n d(V_n, W_n^\infty) < \infty. \quad (\text{resp. } \sup_n d(V_n, W_n^1) < \infty).$$

This has been observed by J.Bourgain and A.Pełczyński. Let us SKETCH THE CONSTRUCTION for the disc-algebra (the H^1-case is considered analogously).

Replace C_A by the direct sum $C_A \oplus \bar{z}\bar{C}_A$ which is isomorphic to $C_A (\bar{z}\bar{C}_A \overset{def}{=} \{\bar{z}\bar{f} : f \in C_A\})$. Then define $I : W_n \longrightarrow C_A \oplus \bar{z}\bar{C}_A$ and $Q : C_A \oplus \bar{z}\bar{C}_A \longrightarrow W_n$ by

$$I(p) = (p, \bar{z}^{(n+1)} p);$$

$$Q(f,g) = K_{n+1} * f + z^{n+1} K_{n+1} * g,$$

where K_n stands for the n-th Fejer kernel. It is easy to verify that I is an isometry, $QI = id_n$ and $\sup_n \|Q\| < +\infty$ ●

The spaces $H^1(\mathbb{T})$ and $H^1(\mathbb{T}^2)$ are non-isomorphic [5]. Moreover no two of the spaces $H^1(\mathbb{T}^n)$ are isomorphic. The last result has been also proved by J.Bourgain.

REFERENCE

5. B o u r g a i n J. The non-isomorphism of H^1 -spaces in one and several variables. - J.Funct.Anal., 1982, 46, p.45-57.

1.5. old

COMPLEMENTED SUBSPACES OF A , H^1 AND H^∞

Per Enflo's counterexample to the approximation problem [1], and subsequent results by Davie [2] and Figiel [3], indicate that an isomorphic classification of all closed subspaces of a Banach Space X (X not isomorphic to a Hilbert space) is probably impossible in the near future. An important and difficult, but not impossible, problem is the classification of the complemented subspaces of X . Because of the recent advances in the study of the Banach space properties of the Disc Algebra A , H^1 , and H^∞ (see [4]), I think we can now give serious consideration to classifying their complemented subspaces. As a first step in the process, I make the following conjecture:

CONJECTURE. A <u>and</u> H^∞ <u>are primary</u>.

A Banach space X is p r i m a r y if whenever $X \approx Y \oplus Z$ then either $X \approx Y$ or $X \approx Z$. In support of the conjecture we will prove that if $A \approx Y \oplus Z$ and if Y is isomorphic to a complemented subspace of $C[0,1]$ then $A \approx Z$. We first use an observation of S.V.Kisljakov which states that if $A \approx Y \oplus Z$ and if Y is isomorphic to a complemented subspace of $C[0,1]$ then Z^* is non-separable. To see this, we let P be a projection of A onto Y and use an argument similar to the proof of corollary 8.5 (e) of [4] to show that $P^*|L^1/H^1_0$ maps weakly Cauchy sequences to norm convergent sequences. If Z^* is separable, it is known that weak and norm convergent sequences in Z^* coincide, and hence it follows that the same is true in L^1/H^1_0 - which is a contradiction. It now follows by corollary 8.5 (b) of [4] that $C[0,1]$ is isomorphic to a complemented subspace of Z , i.e. $Z \approx C[0,1] \oplus W$ for some space W . Since $C[0,1]$ is primary [5], it follows that $Y \oplus C[0,1] \approx C[0,1]$. Hence

$$A \approx Y \oplus Z \approx Y \oplus C[0,1] \oplus W \approx C[0,1] \oplus W \approx Y .$$

Bochkariev [6] has shown that A has a basis consisting of the Franklin system in $L^2_R[0,\pi]$. (Here we are identifying A with the subspace of $C[-\pi,\pi]$ spanned by the characters $\{e^{inx}\}_{n\geq 0}$). If we let H^∞_n be the span of the first n elements of this basis, Delbaen has recently announced that $(\Sigma \oplus H^\infty_n)_{c_0}$ is isomorphic to a complemented subspace of A . This subspace is particularly interesting because it is not isomorphic to A and it is also not isomorphic to a complemented subspace of $C[0,1]$. The complement of this subspace is unknown and identifying it should be the first step in proving (or disproving) the conjecture.

We now outline one approach you might use to try to prove the conjecture. If $X \approx X \oplus X$, then X is primary if and only if X satisfies: (1) If $X \approx Y \oplus Z$, then either Y or Z has a complemented subspace isomorphic to X; and (2) If Y is a Banach space and if X and Y are isomorphic to complemented subspaces of each other, then $X \approx Y$. By Pełczyński's decomposition method, if $X \approx (\sum \oplus X)_E$, where E is c_0 or ℓ^p, $1 \leq p \leq \infty$, then property (1) implies property (2). Therefore, you should first consider the question of Mitjagin [7]: Is A isomorphic to $(\sum \oplus A)_{c_0}$? To give a positive answer to this question, it suffices to show that $(\sum \oplus A)_{c_0}$ is isomorphic to a complemented subspace of A. In this case then, you need to carefully examine the construction of Delbaen. Next, you should try to generalize the technique of [8] to the basis of A, or produce a new basis of A for which the technique works. This approach to the problem has the advantage that it may immediately imply that H^∞ is primary. Since Wojtaszczyk [9] has shown that $H^\infty \approx (\sum \oplus H^\infty)_{\ell^\infty} \approx (\sum \oplus H_n^\infty)_{\ell^\infty}$, if the above approach proves that A is primary, then the technique of [10] should show that H^∞ is primary. As a word of warning concerning the naivety of the conjecture, let us mention that the only complemented subspaces of A which are known are either isomorphic to A, $A \oplus Y$, or to $X \oplus Y$, where X is isomorphic to a complemented subspace of $C[0,1]$ and

$$Y \approx (\sum \oplus E_n)_{c_0}$$

with $\dim E_n < +\infty$ for all $n = 1, 2, \ldots$.

Much less seems to be known about the subspaces of H^1. It is also not known if H^1 is primary. If you want to try to prove that H^1 is primary, you should first consider the question: Is H^1 isomorphic to $(\sum \oplus H^1)_{\ell^1}$? Next, you should look at Billard's basis for H^1 [11]. Since this basis is even more directly related to the Haar system than the basis for A, this question could actually prove to be easier than the others. (Again using the techniques of [8] and [10]).

REFERENCES

1. E n f l o P. A counterexample to the approximation property in Banach spaces. - Acta.Math., 1973, 130, 309-317.
2. D a v i e A.M. The approximation problem for Banach spaces. - Bull.London Math.Soc., 1973, 5, 261-266.
3. F i g i e l T. Further counterexamples to the approximation

problem, dittoed notes.

4. P e ł c z y ń s k i A. Banach spaces of analytic functions and absolutely summing operators. - CBMS Regional Confer. Ser. in Math., 1977, N 30.

5. L i n d e n s t r a u s s J., P e ł c z y ń s k i A. Contributions to the theory of classical Banach spaces. - J. Funct. Anal., 1971, 8, 225-249.

6. Б о ч к а р е в С.В. Существование базиса в пространстве функций, аналитических в круге, и некоторые свойства системы Франклина. - Матем. сб., 1974, 95, № 1, 3-18.

7. М и т я г и н Б.С. Гомотопическая структура линейной группы банахова пространства. - Успехи матем. наук, 1970, 25, № 5, 63-106.

8. A l s p a c h D., E n f l o P., O d e l l E. On the structure of separable $\mathcal{L}_p$ spaces $(1<p<\infty)$. - Stud. Math., 1977, 60, 79-90.

9. W o j t a s z c z y k P. On projections in spaces of bounded analytic functions with applications. - Studia Math., 1979, 65, N 2, 147-173.

10. C a s a z z a P.G., K o t t m a n C., L i n B.L. On some classes of primary Banach spaces. - Canadian J. Math., 1977, 29, N 4, 856-873.

11. B i l l a r d P. Bases dans H^1 et bases de sous espaces de dimension finie dans A. - Proc. Confer. Oberwolfach, August 14-22, 1971, ISNM Vol. 20, Birkhäuser Verlag, Basel and Stuttgart, 1972.

P.G. CASAZZA

Department of Mathematics,
The University of Missouri-
-Columbia, Columbia, Missouri 65211
USA

* * *

COMMENTARY

P. Wojtaszczyk proved that $(\sum\oplus A)_{c_0}\sim A$ and $(\sum\oplus H^1)_{\ell^1}\sim H^1$ when the first edition of the Collection was in preparation. (Now his result is published, cf. [12]). Nevertheless the problem of primariness of A and H^1 seems to remain open.

J. Bourgain [13] has proved that H^∞ is primary (and the same is true for $H^\infty(\mathbb{T}^m)$, m being arbitrary). It is worth mentioning that in Bourgain's proof the relation $H^\infty \approx (\sum_{n\geq 1} W_n^\infty)_\infty$ is used ($W_n^\infty = span\{\mathbb{1}, z, \ldots, z^n\}$, with the sup-norm), rather than $H^\infty \approx (\sum_{n\geq 1} H_n^\infty)_\infty$, as proposed in the text of Problem. The decomposition $H^\infty \approx (\sum_{n=1} W_n^\infty)_\infty$ is due to Bourgain and Pełczyński. The reason why this is valid rests just on the observation about the complemented imbeddings of W_n's into A made in Commentary to Problem 1 4

We end with a quotation from the author's letter to the editors: "The primariness of H^1 seems to be still unknown but with the myriad of new results here the last few years by J.Bourgain, this may be "almost obviously" true".

REFERENCES

12. W o j t a s z c z y k P. Decompositions of H^p-spaces. - Duke Math.J., 1979, 46, N 3, 635-644.
13. B o u r g a i n J. On the primarity in H^∞-spaces. - Preprint.

1.6.
old

SPACES OF HARDY TYPE

A Banach Space E of measurable functions on $[0,2\pi]$ is called a symmetric (or rearrangement invariant) space iff the norm of E is monotone and any two equimeasurable functions have equal norms. ([1], chapter 2). The L^p -spaces $(1\leqslant p\leqslant\infty)$, the Orlicz spaces and the Lorentz spaces can serve as examples. Remind that if the function Ψ is non-decreasing and concave on $[0,2\pi]$, $\Psi(0)=0$, then the Lorentz space $\Lambda(\Psi)$ consists of functions x such that

$$\|x\|_{\Lambda(\Psi)}=\int_0^1 x^*(t)\,d\psi(t)<+\infty ,$$

where x^* is the function non-increasing on $[0,2\pi]$ and equimeasurable with x .

A symmetric space E gives rise to a space of complex functions on $\mathbb{T}$ consisting of functions with **moduli** from E . This space is also denoted by E . By $H(E)$ we denote the set of all functions f analytic in the unit circle $\mathbb{D}$ and satisfying $\|f\|_{H(E)}<\infty$,

$$\|f\|_{H(E)}\overset{def}{=\!=}\sup_{0<r<1}\|f_r\|_E\ ,\quad f_r(\xi)\overset{def}{=\!=}f(r\xi).$$

Though the classical Hardy spaces $H^p=H(L^p)$ have been studied rather well, the theory of general spaces $H(E)$ is only fragmentary.

The set of extreme points of the unit ball of $\Lambda(\Psi)$ is contained in the set of functions $\frac{\chi_e\cdot\varepsilon}{\Psi(me)}$, where $|\varepsilon(t)|\equiv 1$ and χ_e is the characteristic function of a measurable set $e, e\subset[0,2\pi]$. In the case when Ψ is strongly concave these two sets coincide. The following PROBLEM arises natrually: describe the set of extreme points of the unit ball of $H(\Lambda(\Psi))$. Some partial results are contained in [2]. The space $H(\Lambda(\Psi))$ is nothing but H^1 if $\Psi(t)=t$, and coincides with H^∞ , if $\Psi(t)=\operatorname{sign}t$. In these two cases the set of extreme points of the unit ball is well-known, see [3], part 9.

We believe that the solution of the above-mentioned problem will possibly be useful for describing all isometric operators on $H(\Lambda(\Psi))$ Some interesting results on isometric operators on a symmetric space are contained in [4].

REFERENCES

1. Крейн С.Г., Петунин Ю.И., Семёнов Е.М. Интерполяция линейных операторов. М., Наука, 1978.
2. Брыскин И.Б., Седаев А.А. О геометрических свойствах единичного шара в пространствах типа классов Харди. – Зап. научн.семин.ЛОМИ, 1974, 39, 7-16.
3. Hoffman K. Banach spaces of Analytic Functions. Prentice-Hall, Englewood Cliffs, New Jersey, 1962.
4. Зайденберг М.Г. К изометрической классификации симметричных пространств. – Докл.АН СССР, 1977, 234, № 2, 283-286.

E.M.SEMËNOV
(Е.М.СЕМЁНОВ)

СССР 394000 Воронеж,
Воронежский Государственный
Университет

EDITORS' NOTE. Here are some more articles connected with isometries of L^p-spaces of analytic and harmonic functions:

1. Плоткин А.И. Продолжение L^p-изометрий. – Зап.научн. семин.ЛОМИ, 1971, 22, 103-129.
2. Плоткин А.И. Изометрические операторы в L^p-пространствах аналитических и гармонических функций. – Ibid., 1972, 30, 130-145.
3. Плоткин А.И. Алгебра, порожденная операторами сдвига, и L^p-нормы. В кн.: "Функциональный анализ. Выпуск 6. Межвузовский сборник", Ульяновск, 1976, 112-121.
4. Плоткин А.И. Об изометрических операторах в пространствах суммируемых аналитических и гармонических функций. – Докл. АН СССР, 1969, 185, № 5, 995-997.

1.7. BASES IN H^p SPACES ON THE BALL

By $H^p(B)$ we will mean the natural Hardy space of analytic functions on the unit ball of $\mathbb{C}^n$. $A(B)$ denotes the ball algebra of all functions continuous in $\overline{B}$ and analytic in B. We are interested in construction and existence problems for Schauder bases in these spaces.

Let me recall that a sequence of elements $(f_n)_{n=0}^{\infty}$ in a Banach spaces X is a Schauder basis for X if for every x in X there exists a unique sequence of scalars $(a_n)_{n=0}^{\infty}$ such that the series $\sum_{n=0}^{\infty} a_n f_n$ converges to x in the norm of X. The basis is called unconditional if for every x in X the corresponding series $\sum_{n=0}^{\infty} a_n f_n$ is unconditionally convergent.

For the ball algebra $A(B)$ the question of the existence of a Schauder basis is a well known open problem (cf. [1]). This seems to be the most concrete separable Banach space for which this question is still open today. It is known that $A(B)$ does not have an unconditional basis.

For $1 \leqslant p < \infty$ the situation is a little more intriguing. It is a relatively easy task to check that for $1<p<\infty$ the monomials in a correct order form a Schauder basis for $H^p(B)$. However this basis is not unconditional for $p \neq 2$. It was proved in [6] that for $1 \leqslant p < \infty$ $H^p(B)$ is isomorphic as a Banach space to $H^p(\mathcal{D})$, the classical Hardy space on the unit disc. Since unconditional bases for $H^p(\mathcal{D})$ are well known, cf. [2], [4], [5], we get the existence of unconditional bases in $H^p(B)$ for $1 \leqslant p < \infty$. This argument however has one drawback, it is non-constructive, so we pose the following

PROBLEM. <u>Construct an unconditional basis in the space</u> $H^p(B)$ $1 \leqslant p \neq 2 < \infty$.

The most interesting case is $p=1$. There is also an auxiliary question related to this:

<u>Does there exist an orthonormal unconditional basis in</u> $H^1(B)$?

The case $p<1$ is less clear. In $H^p(\mathcal{D})$ we have unconditional bases, cf. [3], [5]. However in several variables the very existence of an unconditional basis in $H^p(B)$, $p<1$ is still open. The proof of isomorphism between $H^1(B)$ and $H^1(\mathcal{D})$ given in [6]

can be extended to $p<1$ (after some technical modifications) provided the following question has positive answer.

QUESTION. Is $H^p(\mathcal{D}),\ p<1$ isomorphic to a complemented subspace of $H^p(B)$?

REFERENCES

1. P e ł c z y ń s k i A. Banach spaces of analytic functions and absolutely summing operators. CBMS regional conference series N°30.
2. Б о ч к а р е в С.В. Существование базиса в пространстве функций, аналитических в круге, и некоторые свойства системы Франклина. - Матем.сборник, I974, 95 (I37), вып.I, 3-I8.
3. S j ö l i n P., S t r o m b e r g J.-O. Basis properties of Hardy spaces. Stockholms Universitet preprint No 19, 1981.
4. W o j t a s z c z y k P. The Franklin system is an unconditional basis in H^1. - Arkiv för Mat., 1982, 20, No 2, 293-300.
5. W o j t a s z c z y k P. H^p -spaces, $p<1$ and spline systems. - Studia Math. (to appear).
6. W o j t a s z c z y k P. Hardy spaces on the complex ball are isomorphic to Hardy spaces on the disc, $1\leqslant p<\infty$. - Annals of Math. (to appear)

P.WOJTASZCZYK

Math.Inst.Polish Acad.Sci.
00-950 Warszawa,
Śniadeckich, 8
POLAND

1.8. SPACES WITH THE APPROXIMATION PROPERTY?

Recall that a Banach space X has the approximation property (a.p.) if for all compact $E \subset X$ and for all $\varepsilon > 0$ there is a bounded linear operator $T : X \longrightarrow X$ such that $\|x - Tx\| < \varepsilon$ when $x \in E$, and such that T has finite rank. Not every Banach space has the a.p. [1] .

<u>Does</u> H^∞ <u>have the a.p.?</u>

Some mild evidence that this might be true comes from the recent result that L^∞ / H^∞ (i.e. BMO) has the a.p. [2] . Another interesting space for which the a.p. is unknown is $W^{k,\infty}(\mathbb{R}^n)$, $n \geq 2$. (When $n = 1$ the answer is easy and positive.) Here

$$W^{k,\infty}(\mathbb{R}^n) = \{ f : D^\alpha f \in L^\infty(\mathbb{R}^n), \quad 0 \leq |\alpha| \leq k \} .$$

REFERENCES

1. E n f l o P. A counter-example to the approximation problem in Banach spaces. - Acta Math., 1973, 130, 309-317.
2. J o n e s P.W. BMO and the Banach space approximation problem. - Institut Mittag-Leffler report No.2, 1983.

PETER W.JONES

Institut Mittag-Leffler
Auravägen 17
S-182 62 Djursholm
Sweden

Usual Address:

Dept. of Mathematics
University of Chicago
Chicago, Illinois 60637
USA

1.9. OPERATOR BLOCKS IN BANACH LATTICES

The operator Q_e of multiplication by the characteristic function of a measurable subset $e \subset [0,1]$ has the unit norm in every functional Banach lattice E on $[0,1]$ (see [1] for a definition).

Associate with every continuous linear operator $T : E \longrightarrow E$ the number

$$\sigma(T, E) = \inf\{\| Q_e T Q_f \|_E : me \cdot mf > 0\}$$

and let

$$\mathcal{D}(E) = \{T \in \mathcal{L}(E,E) : \sigma(T,E) > 0\} .$$

PROBLEM. <u>Under what conditions on E the set $\mathcal{D}(E)$ is empty, i.e.</u> $\sigma(T, E) = 0$ <u>for every linear operator?</u>

This question arose for the first time in [2] (for concrete spaces) in connection with the contractibility problem of linear groups in Banach spaces. In particular an isometry T of L^2 satisfying $\sigma(T, L^2) > 0$ was constructed there. On the other hand, it has been proved [3] that $\mathcal{D}(L^1) = \mathcal{D}(L^\infty) = \emptyset$ and that $\mathcal{D}(L^p) \neq \emptyset$ for $1 < p < \infty$ [4].

Recall now the definition of the Lorentz space $L^{p,q}$ (see [5] for their properties.) For a measurable function x on $[0,1]$ let x^* denote the non-increasing rearrangement of $|x|$. Then

$$\|x\|_{L^{p,q}} = \left(\int_0^1 (x^*(t) \cdot t^{1/p})^q \frac{dt}{t}\right)^{\frac{1}{q}} , \quad 1 < p < \infty, \; 1 \leq q \leq \infty.$$

CONJECTURE. $\mathcal{D}(L^{p,q}) \neq \emptyset$ <u>iff either</u> $1 \leq q \leq p \leq 2$ <u>or</u> $2 \leq p \leq q \leq \infty$.

It is well-known that $(L^{p,q})^* = L^{p',q'}$, where $p' = \frac{p}{p-1}$, $q' = \frac{q}{q-1}$. Therefore without loss of generality it can be assumed that $p \leq 2$. It is also known that $\mathcal{D}(L^{p,\infty}) = \emptyset$, $1 < p < 2$ and that $\bigcap_{q : 1 \leq q \leq p} \mathcal{D}(L^{p,q}) \neq \emptyset$, $1 \leq p \leq \infty$ [6].

For the case $p < q < +\infty$ the situation remains unclear.

Nothing is known about the set $\mathfrak{D}(L^{2,q})$ except $q=2$ when $\mathfrak{D}(L^{2,2})=\mathfrak{D}(L^2)\neq\emptyset$.

The problem of non-emptiness of $\mathfrak{D}(E)$ remains open for Orlicz spaces.

The operators $T=T(p)\in\mathfrak{D}(L^p)$ constructed in [4] depend on p and this is not a mere occasion. The set $\mathfrak{D}(L^{p_1})\cap\mathfrak{D}(L^{p_2})$ is not empty (let $1<p_1<p_2<\infty$ for the definiteness) iff $p_1\leq 2\leq p_2$ [6]. However it is not clear what conditions provide $\mathfrak{D}(E_1)\cap\mathfrak{D}(E_2)=\emptyset$ in the general case.

REFERENCES

1. L i n d e n s t r a u s s J., T z a f r i r i L. Classical Banach Spaces II . Berlin, Springer Verlag, 1979.
2. М и т я г и н Б.С. Гомотопическая структура линейной группы банахова пространства.-Успехи матем.наук,1970,25, № 5, 63-106.
3. E d e l s t e i n I., M i t y a g i n B., S e m e n o v E. The Linear Groups of C and $\mathcal{L}_1$ are Contractible. - Bull.Acad. Polon.Sci., Ser.Math., 1970, 18, N 1.
4. С е м е н о в Е.М., Ц и р е л ь с о н Б.С. Задача о малости операторных блоков в пространствах $\mathcal{L}_p$.- Zeit.Anal. und ihre Anwend., 1983, 2, N 4.
5. K r e i n S.G., P e t u n i n Ju.I., S e m e n o v E.M. Interpolation of Linear Operators. AMS Providence, 1982.
6. С е м е н о в Е.М., Ш т е й н б е р г А.М. Операторные блоки в пространствах $\mathcal{L}_{p,q}$. - Докл.АН СССР. 1983, to appear.

E.M.SEMЁNOV

(Е.М.СЕМЁНОВ)

СССР, 394693
Воронеж, Воронежский
государственный университет

1.10. old SPACES OF ANALYTIC FUNCTIONS (ISOMORPHISMS, BASES)

$\mathcal{O}(\mathfrak{D})$ will denote the space of all functions analytic in the domain $\mathfrak{D}$, $\mathfrak{D}\subset\hat{\mathbb{C}}$. A domain $\mathfrak{D}$ is called standard if $\mathcal{O}(\mathfrak{D})$ is isomorphic (as a linear topological space) to one of three (mutually non-isomorphic) spaces

$$\mathcal{O}_1 \overset{def}{=} \mathcal{O}(\mathbb{D}), \qquad \mathcal{O}_\infty \overset{def}{=} \mathcal{O}(\mathbb{C}), \qquad \mathcal{O}_2 \overset{def}{=} \mathcal{O}_1\times\mathcal{O}_\infty .$$

In [1] the class R of all standard domains was completely described. Moreover in [1] the properties of $\mathfrak{D}\in R$ were found out determining to which particular one of these three spaces the space $\mathcal{O}(\mathfrak{D})$ is isomorphic. These properties involve the structure of the set of all irregular points of the boundary $\partial\mathfrak{D}$ (see [2] for notions from the potential theory). The isomorphic classification of spaces $\mathcal{O}(\mathfrak{D})$ for $\mathfrak{D}$'s not in R remains unknown.

Any domain $\mathfrak{D}(q,r) \overset{def}{=} \hat{\mathbb{C}}\setminus(\bigcup_{j=1}^{\infty} \operatorname{clos}\mathbb{D}(q^j,r_j)\cup\{0\})$, where $\mathbb{D}(a,r)=\{\zeta:|\zeta-a|<r\}$, $q\in(0,1)$, $r=(r_j)_{j\geq1}$ is a monotone sequence of positive numbers with

$$\sum_{j=1}^{\infty}\left(\log\frac{1}{r_j}\right)^{-1}<\infty, \tag{1}$$

does not belong to R (and $\mathfrak{D}(q,r)\in R$ whenever the series in(1) diverges).

CONJECTURE. <u>There exists a continuum of mutually non-isomorphic spaces</u> $\mathcal{O}(\mathfrak{D}(q,r))$.

This conjecture is stated also in [6] (problem 63).

Let us mention - in connection with the open question on the existence of a basis in the space $\mathcal{O}(K)$ of all functions analytic on a compact set K, $K\subset\mathbb{C}$, that this question is open for $K=\mathbb{C}\setminus\mathfrak{D}(q,r)$ as well (under condition (1)), though it was proved [7] for such K that $\mathcal{O}(K)$ has no basis in common with $\mathcal{O}(\mathfrak{D})$, $\mathfrak{D}$ being any regular (in the potential - theoretic sense) neighbourhood of K. From this fact it follows that $\mathcal{O}(K)$ has no basis of the form $\{\sum_{j=0}^{n} P_{jn} z^j\}_{n=0}^{\infty}$.

Let Ω be 1-dimensional open Riemann surface.

(a) We say that Ω is regular iff there exists the Green function $G(\zeta, z)$ with $\lim_{n\to\infty} G(\zeta, z_n) = 0$, $\zeta \in \Omega$, for any sequence (z_n) with no limit point in Ω. Under additional restrictions (for example if Ω is a relatively compact subdomain of another Riemann surface Ω_1) it has been proved that $\mathcal{O}(\Omega)$ and $\mathcal{O}_1$ are isomorphic if Ω is regular (cf.[8] and references therein). Is this true in the general case? The necessity of this condition follows from general results for Stein manifolds ([3] and references therein).

(b) Let Ω be a Riemann surface with the ideal boundary of capacity zero. Is then $\mathcal{O}(\Omega)$ isomorphic to $\mathcal{O}_\infty$? The condition is necessary even in the multidimensional case (unpublished).

(c) The QUESTION about the existence of a basis in $\mathcal{O}(\Omega)$ is solved only under some additional restrictions (even for surfaces satisfying (a) and (b) above [8],[9]).

Clearly $\mathcal{O}(G)$ and $\mathcal{O}(K)$ are non-isomorphic whenever $\mathcal{O}$ is open and K is a compact set $(\mathcal{O}, K \subset \mathbb{C})$.

QUESTION. Which other differences in topological properties of sets $E_1, E_2 (\subset \mathbb{C})$ imply that $\mathcal{O}(E_1)$ and $\mathcal{O}(E_2)$ are non-isomorphic?

Here $\mathcal{O}(E)$ denotes the inductive limit of the net $\{\mathcal{O}(V)\}$ of countably normed spaces $\mathcal{O}(V)$, V running through the set of all open neighbourhoods of E. V.P.Erofeev proved (unpublished) that $\mathcal{O}(\mathbb{D} \cup \alpha) \not\approx \mathcal{O}(\mathbb{D} \cup \beta)$, α and β being an open and a closed subarcs of the unit circle $\partial\mathbb{D}$. It is not known whether $\mathcal{O}(\mathbb{D} \cup \{1\})$ and $\mathcal{O}(\mathbb{D} \cup \beta)$ are isomorphic if β is a closed non-degenerate arc of $\partial\mathbb{D}$.

In [4] a method was proposed to construct common bases for $\mathcal{O}(\mathfrak{D})$ and $\mathcal{O}(K)$, $K \subset \mathfrak{D}$. This method uses a special orthogonal basis common for a pair of Hilbert spaces H_0, H_1 and for the Hilbert scale *) H^α generated by H_0 and H_1 essentially genera-

---

*) The notion of a Hilbert scale introduced by S.G.Krein has a number of important applications to problems of the isomorphic classification of linear spaces and to the theory of bases. We refer to the paper by Б.С.Митягин, Г.М.Хенкин, "Линейные задачи комплексного анализа", Успехи матем. наук, 1971, 26, N 4, 93-152 containing many results concerning spaces of analytic functions, a list of unsolved problems and an extensive bibliography. - Ed.

lizing well-known results of V.P.Erohin about common bases (see,e.g. [4],[3],[7],[8],[9]).

THEOREM ([4],[12],[8]). Let $K \subset \mathcal{D}$, $K = \{t \in \mathcal{D} : |f(t)| \leq \sup_K |f|, \forall f \in \mathcal{O}(\mathcal{D})\}$ and suppose $\mathcal{D} \setminus K$ is a regular domain in $\hat{\mathbb{C}}$ (or a relatively compact domain on a Riemann surface). Then there exist Hilbert spaces H_0, H_1 with

$$H_1 \hookrightarrow \mathcal{O}(\mathcal{D}) \hookrightarrow \mathcal{O}(K) \hookrightarrow H_0 \tag{2}$$

and for all spaces H^α of the corresponding scale

$$\mathcal{O}(\operatorname{clos} \mathcal{D}_\alpha) \hookrightarrow H^\alpha \hookrightarrow \mathcal{O}(\mathcal{D}_\alpha), \tag{3}$$

where $\mathcal{D}_\alpha = \{z \in \mathcal{D} : \omega(\mathcal{D}, K, z) < \alpha\} \cup K$, $\omega(\mathcal{D}, K, z)$ is the harmonic measure of $\partial\mathcal{D}$ with respect to $\mathcal{D} \setminus K$ ([3], p.299). All embeddings in (2) and (3) are continuous. The common orthogonal basis $(e_n)_{n \geq 0}$ of the spaces H_0, H_1 is a common basic in $\mathcal{O}(\mathcal{D})$ and $\mathcal{O}(K)$.

A QUESTION arises: how "far" is it possible to "move apart" the spaces H_0, H_1 satisfying (2) without breaking (3)?

Let $H^\infty(\mathcal{D})$ be a Banach space of all bounded functions analytic in $\mathcal{D}$. We consider Hilbert spaces H_1 with

$$H^\infty(\mathcal{D}) \hookrightarrow H_1 \hookrightarrow \mathcal{O}(\mathcal{D}). \tag{4}$$

The well-known Kolmogorov's problem about the validity of the asymptotic relation

$$\log d_n(A_K^{\mathcal{D}}) \sim -\frac{n}{\mathfrak{C}(K, \mathcal{D})}$$

for the n widths $d_n(A_K^{\mathcal{D}})$ of the compact set $A_K^{\mathcal{D}} \overset{\text{def}}{=} \{f \in H^\infty(\mathcal{D}) : \max_K |f| \leq 1\}$ ($\mathfrak{C}(K, \mathcal{D})$ is the Green's capacity of the compact set K with respect to $\mathcal{D}$) can be reduced to the following

PROBLEM. Describe all domains $\mathcal{D}$ with (4) $\Longrightarrow$ (3) (for a suitable H_0) ([8], see also [11]).

REFERENCES

1. З а х а р ю т а В.П. Пространства функций одного переменного, аналитических в открытых множествах и на компактах. - "Мат.сб.", 1970, 82, № 1, 84-98.
2. Л а н д к о ф Н.С. Основы современной теории потенциала. М., "Наука", 1966.
3. З а х а р ю т а В.П. Экстремальные плюрисубгармонические функции, гильбертовы шкалы и изоморфизм пространств аналитических функций многих переменных, I, II. - В сб.Теория функций, функц. анализ и их прилож., Харьков,1974, № 19, 133-157, № 21, 65-83.
4. З а х а р ю т а В.П. О продолжаемых базисах в пространствах аналитических функций одного и многих переменных. - Сибирск.матем.ж., 1967, 8, № 2, 277-292.
5. Д р а г и л е в М.М., З а х а р ю т а В.П., Х а п л а н о в М.Г. О некоторых проблемах базиса аналитических функций. - В сб.: "Актуальные проблемы науки", Ростов-на-Дону, 1967, 91-102.
6. Unsolved problems. Proceedings of the International Colloqium on Nuclear Spaces and Ideals in Operator Algebras, Warsaw, 1969. Warszawa - Wrocław, 1970, 467-483.
7. З а х а р ю т а В.П., К а д а м п а т т а С.Н. О существовании продолжаемых базисов в пространствах функций, аналитических на компактах. - Мат.заметки, 1980, 27, № 5, 701-713.
8. З а х а р ю т а В.П., С к и б а Н.И. Оценки n-поперечников некоторых классов функций, аналитических на римановых поверхностях. - Мат.заметки, 1976, 19, № 6, 899-911.
9. С е м и г у к О.С. О существовании общих базисов в пространстве аналитических функций на компактной римановой поверхности, Ростов.ун-т, Ростов-на-Дону, 10 с, библ.7 назв. (Рукопись деп. в ВИНИТИ 15 февр. 1977 № 620-77 Деп.) РЖМат 1977, 6 Б 138 Деп.).
10. W i d o m H. Rational approximation and n-dimensional diameter. - J.Approximation Theory, 1972, 5, N 2, 343-361.
11. С к и б а Н.И. Об оценке сверху n-поперечников одного класса голоморфных функций. - В сб."Труды молодых ученых кафедры высшей математики", РИМИ, Ростов-на-Дону, 1978. Депонировано в ВИНИТИ, № 1593-78 Деп.
12. N g u e n T h a n h V a n. Bases de Schauder dans certains espaces de fonctions holomorphes. - Ann.Inst.Fourier (Grenoble), 1972, 22, N 2, 169-253.

V. P. ZAHARIUTA
(В.П.ЗАХАРЮТА)
O. S. SEMIGUK
(О.С.СЕМИГУК)
N. I. SKIBA
(Н.И.СКИБА)

СССР 344711
Ростов-на-Дону
Ростовский государственный
университет

Ростовский Инженерно-строительный
институт

* * *

COMMENTARY BY THE AUTHORS

The problems a), b), c) (in a more general situation, namely for Stein manifolds) have been solved in [13] by a synthesis of results on Hilbert scales of spaces of analytic functions [3] and last results on characterization of power series spaces of finite or infinite type [14], [15].

We formulate one of this results as an example.

Let Ω be a connected Stein manifold. Ω is said to be P-regular if there exists a plurisubharmonic function $u(z)$ such that $u(z) < 0$ in Ω and $u(z_n) \to 0$ for any sequence (z_n) without limit points in Ω.

THEOREM. $A(\Omega) \simeq A(\mathbb{D}^n)$ <u>if and only if</u> Ω <u>is P-regular</u>.

REFERENCES

13. З а х а р ю т а В.П. Изоморфизм пространств аналитических функций. - Докл.АН СССР, 1980, 255, № 1, 11-14.
14. V o g t D. Eine Charakterisierung der Potenzreihenräume vom endlichen Typ und ihre Folgerungen, preprint (to appear in Studia Math.).
15. V o g t D., W a g n e r M.J. Charakterisierung der Unterräume der nuklearen stabilen Potenzreihenräume vom unendlichen Typ, preprint (to appear in Studia Math.).

1.11. ON ISOMORPHIC CLASSIFICATION OF F-SPACES

1. For a given family of positive sequences $\{a_{ip}\}$ let $K(a_{ip})$ be a Köthe space i.e. F is the space of all sequences $x=\{x_n\}_{n\geqslant 1}$ satisfying

$$|x|_p \overset{def}{=} \sum_{n\geqslant 1} |x_n|\, a_{np} < +\infty, \quad p=1,2,\dots . \tag{1}$$

The space $K(a_{ip})$ is endowed with the topology defined by the family of semi-norms (1). It is called a p o w e r (K ö t h e) s p a c e if $a_{ip} = h_p(i)\, a_i$, where $-\infty < h_p(i) \leqslant h_{p+1}(i)$, $h_p(i) - h_1(i) \leqslant C(p) < +\infty$, $i, p \in \mathbb{N}$.
For example the so-called power series spaces

$$E_\alpha(a) \overset{def}{=} K(\exp \alpha_p a_i), \quad \alpha_p \uparrow \alpha, \ -\infty < \alpha \leqslant +\infty, \quad a=(a_i), \tag{2}$$

are power spaces in our sense. $E_\alpha(a)$ is said to be of finite (infinite) type if $\alpha < +\infty$ $(\alpha = +\infty)$.

Consider two classes of power spaces:

1) the class $\mathcal{E}$ of power spaces of the first kind [1],[2]:

$$E(\lambda, a) \overset{def}{=} K(\exp(-\tfrac{1}{p} + \lambda_i p) a_i),$$

2) the class $\mathcal{F}$ of power spaces of the second kind:

$$F(\lambda, a) \overset{def}{=} K(\exp \varphi_p(\lambda_i) a_i),$$

where $\varphi_p(t) \overset{def}{=} -\frac{1}{p} + \min\{\frac{1}{t}, p\}$, $p \in \mathbb{N}$.

Here $a=(a_i)$, $a_i > 0$; $\lambda = (\lambda_i)$, $0 \leqslant \lambda_i \leqslant 1$

in both cases.

If we consider isomorphic spaces as identical, then $\mathcal{E} \cap \mathcal{F}$ consists of spaces (2) and also of their cartesian products; $\mathcal{E}$ contains spaces $E_0(a) \hat{\otimes} E_\infty(b)$ [2] and $\mathcal{F}$ contains spaces of all analytic functions on unbounded n-circular domains in $\mathbb{C}^n$ [3].

PROBLEM 1. Give criteria of isomorphisms:

$$E(\lambda,a)\simeq E(\mu,b) \tag{3}$$

$$F(\lambda,a)\simeq F(\mu,b) \tag{4}$$

in terms of (λ,a), (μ,b).

Articles [1],[2] contain a criterion of isomorphism (3) but under an additional requirement on $E(\lambda,a)$ (note that in AMS translation of [1] in Lemma 4 the important chain of quantifiers „$\forall p' \exists p \forall q \exists q' \forall r' \exists r \forall s \exists s' \exists c \forall t, \tau,\sigma>0$" has been omitted).

Let us formulate one result on isomorphism (4), which somewhat generalizes the result of [3]. Denote by Λ the set of all sequences $\lambda=(\lambda_i)$, $0<\lambda_i\leqslant 1$ such that there exist limits

$$\varphi_\lambda(\theta)=\lim_{n\to\infty}\frac{|\{i:\lambda_i\leqslant\theta,\ i\leqslant n\}|}{n},\quad \theta\in(0,1],$$

and $\varphi_\lambda(\theta)$ is strongly increasing in θ.

THEOREM 1. Let $\lambda,\mu\in\Lambda$ and $a_{2i}\asymp a_i$. Then (4) implies 1) $a_i\asymp b_i$, 2) $\exists c$: $\frac{1}{c}\varphi_\mu\left(\frac{\theta}{c}\right)\leqslant\varphi_\lambda(\theta)\leqslant c\varphi_\mu(c\theta)$, $0<\theta\leqslant 1$.

Note for a comparison that isomorphism (3) takes place for arbitrary $\lambda,\mu\in\Lambda$ whenever the condition $a_i\asymp b_i$ is fulfilled.

Class $\mathcal{F}$ is also of a great interest because the following conjecture seems plausible.

CONJECTURE 1. There exists a nuclear power space of the second kind without the bases quasiequivalence property *).

2. Let X be an F-space with the topology defined by a system of semi-norms $\{|\cdot|_p,\ p\in\mathbb{N}\}$ and $\varphi(t)$ be a convex increasing function on $[1,\infty)$. Denote by $\mathcal{D}_\varphi$ the class of all spaces X such that $\exists p\,\forall q\,\exists m,r,c$:

*) The definition see e.g. in the article Митягин Б.С. Аппроксимативная размерность и базисы в ядерных пространствах. - Успехи матем. наук, I96I, I6, № 4, 63-I32.

$$|f|_q \leq (\varphi(t))^m |f|_p + \frac{c}{t}|f|_r, \quad t \geq 1, \quad f \in X.$$

Classes $\mathcal{D}_\varphi$, being invariant with respect to isomorphisms, are a modified generalization of Dragilev's class $\mathcal{D}_1$ [4] (see also [5]); similar dual classes Ω_φ were considered in Vogt-Wagner [6].

Classes $\mathcal{D}_\varphi$ have been used in [7] to give a positive answer to a question of Zerner. Consider the family $\mathcal{O}$ of all domains $\mathcal{G}$ with a single cusp:

$$\mathcal{G}_\varphi = \{(x,y) \in \mathbb{R}^2 : |y| < \psi(x),\ 0 < x < 1\},$$

ψ being a non-decreasing C^1-function on $[0,1]$, $\psi(0)=0$. Then there exists a continuum of mutually non-isomorphic spaces $C^\infty(\bar{\mathcal{G}})$ with $\mathcal{G} \in \mathcal{O}$. The following theorem clarifies the role of classes $\mathcal{D}_\varphi$ in this problem.

THEOREM 2. $C^\infty(\bar{\mathcal{G}}_\varphi) \in \mathcal{D}_\varphi$ <u>iff there exist</u> $\mu, \gamma > 0$ <u>satisfying</u> $1/\psi(x) \leq \left(\varphi\left(\frac{1}{x^\gamma}\right)\right)^\mu$ <u>for</u> $0 < x < x_0$.

PROBLEM 2. <u>Are all spaces</u> $C^\infty(\bar{\mathcal{G}})$ <u>from the same class</u> $\mathcal{D}_\varphi$ <u>isomorphic or there exists a more subtle</u> (<u>than in Theorem 2</u>) <u>classification of these spaces</u>?

CONJECTURE 2. <u>There exists a modification</u> (<u>apparently very essential one) of Vogt-Wagner's classes</u> Ω_φ <u>which allows to prove the conjecture on the existence of a continuum of mutually non-isomorphic spaces</u> $\mathcal{O}(\mathcal{D}(q,r))$ (see this Collection, Problem 1.10)

REFERENCES

1. З а х а р ю т а В.П. Об изоморфизме и квазиэквивалентности базисов для степенных пространств Кёте.- ДАН СССР, 1975, 221, № 4, 772-774.
2. З а х а р ю т а В.П. Об изоморфизме и квазиэквивалентности базисов для степенных пространств Кёте.-Труды 7-й Дрогобычской матем.школы по функц.анализу, М., 1974.
3. З а х а р ю т а В.П. Обобщенные инварианты Митягина и конти-

нуум попарно неизоморфных пространств аналитических функций.- Функциональный анализ и его приложения, I977, II, № 3, 24-30.

4. З а х а р ю т а В.П. Некоторые линейные топологические инварианты и изоморфизмы тензорных произведений центров шкал.-Известия Северо-Кавказского научного центра высшей школы, I974, 4, 62-64.

5. V o g t D. Charakterisierung der Unterräume von S.-Math.Z., 1977, 155, 109-117.

6. V o g t D., W a g n e r M.J. Charakterisierung der Quotientenräume von S und eine Vermutung von Martineau.-Studia Math. 1980, 67, 225-240.

7. Г о н ч а р о в А.П., З а х а р ю т а В.П. Пространство бесконечно дифференцируемых функций на областях с углами (to appear)

V.P.ZAHARIUTA
(В.П.ЗАХАРЮТА)

СССР, 344711, Ростов-на-Дону,
Ростовский государственный
университет

1.12. WEIGHTED SPACES OF ENTIRE FUNCTIONS

Let $p: \mathbb{C} \to \mathbb{R}$ be a continuous function and define

$$\mathcal{F}_R = \{ f \text{ entire: } \|f\|_r \overset{\text{def}}{=} \sup_{z \in \mathbb{C}} |f(z) \exp(-p(rz))| < \infty, \ \forall r > R \},$$

where $R \in \mathbb{R}_+$. We suppose that $\|\cdot\|_r \leqslant \|\cdot\|_s$ for $r > s > R$ and that $\mathcal{F}_R$ is not trivial. It is easily seen that $\mathcal{F}_R$ is a Fréchet space, the topology of which is strictly stronger than the topology of uniform convergence on the compact subsets of $\mathbb{C}$. With the help of the Riesz representation theorem the dual space of $\mathcal{F}_R$ can be identified with the space of all complex valued measures μ on $\mathbb{C}$ such that

$$\int_{\mathbb{C}} \exp(rp(z))\, d|\mu|(z) < \infty$$

for an $r > R$ (see [5]).

As an example consider $p(z) = |z|^{\alpha}$, for $\alpha > 0$, then $\mathcal{F}_R$ is the space of all entire functions of order α and type R^{α} (see [7]); in this case the monomials $\{z^n\}_{n \geqslant 0}$ constitute a Schauder basis in $\mathcal{F}_R$ and $\mathcal{F}_R$ is topologically isomorphic to the space $\mathcal{H}$ of all holomorphic functions on the disc $\mathbb{D}$ if $R > 0$ and to the space $\mathcal{H}(\mathbb{C})$ of all entire functions if $R = 0$, both spaces $\mathcal{H}(\mathbb{D})$ and $\mathcal{H}(\mathbb{C})$ endowed with the topology of compact convergence. Here, $\mathcal{F}_R$ is also a nuclear Fréchet space (see [7]) and the dual space can be identified with a space of germs of holomorphic functions (Köthe duality [4]). All this can be used to find a solution for interpolation problems such as for instance $f^{(n)}(\lambda_n) = a_n$, $n = 0, 1, 2, \ldots$ in the space $\mathcal{F}_R$ by means of methods from functional analysis (see [3], [1]).

If we take $p(z) = |e^z| = e^x$, $z = x + iy$ (here p is not a function of $|z|$), then the corresponding spaces $\mathcal{F}_R$ do not contain the polynomials and properties similar to the above example are not known. Another example of interest is due to Gel'fand and Shilov [2]:

$$S_{\alpha,A}^{\beta,B} = \{ f \in C^{\infty}(\mathbb{R}) : \sup_{x \in \mathbb{R}} |x^k f^{(q)}(x)| / (\tilde{A}^k \tilde{B}^q k^{k\alpha} q^{q\beta}) < \infty, \ \tilde{A} > A, \tilde{B} > B \}$$

$k, q = 0, 1, 2, \ldots$, where $\alpha, \beta, A, B > 0$ and $\alpha + \beta \geqslant 1$.

In fact, each function $f \in S_{\alpha,A}^{\beta,B}$ can be extended to $\mathbb{C}$ and $S_{\alpha,A}^{\beta,B}$ coincides with the space of all entire functions such that

$$\sup_{z\in\mathbb{C}} |f(z)\exp(a'|x|^{1/\alpha} - b'|y|^{1/(1-\beta)})| < \infty,$$

where $0 < a' < a$, $b' > b > 0$ (see [2]).

The following problems are of special interest if the weight ρ is not a function of $|z|$.

PROBLEM 1. Is it possible to find a representation of the dual space $\mathcal{F}_R'$ of $\mathcal{F}_R$ as a space of certain holomorphic functions or germs of holomorphic functions, analogous to the so called "Köthe-duality" [4] for the space $\mathcal{H}(\mathbb{D})$ or $\mathcal{H}(\mathbb{C})$?

PROBLEM 2: For which weights ρ is the space $\mathcal{F}_R$ nuclear?

Mityagin [6] proved the nuclearity of the spaces $S_{\alpha,A}^{\beta,B}$.

PROBLEM 3. Existence of Schauder bases in $\mathcal{F}_R$.

This problem seems to be quite difficult. If $\mathcal{F}_R$ is nuclear and has a Schauder basis, then $\mathcal{F}_R$ can be identified with a Köthe sequence space (see [7]). If the monomials $\{z^n\}_{n\geq 0}$ constitute a Schauder basis in $\mathcal{F}_R$, as in the first example, then $\mathcal{F}_R$ is a so called power-series-space(see [7]). Let $h \in \mathcal{F}_R$ be an entire function, which is not of the form $az+b$, $a,b \in \mathbb{C}$, then $\mathrm{span}(h^n : n \geq 1) \neq \mathcal{F}_R$: by our assumption on h, there exist two points $z_1, z_2 \in \mathbb{C}$, $z_1 \neq z_2$ with $h(z_1) = h(z_2)$, now set $\mu = \delta_{z_1} - \delta_{z_2}$, where δ_{z_i} denotes the Dirac measures $(i = 1,2)$; then $\mu \in \mathcal{F}_R'$ and $\langle h^n, \mu \rangle = 0 \quad \forall n \in \mathbb{N}$, therefore, by the Hahn-Banach theorem, $\mathrm{span}(h^n : n \geq 1) \neq \mathcal{F}_R$. So, if $\mathcal{F}_R$ does not contain the monomials, $\mathcal{F}_R$ cannot have a Schauder basis of the form $\{h^n\}_{n\geq 1}$. B.A.Taylor [8] constructed an example of a weighted space of entire functions containing the polynomials and the function $\exp(z)$, but where $\exp(z)$ cannot be approximated by polynomials.

REFERENCES

1. Berenstein C.A. and Taylor B.A. A new look

at interpolation theory for entire functions of one variable. - Adv. of Math., 1979, 33, 109-143.

2. G e l ' f a n d I.M. and S h i l o v G.E. Verallgemeinerte Funktionen II, III. VEB Deutscher Verlag der Wissenschaften, Berlin 1962.
3. H a s l i n g e r F. and M e y e r M. Abel - Gončarov approximation and interpolation. - Preprint.
4. K ö t h e G. Topologische lineare Räume. Berlin, Heidelberg, New York, Springer Verlag, 1966.
5. M a r t i n e a u A. Equations différentielles d'ordre infini. - Bull.Soc.Math. de France, 1967, 95, 109-154.
6. М и т я г и н Б.С. Ядерность и другие свойства пространств типа S . - Тр.Москв.матем.о-ва, 1960, 9, 317-328. (Amer.Math. Soc.Transl., 1970, 93, 45-60).
7. R o l e w i c z S. Metric linear spaces. Warsaw, Monografie Matematyczne, 56, 1972.
8. T a y l o r B.A. On weighted polynomial approximation of entire functions. - Pac.J.Math., 1971, 36, 523-539.

F.HASLINGER

Institut für Mathematik
Universität Wien
Strudlhofgasse 4
A-1090 Wien
AUSTRIA

1.13. old LINEAR FUNCTIONALS ON SPACES OF ANALYTIC FUNCTIONS AND THE LINEAR CONVEXITY IN $\mathbb{C}^n$

A domain $\mathcal{D}$ in $\mathbb{C}^n$ is called linearly convex (l.c.) if for each point ζ of its boundary $\partial\mathcal{D}$ there exists an analytic plane $\{z\in\mathbb{C}^n : a_1z_1+\dots+a_nz_n+b=0\}$ passing through ζ and not intersecting $\mathcal{D}$. A set E is said to be approximable from inside (from outside) by a sequence of domains $\mathcal{D}_k$, $k=1,2,\dots$ if $\operatorname{Clos}\mathcal{D}_k\subset\mathcal{D}_{k+1}$ (resp., $\operatorname{Clos}\mathcal{D}_{k+1}\subset\mathcal{D}_k$) and $E=\bigcup_k\mathcal{D}_k$ (resp., $E=\bigcap_k\mathcal{D}_k$). A compact set M is called linearly convex (l.c.) if there exists a sequence of l.c. domains approximating M from the outside. Applications of these notions to a number of problems of Complex Analysis, similar concepts introduced by A.Martineau and references may be found in [1]-[5].

If $\mathcal{D}$ is a bounded l.c. domain with C^2-boundary then every function continuous in $\operatorname{Clos}\mathcal{D}$ and holomorphic in $\mathcal{D}$ has a simple integral representation in terms of its boundary values. The representation follows from the Cauchy-Fantappié formula [7] and is written explicitly in [8],[1],[2]. It leads to a description of the conjugate space of the space $\mathcal{O}(\mathcal{D})$ (resp. $\mathcal{O}(M)$) of all functions holomorphic in a l.c. domain $\mathcal{D}$ (resp. on a compact set M) which can be approximated from inside (from the outside) by bounded l.c. domains with C^2-boundary (see [9] for convex domains and compacta and [2] for linear convex sets; the additional condition on approximating domains imposed in [2] can be removed). Such an approximation is not always possible [6]. This description of the conjugate space is a generalization of well-known results by G.Köthe, A.Grotendieck, Sebastião e Silva, C.L. da Silva Dias and H.G.Tillman for the case $n\geqslant 1$. Let $0\in E$. Then $\tilde{E}=\{w\in\mathbb{C}^n : w_1z_1+\dots+w_nz_n\neq 1$ for every $z\in E\}$ is called the conjugate set and plays the role of "the exterior" in this description. Let

$$\mathcal{D}_m=\{z:\Phi_m(z)<0\},\ 0\in\mathcal{D}_m,\ m=1,2,\dots$$

be the approximating domains specified above, $\Phi_m\in C^2$, $\operatorname{grad}\Phi_m\neq 0$ on $\partial\mathcal{D}_m$. Consider a differential form

$$\omega(u,z)=\frac{(n-1)!}{(2\pi i)^n\langle u,z\rangle^n}\sum_{k=1}^{n}(-1)^{k-1}u_k\,du_1\wedge\dots\wedge du_{k-1}\wedge$$

$$\wedge du_{k+1} \wedge \ldots \wedge du_n \wedge dz_1 \wedge \ldots \wedge dz_n ,$$

where $\langle u, z\rangle = u_1 z_1 + \ldots + u_n z_n$. Let $\tau(\Phi) = (\tau_1(\Phi), \ldots, \tau_n(\Phi))$ where $\tau_k(\Phi) = \Phi'_{z_k} \langle \operatorname{grad} \Phi(z), z\rangle^{-1}$. Every linear continuous functional F on $O(\mathfrak{D})$ (on $O(M)$) has a representation

$$F(f) = \int_{\partial\mathfrak{D}_m} f(z)\varphi(\tau(\Phi_m))\,\omega(\operatorname{grad}\Phi_m(z), z), \tag{1}$$

where $\varphi \in O(\tilde{\mathfrak{D}})$ (respectively, $\varphi \in O(\tilde{M})$), m depends only on φ. Formula (1) establishes an isomorphism between the linear topological spaces $O'(\mathfrak{D})$ and $O(\tilde{\mathfrak{D}})$ (respectively $O'(M)$ and $O(\tilde{M})$).

PROBLEM I. <u>Describe l.c. domains and compact sets which can be approximated from inside (from outside) by bounded l.c. domains with C^2-boundary.</u>

Let $z \in \partial\mathfrak{D}$, $0 \in \mathfrak{D}$ and let $\Gamma(z)$ denote the set of $u \in \mathbb{C}^n$ such that the plane $\{\zeta : \langle u, \zeta\rangle = 1\}$ passes through z and does not intersect $\mathfrak{D}$.

CONJECTURE I. <u>A bounded l.c. domain $\mathfrak{D}$, $0 \in \mathfrak{D}$ with the piecewise smooth boundary $\partial\mathfrak{D}$ admits the approximation incicated in PROBLEM I if and only if the sets $\Gamma(z)$ are connected for all $z \in \partial\mathfrak{D}$.</u>

Let $F \in O'(\mathfrak{D})$ (respectively, $F \in O'(M)$). The function $F_z[(1 - \langle z, w\rangle)^{-n}]$ is called the *Fantappié indicator*; here $z \in \mathfrak{D}$, $w \in \tilde{\mathfrak{D}}$, $0 \in \mathfrak{D}$ (respectively, $z \in M$, $w \in \tilde{M}$, $0 \in M$). The function φ in (1) is the Fantappié indicator of the functional F. A l.c. domain $\mathfrak{D}$ (a compact M) is called **strongly linearly convex**, **if the mapping which establishes the correspondence between functionals and** their Fantappié indicators is an isomorphism of spaces $O'(\mathfrak{D})$ and $O(\tilde{\mathfrak{D}})$ (respectively $O'(M)$ and $O(\tilde{M})$). Similar definition has been introduced by A.Martineau (see references in [9]). Every convex domain or compact is strongly l.c. (see, for example, [9]). At last, the result from [2] discussed above means that the existence of approximation indicated in PROBLEM 1 is sufficient for the strong linear convexity. Strongly l.c. sets have applications in such problems of multidimensional complex analysis, as decompositions of holomorphic functions into series of simplest fractions or into generalized Laurent series, the separation of singularities [1],[2],[5]. That is why the following problem is of interest.

PROBLEM II. Give a geometrical description of strongly linearly convex domains and compact sets.

CONJECTURE 2. A domain (a compact set) is a strongly l.c. set if and only if there exists an approximation of this set indicated in PROBLEM 1.

It was shown in [5] that under some additional conditions, the intersection of any strongly l.c. compact with any analytic line contains only simply connected components. The next conjecture arose in Krasnoyarsk Town Seminar on the Theory of Functions of Several Complex Variables.

CONJECTURE 3. A domain (a compact set) is a strongly l.c. set if and only if the intersection of this set with any analytic line is connected and simply connected.

Let $\mathfrak{D}$ be a bounded l.c. domain with the piecewise smooth boundary $\partial\mathfrak{D}$. The set $\gamma = \{(\zeta, u) \in \mathbb{C}^{2n} : \zeta \in \partial\mathfrak{D},\ u \in \Gamma(\zeta)\}$ is called the Leray boundary of $\mathfrak{D}$. Suppose that γ is a cycle. In this case it can be shown that for any function f holomorphic in $\mathfrak{D}$ and continuous in $\operatorname{Clos}\mathfrak{D}$, we have

$$f(z) = \int_{\gamma} f(\zeta)\omega(u, \zeta - z), \quad z \in \partial\mathfrak{D}. \tag{2}$$

This representation generalizes the integral formula indicated at the beginning of the note to the case of l.c. domains with non-smooth boundaries. If a l.c. domain $\mathfrak{D}$ (a compact set M) can be approximated from inside (from the outside) by l.c. domains whose Leray boundaries are cycles then every linear continuous functional on $O(\mathfrak{D})$ $(O(M))$ can be described by a formula analogous to (1) with $\gamma(\mathfrak{D}_m)$ instead of $\partial\mathfrak{D}_m$. Note that such a domain $\mathfrak{D}$ (a compactum M) is strongly l.c. Therefore the following problem is closely connected with PROBLEM II.

PROBLEM III. Describe bounded l.c. domains whose Leray boundary is a cycle.

This problem is important not only in connection with the description of linear continuous functionals on spaces of functions holomorphic in l.c. domains (on compacta). Formula (2) would have other interesting consequences (cf.[1],[2]).

CONJECTURE IV. The classes of domains in problems I-III coincide.

REFERENCES

1. Айзенберг Л.А. О разложении голоморфных функций многих комплексных переменных на простейшие дроби. - Сиб.мат.ж. 1967, 8, №5, 1124-1142.
2. Айзенберг Л.А. Линейная выпуклость в $\mathbb{C}^n$ и разделение особенностей голоморфных функций. - Bull.Acad.Polon.Sci., Ser.mat., 1967, 15, N 7, 487-495.
3. Айзенберг Л.А., Трутнев В.М. Об одном методе суммирования по Борелю n-кратных степенных рядов. - Сиб. мат.ж. 1971, 12, № 6, 1398-1404.
4. Айзенберг Л.А., Губанова А.С. Об областях голоморфности функций с действительными или неотрицательными тейлоровскими коэффициентами. - Теор.функций, функ.анализ и их прилож., 1972, 15, 50-55.
5. Трутнев В.М. О свойствах функций, голоморфных на сильно линейно выпуклых множествах. - В сб."Некотор.свойства голоморф. функ.мног.компл.перем.", Красноярск, 1973, 139-155.
6. Айзенберг Л.А., Южаков А.П., Макарова Л.Я. О линейной выпуклости в $\mathbb{C}^n$. - Сиб.мат.ж. 1968, 9, № 4, 731-746.
7. Лере Ж. Дифференциальное и интегральное исчисления на комплексном аналитическом многообразии. М., ИЛ, 1961.
8. Айзенберг Л.А. Интегральное представление функций, голоморфных в выпуклых областях пространства $\mathbb{C}^n$. - ДАН СССР, 1963, 151, 1247-1249.
9. Айзенберг Л.А. Общий вид линейного непрерывного функционала в пространствах функций, голоморфных в выпуклых областях $\mathbb{C}^n$. - ДАН СССР, 1966, 166, 1015-1018.

L.A.AIZENBERG
(Л.А.АЙЗЕНБЕРГ)

СССР, 660036, Красноярск
Академгородок
Институт физики СО АН СССР

COMMENTARY BY THE AUTHOR

A solution of Problem II given in [10], [11] shows that Conjecture III is true. The definition of a strong linear convexity (s.l.c) due to Martineau differs from the definition in the text only by the power (-1) (instead of $(-n)$) in the indicatrix formula. The two definitions turned out to be equivalent.

Yu.B.Zelinsky has shown in [12], [13] that the second conditions of Conjecture III and Conjecture I are equivalent. They mean the acyclicity of all sections of the domain by analytic planes of a fixed dimension k, $1 \leq k < n$, and coincide with the s.l.c. [1] These conditions form a precise "complex analogue" of the usual convexity [11], [12]. But the standard convex machinery cannot be generalized to this context. In particular, an ε-contraction of a s.l.c. domain is in general no more s.l.c.

Using the results of [10] one can show that the sections of s.l.c. domains are not too tortuous. This observation yields examples of unbounded s.l.c. domains non-approximable by bounded l.c. domains with smooth boundaries.

REFERENCES

10. З н а м е н с к и й С.В. Геометрический критерий сильной линейной выпуклости. - Функц.анал. и его прил., 1979, т.13, № 3, 83-84.

11. З н а м е н с к и й С.В. Эквивалентность различных определений сильной линейной выпуклости. Международная конференция по комплексному анализу и приложениям. Варна, 20-27 сентября 1981 г., 30.

12. Z e l i n s k y Y.B. On the strongly linear convexity. International conference on complex analysis and applications. Varna, September 20-27, 1981, 198.

13. З е л и н с к и й Ю.Б. О геометрических критериях сильной линейной выпуклости. Доклады Академии Наук СССР, 1981, т.261, № 1, 11-13.

1.14. ON THE UNIQUENESS OF THE SUPPORT OF AN ANALYTIC FUNCTIONAL
old

The symbol $H(E)$ will denote the space of all functions analytic on the (open or compact) set E, $E \subset \mathbb{C}^n$, endowed with the usual topology. Elements of the dual space $H'(\mathcal{D})$ (here and below $\mathcal{D}$ stands for an open set) are called analytic functionals (= a.f.). A.Martineau has introduced the notions of the carrier (porteur) and of the support of an a.f.

A compact set K, $K \subset \mathcal{D}$, is called a carrier of an a.f. T if T admits a continuous extension onto $H(K)$, or equivalently [2] if T is continuously extendable onto $H(\omega)$ for an arbitrary open ω, $\mathcal{D} \supset \omega \supset K$. Every a.f. has at least one carrier.

Let $\mathcal{A}$ be a family of compact subsets of $\mathcal{D}$ such that if $\{A_\alpha\}$ is a subfamily of $\mathcal{A}$ linearly ordered by inclusion then $\bigcap A_\alpha \in \mathcal{A}$. A compact set K, $K \in \mathcal{A}$ is called an $\mathcal{A}$-support of the analytic functional T if K is a carrier of T and K is minimal (with respect to the inclusion relation) among all carriers of T in $\mathcal{A}$. If $\mathcal{D}$ has a fundamental sequence of compact sets from $\mathcal{A}$ then any analytic functional has an $\mathcal{A}$-support but in general the $\mathcal{A}$-support is not unique. It is possible to consider various families of compact subsets of $\mathcal{D}$, e.g. the family of all compact subsets of $\mathcal{D}$, the family of $H(\mathcal{D})$-convex compact sets, the family of all convex compact sets (in this case an $\mathcal{A}$-support is called a convex support or a C-support).

Any analytic functional T on $\mathbb{C}^1$ has a unique C-support but T can have many polynomially convex ($= pc$) supports. (If for example

$$T(f) = \int_0^1 f(z)\,dz, \qquad f \in H(\mathbb{C}^1),$$

then any simple arc connecting 0 and 1 is a pc-support of T.)

PROBLEM. <u>Describe convex compact sets</u> $K (\subset \mathbb{C}^n)$ <u>such that</u> K <u>is the unique</u> C-<u>support of any analytic functional</u> C-<u>supported by</u> K.

C.O.Kiselman [3] has obtained for $n=1$ necessary and sufficient conditions for a compact set to be a unique pc-support. For $n>1$ a compact set with a C^2-boundary is a unique pc-support [4]. Kiselman has proved in [4] that a convex compact set with a smooth boundary is a unique $\mathbb{C}$-support. A stronger result is due to Martineau [5]: a convex compact set K is a unique $\mathbb{C}$-support if any extreme point p of K ($p \in \mathrm{Extr}\, K$) belongs to a unique complex supporting hyperplane (with respect to the complex affine manifold $V(K)$ generated by K).

Our problem is stated for the above two families of compacts only, though it is interesting for other families as well. Using the ideas of Martineau one can prove the following

THEOREM. A convex compact set $K (\subset \mathbb{C}^n)$ is a unique $\mathbb{C}$-support if the set of all its supporting hyperplanes is the closure of the set of hyperplanes h with the following property: $\mathrm{Extr}\, K \cap h$ contains a point lying in a unique complex supporting hyperplane with respect to $V(K)$.

It is probable that the sufficient condition of the theorem is also necessary.

REFERENCES

1. Martineau A. Sur les fonctionneles analytiques et la transformation de Fourier-Borel. - J.Analyse Math., 1963, 9, 1-164.
2. Björk J.-E. Every compact set in $\mathbb{C}^n$ is a good compact set. - Ann.Inst.Fourier, 1970, 20, 1, 493-498.
3. Kiselman C.O. Compact d'unicité pour les fonctionnelles analytiques en une variable. - C.R.Acad.Sci., Paris, 1969, 266, 13, A661-A663.
4. Kiselman C.O. On unique supports of analytic functionals. - Arkiv för Math. 1965, 16, 6, 307-318.
5. Martineau A. Unicité du support d'une fonctionnelle analytique: un théorème de C.O.Kiselman. - Bull.Soc.Math.France, 1968, 92, 131-141.

V.M.TRUTNEV
(В.М.ТРУТНЕВ)

СССР, 660075, Красноярск,
ул.Маерчака 6,
Красноярский государственный
университет

CHAPTER 2

BANACH ALGEBRAS

Thirteen sections of this Chapter can be conventionally divided into three groups. General theory of Banach algebras is represented by Problems 2.1-2.5. This group of problems is connected mainly with the spectral structure of elements of an abstract Banach algebra.

The second group consists of a couple of problems concerning Convolution Measure Algebra followed by a problem on harmonic synthesis in group algebras. The convolution algebra $M(G)$ of all finite Borel measures on a locally compact abelian group G is interesting from many points of view and among them from the spectral one. The subject originates in the classical paper by Wiener and Pitt (Duke Math.J. 1938, 4, N 2, 420-436) and has been intensively studied so far. However, the bulk of all publications on the theme has revealed only different pathologies in the structure of $M(G)$, and the number of "positive" achievements here is not large. J.L.Taylor has calculated the cohomologies of the maximal ideal space of $M(G)$ (see [2] in references of Problem 2.6). G.Brown and W.Moran have described the structural semi-groups of important subalgebras of $M(G)$ (Acta Math., 1974, 132, N 1-2, 77-109). Some years ago B.Host and M.Parreau solved a problem of I.Glicksberg (C.R.Ac.sc. Paris, 1977, 285, 15-17 and Ann.Inst.Fourier, 1978, 28, N 3, 143-164). They described all measures μ whose ideal $\mu * M(G)$ is closed in $M(G)$.

The question of description of Shilov boundary for $M(G)$, which is the subject of Problem 2.6, undoubtedly, is the core problem of the theory. It has been posed by J.L.Taylor and still remains unsolved. Problem 2.7 considers a description of homomorphisms of L-subalgebras of $M(G)$ in the spirit of the well-known Cohen-Rudin theorem. The last problem of the second group, Problem 2.8, deals with the structure of ideals in group algebras.

The third series of questions concerns more visible, but not less mysterious algebras such as the algebra $H^{\infty}(V)$ of all bounded and analytic functions in a domain $V\subset\mathbb{C}$. Note that Problem 2.9 contains an interesting conjecture about the axiomatic description of $H^{\infty}(\mathbb{D})$ in the category of all uniform algebras. Eleven problems are formulated in 2.10 and among them the Corona Problem for $H^{\infty}(V)$ which remained unsolved untill now. We would like to complete the list of references to 2.10 by the following ones. M.F.Behrens (Trans. Amer.Math.Soc., 1971, 161, 359-379) has shown that it is sufficient to solve the Corona Problem for a special class of domains. It is also known that for some of these V the algebra $H^{\infty}(V)$ does not have a corona. New progress has been obtained in a recent paper by L.Carleson (Proc.Conf.Harm.Anal.in Honour of A.Zygmund, Wadsworth Inc., Belmont, California, 1983, 349-372). The classical Hardy algebra $H^{\infty}= H^{\infty}(\mathbb{D})$ does not have a corona but nevertheless the structure of its maximal ideal space remains puzzling. Problem 2.11 is important for the understanding of this structure.

The last two problems of the third group concern the disk algebra, though the question posed in Problem 2.13 is considered in a more general setting.

2.1. old THE SPECTRAL RADIUS FORMULA IN QUOTIENT ALGEBRAS

If A is a complex Banach algebra and $x \in A$ let $r(x)$ denote the spectral radius of x. If I is a proper closed two-sided ideal of A, $x+I$ denotes the coset in the quotient algebra containing x. Clearly, by spectral inclusion, $r(x+I) \leq \inf_{y \in I} r(x+y)$. A is called an S R - a l g e b r a if equality holds in this formula for each x, $x \in A$, and each closed two-sided ideal I of A. The algebra $A(\mathbb{D})$ of all continuous functions on the disk analytic on its interior is not SR [1]. The following algebras are SR [2], [3], [1]: C^*-algebras, H^*-algebras, algebras of compact or Riesz operators, semi-simple dual algebras, semi-simple annihilator algebras and algebras with a dense socle. If A is commutative and has a discrete structure space then A is SR.

QUESTION. Is this true for noncommutative A?

Let A be commutative and let $\hat{A}$ be the Gelfand transform algebra of A and let $\Sigma(A)$ denote the spectrum of A. Then [2] if $\hat{A}$ is dense in $C_0(\Sigma(A))$, A is an SR-algebra. Conversely, if A is a regular SR-algebra, $\hat{A}$ is dense in $C_0(\Sigma(A))$.

QUESTION. Can the condition of regularity be omitted from this hypothesis?

Let A be a C^*-algebra and let x be any element in A and let I be any closed two-sided ideal of A.

QUESTION. Is it true that there always exists $y, y \in I$ (depending on x), such that $r(x+I) = r(x+y)$?

This result is true if $r(x+I) \neq 0$ and it is a corollary of [4] Theorem 3.8. The case $r(x+I) = 0$ is open.

REFERENCES

1. S m y t h M.R.F., W e s t T.T. The spectral radius formula in quotient algebras. - Math.Zeit. 1975, 145, 157-161.
2. M u r p h y G.J., W e s t T.T. Spectral radius formulae. - Proc.Edinburgh Math.Soc.(2), 1979, 22, N 3, 272-275.
3. P e d e r s e n G.K. Spectral formulas in quotient algebras. - Math.Zeit. 148.
4. A k e r m a n n C.A., P e d e r s e n G.K. Ideal perturbations of elements in C^*-algebras. - Math.Scand. 1977, 41, 137-139

G.J.MURPHY
M.R.F.SMYTH T.T.WEST

39 Trinity college
Dublin 2 Ireland

2.2. EXTREMUM PROBLEMS

1. THE GENERAL MAXIMUM PROBLEM. *Given a unital Banach algebra* A , *a compact set* F *in the plane and a function* f *holomorphic in a neighbourhood of* F , *to find*

$$\sup\{|f(a)|: |a|\leqslant 1,\ \sigma(a)\subset F\}.$$

The problem was formulated and solved first [1,2] in the case of

2. THE SPECIAL MAXIMUM PROBLEM. *Let* n *be a natural number,* r *a positive number less than one, to find, among all contractions* T *on* n*-dimensional Hilbert space whose spectral radius does not exceed* r *those (it turns out that there is essentially only one) for which the norm* $|T^n|$ *assumes its maximum.*

This is a particular case of the general problem for

$$A=B(H_n),\qquad F=\{z:|z|\leqslant r\}\ ,\quad f(z)=z^n.$$

The solution of problem 2 was divided into two stages. The first step consists in replacing the awkward constraint that the spectral radius be $\leqslant r$ by a more restrictive one which makes the problem considerably easier: the operator is to be annihilated by a given polynomial. This is

3. THE FIRST MAXIMUM PROBLEM. *Let* n *be a natural number,* p *a polynomial of degree* n *with all roots inside the unit disc. Let* $A(p)$ *be the set of all contractions* $T\in B(H_n)$ *such that* $p(T)=0$. *To find the maximum of* $|T^n|$, *more generally of* $|f(T)|$ *as* T *ranges in* $A(p)$.

We call this maximum $C(p,f)$.

Having solved the first maximum problem we have to solve

4. THE PROBLEM OF THE WORST POLYNOMIAL. *To find, among all polynomials with roots* $\leqslant r$ *in modulus that one for which* $C(p,f)$ *is maximal.*

For the case $f(x)=x^n$ the result of [2] shows that the worst polynomial is $p(z)=(z-r)^n$. The method used in [2] is based on some fairly complicated algebraic considerations and does not extend to functions other than x^n. It is not known whether the worst polynomial for f other than x^n has a root of multiplicity n whether the roots have to be concentrated on the boundary of the disc $|z| \leq r$. Thus it seems useful to study $C(p, f)$ as a function of the roots of p; a recent contribution to Problem 4 is [4]. A list of references up to 1979 is contained in the survey paper [3].

REFERENCES

1. P t á k V. Norms and the spectral radius of matrices. - Czechosl.Math.J. 1962, 87, 553-557.
2. P t á k V. Spectral radius, norms of iterates and the critical exponent. - Lin.Alg.Appl. 1968, 1, 245-260.
3. P t á k V., Y o u n g N.J. Functions of operators and the spectral radius. - Lin.Alg.Appl. 1980, 29, 357-392.
4. Y o u n g N.J. A maximum principle for interpolation in H^∞, Acta Sci.Math.Szeged.

VLASTIMIL PTÁK

Institute of Mathematics
Czechoslovak Academy of Sciences
Žitná 25
11567 Praha 1
Czechoslovakia

2.3. MAXIMUM PRINCIPLES FOR QUOTIENT NORMS IN H^∞

A surprising variety of starting points can lead one to study norms in quotient algebras of H^∞ by a closed ideal. Well known examples are classical complex interpolation [4], canonical models of operators [2] and some problems of optimal circuit design [1]. It also arises from a maximum problem for matrices to which V.Pták was led by considerations relating to numerical analysis. The problem is to estimate the maximum value of $\|\psi(A)\|$, where $\psi \in H^\infty$, over all contractions A of spectral radius at most $r<1$ on n-dimensional Hilbert space (Pták was mainly concerned with the case $\psi(A)=A^m$, some $m \in \mathbb{N}$). An account of this problem is given in [3].

The only known way of handling the spectral constraint is to replace it by the condition $p(A)=0$, for some polynomial p , and then to vary p among the polynomials of degree n having all their zeros in $r(\operatorname{clos}\mathbb{D})$. After some calculations one is led to study the functional

$$F(p) = \|\psi + pH^\infty\|_{H^\infty/pH^\infty}$$

as a function of $p \in H^\infty$ for fixed ψ . In particular, as p varies over the class of polynomials described above, <u>at what p does F attain its maximum</u>? The result we should like to prove is that F attains its maximum at a polynomial of the form $p(z)=(z-\varepsilon r)^n$ for some $\varepsilon \in \mathbb{C}$ with $|\varepsilon|=1$. Pták proved this was so in the case $\psi(z)=z^n$, and I proved that if ψ is a Blaschke product of degree n then all the zeros of an extremal polynomial p have modulus r: in fact, F is then the composition of a strictly increasing function and a plurisubharmonic function of the zeros of p [5]. Nothing, is known, however, about the most interesting case,

$$\psi(z) = z^m \text{ with } \quad m > n .$$

To formulate the problem concisely, let us say that the *maximum principle holds for* $f : \Omega \subset \mathbb{C} \to \mathbb{R}$ if for any compact set $K \subset \Omega$, the supremum of f on K is attained at some point of the boundary of K relative to Ω . And if M is a complex manifold, we shall say that the maximum principle holds for $F : M \to \mathbb{R}$ if, for any open set $\Omega \subset \mathbb{C}$ and any analytic function $G : \Omega \to M$, the maximum principle holds for $F \circ G$.

PROBLEM. Let $\psi \in H^\infty$ and let $F : \mathbb{D}^n \to \mathbb{R}$ be defined by

$$F(\alpha_1, \dots, \alpha_n) = \|\psi + \varphi H^\infty\|_{H^\infty/\varphi H^\infty}$$

where

$$\varphi(z) = \prod_{i=1}^{n} (z - \alpha_i) .$$

Does the maximum principle hold for F ?

REFERENCES

1. H e l t o n J.W. Non-Euclidean functional analysis and electronics. - Bull.Amer.Math.Soc., 1982, 7, 1-64
2. Н и к о л ь с к и й Н.К. Лекции об операторе сдвига. Москва, Наука, 1980.
3. P t á k V., Y o u n g N.J. Functions of operators and the spectral radius. - Linear Algebra and its Appl., 1980, 29, 357-392
4. S a r a s o n D. Generalized interpolation in H^∞. - Trans.Amer. Math.Soc., 1967, 127, 179-203.
5. Y o u n g N.J. A maximum principle for interpolation in H^∞. - Acta Sci.Math., 1981, 43, N 1-2, 147-152.

N.J.YOUNG

Mathematics department
University Gardens
Glasgow G128QW
Great Britain

2.4. OPEN SEMIGROUPS IN BANACH ALGEBRAS

Let A be a complex Banach algebra with identity, not necessarily commutative. Let S be some open multiplicative semigroup in A. For an element a in A let $d(a)$ denote the distance from the point a to the closed set $A \setminus S$ (in other words, it is the radius of the largest open ball centred at a and contained in S), and let $r(a)$ be the supremum of all $\varepsilon \geqslant 0$ such that the elements $a - \lambda \cdot 1$ belong to S for $|\lambda| < \varepsilon$ (that is the radius of the largest open disk centred at a and contained in the intersection of S with the subspace spanned by a and 1). So we clearly have $r(a) \geqslant d(a)$. For a variety of particular semigroups S we know that the formula

$$r(a) = \lim_{n \to \infty} d(a^n)^{1/n}$$

is valid for every a in A. We list below the most important cases.

First of all, if $S = G(A)$, the group of invertible elements, the result follows from the spectral radius formula. Second, the formula is true when S is the semigroup of left (or right) invertible elements of A, cf. [6]. Third, it also holds when S is the complement of the set of left (right) topological divisors of zero in A, cf. [1].

Next, the formula is true for various semigroups of the algebra $A = B(X)$ of bounded linear operators on a Banach space. In the case when S is the semigroup of surjective (or bounded from below) operators on X it was obtained in [1] by an analytic argument (in fact, this is equivalent to the third case mentioned before). Using an additional geometric device these results were applied in [6] to prove the above formula for the semigroup S of (upper or lower) semi-Fredholm operators on X, and hence it follows for the semigroup of Fredholm operators as well. In these cases the distance $d(T)$ admits other natural interpretations, namely, it coincides with (or is related to) certain geometric characteristics of the operator T (like the surjection modulus, the injection modulus, the essential minimum modulus [5], etc.).

In each of the cases listed above an individual approach was needed to find a proof. The difficult steps are of an analytic character, based on the theorems of G.R.Allan (1967) and J.Leiterer (1978) on analytic vector-valued solutions of linear equations depending analytically on a parameter. The main idea is well demonstrated in [1] though in that case a combinatorial argument is also available [2]. So there seems to be some motivation for investigating the problem in general to seek a theorem which would contain all these particular results.

Let us give some warnings. Let $S = G_1(A)$ be the principal component of the set of invertible elements in A. There are (non-commutative) Banach algebras A for which the group $G(A)/G_1(A)$ is finite, but not trivial, cf.[4],[3]. In such a situation for every invertible element x not in $G_1(A)$ one can find a positive integer k such that x^k is in $G_1(A)$. Then we have $\tau(x) = 0$ but $\tau(x^k) > 0$ so that the formula cannot be true for this S and all a in A. Moreover, it is easy to see that $\lim_{n\to\infty} d(x^n)^{1/n}$ cannot exist for these x.

Let A be a commutative Banach algebra and let S be the open semigroup of all elements whose spectra are contained in the open unit disk. In this case we have $\tau(x) = d(x)$ for every x in S but $\lim_{n\to\infty} d(x^n)^{1/n} = 1 \geqslant \tau(x)$, and the last inequality may be strict. We arrive at the same conclusion if we replace the unit disk in the preceding definition by a q-multiple of it, with $0 < q < 1$. This suggests some analogy between our problem and the classical formula for the radius of convergence of a power series (the formula does not give the radius of the disk where we are considering the function but the radius of the disk where the given function can naturally be defined).

Thus some additional conditions should be imposed on S in general. For instance, the property "if $ab = ba$ belongs to S then both a and b are in S " is shared by most of the semigroups for which the problem is solved in the affirmative. This condition ensures, by the way, that $\tau(x^n) = \tau(x)^n$ for all x in A and $n = 1, 2, \dots$ (we note that $\tau(x^n) \geqslant \tau(x)^n$ is always true). Is it important that the identity element be in S, or that S be connected, or "maximal"? Does $\lim_{n\to\infty} d(a^n)^{1/n}$ exist for a in an arbitrary open semigroup S ? What is the meaning of it in general?

I should like to thank Tom Ransford for a valuable discussion on this topic.

REFERENCES

1. M a k a i E., Jr., Z e m á n e k J. The surjectivity radius, packing numbers and boundedness below of linear operators. - Integr.Eq.Oper.Theory, 1983, 6.
2. M ü l l e r V. The inverse spectral radius formula and removability of spectrum.
3. P a u l s e n V. The group of invertible elements in a Banach algebra. - Colloq.Math., 1982, 47.
4. Y u e n Y. Groups of invertible elements of Banach algebras. - Bull.Amer.Math.Soc., 1973, 79.
5. Z e m á n e k J. Geometric interpretation of the essential minimum modulus. - Operator Theory: Advances and Applications, vol.6 p.225-227. Birkhäuser Verlag, Basel, 1982.
6. Z e m á n e k J. The semi-Fredholm radius of a linear operator. - To appear.

JAROSLAV ZEMÁNEK

Institute of Mathematics
Polish Academy of Sciences
00-950 Warszawa, P.O. Box 137
Poland

2.5. HOMOMORPHISMS FROM C^*-ALGEBRAS

Let A and B be Banach algebras. A basic AUTOMATIC CONTINUITY PROBLEM is to give algebraic conditions on A and B which ensure that each homomorphism from A into B is necessarily continuous.

An important tool in investigations of this problem is the separating space: if $\theta : A \to B$ is a homomorphism, then the *separating space* of θ is

$$\mathfrak{S}(\theta) = \{ b \in B : \text{ there is a sequence } (a_n) \subset A \text{ with } a_n \to 0 \text{ and } \theta a_n \to b \}.$$

Of course, θ is continuous if and only if $\mathfrak{S}(\theta) = \{0\}$. The basic properties of $\mathfrak{S}(\theta)$ are described in [6].

Consider the GENERAL QUESTION: <u>if</u> $b \in \mathfrak{S}(\theta)$, <u>what can one say about</u> $\sigma(b)$, <u>the spectrum of</u> b <u>in the Banach algebra</u> B ?

First let us note that, if $b \in \mathfrak{S}(\theta)$, then $b \in \ker \varphi$ for each character φ on B. For such a character φ is necessarily continuous, the character $\varphi \circ \theta$ is continuous on A, and so $\varphi(b) = \lim \varphi(\theta a_n) = \lim (\varphi \circ \theta)(a_n) = 0$. Thus, if B is commutative, it follows that $\sigma(b) = \{0\}$.

Is the same result true in the non-commutative case? An element b of a Banach algebra is a *quasi-nilpotent* if $\sigma(b) = \{0\}$, and so our question is the following.

QUESTION 1. <u>Let</u> $\theta : A \longrightarrow B$ <u>be a homomorphism, and let</u> $b \in \mathfrak{S}(\theta)$. <u>Is</u> b <u>necessarily a quasi-nilpotent element of</u> B ?

It can be shown that $\sigma(b)$ is always a connected subset of $\mathbb{C}$ containing the origin (see [6 , 6.16]), but nothing further seems to be known in general.

The question was raised as Question 5' in [3], and it is shown there that the question is equivalent to the following. Let $\theta : A \to B$ be a homomorphism, and suppose that $\overline{\theta(A)}$ is semi-simple. <u>Is</u> θ <u>necessarily continuous</u>?

It is shown by Aupetit in [2] that, if A and B are unital Banach algebras, if $\theta : A \to B$ is a homomorphism, and if $b \in \mathfrak{S}(\theta)$ then $\nu(\theta a) \leq \nu(b + \theta a)$ for all $a \in A$ (Here, ν denotes the spectral radius.) Thus, if $b \in \mathfrak{S}(\theta) \cap \theta(A)$, then $\nu(b) = 0$ and b is a quasi-nilpotent. However, it is not in general true that the set of quasi-nilpotents in a Banach algebra is closed (see [1]), and so we cannot immediately conclude from this result that each $b \in \mathcal{S}(\theta)$ is quasi-nilpotent.

Quite probably, there is a counter-example to Question 1. However, let us concentrate on the case in which both A and B are C^*-algebras.

QUESTION 2. If, in Question 1, both A and B are C^*-algebras, can we then conclude that b is necessarily quasi-nilpotent?

The continuity ideal of a homomorphism $\theta : A \longrightarrow B$ is the set

$$\mathcal{J}(\theta) = \{a \in A : \theta(a)\mathcal{S}(a) = \mathcal{S}(a)\theta(a) = \{0\}\}.$$

It was proved by Johnson ([4], see[6, 12.2]) that, if A is a C^*-algebra, then $\mathcal{J}(\theta)$ is a two-sided ideal in A and that its closure $\overline{\mathcal{J}(\theta)}$ has finite condimension in A. Next, Sinclair ([5, Theorem 4.1]) showed that, if A and B are both C^*-algebras, and if $\theta : A \longrightarrow B$ is a homomorphism with $\overline{\theta(A)} = B$ then $\theta \,|\, \overline{\mathcal{J}(\theta)}$ can be decomposed as $\mu + \lambda$, where μ is a continuous homomorphism and $\lambda : \overline{\mathcal{J}(\theta)} \longrightarrow \mathcal{S}(\theta)$ is a discontinuous homomorphism (or $\lambda = 0$). Now $\overline{\mathcal{J}(\theta)}$ and $\mathcal{S}(\theta)$ are closed ideals in A and B, respectively, and so both are C^*-algebras. Moreover, the range $\lambda(\overline{\mathcal{J}(\theta)})$ is a dense subalgebra of $\mathcal{S}(\theta)$ and so, by our above remarks, consists of quasi-nilpotent elements. Thus, $\mathcal{S}(\theta)$ is a C^*-algebra with a dense subalgebra consisting of quasi-nilpotents. No such C^*-algebra is known, and I would like it to be true that no such C^*-algebra exists. So we come to the sharpest form of our original question.

QUESTION 3. Is there a C^*-algebra (other than $\{0\}$) which has a dense subalgebra consisting of quasi-nilpotent elements?

If no such C^*-algebra exists, then the homomorphism λ in Sinclair's theorem must be zero, and so the element b in Question 2 must indeed be quasi-nilpotent.

REFERENCES

1. A u p e t i t B. Propriétés spectrales des algèbres de Banach. - Lect.Notes Math., 1979, 735, Springer-Verlag.

2. A u p e t i t B. The uniqueness of the complete norm topology in Banach algebras and Banach-Jordan algebras. - J.Functional Analysis, 1982, 47, 1-6.
3. D a l e s H.G. Automatic continuity: a survey. - Bull.London Math.Soc., 1978, 10, 129-183.
4. J o h n s o n B.E. Continuity of homomorphisms of algebras of operators II. - J.London Math.Soc.(2),1969, 1, 81-84.
5. S i n c l a i r A.M. Homomorphisms from C^*-algebras. - Proc. London Math.Soc., (3), 1974, 29, 435-452; Corrigendum 1976, 32, 322
6. S i n c l a i r A.M. Automatic continuity of linear operators. - London Math.Soc.Lecture Note Series, 21, C.U.P., Cambridge,1976.

H.G.DALES

School of Mathematics,
University of Leeds,
Leeds LS2 9JT.
Great Britain

2.6. old ANALYTICITY IN THE GELFAND SPACE OF THE ALGEBRA OF $L^1(\mathbb{R})$ MULTIPLIERS

We shall be concerned with spectral properties of the Banach algebra of those bounded linear operators on $L^1(\mathbb{R})$ which commute with translations. However, it is convenient to represent the action of each operator by convolution so that the object of study becomes the algebra $M(\mathbb{R})$ of bounded regular Borel measures on $\mathbb{R}$. The general problem to be considered is the classification of the analytic structure of the Gelfand space, Δ, of $M(\mathbb{R})$ despite the fact that Δ is sometimes regarded as the canonical example of a "horrible" maximal ideal space from the point of view of complex analysis (cf. [1] p.9). Some encouraging progress has been made in recent years and it will be possible to pose some specific questions which should be tractable.

We refer to Taylor's monograph, [2], for a survey of work up to 1973 (Miller's conjectured characterization of the Gleason parts of Δ has since been verified in [3]) and for further details concerning general theory of convolution measure algebras. In particular, we follow Taylor in representing Δ as the semigroup of continuous characters on a compact semigroup $\mathcal{S}$ (the so-called *structure semigroup* of $M(\mathbb{R})$) and in transferring measures in $M(\mathbb{R})$ to measures on $\mathcal{S}$. In this formulation an element f of Δ acts as a homomorphism according to the rule

$$f(\mu)=\int_{\mathcal{S}} f(s)\,d\mu(s).$$

Every member f of Δ then has a canonical polar decomposition, $f=|f|h$, where $|f|$, $h\in\Delta$ and h has idempotent modulus. If f itself does not have idempotent modulus (a possibility which corresponds to the Wiener-Pitt phenomenon and was first noted by Šreider [4]) then the map $z\to|f|^z h$, for $\mathrm{Re}(z)>0$, demonstrates analyticity in Δ. From that observation Taylor showed that the Šilov boundary ∂ of $M(\mathbb{R})$ is contained in clos θ, where

$$\theta=\{f\in\Delta:|f|=|f|^2\}.$$

He posed the converse question which is still unresolved. Subsequent work tends to suggest a negative answer so that we propose

CONJECTURE 1. $\theta\setminus\partial\neq\emptyset$.

It should be noted that the result $\partial\subset$ clos θ remains valid for abstract convolution measure algebras and that it is very easy to find convolution measure algebras for which $\theta\setminus\partial\neq\emptyset$. It is also possible to find natural L-subalgebras of $M(\mathbb{R})$, itself, for which the corresponding conjecture is true. (An L-subalgebra is a closed subalgebra A which contains all measures absolutely continuous with respect to any measure in A). Thus, a disproof would depend on not only a new phenomenon peculiar to $M(\mathbb{R})$ but one which is specific to the full algebra. In addition we established a weak form of the conjecture in [5] by showing that a certain idempotent $\mathbb{1}_d$ (see below) fails to be a strong boundary point for $M(\mathbb{R})$ (although it is a strong boundary point for the L-subalgebra of discrete measures). It should also be noted that Johnosn, [6], proved that $\Delta\setminus\partial\neq\emptyset$ but the techniques used to prove this result and its subsequent refinements depend essentially on the use of elements lying outside θ. A natural strategy is to embed $M(\mathbb{R})$ in a suitable super-algebra and prove the impossibility of extension of an appropriate homomorphism.

It appears to be almost as difficult to exhibit, in the opposite direction, large numbers of elements of θ which DO belong to ∂. Before we describe some progress in this direction let us introduce the notation $\mathbb{1}$ for the unit function in Δ and the notation $\mathbb{1}_d$ for the homomorphism given by

$$\mathbb{1}_d(\mu)=\int_{\mathbb{R}} d\mu_d ,$$

where μ_d is the discrete part of μ. ($\mathbb{1}_d$ plays the role of the unit function for the subalgebra of discrete measures, which can be regarded as $M(\mathbb{R}_d)$, where $\mathbb{R}_d$ is the discrete real line).

Let us define a partial order on Δ by saying that $f\leq g$ if

$$|f(s)|^2\leq g(s)\,\overline{f(s)} \qquad (s\in S).$$

We have shown in [7] that maximal elements are members of the Šilov boundary.

THEOREM 1. (i) <u>If</u> f <u>is maximal in</u> Δ <u>then</u> f <u>is a strong boundary point</u>.

(ii) <u>If</u> $|f|$ <u>is maximal in</u> $\Delta\setminus\{\mathbb{1}\}$ <u>then</u> f <u>belongs to</u> ∂. <u>If</u>, <u>moreover</u>, <u>the</u> L-<u>subalgebra</u>

$$\{\mu\in M(\mathbb{R}) : |f|\mu=\mu\}$$

<u>is countably generated then</u> f <u>is a strong boundary point</u>.

It is obvious that maximal elements belong to θ but not entirely trivial that there are many examples other than those homomorphisms induced by continuous characters of $\mathbb{R}$ (viz. extensions of non-zero homomorphisms of $L^1(\mathbb{R})$). To see that this is the case consider in connexion with (i), homomorphisms which are induced on the discrete measures by discontinuous characters, and in connexion with (ii), homomorphisms which annihilate some fixed member of $L^1(\mathbb{R})$.

The additional hypothesis in (ii) does not correspond to a specific obstruction and merely reflects the constructive nature of our proof. We have avoided a similar difficulty in (i) by an appeal to Rossi's local peak set theorem and it seems plausible that a similar device should be available here. A proof which reduced the uncountably generated case to the countably generated case by pure measure algebra techniques would be particularly interesting since this species of difficulty often arises. In any event we propose

CONJECTURE 2. _If_ f _is maximal in_ $\Delta\setminus\{\mathbb{1}\}$ _then_ f _is a strong boundary point_.

It would be useful to determine for specific subclasses of θ whether or not the elements are strong boundary points. The result that $\mathbb{1}_d$ is the centre of an analytic disc was extended in [8] to cover the case of the idempotent corresponding to any single generator Raikov system. On the other hand we show in [7] that $\mathbb{1}_d$ is accessible in the sense that it is the infimum of those maximal elements of $\Delta\setminus\{\mathbb{1}\}$ below which it lies. It is natural to expect that both results extend, although we feel that present techniques would require substantial development to prove

CONJECTURE 3. _The idempotents corresponding to proper Raikov systems are accessible but fail to be strong boundary points_.

We have chosen to present these problems from the standpoint of the development of the general theory. From a practical position the most useful results are those which exhibit classes of homomorphisms which belong to the Šilov boundary of L-subalgebras of $M(\mathbb{R})$ – because such results give information on spectral extension. In fact, THEOREM 1 is of this type because it remains valid for arbitrary convolution measure algebras (provided the technical hypothesis that $\mathbb{1}$ is a critical point is added to part (ii)). Variants of that theorem with the weaker conclusion that f belongs to the Šilov boundary but valid for a larger class of f would be of considerable interest.

REFERENCES

1. G a m e l i n T.W. Uniform Algebras. New Jersey, Prentice-Hall, 1969.

2. T a y l o r J.L. Measure Algebras. CBMS Regional confer.ser.

math., 16, Providence, Amer.Math.Soc., 1973.

3. Brown G., Moran W. Gleason parts for measure algebras. - Math.Proc.Camb.Phil.Soc., 1976, 79, 321-327.

4. Шрейдер Ю.А. Об одном примере обобщённого характера. - Матем.сб., 1951, 29, № 2, 419-426.

5. Brown G., Moran W. Point derivations on $M(G)$. - Bull.Lond.Math.Soc., 1976, 8, 57-64.

6. Johnson B.E. The Šilov boundary of $M(G)$. - Trans.Amer.Math.Soc., 1968, 134, 289-296.

7. Brown G., Moran W. Maximal elements of the maximal ideal space of a measure algebra. - Math.Ann., 1979/80, 246, N 2, 131-140.

8. Brown G., Moran W. Analytic discs in the maximal ideal space of $M(G)$. - Pacif.J.Math., 1978, 75, N 1, 45-57.

GAVIN BROWN

University of New South Wales
Sydney, Australia

WILLIAM MORAN

University of Adelaide
Adelaide, Australia

2.7. old ON THE COHEN-RUDIN CHARACTERISATION OF HOMOMORPHISMS OF MEASURE ALGEBRAS

Let $L(\mathbb{T})$ be the Lebesgue space and $M(\mathbb{T})$ the set of all bounded regular Borel measures on the unit circle $\mathbb{T}$. $M(\mathbb{T})$ is a commutative Banach algebra with the convolution product and the norm of total variation, and $L(\mathbb{T})$ is embedded in $M(\mathbb{T})$ as a closed ideal. A subalgebra N of $M(\mathbb{T})$ is said to be L - subalgebra if it is a closed subalgebra of $M(\mathbb{T})$ and $\mu\in N$ and $\nu\ll\mu$, that is, ν is absolutely continuous with respect to μ, implies $\nu\in N$.

Let $\Delta'(N)$ be the set of all homomorphisms of N to the complex numbers (which might be trivial). Then, by Yu. Šreider [1], for every ψ, $\psi\in\Delta'(N)$, there corresponds a unique generalized character $\{\psi_\mu:\mu\in N\}$ or zero system such that

$$\psi(\mu)=\int_{\mathbb{T}}\psi_\mu(t)\,d\mu(t),\quad \mu\in N .$$

In the following we shall use the same notation ψ for $\{\psi_\mu\}$. A generalized character $\psi=\{\psi_\mu:\mu\in N\}$ satisfies, by definition,

(i) $\psi_\mu\in L^\infty(|\mu|)$ and $\mu-\operatorname{ess\,sup}|\psi_\mu|>0$

(ii) $\psi_\mu=\psi_\nu$ ν-a.e. if $\nu\ll\mu$

(iii) $\psi_{\mu*\nu}(s+t)=\psi_\mu(s)\psi_\nu(t)$ $\mu\times\nu$ - a.e. (s,t).

Let Ψ be a homomorphism of N to $M(\mathbb{T})$. Then the mapping $\nu\to(\Psi_\nu)^\wedge(n)$, $\nu\in N$, defines a homomorphism for every integer n, where „∧” denotes the Fourier-Stieltjes transform

$$\hat{\mu}(n)=\int_{\mathbb{T}}e^{-int}\,d\mu(t).$$

Thus there exists a generalized character $\psi(n)=\{\psi_n(n,t):\nu\in N\}$ or zero system such that

$$(\Psi\nu)^\wedge(n)=\int_{\mathbb{T}}\psi_\nu(n,t)\,d\nu(t),\quad n\in\mathbb{Z} . \tag{1}$$

Let $\{a_n\}_{n\geq 0}$ be a sequence of integers such that $a_n\geq 2$ and $a_n>2$ for infinitely many n. Put

$$d_n=2\pi\prod_{m=1}^{n}a_m^{-1} .$$

Let

$$\mu=\mathop{*}_{n=1}^{\infty}\frac{1}{2}(\delta(0)+\delta(d_n))$$

be a Bernoulli convolution product, where $\delta(a)$ is a Dirac measure concentrated on a point a. We fix such a μ and denote by $N(\mu)$ the smallest L-subalgebra containing μ.

THEOREM ([2], [3]). Let M be L-subalgebra $L(\mathbb{T})$ or $N(\mu)$ and ψ be a homomorphism of M to $M(\mathbb{T})$. Suppose

(A) $|\psi(n)|^2=|\psi(n)|$ i.e. $|\psi_\nu(n,t)|^2=|\psi_\nu(n,t)|$ ν-a.e. for all n in $\mathbb{Z}$ and ν in M.

Then we have

(a) a positive integer m and a finite subset $R=\{n_{m+1}, n_{m+2}, \ldots, n_\ell\}$ of $\mathbb{Z}$,

(b) $\varphi_j \in \Delta'(M)$ $(j=1,2,\ldots,\ell)$

(c) $\pi_j \in \Delta'(M)$ with $|\pi_j|^2=|\pi_j|$ $(j=1,2,\ldots,m)$ such that

$$\psi(n)=\begin{cases} \varphi_j, & n=n_j\in R \\ \sum_{j=1}^{m} \pi_j^n \varphi_j \; C_{m\mathbb{Z}+j}(n) & , \end{cases} \tag{2}$$

where C_E denotes the characteristic function of the set E.

Conversely if $\{\psi(n)\}$ is a sequence in $\Delta'(M)$ satisfying (A), (a), (b) and (c), then the mapping Ψ given by (1) is a homomorphism of M to $M(\mathbb{T})$.

When $M=L(\mathbb{T})$, then $\Delta'(M)=\{e^{int}: n\in\mathbb{Z}\}\cup\{0\}$ and the condition (A) is obviously satisfied. For this case the theorem is due to W. Rudin. The other case, when $M=N(\mu)$, the theorem is proved by S. Igari - Y.Kanjin. Since $L(\mathbb{T})$ is an ideal, our theorem holds good for

$$M=L(\mathbb{T})\oplus N(\mu).$$

We remark that we cannot expect the conditions (a), (b) and (c) without the hypothesis (A) (cf. [2]).

PROBLEM 1. For what kind of L-subalgebra M does the above theorem hold good?

PROBLEM 2. Let $M=M(\mathbb{T})$ and Ψ be a homomorphism of $M(\mathbb{T})$ to $M(\mathbb{T})$. Let $\{\psi(n)\}$ be a sequence of $\Delta'(M(\mathbb{T}))$ given by (1) and assume that $\{\psi_\nu(n)\}$ satisfies the condition (A) for a measure ν. Then, characterize ν such that the conditions (a), (b) and (c) hold for $\{\psi_\nu(n)\}$.

REFERENCES

1. Ш р е й д е р Ю.А. Строение максимальных идеалов в кольцах мер со сверткой. - Матем.сб., 1950, 27, 297-318.
2. R u d i n W. The automorphisms and the endomorphisms of the group algebra of the unit circle. - Acta Math., 1976, 95, 39-56.
3. I g a r i S., K a n j i n Y. The homomorphisms of the measure algebras on the unit circle. - J.Math.Soc.Japan, 1979, 31, N 3, 503-512.

SATORU IGARI

Mathematical Institute
Tohôku University Sendai 980, Japan

2.8. TWO PROBLEMS CONCERNING SEPARATION OF IDEALS IN GROUP ALGEBRAS

All algebras in this paper are commutative complex regular Banach algebras. In such algebras all ideals consist of joint topological divisors of zero, i.e. if I is a (not necessarily closed) ideal in a regular Banach algebra, then there is a net (z_α) of elements of the algebra in question, which does not tend to zero, but $\lim_\alpha z_\alpha x = 0$ for all x in I (cf.[1]). In this case we say that the net (z_α) annihilates the ideal I. We say that an ideal $I \subset A$ has the separation property if for each x in $A \setminus I$ there is a net $(z_\alpha) \subset A$ annihilating I and such that the net $z_\alpha x$ does not tend to zero. It can be shown that in this case there exists one net (z_α) which works for all elements x in $A \setminus I$, and, in fact, $I = \{x \in A : z_\alpha x \to 0\}$ (cf.[2]). In case when there exists such a bounded net we say that the ideal I has the bounded separation property. An ideal with bounded separation property is necessarily closed. If A is a regular Banach algebra and F is a closed non-void subset of its maximal ideal space, then both the maximal and the minimal (non-closed) ideal with the hull F have the separation property. However the bounded separation property may fail for the minimal closed ideal with the given hull, even if it possesses the separation property. It is also possible to construct a closed ideal in a regular Banach algebra which has bounded separation property and it is different from the intersection of all maximal ideals containing it (the question whether it is possible was stated as a problem in the paper [2], but the construction of suitable example is rather easy: we take as the algebra A the algebra of all continuous functions on the unit interval possesing the derivative at 0, and provide it with the norm $\|x\| = |x|_\infty + |x'(0)|$. The ideal in question is then $I_0 = \{x \in A : x(0) = x'(0) = 0\}$). Thus the nets provide a tool for separation and description of ideals. It is particularly interesting whether this tool works for the group algebras. In this context we pose the following problems.

PROBLEM 1. Let I be a closed ideal in $L_1(G)$ for an LCA group G. Does I possess the separation property?

PROBLEM 2. Does there exist an LCA group G and a closed ideal I in $L_1(G)$ which has the bounded separation property and is not of the form (*) $I = \bigcap \{M \in \mathfrak{M}(A) : I \subset M\}$?

In fact we do not know any example of a closed ideal in a group algebra which has separation property and is not of the form (*) .

REFERENCES

1. Żelazko W. On a certain class of non-removable ideals in Banach algebras. - Studia Math. 1972, 44, 87-92.
2. Żelazko W. On domination and separation of ideals in commutative Banach algebras. - Studia Math. 1981, 71, 179-189.

WIESŁAW ŻELAZKO

Math.Inst.Polish Acad.Sc.,00-950
Warszawa, Śniadeckich 8
POLAND

2.9.
old

POLYNOMIAL APPROXIMATION

Let A be a uniformly closed algebra of continuous functions on a compact Hausdorff space S. Assume that A contains constants, that every continuous linear multiplicative functional on A is of the form $f \to f(s)$ for a unique element s of S, and that every element of A whose reciprocal belongs to A is of the form $\exp f$ for an element f of A.

Let σ be a positive measure on the Borel subsets of S, whose support contains more than one point, such that the closure of A in $L^\infty(\sigma)$, considered in its weak topology induced by $L^1(\sigma)$, contains no nonconstant real element. Assume that the functions of the form $f+\bar{g}$ with f and g in A are dense in $L^\infty(\sigma)$ in the same topology. It is CONJECTURED <u>that the closure of A in $L^\infty(\sigma)$ is isomorphic to the algebra of functions which are bounded and analytic in the unit disk</u>.

Positive measures on the Borel subsets of S are considered in the weak topology induced by the continuous functions on S. Two positive measures μ and ν are said to be e q u i v a l e n t (with respect to A) if the identity

$$\int f\,d\mu = \int f\,d\nu$$

holds for every element f of A. The closure of the set of measures which are absolutely continuous with respect to μ and equivalent to μ is a compact convex set, which is the closed convex span of its extreme points. Extremal measures are characterized by the density of the functions of the form $f+\bar{g}$, with f and g in A, in $L^1(\mu)$. Let μ be an extremal measure and let B be the weak closure in $L^\infty(\mu)$ of the functions of the form $f+\bar{g}$ with f and g in A. It is CONJECTURED <u>that the quotient Banach space $L^\infty(\mu)/B$ is reflexive</u>.

For equivalent positive measures μ and ν, define μ to be l e s s t h a n o r e q u a l t o ν if the inequality

$$\int \log|f|\,d\mu \leq \int \log|f|\,d\nu$$

holds for every element f of A. If μ is an extremal measure, it is CONJECTURED that a greatest element σ exists in (the closure of) the set of measures which are absolutely continuous with respect to μ and equivalent to μ. It is CONJECTURED that the functions of the form $f+\bar{g}$ with f and g in A are weakly dense in $L^\infty(\sigma)$.

REFERENCES

1. de Branges L., Trutt D. Quantum Cesàro operators. - In: Topics in functional analysis (essays dedicated to M.G.Krein on the occasion of his 70th birthday), Advances in Math., Suppl.Studies, 3, Academic Press, New York, 1978, pp.1-24.
2. de Branges L. The Riemann mapping theorem. - J.Math. Anal.Appl., 1978, 66, N 1, 60-81.

L.DE BRANGES

Purdue University
Department of Math.
Lafayette, Indiana 47907
USA

2.10. old PROBLEMS PERTAINING TO THE ALGEBRA OF BOUNDED ANALYTIC FUNCTIONS

Here is a list of problems concerning my favorite algebra, the algebra $H^\infty(V)$ of bounded analytic functions on a bounded open subset V of the complex plane. Some of the problems are old and well-known, while some have arisen recently. We will restrict our discussion of each problem to the barest essentials. For references and more details, the reader is referred to the expository account [1], where a number of these same problems are discussed. The maximal ideal space of $H^\infty(V)$ will be denoted by $\mathfrak{M}(V)$, and V will be regarded as an open subset of $\mathfrak{M}(V)$. The grandfather of problems concerning $H^\infty(V)$ is the following.

PROBLEM 1 (CORONA PROBLEM). <u>Is</u> V <u>dense in</u> $\mathfrak{M}(V)$?

The Corona Theorem of L.Carleson gives an affirmative answer when V is the open unit disc $\mathbb{D}$.

In the cases in which $\mathfrak{M}(V)$ has been described reasonably completely, there are always analytic discs in $\mathfrak{M}(V)\setminus V$, but never a higher dimensional analytic structure.

PROBLEM 2. <u>Is there always an analytic disc in</u> $\mathfrak{M}(V)\setminus V$? <u>Is there ever an analytic bidisc in</u> $\mathfrak{M}(V)$?

The Shilov boundary of $H^\infty(V)$ will be denoted by $Ш(V)$. A function f, $f\in H^\infty(V)$, is i n n e r if $|f|=1$ on $Ш(V)$. There is a plethora of inner functions in $H^\infty(V)$, but the following question remains unanswered.

PROBLEM 3. <u>Do the inner functions separate the points of</u> $\mathfrak{M}(V)$?

An affirmative answer in the case of the unit disc was obtained by K.Hoffman, R.G.Douglas, and W.Rudin [2, p.316].

The Shilov boundary $Ш(V)$ is extremely disconnected. Its Dixmier decomposition takes the form $Ш(V)=T\cup Q$, where T and Q are closed disjoint sets, $C(T)\cong L^\infty(\nu)$ for a normal measure on T, and Q carries no nonzero normal measures. The next problem is to identify the normal measure ν. There is a natural candidate at hand. Let λ_z be the "harmonic measures" on $\mathfrak{M}(V)$. These are certain naturally-defined probability measures on $\mathfrak{M}(V)\setminus V$ that satisfy $f(z)=\int f\,d\lambda_z$ for $f\in H^\infty(V)$.

PROBLEM 4. Can the normal measure ν on $\mathbb{T}$ be taken to be the restriction of harmonic measure to $\mathbb{T}$?

There are a number of problems related to the linear structure of $H^\infty(V)$. It is not known, for instance, whether $H^\infty(V)$ has the approximation property, even when V is the unit disc. As a weak-star closed subalgebra of $L^\infty(dxdy\,|\,V)$, $H^\infty(V)$ is a dual space. The following problem ought to be accessible by the same methods used to study $Ш(V)$.

PROBLEM 5. Does $H^\infty(V)$ have a unique predual?

T.Ando [3] and P.Wojtaszczyk [4] have shown that any Banach space B with dual isometric to $H^\infty(\mathbb{D})$ is unique (up to isometry). However, Wojtaszczyk shows that various nonisomorphic B 's have duals isomorphic to $H^\infty(\mathbb{D})$. An extension of the uniqueness result is obtained by J.Chaumat [5].

The weak-star continuous homomorphisms in $\mathfrak{M}(V)$ are called d i s t i n g u i s h e d h o m o m o r p h i s m s. The evaluations at points of V are distinguished homomorphisms, and there may be other distinguished homomorphisms. Related to Problem 2 is the following.

PROBLEM 6. Does each distinguished homomorphism lie on an analytic disc in $\mathfrak{M}(V)$?

The coordinate function z extends to a map $Z: \mathfrak{M}(V) \to \overline{V}$. If $\zeta \in \partial V$, then the fiber $\mathfrak{M}_\zeta = Z^{-1}(\{\zeta\})$ contains at most one distinguished homomorphism.

PROBLEM 7. Suppose there is a distinguished homomorphism φ , $\varphi \in \mathfrak{M}_\zeta$ and suppose Γ is an arc in V terminating at ζ. If $f, f \in H^\infty(V)$ has a limit along Γ, does that limit coincide with $f(\varphi)$?

J.Garnett [6] has obtained an affirmative answer when Γ is appropriately smooth.

The next problem is related to Iversen's Theorem on cluster values, and to the work in [7]. Define $Ш_\zeta = Ш(V) \cap \mathfrak{M}_\zeta$. Denote by $R(f,\zeta)$ the range of f , $f \in H^\infty(V)$, at ζ , $\zeta \in \partial V$, consisting of those values assumed by f on a sequence in V tending to ζ . An abstract version of Iversen's Theorem asserts that $f(Ш_\zeta)$ includes the topological boundary of $f(\mathfrak{M}_\zeta)$, so that $f(\mathfrak{M}_\zeta) \setminus f(Ш_\zeta)$ is open in $\mathbb{C}$. The problem involves estimating the defect of $R(f,\zeta)$ in $f(\mathfrak{M}_\zeta) \setminus f(Ш_\zeta)$.

PROBLEM 8. <u>If every point of ∂V is an essential singularity for some function in $H^\infty(V)$</u>, <u>does</u> $f(\mathfrak{M}_\zeta)\setminus[f(\text{Ш}_\zeta)\cup R(f,\zeta)]$ <u>have zero logarithmic capacity for each</u> f, $f\in H^\infty(V)$?

The remaining problems pertain to the algebra $H^\infty(R)$, where R is a Riemann surface. We assume that $H^\infty(R)$ separates the points of R. Then there is a natural embedding of R into $\mathfrak{M}(R)$.

PROBLEM 9. <u>Is the natural embedding</u> $R\longrightarrow\mathfrak{M}(R)$ <u>a homeomorphism of</u> R <u>and an open subset of</u> $\mathfrak{M}(R)$?

PROBLEM 10. <u>If</u> Γ <u>is a simple closed curve in</u> R <u>that separates</u> R, <u>does</u> Γ <u>separate</u> $\mathfrak{M}(R)$?

The preceding problem arises in the work of M.Hayashi [8], who has treated Widom surfaces in some detail. For this special class of surfaces, Hayashi obtains an affirmative answer to the following problem.

PROBLEM 11. <u>Is</u> $\text{Ш}(R)$ <u>extremely disconnected</u>?

REFERENCES

1. G a m e l i n T.W. The algebra of bounded analytic functions. - Bull.Amer.Math.Soc., 1973, 79, 1095-1108.
2. D o u g l a s R.G., R u d i n W. Approximation by inner functions. - Pacif.J.Math., 1969, 31, 313-320.
3. A n d o T. On the predual of H^∞. - Special issue dedicated to Władisław Orlicz on the occasion of his seventy-fifth birthday. Comment.Math.Special Issue, 1978, 1, 33-40.
4. W o j t a s z c z y k P. On projections in spaces of bounded analytic functions with applications. - Studia Math., 1979, 65, N 2, 147-173.
5. C h a u m a t J. Unicité du prédual. - C.R.Acad.Sci.Paris, Sér A-B, 1979, 288, N 7, A411-A414.
6. G a r n e t t J. An estimate for line integrals and an application to distinguished homomorphisms. - Ill.J.Math., 1975, 19, 537-541.
7. G a m e l i n T.W. Cluster values of bounded analytic functions. - Trans.Amer.Math.Soc., 1977, 225, 295-306.
8. H a y a s h i M. Linear extremal problems on Riemann surfaces, preprint.

T.W.GAMELIN

Dept. of Math.UCLA,
Los Angeles, CA 90024, USA

2.11. old SETS OF ANTISYMMETRY AND SUPPORT SETS FOR $H^\infty + C$.

Let X be a compact Hausdorff space and A a closed subalgebra of $C(X)$ which contains the constants and separates the points of X. A subset S of X is called a set of antisymmetry for A if any function in A which is real valued on S is constant on S. This notion was introduced by E.Bishop [1] [*] (see also [2]), who established the following fundamental results: (i) X can be written as the disjoint union of the maximal sets of antisymmetry for A; the latter sets are closed. (ii) If S is a maximal set of antisymmetry for A, then the restriction algebra $A|S$ is closed. (iii) If f is in $C(X)$ and $f|S$ is in $A|S$ for every maximal set of antisymmetry S for A, then f is in A.

A closed subset of X is called a *support set for* A if it is the support of a representing measure for A (i.e., a Borel probability measure on X which is multiplicative on A). It is trivial to verify that every support set for A is a set of antisymmetry for A. However, there is in general no closer connection between these two classes of sets. This is illustrated by B.Cole's counterexample to the peak point conjecture [3, Appendix], which is an algebra $A \neq C(X)$ such that X is the maximal ideal space of A and such that every point of X is a peak point of A. For such an algebra, the only support sets are the singletons, but not every set of antisymmetry is a singleton (by (iii)).

The present problem concerns a naturally arising algebra for which there does seem to be a close connection between maximal sets of antisymmetry and support sets. However, the evidence at this point is circumstantial and the precise connection remains to be elucidated. Let L^∞ denote the L^∞-space of Lebesgue measure on $\mathbb{T}$. Let H^∞ be the space of boundary functions on $\mathbb{T}$ for bounded holomorphic functions in $\mathbb{D}$, and let C denote $C(\mathbb{T})$. It is well known that $H^\infty + C$ is a closed subalgebra of L^∞ [4], so we may identify it, under the Gelfand transformation, with a closed subalgebra of $C(M(L^\infty))$, where $M(L^\infty)$ denotes the maximal ideal space of L^∞ (with its Gelfand topology). In what follows, by a set of antisymmetry or a support set, we shall mean these notions for the case $X = M(L^\infty)$ and $A =$ (the Gelfand transform of) $H^\infty + C$. Also, we shall identify the functions in L^∞ with their Gelfand transforms.

---

[*] See the note at the end of the section. - Ed.

The first piece of evidence for the connection alluded to above is the following result from [5]: If f is in L^∞ and $f|S$ is in $(H^\infty+C)|S$ for each support set S, then f is in $H^\infty+C$. This is an ostensible improvement of part (iii) of Bishop's theorem in the present special situation. It is natural to ask whether it is an actual improvement, or whether it might not be a corollary to Bishop's theorem via some hidden connection between maximal sets of antisymmetry and support sets. The proof of the result is basically classical analysis and so offers no clues about the latter question. The question is motivated, in part, by a desire to understand the result from the viewpoint of abstract function algebras.

A second piece of evidence comes from [6], where a sufficient condition is obtained for the semi-commutator of two Toeplitz operators to be compact. The condition can be formulated in terms of support sets, and it is ostensible weaker than an earlier sufficient condition of Axler [7] involving maximal sets of antisymmetry. Again, it is natural to ask whether the newer result is really an improvement of the older one, or whether the two are actually equivalent in virtue of a hidden connection between maximal sets of antisymmetry and support sets. As before, the proof offers no clues.

As a final bit of evidence one can add the following unpublished results of K.Hoffman: (1) If two support sets for $H^\infty+C$ intersect, then one of them is contained in the other; (2) There exist maximal support sets for $H^\infty+C$.

All of the above makes me suspect that each maximal set of antisymmetry for $H^\infty+C$ can be built up in a "nice" way from support sets. It would not even surprise me greatly to learn that e a c h m a x i m a l s e t o f a n t i s y m m e t r y i s a s u p p o r t s e t. At any rate, there is certainly a connection worth investigating.

REFERENCES

1. B i s h o p E. A generalization of the Stone-Weierstrass theorem. - Pacif.J.Math., 1961, 11, 777-783.
2. G l i c k s b e r g I. Measures orthogonal to algebras and sets of antisymmetry. - Trans.Amer.Math.Soc.,1962, 105, 415-435.
3. B r o w d e r A. Introduction to Function Algebras. New York, W.A.Benjamin, Inc., 1969.
4. S a r a s o n D. Algebras of functions on the unit circle. - Bull.Amer.Math.Soc.,1973, 79, 286-299.

5. S a r a s o n D. Functions of vanishing mean oscillation. - Trans.Amer.Math.Soc.,1975, 207, 391-405.

6. A x l e r S., C h a n g S.-Y., S a r a s o n D. Products of Toeplitz operators. - Int.Equat.Oper.Theory, 1978, 1, N 3, 285-309.

7. A x l e r S. Doctoral Disseration. University of California, Berkeley, 1975.

DONALD SARASON

University of California,
Dept.Math., Berkeley,
California, 94720, USA

EDITORS' NOTE: The notion of a set of antisymmetry was introduced by G.E.Shilov as early as in 1951. He has proved the first theorem about representation of a maximal ideal space of a uniform algebra as a union of sets of antisymmetry (see Chapter 8 of the monograph И.М.Гельфанд, Г.Е,Шилов, Д.А.Райков, "Коммутативные нормированные кольца", М., Физматгиз, I958).

* * *

COMMENTARY BY THE AUTHOR

The structure of the maximal sets of antisymmetry for $H^\infty+C$ remains mysterious, although a little progress has occurred. P.M.Gorkin in her dissertation [8] has the very nice result that $M(L^\infty)$ contains singletons which are maximal sets of antisymmetry for $H^\infty+C$. Such singletons are of course also maximal support sets. The author's paper [9] contains a result which is probably relevant to the problem.

REFERENCES

8. G o r k i n P.M. Decompositions of the maximal ideal space of L^∞, Doctoral Dissertation, Michigan State University, East Lansing, 1982.

9. S a r a s o n D. The Shilov and Bishop decompositions of $H^\infty+C$. - Conference on Harmonic Analysis in Honour of Antoni Zygmund, vol.II, pp.461-474, Wadsworth, Belmont, CA, 1983.

2.12. old SUBALGEBRAS OF THE DISK ALGEBRA

Let A denote the disk algebra, i.e. the algebra of all functions continuous on $\operatorname{clos}\mathbb{D}$ ard analytic on $\mathbb{D}$. Fix functions f and g in A. We denote by $[f,g]$ the closed subalgebra of A generated by f and g, i.e. the closure in A of the set of all functions

$$\sum_{n,m=0}^{N} c_{n,m} f^n g^m, \; c_{n,m} \text{ constants.}$$

We ask: w h e n d o e s $[f,g]=A$?

Necessary conditions are

1) f, g together separate points of $\operatorname{clos}\mathbb{D}$.
2) For each a in $\mathbb{D}$, either $f'(a)\neq 0$ or $g'(a)\neq 0$.

1) and 2) together are not sufficient for $[f,g]=A$. Some regularity condition must be imposed on the boundary. We assume

3) f,g are smooth on $\mathbb{T}$, i.e. the derivatives f' and g' extend continuously to $\mathbb{T}$.

1), 2), 3) are not yet sufficient conditions. We add

4) For each a on $\mathbb{T}$, either $f'(a)\neq 0$ or $g'(a)\neq 0$.

In [1] R.Blumenthal showed

THEOREM 1. 1), 2), 3) <u>and</u> 4) <u>together are sufficient for</u> $[f,g]=A$.

Related results are due to J.-E.Björk, [2], and to Sibony and the author, [3].

Condition 4) is, however, not necessary, since for instance $[(z-1)^2, (z-1)^3]=A$ and conditions 1), 2), 3) hold here while 4) is not satisfied.

The problem arises to give a condition that replaces 4) which is both necessary and sufficient for $[f,g]=A$. In the special case $f=(z-1)^3$ this problem has been solved by J.Jones in [4] and his result is the following: let W^+ and W^- be the two subregions of $\operatorname{clos}\mathbb{D}$ which are identified by the map $(z-1)^3$. Put

$$z^* = 1+e^{\frac{2\pi i}{3}}(z-1).$$

Then for z in W^+, z^* lies in W^- and $(z-1)^3$ identifies z and z^*. Let χ be an inner function on W^+ whose only singularity is at $z=1$. Then for some t, $t>0$,

$$\chi(z) = \exp\left\{ t\left(1+\frac{1}{(z-1)e^{\pi i/3}}\right)^3 \right\}. \tag{1}$$

THEOREM 2 ([5]). *Let* g *be a function in* A *such that* $f=(z-1)^3$ *and* g *together satisfy 1), 2), 3). Then* $[f,g] \neq A$ *if and only if for some* χ *of the form (1),*

$$|g(z) - g(z^*)| \leq K|\chi(z)| \tag{2}$$

for all z *in* W^+, *where* K *is some constant.*

We propose two problems.

PROBLEM 1. *Prove an analogue of Theorem 2 for the case when* f *is an arbitrary function analytic in an open set which contains* $\operatorname{clos}\mathbb{D}$ *by finding a condition to replace (2) which together with* 1), 2), 3) *is necessary and sufficient for* $[f,g] \neq A$.

Furthermore, condition (2) implies that the Gleason distance from z to z^*, computed relative to the algebra $[f,g]$, approaches 0 rapidly as $z \to 1$, and so is inequivalent to the Gleason distance computed relative to the algebra A. Let B denote a closed subalgebra of A which separates the points of $\operatorname{clos}\mathbb{D}$ and contains the constants. Let ρ_B denote the Gleason distance induced on $\operatorname{clos}\mathbb{D}$ by B, i.e.

$$\rho_B(z_1,z_2) = \sup_{\substack{\varphi\in B\\ \|\varphi\|=1}} |\varphi(z_1)-\varphi(z_2)|; \quad |z_1|\leq 1, \ |z_2|\leq 1.$$

Let ρ denote the Gleason distance on $\operatorname{clos}\mathbb{D}$ induced by A.

PROBLEM 2. *Assume that*

(a) *The maximal ideal space of* B *is the disk* $\operatorname{clos}\mathbb{D}$.

(b) *There exists a constant* K, $K>0$, *such that*

$$\rho_B(z_1,z_2) \geq K\rho(z_1,z_2); \quad |z_1|\leq 1, \ |z_2|\leq 1.$$

Show that then $B = A$.

REFERENCES

1. Blumenthal R. Holomorphically closed algebras of analytic functions. - Math.Scand.,1974, 34, 84-90.
2. Björk J.-E. Holomorphic convexity and analytic structures in Banach algebras. - Arkiv för Mat.,1971, 9, 39-54.
3. Sibony N., Wermer J. Generators for $A(\Omega)$. - Trans.Amer.Math.Soc.,1974, 194, 103-114.
4. Jones J. Generators of the disc algebra (Dissertation), Brown University, June, 1977.
5. Wermer J. Subalgebras of the disk algebra. - Colloque d'Analyse Harmonique et Complexe, Univ.Marseille I, Marseille, 1977.

J.WERMER

Brown University
Department of Math.
Providence, R.I., 02912
USA

2.13. A QUESTION INVOLVING ANALYTIC FAMILIES OF OPERATORS

Let A be a uniform algebra with Shilov boundary $\mathfrak{X}$. (The case when A is the disk algebra and $\mathfrak{X}=\mathbb{T}$ is an interesting example for this purpose.) Suppose we are given a linear operator S which maps A into $C(\mathfrak{X})$ and has small norm. Suppose further that the image $(I+S)(A) \subseteq C(\mathfrak{X})$ is a subalgebra (here I is the inclusion of A into $C(\mathfrak{X})$).

QUESTION: Is there an analytic family of linear operators $S(z)$ defined for z in $\bar{\mathbb{D}}$ so that $(I+S(z))(A)$ is a subalgebra of $C(\mathfrak{X})$ for each z in $\bar{\mathbb{D}}$ and so that $S(1)=S$?

The hypotheses are related to questions of deformation of the structure of A, see [2] for details and examples. In cases where $S(z)$ can be obtained the differential analysis of $S(z)$ connects the deformation theory of A with the cohomology of A. (See [1]). For instance, $S'(0)$ would be a continuous derivation of the algebra A into the A-module $C(\mathfrak{X})/A$. Such considerations lead rapidly to questions about operators on spaces such as VMO(=$C(\mathbb{T})$/disk algebra). (See [3] for an example.)

REFERENCES

1. J o h n s o n B.E. Low Dimensional Cohomology of Banach Algebras. - Proc.Symp.Pure Math. 38, 1982, part 2, 253-259.
2. R o c h b e r g R. Deformation of Uniform Algebras. - Proc. Lond.Math.Soc. (3) 1979, 39, 93-118.
3. R o c h b e r g R. A Hankel Type Operator Arising in Deformation Theory. - Proc.Symp.Pure Math. 35,1979, Part I, 457-458.

RICHARD ROCHBERG

Washington University, Box 1146
St.Louis, MO 63130
USA

CHAPTER 3

PROBABILISTIC PROBLEMS

The problems assembled in this Chapter are of probabilistic origin but are more or less closely connected with Spectral Function Theory. Nowadays, probabilistic methods are increasingly applied in Harmonic Analysis. Many such examples can be found, for instance, in the book by J.-P.Kahane "Some random series of functions".

The theory of stationary Gaussian processes bridges Probability and Function Theory. Moreover it supplies Function Theory with many interesting problems. This does not exhaust, of course, all connections between the two theories. Recall, for example, the traditional application of Fourier integrals to the investigation of probability distributions, or the martingale theory of Hardy classes, or the Brownian motion which is a traditional source of counterexamples in classical Fourier analysis as well as a powerful tool for the study of boundary problems. However, we would like to emphasize the circle of "spectral" ideas arising in the theory of stationary Gaussian processes (see 3.1, 3.2, 3.3, 3.4 below). These processes are linked with Spectral Function Theory by the concept of filter (i.e. the convolution operator $f \mapsto K*f$). N.Wiener systematically and successfully applied this concept, originating in engineering, to various purely mathematical problems. Almost any stationary Gaussian process can be considered as a response of some filter to a "white noise". All statistical information being contained in the "white noise", the problem is reduced to the study of its redistribution under the action of a given filter. This is, maybe, one of reasons why Hardy classes, entire functions, etc. - in other words almost all the tools used now in Function Theory - are so important for some purely probabilistic papers.

The questions posed in 3.1 can hardly be considered as concrete problems. Rather, they indicate possible directions of investigation and the interested reader may consult the excellent book [1] (see References in 3.1).

In contradistinction to Problem 3.1 the "old" Problem 3.2 contained a series of analytic questions of which one is solved (see the Commentary).

The first part of Problem 3.7 has also been solved, whereas, all aspects of Problem 3.6 remain open.

In this edition this Chapter has been enlarged by three Problems. Problem 3.3 deals with the Hilbert space geometry of Past and Future. The questions posed there concern also the theory of Hankel operators. Problem 3.5 outlines a new field for dilation theory in the theory of Markov processes. Problem 3.4 deals with limit theorems.

We conclude by indicating some Problems from the remaining ("deterministic") part of the Collection. In 8.4 approximation by trigonometrical polynomials with bounded spectra is discussed. The probabilistic interpretation is well-known (see e.g. the above metnioned book by Dym and McKean, [1] in 3.1). Chapter 6 has some relations with Probability as mentioned above. Sarason's result cited as Theorem 1 in 3.2 has very much in common with the contents of that Chapter.

The years since the first edition have been marked by closer and clearer connections among the elements of the triad "Function Theory - Operator Theory - Probability". This tendency (partly reflected in the operator-theoretic item 3.5) is well illustrated by the papers [5], [9] cited in the Commentary to Problem 3.2

3.1.
old

SOME QUESTIONS ABOUT HARDY FUNCTIONS

The theory of Gaussian-distributed noise leads to a variety of substantial mathematical questions about Hardy functions. I will put the questions in a purely mathematical way; the reader is referred to [1] for the statistical interpretation and/or additional information.

1. Let Δ , $\Delta \geq 0$, be summable on the line and let

$$\int_{\mathbb{R}} \frac{\log \Delta}{1+x^2}\, dx = -\infty .$$

Then the exponentials e^{ixt}, $t \leq 0$, span $L^2(\mathbb{R}, \Delta dx)$, but how is e^{ixT} for fixed $T > 0$ e f f i c i e n t l y approximated (by these functions)? See [1], § 4.2.

2. Let h , $h \in H^2_+$, be outer, let T , $T > 0$, be fixed, and let

$$K(x) = \frac{e^{-ixT}}{h(x)} \int_T^{\infty} e^{ixt}\, \check{h}(t)\, dt ,$$

$\check{h}(t)$ being the inverse transform $\frac{1}{2\pi}\int_{\mathbb{R}} e^{-ixt} h(x)\, dx .$

What can be said about $\check{K}$? $\|\check{K}\|_2$ cannot be $< \infty$ for all small T ; also, $\check{K}$ can be enormously singular; see [1]. §4.4.

3. Let h , $h \in H^2_+$, be outer. The QUESTION is to explain what makes the phase function h^*/h the ratio of two inner functions or the reciprocal of an inner function; see [1], §4.6. $e^{ixT} h^*/h$ is itself an inner function if and only if h is integral of exponential type $\leq T$.

4. Let h , $h \in H^2_+$, be outer. When does h^*/h belong to the span of $e^{ixt} H^\infty$, $t \leq 0$, in L^∞ ? See [1], §4.12.

5. The following conditions are equivalent for outer h , $h \in H^2_+$: a) $e^{2ixT}\, h^*/h$ is the ratio of a function of class H^2_+ and a function of class H^2_- ; b) $\int_{\mathbb{R}} |f/h|^2 dx < \infty$ for

some integral function f of exponential type $\leq T$;

c) $\left\| e^{2ixt} \frac{h^*}{h} \frac{x+i}{x-i} - k \right\|_\infty \leq 1$

<u>for some</u> k, $k \neq 0$, $k \in H_+^\infty$; see [1], §4.13. <u>What can be said about such functions</u> h ? Note that b) is a problem of "multiplying down" the function $1/h$ in the style of [2]. <u>What outer function satisfy</u> a), b), c) <u>for every</u> T , $T > 0$? <u>for no</u> T , $T > 0$? Note that h cannot satisfy c) for T , $T = 0$.

6. The phase function h^*/h is ubiquitous. <u>What can be said about it for the general outer function</u> h , $h \in H_+^2$?

REFERENCES

1. D y m H., M c K e a n H.P. Gaussian Processes, Function Theory, and the Inverse Spectral Problem. New York, Academic Press, 1976.
2. B e u r l i n g A., M a l l i a v i n P. On Fourier transforms of measures with compact support. - Acta Math. 1962, 107, 291-302.

H.P.MCKEAN

New York University,
Courant Institute of Mathematical Sciences,
251 Mercer Street,
New York, N.Y. 10012, USA

* * *

COMMENTARY

In connection with section 4 see the COMMENTARY to Problem 3.2. The question discussed in section 5 is related to the paper [3].

REFERENCE

3. K o o s i s P. Weighted quadratic means of Hilbert transforms. - Duke Math.J., 1971, 38, N 3,609-634.

3.2. old SOME ANALYTICAL PROBLEMS IN THE THEORY OF STATIONARY STOCHASTIC PROCESSES

1. Let $\xi(t)$ be a stationary Gaussian process with discrete or continuous time (see [1] for the definitions of basic notions of stochastic processes used here). Denote by $f(\lambda)$ the spectral density of ξ (in case of discrete time f is a non-negative integrable function on the unit circle $\mathbb{T}$; in case of continuous time $\mathbb{T}$ is replaced by the real line $\mathbb{R}$).

Let $L(f)$ be a Hilbert space of functions on $\mathbb{T}$ (or $\mathbb{R}$) with the inner product

$$(\varphi,\psi)_f=\int\varphi(\lambda)\,\overline{\psi(\lambda)}\,f(\lambda)\,d\lambda.$$

For $\tau\geqslant 0$, let $L^+_\tau(f)$ (resp. $L^-_\tau(f)$) be the subspaces of $L(f)$ genetated by exponentials $e^{it\lambda}$ with $t\geqslant\tau$ (resp. $t\leqslant-\tau$). Let $\mathcal{P}^+_\tau$ and $\mathcal{P}^-_\tau$ be the orthogonal projections onto $L^+_\tau(f)$ and $L^-_\tau(f)$. Consider the operators

$$B_\tau=\mathcal{P}^-_\tau\,\mathcal{P}^+_0\,\mathcal{P}^-_\tau .$$

These positive selfadjoint operators were introduced into the theory of stochastic processes in [2]. Many characteristics of processes can be expressed in their terms. In particular, important classes of Gaussian processes correspond to the following conditions on B_τ :

a) B_τ is compact for all (sufficiently large in case of continuous time) τ ;

b) B_τ is nuclear for all (sufficiently large in case of continuous time) τ .

Since the finite-dimensional distributions of a Gaussian process are completely determined by its spectral density f , it would be desirable to describe properties of B_τ in terms of f .

2. __Processes with discrete time__.

THEOREM 1 ([3]). __The operators__ B_τ __are compact if and only if the spectral density__ f __can be represented in the form__

$$f(\lambda)=|P(e^{i\lambda})|^2\exp\left(u(\lambda)+\tilde{v}(\lambda)\right).$$

Here P is a polynomial with roots on the unit circle and the functions u and v are continuous.

THEOREM 2 (I.A.Ibragimov, V.N.Solev, cf.[1]). The operators B_τ belong to the trace class if and only if

$$f(\lambda) = |P(e^{i\lambda})|^2 e^{a(\lambda)}.$$

Here P stands for a polynomial with roots on $\mathbb{T}$ and

$$a(\lambda) \sim \sum a_j e^{ij\lambda}, \quad \sum |a_j|^2 |j| < \infty.$$

PROBLEM 1. Under what conditions on spectral density f do the operators B_τ belong to the class $\mathfrak{S}_p$, $1 \leqslant p \leqslant \infty$, i.e. for what f

$$\sum \lambda_{j\tau}^p < \infty,$$

where $\lambda_{j\tau}$ are eigen-values of B_τ ?

Theorems 1 and 2 deal with the extreme cases $p = \infty$, 1.

THEOREM 3 (I.A.Ibragimov, see [1]). The estimate

$$|B_\tau| = O(\tau^{-r-\alpha}) \qquad \text{for } \tau \to \infty$$

(r is an integer, $0 < \alpha < 1$) holds if and only if

$$f(\lambda) = |P(e^{i\lambda})|^2 e^{g(\lambda)},$$

where P is a polynomial, g is an r-time differentiable function and $g^{(r)}$ satisfies the Lipschitz condition of order α .

The value $|B_\tau|$ is exponentially small for $\tau \to \infty$ if and only if the spectral density f is an analytic function.

3. **Processes with continuous time.** Nothing similar to

theorem 3 is known in that case.

PROBLEM 2. Under what conditions does the value $|B_\tau|$ decrease with power or exponential rate as $\tau \longrightarrow +\infty$?

THEOREM 4 (I.A.Ibragimov, [1]). Let $f(\lambda)=|\Gamma(\lambda)|^{-2}$, where Γ is an entire function of exponential type with roots $z_1, z_2, \dots$.

Then $\lim_{\tau\to+\infty} |B_\tau| = 0$ if and only if

1. $$\frac{\log(|\Gamma(\lambda)|}{1+\lambda^2} \in L^1(\mathbb{R})$$

2. $$\sup_{-\infty<\lambda<\infty} \sum_j \left| \operatorname{Im} \frac{1}{\lambda - z_j} \right| < \infty .$$

PROBLEM 3. Investigate the case when $f(\lambda)=\dfrac{|\Gamma_1(\lambda)|^2}{|\Gamma_2(\lambda)|^2}$ and Γ_1 , Γ_2 are entire functions of exponential type.

This problem is essential for the analysis of the operator B_τ in the multivariate case [4].

Note in conclusion that Problem 1 can be easily reformulated for continuous time.

REFERENCES

1. Ибрагимов И.А., Розанов Ю.А. Гауссовские случайные процессы, М., "Наука", 1970.
2. Гельфанд И.М., Яглом А.М. О вычислении количества информации о случайной функции, содержащейся в другой такой функции. - Успехи матем.наук, 1957, XII, I, 3-52.
3. Sarason D. An addendum to Past and Future. - Math.Scand., 1972, 30, 62-64.
4. Ибрагимов И.А. О полной регулярности многомерных стационарных процессов. - Докл.АН СССР, 1962, 162, № 5.

I.A.IBRAGIMOV (И.А.ИБРАГИМОВ) СССР, 191011, Ленинград,
V.N.SOLEV (В.Н.СОЛЕВ) Фонтанка 27, ЛОМИ АН СССР

* * *

COMMENTARY

Problem 1 has been solved by V.V.Peller in [5]: $B_\tau \in \gamma_p$ iff $f = |P|^2 e^a$ where P is a polynomial with roots on $\mathbb{T}$ and a belongs to the Besov class $B_{2p}^{1/2p}$.

Things are more complicated in case of continuous time Let h be the outer function in $\mathbb{C}_+$ satisfying $f = |h|^2$ on $\mathbb{R}$ and let $A_1 = clos_{L^\infty(\mathbb{R})}(\bigcup_{\tau>0} e^{-i\tau x}\cdot H^\infty)$. Then it is easy to show that

$$\|B_\tau\|^{1/2} = \|\mathcal{P}_0^+ \mathcal{P}_\tau^-\| = dist_{L^\infty(\mathbb{R})}(\frac{\bar h}{h}, e^{-i\tau x}\cdot H^\infty) ,$$

where H^∞ stands for the usual Hardy algebra in $\mathbb{C}_+$, and therefore a process satisfies the strong mixing condition iff $\bar h/h \in A_1$. The structure of the Douglas algebra A_1 in contrast with that of $H^\infty + C(\mathbb{T})$ is very complicated. This is the main reason of troubles arising in the investigation of the continuous time case. Many results valid for processes with discrete time could be easily extended to processes with continuous time, if the following factorization were true

$$\frac{\bar h}{h} = u \cdot I ,$$

where u is an invertible unimodular function in A_1 and I is an inner function in H^∞ . Being valid in $H^\infty + C$, see [6], this factorization, unfortunately, does not hold in A_1 [7].

The factorization theorem in $H^\infty + C$ can be applied to a description of γ_∞-regular Gaussian processes with continuous time, i.e. the processes with compact B_τ , $\tau \geq a > 0$. Let C denote the space of all continuous functions on $\mathbb{R}$ having a finite limit at infinity.

THEOREM (Hruščëv - Peller). A stationary Gaussian process $\{X_t\}_{t\in\mathbb{R}}$ with spectral measure Δ is γ_∞-regular iff

$$d\Delta_x = |F|^2 e^{u+\tilde v}\, dx ,$$

where $u, v \in C$, F is an entire function of exponential type such that $|F|^2 e^{u+\tilde v} \in L^1$.

PROOF (SKETCH). Let h be an outer function in H^2 satisfying $d\Delta_X = |h|^2 dx$. Then $B_\tau \in \mathfrak{S}_\infty$, $\tau \geq a$ iff $\frac{\bar h}{h} e^{i\sigma x} \in H^\infty + C$ (see e.g. [9]), which in turn by the factorization theorem of T.Wolff is equivalent to $\frac{\bar h}{h} e^{i\sigma x} = e^{i(v+\tilde u)} \cdot B$, where $u, v \in C$ and B is a Blaschke product in $\mathbb{C}_+$. Consider an auxiliary outer function h_0 defined by $\log|h_0(x)| = \frac{1}{2}\{u(x) - \tilde v(x)\}$, $x \in \mathbb{R}$. Let $h_1 = \frac{h}{h_0}$. Then clearly $\frac{\bar h_1}{h_1} e^{i\sigma x} = B$, which implies that

$$\bar h_1 e^{i\sigma x} = B h_1 = F$$

is a restriction to $\mathbb{R}$ of an entire function of exponential type.

Conversely, suppose $|h|^2 = |F|^2 e^{u+\tilde v} \in L^1$ with an entire function F of exponential type. Clearly, F belongs to the Cartwright class and it can be replaced by an entire function of exponential type being an outer function in $\mathbb{C}_+$ of the same modulus on $\mathbb{R}$. It follows that on $\mathbb{R}$

$$\frac{\bar h}{h} = c\,\frac{\bar F}{F} \cdot e^{i(v-\tilde u)}, \qquad |c| = 1 .$$

Since F is of exponential type, $\frac{\bar F}{F} e^{i\sigma z} \in H^\infty$ for some $\sigma > 0$. Therefore $\frac{\bar h}{h} e^{i\sigma z} \in H^\infty + C$. ●

Some sufficient conditions for the strong mixing were obtained in [8]. See also [9] for a brief introduction to the subject

REFERENCES

5. П е л л е р В.В. Операторы Ганкеля класса $\mathfrak{S}_p$ и их приложения (рациональная аппроксимация, гауссовские процессы, проблема мажорации операторов). - Матем.сборн., 1980, 113, № 4; 538-581.
6. W o l f f T. Two algebras of bounded functions. - Duke Math J, 1982, 49, N 2, 321-328.
7. S u n d b e r g C. A counterexample in $H^\infty + BUC$. - 1983 (May-June), preprint.
8. H a y a s h i E. The spectral density of the strong mixing stationary Gaussian process. - 1981, preprint.
9. П е л л е р В.В., Х р у щ ё в С.В. Операторы Ганкеля, наилучшие приближения и стационарные гауссовские процессы. - Успехи матем. наук, 1982, 37, № 1, 53-124.

3.3. MODULI OF HANKEL OPERATORS, PAST AND FUTURE

A discrete, zero-mean, stationary Gaussian process is a sequence $\{X_n\}_{n\in\mathbb{Z}}$ in the real L^2 space of a probability measure P, such that $\mathbb{E}X_n = \int X_n dP = 0,\ n\in\mathbb{Z}$; $\mathbb{E}(X_n X_m) = Q(n-m)$ i.e. depends only on $n-m$; every function in the linear span of the functions X_n has a Gaussian distribution. In prediction theory the Past is associated with the closed linear span (over $\mathbb{C}$) G_p of X_k, $k<0$, and the Future with the span G_f of X_k, $k\geqslant 0$. These closed subspaces are usually considered as subspaces of the complex Hilbert space G spanned by the whole sequence $\{X_k\}_{k\in\mathbb{Z}}$.

Our problem concerns the description of all possible positions in G of the Future G_f with respect to the Past G_p.

The sequence $\{Q(n)\}_{n\in\mathbb{Z}}$ being positive definite, there exists a finite positive Borel measure μ on $\mathbb{T}$ satisfying $Q(n) = \hat{\mu}(n),\ n\in\mathbb{Z}$. The measure μ is called *the spectral measure* of $\{X_n\}_{n\in\mathbb{Z}}$. Clearly the mapping Φ defined by $\Phi X_n = z^n$, $n\in\mathbb{Z}$, can be extended to a unitary operator from G to $L^2(\mu)$. To avoid technical difficulties we consider henceforth all stationary sequences $\{X_n\}_{n\in\mathbb{Z}}$ in G, not necessarily real. A stationary sequence is unitarily equivalent to a Gaussian process iff the spectral measure μ is invariant under the transform $z\mapsto\bar{z}$ of $\mathbb{T}$.

Consider the set of all triples $(A, B, \mathcal{H})$ where A and B are closed subspaces of the complex separable infinite-dimensional Hilbert space $\mathcal{H}$ such that $\operatorname{clos}(A+B) = \mathcal{H}$. The triples $(A_1, B_1, \mathcal{H}_1)$ and $(A_2, B_2, \mathcal{H}_2)$ are said to be equivalent if there exists an isometry V of $\mathcal{H}_1$ onto $\mathcal{H}_2$ satisfying $VA_1 = A_2$, $VB_1 = B_2$. Let $\mathcal{T}$ be the set of all equivalence classes with respect to the introduced equivalence relation.

PROBLEM 1. <u>Which classes in $\mathcal{T}$ contain at least one element (G_p, G_f, G) corresponding to a stationary sequence $\{X_n\}_{n\in\mathbb{Z}}$</u> ?

The class $\mathcal{T}$ admits a more explicit description. Let $\mathcal{P}_A$ denote the orthogonal projection onto the subspace A. Each triple $t = (A, B, \mathcal{H})$ defines the selfadjoint operator $\mathcal{P}_A \mathcal{P}_B \mathcal{P}_A$ and the numbers

$$n_+(t) = \dim(A^\perp \cap B), \quad n_-(t) = \dim(A\cap B^\perp).$$

LEMMA. <u>A triple</u> $t_1 = (A_1, B_1, \mathcal{H})$ <u>is equivalent to</u> $t_2 = (A_2, B_2, \mathcal{H}_2)$ <u>iff</u> $\mathcal{P}_{B_1}\mathcal{P}_{A_1}\mathcal{P}_{B_1}$ <u>and</u> $\mathcal{P}_{B_2}\mathcal{P}_{A_2}\mathcal{P}_{B_2}$ <u>are unitarily equivalent and</u> $n_\pm(t_1) = n_\pm(t_2)$.

SKETCH OF THE PROOF. We may assume without loss of generality that $\mathcal{H}_1 = \mathcal{H}_2 = \mathcal{H}$, $B_1 = B_2 = B$, $\mathcal{P}_B\mathcal{P}_{A_1}\mathcal{P}_B = \mathcal{P}_B\mathcal{P}_{A_2}\mathcal{P}_B$ (note that $\dim B_1 = \dim B_2$ under the assumption of the lemma). Given a subspace C in $\mathcal{H}$ let $C^\perp = \mathcal{H} \ominus C$. Consider the partial isometries V_1, V_2 determined by the polar decompositions

$$\mathcal{P}_{A_i}\mathcal{P}_B = V_i(\mathcal{P}_B\mathcal{P}_{A_i}\mathcal{P}_B)^{\frac{1}{2}}, \quad i = 1,2. \tag{1}$$

Let $\mathcal{U}$ be an operator on $\mathcal{H}$ defined by

$$\mathcal{U}x = V_2V_1^*x, \quad x \in A_1 \ominus B^\perp; \quad \mathcal{U}x = x, \quad x \in B;$$

$\mathcal{U}|A_1 \cap B^\perp$ is an arbitrary unitary operator from $A_1 \cap B^\perp$ onto $A_2 \cap B^\perp$. It is easy to see that $\mathcal{U}$ is defined correctly (if $x \in A_1 \cap B$ then $V_2V_1^*x = x$). Also clearly $\mathcal{U}$ maps isometrically A_1 onto A_2. It remains to verify that $\mathcal{U}$ is a unitary operator on $\mathcal{H}$. Clearly it suffices to show that $(\mathcal{U}x, y) = (x, y)$ for $x \in A_1$, $y \in B$. If $x \in A_1 \cap B^\perp$, this is evident. Now we can consider only vectors x of the form $x = \mathcal{P}_{A_1}z$, $z \in B$. It follows from (1) that

$$(\mathcal{U}x,y) = (V_2V_1^*\mathcal{P}_{A_1}\mathcal{P}_B z, y) = (V_2(\mathcal{P}_B\mathcal{P}_{A_2}\mathcal{P}_B)^{1/2}z, y) = (\mathcal{P}_B\mathcal{P}_{A_1}\mathcal{P}_B z, y) = (x,y). \quad \bullet$$

For every $k, m \in \mathbb{Z}_+ \cup \{\infty\}$ and for every selfadjoint operator T on a Hilbert space $\mathcal{H}_1$ such that $\mathbb{O} \leq T \leq I$, $\operatorname{Ker} T = \{\mathbb{O}\}$ there exists $t = (A, B, \mathcal{H}) \in \mathcal{T}$ satisfying $n_+(t) = k$, $n_-(t) = m$ and such that $\mathcal{P}_B\mathcal{P}_A\mathcal{P}_B | B \ominus A^\perp$ is unitarily equivalent to T.

Indeed, without loss of generality $n_\pm(t) = 0$. By the well-known Naimark theorem in $\mathcal{H}_1 \oplus \mathcal{H}_1$ there exists a projection $\mathcal{P}_A$ defined by

$$\mathcal{P}_A = \begin{pmatrix} T & T^{\frac{1}{2}}(I-T)^{\frac{1}{2}} \\ T^{\frac{1}{2}}(I-T)^{\frac{1}{2}} & I-T \end{pmatrix}.$$

Put $B = \mathcal{H}_1 \oplus \{\mathbb{O}\}$ and $\mathcal{H} = \operatorname{clos}(A + B)$. Then clearly $\mathcal{P}_B \mathcal{P}_A \mathcal{P}_B \mid B = T$.

By Szegö's alternative either $G_p = G_f = G$ or $G_p \neq G$ and then $d\mu = |h|^2 dm + d\mu_s$, where h is an outer function in H^2 and μ_s is a singular measure on $\mathbb{T}$. We have $\Phi^{-1}(L^2(\mu_s)) \subset G_p \cap G_f$ and therefore only the case $\mu_s \equiv 0$ is interesting.

Recall that for a bounded function φ on $\mathbb{T}$ the Hankel operator H_φ on H^2 is defined by $H_\varphi f = (I - \mathbb{P}_+)\varphi f$, where $\mathbb{P}_+$ is the orthogonal projection from L^2 onto H^2.

Consider the Hankel operator $H_{\bar{h}/h}$ with the unimodular symbol $\varphi = \bar{h}/h$. It is easy to see that $\mathcal{P}_{G_f} \mathcal{P}_{G_p} \mathcal{P}_{G_f} \mid G_f$ is unitarily equivalent to $H^*_{\bar{h}/h} H_{\bar{h}/h}$ (see [1], Lemma 2.6).

The modulus of an operator T on Hilbert space is the selfadjoint non-negative operator $(T^*T)^{1/2}$. Problem 1 is therefore intimately connected with the problem of description of the moduli of Hankel operators up to the unitary equivalence.

PROBLEM 2. <u>Which operator can be the modulus of a Hankel operator</u>?

There are two n e c e s s a r y c o n d i t i o n s for an operator to be the modulus of a Hankel operator which imply evident restrictions on triples equivalent to (G_p, G_f, G).

For any $\varphi \in L^\infty$ the operator $(H^*_\varphi H_\varphi)^{1/2}$ <u>is not invertible</u> because $\lim_{n\to\infty} \| H_\varphi z^n \| = 0$. It follows that G_f does contain an orthogonal basis $\{e_n\}_{n \geq 0}$ with $\lim_n \| \mathcal{P}_{G_0} e_n \| = 0$ (i.e. an orthogonal sequence "almost independent" with respect to the Past) provided $G_p \neq G_f$.

The kernel $\operatorname{Ker}(H^*_\varphi H_\varphi)^{1/2}$ is <u>either trivial or infinite dimensional</u>. Indeed, being invariant under multiplication by z by Beurling's theorem, it is either trivial or equal to θH^2 for an inner function θ.

Note that for $t = (G_p, G_f, G)$ we always have $n_+(t) = n_-(t)$. Indeed passing to the spectral representation Φ we see that $J: f \mapsto \bar{z}\bar{f}$ is an isometry of G (over $\mathbb{R}$) with $J G_p = G_f$, $J G_f = G_p$. So we have either $n_+(t) = n_-(t) = 0$ or $n_+(t) = n_-(t) = \infty$.

Under the a priori assumption that the angle between G_p and G_f is positive Problem 1 can be in fact reduced to Problem 2.

Indeed, if the angle between G_p and G is positive then by the Helson-Szegö theorem the spectral measure μ is absolutely continuous and $\mathcal{P}_{G_f}\mathcal{P}_{G_p}\mathcal{P}_{G_f}$ is unitarily equivalent to $H^*_{\bar{h}/h}H_{\bar{h}/h}$, $d\mu=|h|^2dm$. On the other hand if $R=H^*_\varphi H_\varphi$ and $\|R\|<1$ then there exists an outer function h in H^2 with $H_\varphi=H_{\bar{h}/h}$ (see [2]). Put $\mathcal{H}=L^2(|h|^2)$, $B=span_{L^2(|h|^2)}\{z^n : n\geqslant 0\}$, $A=span_{L^2(|h|^2)}\{z^n : n\leqslant 0\}$. Clearly $\{z^n\}_{n\in\mathbb{Z}}$ is a stationary sequence with the Future B and the Past A.

It follows from the above considerations that in the case of non-zero angle between A and B the problem of the existence of a stationary sequence with the Future B and the Past A can be reduced to the existence of a Hankel operator whose modulus is unitarily equivalent to $\mathcal{P}_B\mathcal{P}_A\mathcal{P}_B$.

In connection with Problems 1 and 2 we can propose two conjectures.

CONJECTURE 1. <u>Let</u> $\{s_n\}_{n\geqslant 0}$ <u>be a non-increasing sequence of positive numbers and let</u> $\lim_n s_n = 0$. <u>Then there exists a Hankel operator whose singular numbers</u> *) $s_n(H_\varphi)$ <u>satisfy</u>

$$s_n(H_\varphi) = s_n\,,\quad n\geqslant 0\,.$$

CONJECTURE 2. <u>Let</u> T <u>be a compact selfadjoint operator such that</u> $Ker\,T$ <u>is either trivial or infinite dimensional</u>. <u>Then there exists a Hankel operator</u> H_φ <u>satisfying</u> $T=(H^*_\varphi H_\varphi)^{1/2}$.

It can be shown that the last conjecture is equivalent to the following one.

CONJECTURE 2'. <u>Given a triple</u> $t=(A,B,\mathcal{H})\in\mathcal{T}$ <u>such that</u> $\mathcal{P}_A\mathcal{P}_B\mathcal{P}_A$ <u>is compact and</u> $n_+(t)=n_-(t)$ <u>is either</u> 0 <u>or</u> ∞ <u>there exists a stationary sequence</u> $\{X_n\}_{n\in\mathbb{Z}}$ <u>in</u> $\mathcal{H}$ <u>whose Future is</u> B <u>and Past is</u> A.

We can also propose the following qualitative version of Conjecture 1.

THEOREM. <u>Let</u> $\varepsilon>0$ <u>and</u> $\{s_n\}_{n\geqslant 0}$ <u>be a non-increasing sequence of positive numbers. Then there exists a Hankel operator</u> H_φ <u>satisfying</u>

$$\frac{1}{1+\varepsilon}\,s_n\leqslant s_n(H_\varphi)\leqslant(1+\varepsilon)\,s_n\,,\quad n\geqslant 0.$$

*) See def. of singular numbers in [3].

PROOF. Let b be an interpolating Blaschke product having zeros $\{\zeta_n\}_{n\geq 0}$ with the Carleson constant δ (see e.g. [4]). Consider the Hankel operators of the form $H_{f\bar{b}}$, $f \in H^\infty$. Then we have (see [4], Ch.VIII)

$$b \, H_{f\bar{b}} P_b = f(S_b), \tag{2}$$

where S_b is the compression of the shift operator S to $K_b \overset{def}{=} H^2 \ominus bH^2$, $S_b g = P_b z g$, $g \in K_b$, P_b is the orthogonal projection onto K_b, b is multiplication by b, $f(S_b) \overset{def}{=} P_b f g$, $g \in K_b$.
Since b is an interpolating Blaschke product, there exists a function f in H^∞ satisfying $f(\zeta_n) = s_n$, $n \geq 0$.

It follows from (2) that $s_n(H_{f\bar{b}}) = s_n(f(S_b))$, $n \geq 0$.
Consider the vectors $e_n = \frac{b}{z-\zeta_n} |\zeta_n| (1-|\zeta_n|^2)^{1/2}$. We have $f(S_b) e_n = f(\zeta_n) e_n$ (see [4], Ch.VI) and there exists an invertible operator V on K_b such that the sequence $\{Ve_n\}_{n\geq 0}$ is an orthogonal basis of K_b, moreover if $1-\delta$ is small enough then we can choose V so that $\|V\| \cdot \|V^{-1}\| \leq 1+\varepsilon$ (see [4], Ch.VII). The result follows from the obvious estimates

$$\frac{1}{\|V\| \cdot \|V^{-1}\|} s_n \leq s_n(f(S_b)) \leq \|V\| \cdot \|V^{-1}\| s_n . \quad \bullet$$

Conjecture 1 can be interpreted in terms of rational approximation. It follows from the theorems of Nehari and Adamian-Arov-Krein (see [1]) that for a function f in $BMO_A \overset{def}{=} \mathbb{P}_+ L^\infty$ we have

$$s_n(\Gamma_\varphi) = r_n(\varphi) \overset{def}{=} dist_{BMO_A}(\varphi, \mathcal{R}_n) ,$$

where Γ_φ is the operator on ℓ^2 with the matrix $\{\hat{\varphi}(n+k)\}_{n,k\geq 0}$, $\mathcal{R}_n$ is the set of rational functions with at most n poles outside $clos\, \mathbb{D}$ (including possible poles at ∞) counting multiplicities. Conjecture 1 is equivalent to the following one.

CONJECTURE 1'. <u>Let</u> $\{r_n\}_{n\geq 0}$ <u>be a non-increasing sequence,</u> $\lim_{n\to\infty} r_n = 0$. <u>Then there exists</u> f <u>in</u> BMO_A <u>such that</u> $r_n(f) = r_n$, $n \geq 0$.

If the conjecture is true then it would give an analogue of the well-known Bernstein theorem [5] for polynomial approximation. Note in this connection that Jackson-Bernstein type theorems for rational approximation in the norm BMO_A were obtained in [5], [7], [8].

We are grateful to T.Wolff for valuable discussions.

REFERENCES

I. П е л л е р В.В., Х р у щ ё в С.В. Операторы Ганкеля, наилучшие приближения и стационарные гауссовские процессы. - Успехи матем. наук, I982, 37, № I, 53-I24.

2. А д а м я н В.М., А р о в Д.З., К р е й н М.Г. Бесконечные Ганкелевы матрицы и обобщенные задачи Каратеодори-Фейера и И.Шура. - Функц. анал. и его прил., I968, 2, № 4, I-I7.

3. Г о х б е р г И.Ц., К р е й н М.Г. Введение в теорию линейных несамосопряженных операторов. М., "Наука", I965.

4. Н и к о л ь с к и й Н.К. Лекции об операторе сдвига, М., "Наука", I980.

5. Б е р н ш т е й н С.Н. Об обратной задаче теории наилучшего приближения непрерывных функций. - Собрание сочин., т.2, Изд-во АН СССР, I954, 292-294.

6. П е л л е р В.В. Операторы Ганкеля класса $\mathfrak{S}_p$, и их приложения (рациональная аппроксимация, гауссовские процессы, проблема мажорации операторов).- Матем. сборник, I980, II3, № 4, 538-58I.

7. P e l l e r V.V. Hankel operators of the Schatten - von Neumann class $\mathfrak{S}_p$, $0<p<1$. - LOMI Preprints, E-6-82, Leningrad, 1982.

8. S e m m e s S. Trace ideal criteria for Hankel operators, $0<p<1$. Preprint, 1982.

S.V.HRUŠČËV
(С.В.ХРУЩЁВ)

V.V.PELLER
(В.В.ПЕЛЛЕР)

СССР, I9IOII, Ленинград
Фонтанка 27, ЛОМИ

3.4. SOME PROBLEMS RELATED TO THE STRONG LAW OF LARGE NUMBERS FOR STATIONARY PROCESSES

Let $(\xi_k : k \in \mathbb{Z})$ be a process in $L^2(\Omega, \mathcal{A}, P)$ stationary in the wide sense with $\mathbb{E}\xi_k = 0$ and $\mathbb{E}|\xi_k|^2 = \sigma^2 > 0$.

Denote the correlation function of the process by

$$R(n) = \mathbb{E}\bar{\xi}_k \xi_{k+n} \quad \forall (k, n)$$

and let

$$\xi_k = \int_{-\pi}^{\pi} e^{ik\lambda} Z(d\lambda)$$

be its spectral representation. Here $Z(d\lambda)$ stands for the stochastic spectral measure of the process (ξ_k) ; $Z(d\lambda)$ is a process with orthogonal increments.

It is well-known that the strong law of large numbers (h e r e a f t e r a b b r e v i a t e d a s SLLN) holds for all processes stationary in the strict sense, that is, the limit of the means $\sigma_n = n^{-1} \sum_{k=1}^{n} \xi_k$ exists a.e. But there exist processes stationary in the wide sense such that the means σ_n converge in $L^2(\Omega)$ and diverge a.e. (see [1], [2]). SLLN criteria are given in [2].

THEOREM (Gaposhkin [2]). In the above notation

$$\lim_{n\to\infty} \Big[\sigma_n - \int_{|\lambda| \le 2^{-[\log_2 n]}} Z(d\lambda)\Big] = 0 \tag{1}$$

Thus SLLN holds iff the limit

$$\lim_{m\to\infty} \int_{|\lambda| \le 2^{-m}} Z(d\lambda) \tag{2}$$

exists a.e.

The theorem implies the following: if

$$R(n)=O((\log\log n)^{-2-\varepsilon})\quad n\to\infty \tag{3}$$

then SLLN holds provided $\varepsilon>0$, while for $\varepsilon=0$ it does not hold in general.

In all known counterexamples

$$\sup_k \mathbb{E}|\xi_k|^p=\infty \quad (p>2).$$

PROBLEM 1. Is the condition

$$\exists p>2:\ \sup_k \mathbb{E}|\xi_k|^p<\infty$$

(may be with the supplementary condition $\lim R(n)=0$) sufficient for the SLLN?

PROBLEM 2. Is the condition

$$\sup_k \|\xi_k\|_\infty<\infty$$

(may be with the above supplementary condition) sufficient for the SLLN?

PROBLEM 3. If the answers to problems 1 and 2 are negative, we may ask: are there stationary processes (ξ_k) satisfying

$$\sup \|\xi_k\|_\infty<\infty,\ R(n)=O((\log\log n)^{-2})$$

while SLLN does not hold? Or condition (3) can be relaxed for L^∞-bounded processes?

All processes stationary in the strict sense obey SLLN, and so the Theorem implies the existence of limit (2) as well.

PROBLEM 4. Why does limit (2) exist for stationary (in the strict sense) processes?

Analogous problems are of interest not only for unitary operators determining stationary processes but also for normal operators in $L^2(X)$ (see an ergodic theorem of this kind in [3]). Here is one of possible problems in this direction.

PROBLEM 5. Let T be a normal operator in $L^2(X,\mu)$, μ being a σ-finite measure. Suppose $\|T\|=1$, $f\in L^2(X)$, $\sup_k |T^k f(x)|\le C$ a.e. Does

$$\lim_{n\to\infty} n^{-1}\sum_{k=1}^{n} T^k f(x)$$

exist a.e.?

REFERENCES

1. Blanc-Lapierre A., Tortrat A. Sur la loi forte des grands nombres. - C.r.Acad.sci. Paris, 1968, 267 A, 740-743.
2. Гапошкин В.Ф. Критерии усиленного закона больших чисел для классов стационарных в широком смысле процессов и однородных случайных полей - Теор.вероятн. и ее прим., 1977, 22, № 2, 295-319.
3. Гапошкин В.Ф. Об индивидуальной эргодической теореме для нормальных операторов в L_2. - Функц.анализ и его прил., 1981, 15, № 1, 18-22.

V.F.GAPOSHKIN
(В.Ф.ГАПОШКИН)

СССР, 103055, Москва
ул.Образцова, 15,
Московский институт инженеров
железнодорожного транспорта

3.5. THE THEORY OF MARKOV PROCESSES FROM THE STANDPOINT OF THE THEORY OF CONTRACTIONS

Let (X,μ) be a Lebesgue space. A contraction P on $L^2(X,\mu)$ is called a M a r k o v o p e r a t o r if P is order-positive and preserves the constants. In other words P is Markov if $P\mathbb{1}=\mathbb{1}$, $P^*\mathbb{1}=\mathbb{1}$ and P is positive. The integral representation of such an operator is given by a bistochastic measure ν on $X\times X$: $\nu\geqslant 0$, $\nu(X\times X)=1$

$$(Pf)(x)=\int\limits_X f(y)\nu(x,dy),\ \iint\limits_{AX}\nu(x,dy)dx=\mu(A),\ \iint\limits_{XB}\nu(dx,y)dy=\mu(B).$$

Markov operators form a convex semigroup with a zero (=projection onto the constants) and with a unit. This is a functional equivalent of the semigroup of multivalued maps, admitting an invariant measure, of (X,μ) onto itself. A detailed account of an analogous view-point see in [1].

A Markov operator gives rise in a natural way to a stationary Markov process with the state space X , the initial measure μ and the two-dimensional distribution ν (see above). In the space $\mathfrak{X}=\prod\limits_{-\infty}^{\infty}X_i$ $(X_i\equiv X)$ of realizations of the process a Markov measure $M_\nu=M$ appears. The left shift T in $(\mathfrak{X},M)$ generates a unitary operator U_T on $L^2(\mathfrak{X},M)$, a unitary dilation of P (non-minimal in general).

<u>The main problem of the theory of Markov processes is the investigation of metric properties of the shift T in terms of the Markov operator P</u>. The classical theory virtually used spectral properties of P only. This is insufficient for metric problems, P being nonselfadjoint.

Modern tools of the theory of contractions seem not have been used for this aim and we want to draw attention to this point (see also [1]). The connection between the contractions theory, their dilations, the scattering theory on the one hand and Markov processes theory on the other can be usefully applied in both directions.

1. PROBLEMS ABOUT PAST. It is easy to check that a Markov process is forward (back) mixing in the sense of Kolmogorov iff P be-

longs to the class $C_{0\cdot}$ (resp. $C_{\cdot 0}$), see notation in [2]. The opposite class C_1 includes two subcases. The first one is of no interest and corresponds to an isometric P and to deterministic processes. The second one, namely, the case of a completely non-isometric contraction, is very interesting.Its very existence is far from being obvious (for Markov operators), an example was given by M.Rosenblatt [3]. An important theorem (see [2]) asserts that the corresponding process, being non-deterministic, is quasisimilar to a diterministic one. Our PROBLEM is as follows: use the technique of the theory of contractions to study mixing criteria of various kinds, deterministic and quasideterministic, the exactness [1], the bernoullity etc. A powerful tool for these topics is the characteristic function of a Markov contraction. No adequate metric analogue of this notion seems to be found (e.g. how can one connect this function with the bistochastic measure ν?)

2. NON-LINEAR DILATIONS. The theory of Markov processes imlicitly includes some constructions unfamiliar in the theory of contractions. We have already mentioned that a unitary operator U_T acting in $L^2(\mathfrak{X}, M)$ is not the minimal dilation of the Markov operator P. The minimal dilation can be easily described in these terms. It coincides with the restriction of U_T to $\mathcal{L} = span\{L^2(\mathfrak{X}_i): i \in \mathbb{Z}\}$, $L^2(\mathfrak{X}_i)$ being the subspace of $L^2(\mathfrak{X}, M)$ consisting of functions depending on the i-th coordinate of $\{x_n\} \in X$ only. The subspace $\mathcal{L}$ is the subspace of all linear functionals of realizations ("one-particle" subspace). Thus the theory of minimal dilations corresponds to the linear theory of Markov processes whereas the dilation U_T has to be interpreted as a "non-linear" one (clearly U_T is a linear operator acting on non-linear functionals of realizations).

The investigation of the pair (P, U_T) ("a Markov operator plus a non-linear dilation") is of interest for the theory of contractions connecting it with methods and notions of the metric theory of processes (mixing, bernoullity etc.) E.g. the problem of the isomorphism of two Markov processes is analogous with the problem of existence of the wave operator in scattering theory. The enthropy yields an invariant of the dilation etc. It would be interesting to define the non-linear dilation for an arbitrary (non-positive) contraction.

3. C^*-ALGEBRA GENERATED BY MARKOV OPERATORS. Let us mention a more special problem: to describe the C^*-envelope of the set of all Markov operators. This algebra does not coincide with the algebra of all operators. (G.Lozanovsky gave a nice (unpublished) example: the distance between the Fourier transform as an operator in $L^2(\mathbb{R})$ and the set of all regular operators (= differences of positive operators) is one). It seems likely that a direct description of elements of this algebra can be given in terms of the order. This C^*-algebra plays an important rôle in the theory of gruppoids.

REFERENCES

1. В е р ш и к А.М. Многозначные отображения с инвариантной мерой (полиморфизмы) и марковские операторы. - Зап.научн.семин.ЛОМИ, 1977, 72, 26-61.
2. Sz.-N a g y B., F o i a ş C. Analyse harmonique des opérateurs de l'espace de Hilbert. Budapest, Acad.Kiadó, 1967.
3. R o s e n b l a t t M. Stationary Markov Processes. Berlin, 1971.

A.M.VERSHIK
(А.М.ВЕРШИК)

СССР, 198904, Петродворец,
Математико-механический
факультет Ленинградского
университета

3.6. old EXISTENCE OF MEASURES WITH GIVEN PROJECTIONS

Let F denote a measurable subset of the unit cube $Q \subset \mathbb{R}^d$, π_k being the canonical operator of projection onto the k-th axis, $k = 1, \ldots, d$, and X_k the side of Q situated on the k-th axis. Consider the linear operator π transforming every finite measure m on F into the system $(m_1, m_2, \ldots, m_d)$ of its marginals, $m_k A \overset{def}{=} m(\pi_k^{-1} A)$, $A \subset X_k$. It is often of importance to know whether a given system $(m_1, \ldots, m_d)$ belongs to the image under π of a natural class of probability measures on F. Some partial results are known, mostly for $d=2$. E.g. for a class $\mathcal{F}$ of subsets F of Q defined in terms of measure spaces (X_k, m_k), $k = 1, 2$, an existence criterion of a probability measure on F with marginals (m_1, m_2) is as follows: the required measure exists iff no subset of $X_1 \times X_2$ of the form $F \cap (A \times B)$, where $A \subset X_1$, $B \subset X_2$ and $m_1 A + m_2 B > 1$, is a union of subsets N_1, N_2 with $m_1(\pi_1 N_1) = m_2(\pi_2 N_2) = 0$ (see [1]; a smaller class of closed set was considered in [2]). The class $\mathcal{F}$ is in particular characterized by the following property: for every $F \in \mathcal{F}$ the set of all measures on F with given marginals is compact in the topology of convergence on sets of the form $A \times B$. A similar condition (where the non-decomposability of sets $F \cap (A \times B)$ is replaced by $mes_2(F \cap (A \times B)) > 0$) is a criterion of existence of a probability measure on F with marginals (m_1, m_2) subordinated to the Lebesgue type (i.e. absolutely continuous with respect to the Lebesgue measure mes_2), [1].

Analogous conditions fail to be sufficient for $d > 2$, and the corresponding criterion is unknown. For $d=2$ it is not known whether the Lebesgue type can be replaced by any other type in the last sentence of the previous paragraph. For $d>2$ there is no existence criterion for a positive measure on the cube Q with given marginals whose density function with respect to Lebesgue measure is majorized by the density function of a given probability measure on Q (see the discussion in [1]).

REFERENCES

1. Судаков В.Н. Геометрические проблемы теории бесконечно-

мерных вероятностных распределений. - Труды МИАН, I4I, М.-Л., Наука, I976. (Proc. of the Steklov Inst. of Math., 1979, issue 2).

2. S t r a s s e n V. Probability measures with given marginals. - Ann.Math.Stat., 1965, 36, N 2, 423-439.

V.N.SUDAKOV
(В.Н.СУДАКОВ)

СССР, I9I0II, Ленинград,
Фонтанка 27, ЛОМИ

* * *

COMMENTARY BY THE AUTHOR

Consider a finite or countable family of probability distributions $\{m_k, k \in K\}$. The answer to the question as to whether there exists such a family $\{\xi_k, k \in K\}$ of random variables, each ξ_k being distributed according to m_k, that for every pair (k_1, k_2) the equality

$$\varkappa(m_{k_1}, m_{k_2}) \stackrel{def}{=} \min\left\{\iint |x-y|\, dm : m\pi_X^{-1} = m_{k_1},\ m\pi_Y^{-1} = m_{k_2}\right\} = \| \xi_{k_1} - \xi_{k_2} \|_{L^1}$$

holds, depends on existance of a probability measure m with marginals m_k on the set

$$F = \{x \in \mathbb{R}^k : U_{k_1k_2}(x_{k_1}) - U_{k_1k_2}(x_{k_2}) = |x_{k_1} - x_{k_2}| \ \ \forall k_1, k_2 \in K\} .$$

Here $\varkappa$ is the Kantorovich distance and $U_{k_1k_2}$ stands for related potential function (see e.g. [1]):

$$U_{k_1k_2}(\lambda) = \int sign(m_{k_1}(-\infty, t] - m_{k_2}(-\infty, t])\, dt .$$

One can show that for such a special type of subsets there always exists a measure with given marginals $\{m_k, k \in K\}$, so that the family $\{\xi_k\}$ under discussion does exist.

I am grateful to B.I.Berg for stimulating discussions.

3.7. old ON THE FOURIER TRANSFORM OF THE INDICATOR OF A SET IN $\mathbb{R}^n$ OF FINITE LEBESGUE MEASURE

Consider a set $E \subset \mathbb{R}^n$ of finite and non-zero Lebesgue measure: $0<|E|<\infty$. The function χ_E, $\chi_E=\begin{cases} 1, & x\in E \\ 0, & x\notin E \end{cases}$ is called the indicator of E. Set

$$\hat{\chi}_E(t)=\int_{\mathbb{R}^n} e^{itx}\chi_E(x)\,dx ,$$

the Fourier transform of χ_E. We ask whether there is a set E, $0<|E|<\infty$ such that $\hat{\chi}_E$ vanishes on an open non-empty set A, $A\subset\mathbb{R}^n$.

Note that if E is bounded then $\hat{\chi}_E$ is analytic in $\mathbb{R}^n$ and therefore cannot vanish on an open non-empty set. Some other similar cases are considered in the author's paper [1]. If it turns out that there exists such a set E, $0<|E|<\infty$ with $\hat{\chi}_E$ vanishing on an open non-empty set, the SECOND PROBLEM will be to describe all sets E with this property.

These questions are related to the uniqueness problem for a finite Borel measure μ (in $\mathbb{R}^n$) with prescribed values $\mu(B_0+y)$ $(y\in\mathbb{R}^n)$, B_0 being a given Borel set, $0<|B_0|<\infty$. It follows from [1] that if there is no open non-empty $A\subset\mathbb{R}^n$ with $\hat{\chi}_{B_0}|A\equiv 0$, then such a μ is unique. And, conversely, it is not unique provided A does exist.

REFERENCE

1. Сапогов Н.А. Об одной проблеме единственности для конечных мер в евклидовых пространствах. - Зап.научн.семин.ЛОМИ, 1974, 41, 3-13.

N.A.SAPOGOV
(Н.А.САПОГОВ)

СССР, 191011, Ленинград
Фонтанка 27, ЛОМИ

* * *

COMMENTARY

The first problem has been solved by Kargayev [2]. Sets E, $E\subset\mathbb{R}$, with $0<|E|<\infty$ and $\hat{\chi}_E$ vanishing on an interval DO EXIST. Moreover, it is shown in [2] that given numbers $a, b, 0<a<b<\pi$, and an even function $k:\mathbb{R}\longrightarrow[1,+\infty)$ increasing on $[0,+\infty)$ and such that

$$k(x+y)\leq Ck(x)k(y) \qquad (x,y\in\mathbb{R})$$

$$\int^{+\infty}\frac{\log k(x)}{x^2}\,dx<\infty\ , \quad \sum_{n\in\mathbb{Z}}\frac{1}{k(n)}<+\infty\ ,$$

there is a sequence $\{[a_n, a_n+h_n]\}_{n=1}^{\infty}$ of disjoint segments satisfying

$$\frac{C_1}{k(n)}\leq h_n\leq\frac{C_2}{k(n)},\ |n-a_n|\leq\frac{C_2}{k(n)} \qquad (n\in\mathbb{Z}\ ;\ 0<C_1<C_2),$$

and $\hat{\chi}_E\,|\,(a,b)\equiv 0, \qquad E\overset{def}{=\!=}\bigcup_{n\in\mathbb{Z}}[a_n, a_n+h_n]$.

REFERENCE

2. К а р г а е в П.П. Преобразование Фурье характеристической функции множества, исчезающее на интервале. - Матем.сб., I982, II7, № 3, 397-4II.

CHAPTER 4

OPERATOR THEORY

Due to founders of the Spectral Operator Theory the word "operator" became almost inseparable from the word "selfadjoint". This connection was so tight that even now it is still common to speak on "NON-selfadjoint operators" as though forgiving general operators for the absence of appreciable intrinsic structure. This kind of inferiority complex is being overcome nowadays under pressure from Physics (recall complex poles of resolvent on "the non-physical sheet" in the resonance scattering) and under the influence of the increasing power of Analysis. "Analysis" means here mainly "Complex Analysis", and the above tendency may be well illustrated,in particular, by operator-theoretic problems of this book (first of all problems in Chapters 4, 5, 7). Almost all of them are related to the spectral theory tending to blend with Complex Analysis and, in any case ,to borrow from it something more significant than Cauchy formula or Liouville and Stone-Weierstrass theorems, tools not exceeded by the classical approach. This blend is probably the most characteristic feature of the present-day theory, or at least of its parts close to this book. The first steps of this mutual penetration were made in thirties and forties (Stone, Wold, Plessner, M.Krein, Livshic).

The new spectral theory begins with working out convenient functional models whereas in classical analysis such a model was often the end of investigation. Many problems in this Chapter are related to the multiplication operator $f \mapsto zf$ whose restrictions and compressions to suitable subspaces yield models we have just mentioned. One of the most popular models can be described in terms of the so-called characteristic function of the operator (the Szőkefalvi-Nagy - Foiaş model and its generalizations). This model reduces spectral problems to the investigation of boundary properties of vector-valued functions of the Nevanlinna class. The questions posed in Problems 4.8-4.20 exhibit distinctly enough the present state of affairs which can be summarized as follows.

Almost all achievements of the H^p-theory have been exploited (free interpolation, the delicate multiplicative structure of H^p-functions, Corona Theorem etc.) and a new "operator-valued" Function Theory is needed now. Its contents are essentially non-scalar though this fact is often disguised by formally dimension-invariant statements. Therefore the new progress requires not only new efforts in the spirit of the standard H^p- theory, but rather creating a kind of special non-commutative intuition. Vivid examples are Problem 4.12 and the Halmos-Lax theorem describing invariant subspaces of the multiple shift. This theorem is deciphered (for a very special situation) in Problem 4.14. As to Problem 4.12, its seeming simplicity conceals many interesting concrete realizations. Namely, Problem 4.12 includes as particular cases the principal question of Problem 4.10 and the matrix generalization of the Corona Problem discussed in the Commentary. Many problems of the "vector-valued" function theory admit interesting scalar interpretations (as, for instance, Problems 4.4, 4.9, 4.12). Other problems make sense only when the space of values is multidimensional.

It is not easy to classify rapidly growing Operator Theory with its intertwining ramifications. The same is true even for its parts presented here. However, we tried to group the problems in accordance with the intrinsic logic of the subject which we understand approximately as follows.

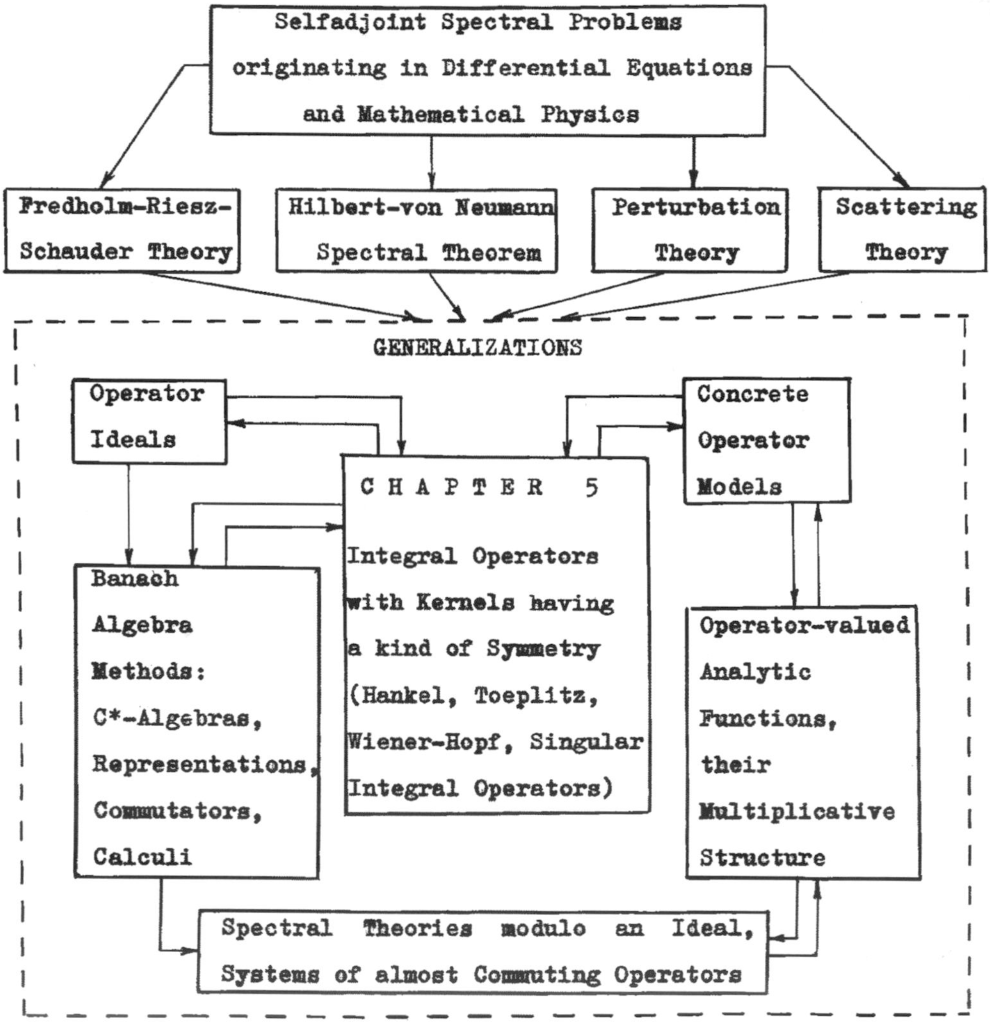

Of course, the scheme leaves aside a great number of links existing in Operator Theory as well as connections with other branches of Mathematics. But the main purpose of the scheme is to explain the following arrangement of Problems inside the Chapter.

I. Selfadjoint spectral theory, including Perturbation Theory and Scattering Theory, and problems they generate (NN 4.1-4.8, 4.15). This group represents so well-known domains of Operator Theory that we do not risk commenting it.

II. Functional Models, Characteristic Function and other operator analytic functions (4.8-4.20, 4.4). Being of quite different origins, these Problems mostly can be reduced to the investigation of the multiplicative structure of operator-functions and they need certain development of operator (or matrix) analogues of H^p-techniques. Problems 4.12, 4.15, 4.20, 4.21, 4.24 are directly related to the themes of the next Chapter 5.

III. Banach Algebra Methods are a common feature of Problems 4.22-4.29 and they also play an important role in Problems 4.30-4.36. We mean here various aspects of "algebraization" as an alternative approach to generalizations of the Spectral Theorem and to interpretations of spectral nature of non-commuting objects. These aspects are C*-algebras, calculi, symbols, theories à la Fredholm-Riesz-Schauder etc.

IV. "Near-normal" (i.e. hypo-, semi-, sub-, quasi-,...normal) operators as a particular case of the perturbation theory for families $(A_1, \ldots, A_n)$ of commuting selfadjoint operators (namely, the case $n=2$). Such operators are the subject of problems 4.30-4.36 (and 4.37 - from the technical point of view). An analogy with the classical Perturbation Theory may be drawn as follows: the classical smallness of "the imaginary part" $\mathfrak{Im}\, A = (A - A^*)/2i$ is replaced here by the smallness of the selfcommutator $[A, A^*] = AA^* - A^*A$, the accretivity condition $\mathfrak{Im}\, A \leqslant 0$ by the hyponormality $[A, A^*] \leqslant 0$ etc. The new situation is in a sense "two-dimensional", the operator A being viewed as a pair $(\mathcal{Re}\, A, \mathfrak{Im}\, A)$. This leads to essential complications in comparison with the "one-dimensional" (i.e. the selfadjoint) case. Some quite simple "one-dimensional" problems look rather difficult in the "two-dimensional" setting, as for example the stability of continuous spectrum or the solvability of equation $\mathcal{I}m A = K$ for $K \in \mathcal{V}$ ($\mathcal{V}$ being a given operator ideal). These problems likely require a hard analysis of operator algebras and finding out of algebraic obstacles to solve the equation $[A, A^*] \in \mathcal{V}$ in the form $A = N + S$ where $[N, N^*] = 0$, $S \in \mathcal{V}$. Problems of this direction are represented rather distinctly by items 4.30-4.38. Problems 4.39, 4.40 deal with interesting concrete questions (circular symmetry of spectra of endomorphisms, quasinilpotency of indefinite integration) which can hardly be placed at a definite point of our scheme On the other hand, parts of Operator Theory indicated in the scheme, are included in Chapters 5 and 7. Let us enlist some close cross-links between problems of this Chapter and of other Chapters.

Chapter 2 (Banach Algebras): Problem 2.13 could be placed among items 4.8-4.20; Problems 2.1, 2.4 are related to 4.39. Finally, Problems 2.2 and 2.3 deal with a subtle behaviour of the norms $\|T^n\|, n \geq 1$, and therefore find a responce in 4.38.

Chapter 9 (Uniqueness). Problem 9.4 admits a clear operator-theoretic interpretation in the spirit of Problems 4.31-4.36, and it is recommended to read Problem 9.6 and Problem 4.4 simultaneously.

Numerous problems on multipliers dispersed throughout the volume are related by their common origins and intrinsic connections (4.21, 4.25, 4.26, 2.6, 10.3, 9.9, 10.8).

Unfortunately (or fortunately - taking into account inevitable volume restrictions) many parts of Operator Theory are not mentioned here. For example, the invariant subspace problem is not presented explicitly. But it is alluded to, incidentally, in constructions of items 4.8, 4.13, 4.22 whereas other problems of this "descriptive" direction of Operator Theory have to wait for another Problem Book. The same fate (i.e. complete oblivion) is shared by many consumers of Operator Theory. The reader will find in our Book neither "pseudodifferential operator", nor "integral Fourier operator", nor "operator K-theory"....

4.1. BOUNDEDNESS OF CONTINUUM EIGENFUNCTIONS AND THEIR RELATION TO SPECTRAL PROBLEMS

We will describe a set of problems for matrices acting on $\ell^2(\mathbb{Z})$. There are analogous problems for $\ell^2(\mathbb{Z}^\nu)$ and for suitable elliptic operators on $L^2(\mathbb{R}^\nu)$. Let A be a bounded self-adjoint operator on $\ell^2(\mathbb{Z})$ whose matrix elements obey $a_{ij} \equiv (\delta_i, A\delta_j) = 0$ if $|i-j| \geq K$. A fundamental result asserts the existence of a measure $d\rho(E)$, a function $n(E)$ taking the values $0, 1, \ldots, \infty$ (infinity allowed) with $n(E) \geq 1$ $(d\rho)$-a.e. E and $n(E) = 0$ if $E \notin \operatorname{supp} \rho$ and for each E, $n(E)$ linearly independent sequences $u_\alpha(E; n)$; $\alpha = 1, \ldots, n(E)$ (not necessarily in ℓ^2) so that (a) $|u_\alpha(E;n)| \leq C(1+|n|)$ (b) $\sum a_{ij} u_\alpha(E;j) = E\, u_\alpha(E;i)$;

(c) Let $\mathcal{H}' = L^2(\mathbb{R}; \mathbb{C}^{n(E)}; d\rho)$, i.e. functions, f, on $\mathbb{R}$ with $f(E)$ having values in $\mathbb{C}^{n(E)}$ (where $\mathbb{C}^\infty = \ell^2$) and let C_0 denote sequences in $\ell^2(\mathbb{Z})$ of compact support. Define U taking C_0 into $\mathcal{H}'$ by $(Ug)_\alpha(E) = \sum_m \overline{u_\alpha(E;m)}\, g(m)$. Then U extends to a unitary map of $\ell^2(\mathbb{Z})$ onto $\mathcal{H}'$; (d) $U(Ag) = E(Ug)$.

These continuum eigenfunction expansions are called BGK expansions in [1] in honor of the work of Berezanskii, Browder, Gårding, Gel'fand and Kac, who developed them in the context of elliptic operators. See [1,2,3] for proofs. These expansions don't really contain much more information than the spectral theorem. The most significant additional information concerns the boundedness properties of u; see [4,5] for applications.

Actually, the general proofs show that $(1+|n|)$ in part (a) can be replaced by $(1+|n|)^\alpha$ for any $\alpha > \frac{1}{2}$. Indeed, one shows that for any $g \in \ell^2$, one can arrange that for $(d\rho)$-a.e. E $g(\cdot)\, u_\alpha(E, \cdot) \in \ell^2$. If one could arrange a set, S, of good E's where $gu \in \ell^2$ for all $g \in \ell^2$ with $\rho(\mathbb{R} \setminus S) = 0$, then on S, $u \in \ell^\infty$. This leaves open:

QUESTION 1. <u>Is it true that for</u> $(d\rho)$-<u>a.e.</u> E, <u>each</u> $u_\alpha(E, \cdot)$ <u>is bounded</u>?

There is a celebrated counterexample of Maslov [6] to the boundedness in the one dimensional elliptic case. As explained in [1], Maslov's analysis is wrong, and it is not clear whether his example has bounded u's a.e. We believe the answer to question 1 (and all other yes/no questions below) is affirmative, but for what we have to say below, a weaker result would suffice:

QUESTION 2. Is it at least true that for $(d\rho)$-a.e. E and all α : $\frac{1}{2N+1} \sum_{|n|\leq N} |u_\alpha(E,n)|^2$ is bounded?

QUESTION 3. Is it true that

$$\lim_{N\to\infty} \frac{1}{2N+1} \sum_{|n|\leq N} |u_\alpha(E,n)|^2 = k(\alpha,E)$$

exists? The $\overline{\lim}$ we will denote by $\bar{k}(\alpha,E)$.

Given a subset M, of $\{(E,\alpha) : E \in \mathbb{R}, \alpha \leq N\}$ we define

$$(P(M)g)(n) = \sum_\alpha \int_{\{E:(E,\alpha)\in M\}} u_\alpha(E,n)(Ug)_\alpha(E)d\rho(E)$$

where a suitable limit in mean may need to be taken. Define

$$M_1 = \{(E,\alpha) : u_\alpha(E,\cdot) \in \ell^2\}$$

$$M_2 = \{(E,\alpha) : k(\alpha,\bar{E}) = 0 \text{ but } (E,\alpha) \notin M_1\}$$

$$M_3 = \{(E,\alpha) : \bar{k}(\alpha,E) \neq 0\} .$$

Obviously, $P(M_1)$ is the projection onto the point spectrum of A.

QUESTION 4. Is it true that $P(M_2)$ is the projection onto the singular continuous space of A and $P(M_3)$ the projection onto the absolutely continuous spectrum of A ?

Among other things this result would imply that in the Jacobi case (where the number K of the third sentence in this note is 2), the singular spectrum is simple.

In higher dimensions, one can see situations where A separates (i.e. $\ell^2(Z^\nu) = \ell^2(Z^{\nu_1}) \otimes \ell^2(Z^{\nu_2})$ and $A = A_1 \otimes I + I \otimes A_2$) where A has a.c. spectrum with eigenfunctions decaying in ν_2 dimensions but of plane wave form in the remaining ν_1-dimensions.

One can also imagine a.c. spectrum from combining singular spectrum for A_1 and A_2. In either case $k=0$ for lots of continuum a.c. eigenfunctions.

QUESTION 5. Is there a sensible (i.e. not obviously false) version of Question 4 in the multidimensional case?

There are examples [7] of cases where A has only point spectrum but there is an eigenfunction with $k(d,E)>0$ (since it occurs on a set of ρ-measure zero, it isn't a counterexample to a positive answer to Question 4). Does the second part of Question 4 have a positive converse?

QUESTION 6. Is it true that if $Au=Eu$ has a bounded eigenfunction with $\bar{k}>0$ for a set, Q , of E's of positive Lebesgue measure, then A has some a.c. spectrum on Q ?

QUESTION 7. What is the proper analog of Question 6 for singular continuous spectrum?

REFERENCES

1. S i m o n B. Schrödinger semigroups. - Bull.Amer.Math.Soc., 1982, 7, 447-526.
2. Б е р е з а н с к и й Ю.М. Разложение по собственным функциям самосопряженных операторов. Киев, Наукова думка, 1965 (Transl.Math. Mono., v.17, Amer.Math.Soc., Providence, R.I., 1968).
3. К о в а л е н к о В.Ф., С е м ё н о в Ю.А. Некоторые вопросы разложения по обобщенным собственным функциям оператора Шредингера с сильно сингулярными потенциалами. - Успехи мат.наук, 1978, 33, вып.4, 107-140 (Russian Math.Surveys,1978,33, 119-157).
4. P a s t u r L. Spectral properties of disordered systems in one-body approximation. - Comm.Math.Phys.,1980, 75, 179.
5. A v r o n J., S i m o n B. Singular continuous spectrum for a class of almost periodic Jacobi matrices. - Bull.Amer.Math.Soc., 1982, 6, 81-86.
6. М а с л о в В.П. Об асимптотике обобщенных собственных функций уравнения Шредингера. - Успехи мат.наук,1961,16, вып.4, 253-254.
7. S i m o n B., S p e n c e r T. unpublished.

BARRY SIMON

Departments of Mathematics and Physics
California Institute of Technology
Pasadena, California 91125 USA

4.2. SCATTERING THEORY FOR COULOMB TYPE PROBLEMS

1. Let self-adjoint operators A and A_0 act in a Hilbert space H and suppose the spectrum of A_0 is absolutely continuous. Suppose further that there exists a unitary operator function $W_0(t)$ satisfying the conditions: $W_0(t)W_0(\tau)=W_0(\tau)W_0(t)$, $W_0(t)A_0 = A_0 W_0(t)$, $\underset{t\to\infty}{s-\lim} W_0^*(t+\tau)W_0(t) = E$ and there exist limits

$$\underset{t\to\pm\infty}{s-\lim}\ exp(iAt)exp(-iA_0t)W_0(t)=U_\pm(A,A_0) \tag{1}$$

$$\underset{t\to\pm\infty}{s-\lim}\ W_0^{-1}(t)exp(iA_0t)exp(-iAt)P=U_\pm(A_0,A) \tag{2}$$

where P is the orthogonal projection onto the absolutely continuous part of A. For the generalized wave operators $U_\pm$ [1-5] the equality $AU_\pm(A,A_0)=U_\pm(A,A_0)A_0$ holds. The factor $W_0(t)$ is not uniquely defined. The factor $\widetilde{W}_0(t)=W_0(t)V_\pm\ (t \gtrless 0)$ can be used too when obvious requirements to $V_\pm$ are fulfilled. Due to this ambiguity the naturally looking definition

$$S(A,A_0)=U_+^*(A,A_0)U_-(A,A_0) \tag{3}$$

of the scattering operator becomes senseless since $\widetilde{S}(A,A_0) = V_+^*S(A,A_0)V_-$ is in fact an arbitrary unitary operator commuting with A_0.

PROBLEM 1. <u>Find physically motivated normalization of $W_0(t)$ when $t\to\pm\infty$, removing the non-uniqueness in definition of the scattering operator.</u>

This problem has been solved for the scattering with the Coulomb main part, i.e. when

$$Af=-\frac{d^2f}{dr^2}+\left[\frac{\ell(\ell+1)}{r^2}-\frac{2z}{r}+q(r)\right]f,\quad A_0f=-\frac{d^2f}{dr^2}+\frac{\ell(\ell+1)}{r^2}f \tag{4}$$

where $0 < r < \infty$, $\ell > 0$. For the system (4) the factor $W_0(t)$ is of the form [1-2] : $W_0(t) = \exp[i(\operatorname{sign} t)\ln|t|\, z/\sqrt{A_0}]$.
In [2] it is proved that with such $W_0(t)$ the scattering operator (3) coincides with the results of the stationary scattering theory. This fact suggests that the normalization of $W_0(t)$ is physically reasonable. A similar problem has been solved for the Dirac equations with the Coulomb main part [5]. Consider the system

$$Af = -\frac{d^2 f}{dx^2} + q(x)f\,,\ A_0 f = -\frac{d^2 f}{dx^2},\ -\infty < x < \infty \qquad (5)$$

where $q(x) \approx \alpha_{\pm}/x,\ x \to \pm\infty$

PROBLEM 2. <u>Find</u> $W_0(t)$ <u>and solve Problem 1 for the system</u> (5).

When $\alpha_+ = -\alpha_-$ the system (5) can be reduced to the Coulomb case, when $\alpha_+ = \alpha_-$ it is considered in [2]. The case $\alpha_- = 0$, $\alpha_+ \neq 0$ is of importance in a number of physical problems [6-7].

2. The Coulomb interaction of n particles is described by the system

$$A = -\sum_{k=1}^{n} \Delta_k/2m_k + \sum_{1 \le k < \ell \le n} z_{k,\ell}/r_{k,\ell}\,, \quad A_0 = -\sum_{k=1}^{n} \Delta_k/2m_k \qquad (6)$$

where r_k is the radius-vector of the k-th particle, $r_{k,\ell} = |r_k - r_\ell|$. Taking bound states into account leads to the transition from A_0 to the extended operator $\tilde{A}_0$ [8]. The operators $\tilde{A}_0$ and $\tilde{W}_0(t)$ for the system (6) have been constructed in [9]. The construction is effective when $n=3$. In [9] the existence of $U_{\pm}(A, \tilde{A}_0)$ is proved.

PROBLEM 3. <u>Prove the existence of</u> $U_{\pm}(\tilde{A}_0, A)$ <u>for the system (6), i.e. the completeness of the corresponding wave operators.</u>

3. We shall consider the non-standard inverse

PROBLEM 4. <u>Let operator</u> A_0 <u>and generalized scattering operator</u> S <u>be known. Recover the operator</u> A <u>and the corresponding</u> $W_0(t)$, <u>find classes where the problem has one and only one solution.</u>

Consider a model example, namely

$$A_0 f = x f(x), \quad A f = x f(x) + i\Big[\int_a^x f(t)\rho(t)dt - \int_x^b f(t)\rho(t)dt\Big]\rho(x) \tag{7}$$

where $\rho(x)$ belongs to the Hölder class and $\rho(x)>0$, $x \in (a,b)$. There is an effective solution of the "direct" problem (the construction of $W_0(t)$ and of S) for the system (7) [2]:

$$W_0(t) f = f(x) \exp[i\rho^2(x) \ln|t|], \quad Sf = f(x)s(x), \quad s(x) = \psi_-(x)/\psi_+(x) \tag{8}$$

where $\psi_\pm(x) = \lim_{y \to \pm 0} \psi(x+iy)$ and

$$\psi(z) = [\Phi(z) + \Phi^{-1}(z)]/2, \quad \Phi(z) = \exp\Big[\frac{i}{2}\int_a^b \frac{\rho^2(u)}{u-z} du\Big]. \tag{9}$$

Suppose in addition that

$$\int_a^b [\rho^2(t)/(t-a)]dt < \pi, \quad \int_a^b [\rho^2(t)/(b-t)]dt < \pi.$$

Then operator A has no discrete spectrum. The formulae (8), (9) give an effective solution of Problem 4 for system (7) by the factorization method, i.e. $\Phi(z)$ and $\rho(x)$ are found by $S(x)$ and hence A and $W_0(t)$. Some problems with non-local potentials and the case (5) with $\alpha_+ = \alpha_-$ can be reduced to the system (7).

4. Now we come to s t a t i o n a r y i n v e r s e p r o b l e m s. When $z > 0$ the discrete spectrum of the operator A (see (4)) is described by the Ritz half-empiric formula

$$\lambda_n = -z^2/[n + l + \delta_l + \varkappa(n,l)]^2, \quad \varkappa(n,l) \to 0, \quad n \to \infty;$$

its proof is given in [10]. The number δ_l is called a quantum defect of the discrete spectrum. The same number δ_l serves as a deviation measure of the operator (4) from the case of hydrogen nucleus $[q(r) = 0]$.

PROBLEM 5. Find a method to recover the potential $q(r)$ from δ_ℓ $(\ell = 0, 1, 2, \ldots)$.

Here the representation of δ_ℓ by solutions of the Schrödinger equation [10] and the transformation operator [11] can be useful. The definition of the quantum defect δ_K $(K = \pm 1, \pm 2, \ldots)$ is introduced in [12] for the Dirac radial equation too

$$\frac{d}{dr}\psi\begin{bmatrix} 0 & 1 \\ -1 & 0 \end{bmatrix} - \psi H(r) - \lambda\psi = 0, \quad H(r) = \begin{bmatrix} \frac{2z}{r} + q(r) - m, & -\frac{K}{r} \\ -\frac{K}{r}, & \frac{2z}{r} + q(r) + m \end{bmatrix} \tag{10}$$

$$\psi = [\psi_1, \psi_2]$$

So problem 5 can be formulated for the case (10) too. Note that the classic inverse problem for the Dirac equation has not yet been solved even when the Coulomb member is missing. The systems of the Dirac type have been thoroughly investigated when (see (10))

$$H(r) = \begin{bmatrix} V(r) - m & W(r) \\ W(r) & V(r) + m \end{bmatrix}$$

where $V(r)$, $W(r)$ are real functions from $L(0, \infty)$. The peculiarity of the system (10) is defined by the fact that the element $W(r)$ is known $[W(r) = -K/r \in L(0, \infty)]$. It is therefore perhaps not necessary to use the scattering (or spectrum) data over the whole energy interval $-\infty < \lambda < \infty$. We come to a peculiar half-inverse problem both for the Coulomb and non-Coulomb case.

PROBLEM 6. Let $z, m, W(r) = -K/r$ $(K = \pm 1, \pm 2, \ldots)$ be known. Reconstruct the element $V(r) = 2z/r + q(r)$ by the scattering (or spectrum) data, belonging to the energy interval $0 \leqslant \lambda < \infty$.

The model half-inverse problem has been solved in [13].

REFERENCES

1. D o l l a r d J. Asymptotic convergence and Coulomb interaction. - J.Math.Phys.,1964, 5, 729-738.
2. С а х н о в и ч Л.А. Обобщенные волновые операторы. - Матем.сб., 1970, 81, № 2, 209-227.
3. С а х н о в и ч Л.А. Обобщённые волновые операторы и регуляризация ряда теорий возмущений. - Теор. и матем.физика, 1970, 2, № 1, 80-86.
4. Б у с л а е в В.С., М а т в е е в В.Б. Волновые операторы для уравнения Шрёдингера с медленно убывающим потенциалом. - Теор. и матем.физика, 1970, 2, 367-376.
5. С а х н о в и ч Л.А. Принцип инвариантности для обобщенных волновых операторов. - Функц.анализ и его прилож., 1971, 5, № 1, 61-68.
6. Туннельные явления в твердых телах. МИР, 1973.
7. Б р о д с к и й А.М., Г у р е в и ч А.Ю. Теория электронной эмиссии из металлов, 1973.
8. Ф а д д е е в Л.Д., Математические вопросы квантовой теории рассеяния для системы трех частиц. - Тр.Матем.ин-та им. В.А.Стеклова, 1963, т.69.
9. С а х н о в и ч Л.А. Об учете всех каналов рассеяния в задаче n тел с кулоновским взаимодействием. - Теор. и матем.физика, 1972, 13, № 3, 421-427.
10. С а х н о в и ч Л.А. О формуле Ритца и квантовых дефектах спектра радиального уравнения Шредингера. - Изв.АН СССР, сер.матем., 1966, 30, № 6, 1297-1310.
11. К о с т е н к о Н.М. Об одном операторе преобразования. - Изв. высш.уч.зав., Математика, 1977, 9, 43-47.
12. С а х н о в и ч Л.А. О свойствах дискретного и непрерывного спектров радиального уравнения Дирака. - Докл.АН СССР, 1969, 185, № 1, 61-64.
13. С а х н о в и ч Л.А. Об одной полуобратной задаче. - Успехи матем.наук, 1963, 18, № 3, 199-206.

L.A.SAHNOVICH
(Л.А.САХНОВИЧ)

СССР, 270000, Одесса,
Электротехнический институт
связи им.А.С.Попова

4.3. old

A QUESTION OF POLYNOMIAL APPROXIMATION ARISING IN CONNECTION WITH THE LACUNAE OF THE SPECTRUM OF HILL'S EQUATION

Let $Q = -\frac{d^2}{dx^2} + q(x)$ be a Hill's operator with $q \in C_1^\infty$, the class of real infinitely differentiable functions of period 1. The spectrum determined by the periodic and anti-periodic solutions $Qf = \lambda f$, $0 \le x < 1$ comprise a simple (periodic) ground state λ_0 followed by separated pairs $\lambda_{2n-1} < \lambda_{2n}$, $n = 1, 2, \dots$ of alternately anti-periodic and periodic eigenvalues increasing to $+\infty$, the equality or inequality signifying the dimensionality (=1 or 2) of the eigenspace; see [1]. The intervals $[\lambda_{2n-1}, \lambda_{2n}]$, $n = 1, 2, \dots$ are the l a c u n a e of the spectrum of Q in $L^2(\mathbb{R})$. Hochstadt [2] proved that the infinite differentiability of q is reflected in the rapid vanishing of the lengths ℓ_n of the lacunae $[\lambda_{2n-1}, \lambda_{2n}]$ as $n \uparrow +\infty$. Trubowitz [3] proved that the real analyticity of q is equivalent to $\ell_n \le a e^{-bn}$, $n \uparrow \infty$. A comparison between ℓ_n and $\hat{q}(n) = \int_0^1 e^{2\pi i n x} q(x)\,dx$ springs to mind. Interest in sharpening these results arises in connection with the following geometrical problem.

Let M, $M \subset C_1^\infty$, be the class of functions giving rise to a fixed periodic and anti-periodic spectrum $\lambda_0 < \lambda_1 \le \lambda_2 < \lambda_3 \le \lambda_4 < \dots$ and let g, $g \le \infty$, be the number of pairs of simple eigenvalues $\lambda_{2n-1} < \lambda_{2n}$. M is a compact g-dimensional torus identifiable as the real part of the Jacobi variety of the hyperelliptic curve of genus g, $g \le \infty$, with branch points over the real spectrum, augmented by the point at ∞; see [4] and [5] for $g < \infty$, and [6] for $g = \infty$. M admits a family of transitive commuting (iso-spectral) flows expressible in Hamiltonian form as $\frac{\partial q}{\partial t} = Xq$ with $Xq = (\operatorname{grad} \lambda)'$, prime signifying $\frac{d}{dx}$, in which λ is a simple eigenvalue of Q and grad λ is the functional gradient $\partial\lambda/\partial q(x) =$ the square of the normalized eigenfunction $[f(x)]^2$. The l o c a l flows: $\partial q/\partial t = X_1 q = \partial q/\partial x$ (translation), $X_2 q = 3q\frac{\partial q}{\partial x} - \frac{1}{2}\frac{\partial^3 q}{\partial x^3}$ (Korteweg-de Vries), e t c. are more familiar. The latter belong to the span of the former, but can be expressed in an independent fashion $[X_m q = (\operatorname{grad} H_m)']$ via the rule $X_m q = (q\mathfrak{D} + \mathfrak{D}q - \frac{1}{2}\mathfrak{D}^3)\operatorname{grad} H_{m-1}$ starting from $H_{-1} = 1$; for example,

$$H_0=\int_0^1 q\,dx\,,\quad H_1=\int_0^1 \frac{q^2}{2}\,dx\,,\quad H_2=\int_0^1 \left(\frac{q'^2}{4}+\frac{q^3}{2}\right)dx\,.$$

THE GEOMETRICAL QUESTION <u>is to decide if the local vector fields</u> X_1, X_2, <u>etc. span the tangent space of</u> M <u>at each point</u>. This is always the case if $g<\infty$; see [5] or [7]. McKean-Trubowitz [6] make the question precise for $g=\infty$ and prove the following necessary and sufficient condition. Let F be the space of sequences $f(\lambda_{2n})$, $n=1,2,3,\ldots$, with quadratic form $\|f\|^2=\sum_{n\geq 1} \ell_n |f(\lambda_{2n})|^2<\infty$.
Let P be the subspace $f(\lambda_{2n})=p(\lambda_{2n})$, $n=1,2,3,\ldots$, with p a polynomial. T h e n t h e s p a n n i n g o f t h e l o c a l v e c t o r f i e l d s t a k e s p l a c e i f a n d o n l y i f P s p a n s F . The condition is met if q is real analytic $(\ell_n<ae^{-bn},\ n\uparrow+\infty)$. It is known that $\lambda_{2n}\sim n^2\pi^2+ +c_0+c_1n^{-2}+c_2n^{-4}+\ldots\ (n\uparrow+\infty)$, permitting the application of a result of Koosis [8] in the case of purely simple spectrum to verify that the spanning takes place in that circumstance only if $\sum_{n\geq 1}\frac{\log \ell_n}{n^2}=-\infty$.

Contrariwise, the spanning c a n n o t t a k e p l a c e if q vanishes on an interval; in fact, if the local vector fields span the tangent space, then the associated gradients $\partial H/\partial q$ span the normal space of M , and the two together (tangent and normal) fill up the whole of the ambient space, taken to be $L^2(0,1)$, which is impossible since X_q and $\partial H/\partial q$ are universal polynomials in q , q', q'' , etc. without constant term and consequently vanish on the same interval as q . I t s e e m s l i k e l y t h a t t h e s p a n n i n g b e c o m e s c r i t i c a l i n t h e v i c i n i t y o f q u a s i - a n a l y t i c q .
The same questions arise for Q on the line with $q\in C_\downarrow^\infty$, the class of infinitely differentiable functions of rapid decay at $\pm\infty$. The rate of vanishing of the lacunae is replaced by the rate of decay of the reflection coefficient $s_{12}(k)$ of Faddeev [9], e.g., $q\in C_\downarrow^\infty$ is reflected in $s_{12}\in C_\downarrow^\infty$, while the analyticity of q in a horizontal strip is reflected in $|s_{12}(k)|<ae^{-b|k|}$, $k\to\pm\infty$;
see [10]. The torus M is now replaced by $g\leq\infty$ - dimensional cy-

linder specified by fixing $|s_{12}|$ and finite number of bound states (negative simple eigenvalues) $-\kappa_n^2\ (n=1,\dots,g)$, and the vector fields $Xq=(\mathrm{grad}\,H)'$ determined from $H=|s_{12}(\kappa)|$, $\kappa\in\mathbb{R}$, or from $H=-\kappa_n^2\ (n=1,\dots g)$ presumably span the tangent space; see [11] for preliminary information. The local vector fields $X_1 q=q'$, $X_2 q=3qq'-\frac{1}{2}q'''$ operate as before, and the question is the same as before: do they span the tangent space of M ? The technical clarification of the question is a necessary part of any discussion.

REFERENCES

1. Magnus W., Winkler. Hill's Equation, New York, Interscience-Wiley, 1966.
2. Hochstadt H. Function-theoretic properties of the discriminant of Hill's equation. - Math.Zeit.,1963, 82, 237-242.
3. Trubowitz E. The inverse problem for periodic potentials. - Comm.Pure Appl.Math.,1977, 30, 321-337.
4. Дубровин Б.А., Новиков С.П. Периодическая задача для уравнений Кортевега-де Фриза и Штурма-Лиувилля. Их связь с алгебраической геометрией. - Докл.АН СССР, 1974, 219, 3, 531-- 534.
5. McKean H.P., P. van Moerbeke. The spectrum of Hill's equation. - Invent.Math.,1975, 30, 217-274.
6. McKean H.P., Trubowitz E. Hill's operator and hyperelliptic function theory in the presence of infinitely many branch points. - Comm.Pure Appl.Math.,1976, 29, 143-226.
7. Lax P. Periodic solutions of the KdV equation. - Comm.Pure Appl.Math.,1975, 28, 141-188.
8. Koosis P. Weighted polynomial approximation on arithmetic progressions of intervals or points. - Acta Math.,1966, 116, 223-277.
9. Фаддеев Л.Д. Свойства S-матрицы одномерного уравнения Шредингера. - Тр.Матем.ин-та АН СССР, 1964, 73, 314-336.
10. Deift P., Trubowitz E. Inverse scattering on the line. - Comm.Pure Appl.Math., 1979, 32, N 2, 121-251.
11. McKean H.P. Theta functions, solutions, and singular curves. (Proc.Conf., Park City, Utah., 1977, 237-254), Lecture Notes in Pure and Appl.Math., 48, Dekker, New York, 1979.

H.P.MCKEAN New York University. Courant Institute of Mathematical Sciences, 251 Mercer Street, New York, N.Y. 10012, USA

4.4 old ZERO SETS OF OPERATOR FUNCTIONS WITH A POSITIVE IMAGINARY PART

Let E be a separable Hilbert space, M be a function analytic in the unit disc $\mathbb{D}$, taking values in the space of bounded operators on E and continuous up to the boundary of $\mathbb{D}$. Suppose also that

$$M(\zeta) = I + C(\zeta), \quad \zeta \in \operatorname{clos}\mathbb{D},$$

where $C(\zeta)$ is a compact operator on E. We also assume that the following properties are satisfied:

A) For a modulus of continuity ω the inequality

$$|M(\zeta) - M(\zeta')| \leqslant \omega(|\zeta - \zeta'|)$$

holds with $\zeta, \zeta' \in \mathbb{D}$.

B) M has a positive imaginary part in $\mathbb{D}$:

$$\operatorname{Im} M(\zeta) \overset{\text{def}}{=\!=} \frac{M(\zeta) - M(\zeta)^*}{2i} > \mathbb{0}, \quad \zeta \in \mathbb{D}$$

A point ζ in $\operatorname{clos}\mathbb{D}$ will be called a *root of* M if

$$\inf_{\|e\|=1} \|M(\zeta)e\| = 0.$$

Since $I - M(\zeta)$ is compact, for any root ζ there exists $e \in E$ such that $M(\zeta)e = \mathbb{0}$. It is not hard to verify that the roots of a function with a positive imaginary part can lie only on $\mathbb{T}$. Denote the set of all roots of M by Λ and let $m\Lambda_\delta$ be the Lebesgue measure of its δ-neighbourhood in $\mathbb{T}$.

CONJECTURE 1. <u>Under hypotheses A, B the inequality</u> $m\Lambda_\delta \leqslant C\omega(\delta)$ <u>holds for a positive constant</u> C.

It seems to be natural to weaken hypothesis A and replace it by the following one taking into account the behaviour of M only near the set Λ.

A') $|M^{-1}(z)|^{-1} \leqslant \omega(\operatorname{dist}(z, \Lambda))$.

CONJECTURE 2. <u>Under hypotheses</u> A', B <u>the inequality</u>

$$m\Lambda_\delta \leqslant c\,\omega(\delta)$$

<u>holds for a positive constant</u> C .

Let us note that the validity of any of CONJECTURES 1 or 2 with $\omega(\delta) \leqslant const\cdot\delta$ would imply that Λ is finite.

The above CONJECTURES agree with known results of operator theory and complex analysis. Their proof would permit us to describe the structure of the singular and discrete spectra of perturbed operators in terms of "relative smoothness" of perturbation. To indicate links let us point out a situation of perturbation theory where the questions of such sort arise.

Let H be a Hilbert space, A° , V be self-adjoint operators on H , $V \geqslant 0$,

$$A = A^\circ + V,$$

$E = clos\, VH$. It follows immediately from the second resolvent identity that the following relation between the resolvents of the original and perturbed operators holds

$$(I_E - V^{1/2} R_z V^{1/2})(I_E + V^{1/2} R_z^\circ V^{1/2}) = I_E\,, \ \operatorname{Im} z > 0\,. \tag{1}$$

Here $R_z = (A - zI)^{-1}$, $R_z^\circ = (A^\circ - zI)^{-1}$. The function M defined by

$$M(z) = I_E + V^{1/2} R_z^\circ V^{1/2}, \quad M : E \to E$$

has a positive imaginary part. The perturbation V will be called r e l a t i v e l y s m o o t h if for some spectral representation τ of A°

$$H \xrightarrow{\ \tau\ } L^2(\mathbb{R})$$

$$A^\circ f \longrightarrow x f(x)$$

an operator with a smooth kernel corresponds to V . The perturbed operator A in this representation coincides with the so-called Friedrichs's model (see for example [1]),

$$Af \xrightarrow{\ \tau\ } xf + \int v(x,\xi)\, f(\xi)\, d\xi\,. \tag{2}$$

The problem of investigation of the singular spectrum σ_s and the discrete spectrum σ_d can be reduced (see [1]) to the investigation of the zero set of the corresponding operator-valued function M, since $\sigma_s \cup \sigma_d \subset \Lambda$ according to (1). Just in this way in [1] a theorem is proved which claims that if the kernel function v has a good behaviour at infinity and satisfies the smoothness condition $v \in \mathrm{Lip}\,\alpha$, $\alpha > 1/2$ then the discrete spectrum is finite and the singular spectrum is empty. The crucial point of the proof of this result is the fact that M has better smoothness properties at the points of Λ in comparison with v and so A') is satisfied for M with $\omega(\delta) = c\delta$.

Investigation of the one-dimensional Friedrichs's model with the kernel of class $\mathrm{Lip}\,\alpha$, $\alpha < 1/2$ (see [2]) shows that in this case the same phenomenon takes place.

THEOREM [2]. <u>If</u> $v(x,\xi) = \varphi(x)\varphi(\xi)$, $0 < x, \xi < 1$, $\varphi(0) = \varphi(1) = 0$, $\varphi \in \mathrm{Lip}\,\alpha$, $\alpha < 1/2$, <u>then the function</u> M,

$$M(z) = \int_{-1}^{1} \frac{|\varphi(t)|^2}{t-z}\,dt, \quad \mathrm{Im}\,z > 0,$$

<u>satisfies A') with</u> $\omega(\delta) = c\delta^{2\alpha}$, <u>the singular and discrete spectra of the operator</u> A <u>on</u> $L^2(-1,1)$ <u>defined by</u>

$$Af = xf(x) + \int_{-1}^{1} \varphi(x)\varphi(t) f(t)\,dt$$

<u>are contained in the set</u> Λ <u>of zeros of</u> M <u>and</u>

$$m\Lambda_\delta \leqslant c\delta^{2\alpha}, \quad \delta > 0.$$

It is proved also in [2] that the above theorem is precise (in a sense).

The tool of [2] is the scalar analogue of CONJECTURE 2 proved in [3]. As to CONJECTURE 1 in the scalar case, apparently it can be considerably strengthened. See in more details 9.6 of the present volume.

We do not venture to formulate so fine conjectures in the multi-dimensional case.

REFERENCES

I. Ф а д д е е в Л.Д. О модели Фридрихса в теории возмущений. - Труды Матем.ин-та АН СССР им.В.А.Стеклова, I964, 30, 33-75.
2. П а в л о в Б.С., П е т р а с С.В. О сингулярном спектре слабо возмущенного оператора умножения. - Функц.анал. и его прил., I970, 4, № 2, 54-6I.
3. П а в л о в Б.С. Теорема единственности для функций с положительной мнимой частью. - В кн.: Проблемы матем.физики, ЛГУ, I970, II8-I24.

L.D.FADDEEV (Л.Д.ФАДДЕЕВ) СССР, I9I0II, Ленинград Фонтанка, 27, ЛОМИ

B.S.PAVLOV (Б.С.ПАВЛОВ) СССР, I98904, Ленинград Петродворец, Физический факультет Ленинградского университета

* * *

COMMENTARY

The following progress is obtained in [4]:

THEOREM. Let $C(\zeta)\in\mathfrak{S}_1\ (\forall\zeta\in\mathbb{D})$ and let $M(\zeta_0)$ be invertible at some (and then at any) point $\zeta_0\in\mathbb{D}$. Then

$$mF_\delta\leqslant c\cdot c_\alpha\cdot\delta^\alpha,$$

where

$$F=\{w\in\mathbb{T}:\|M^{-1}(\zeta)-I\|^{-1}_{\mathfrak{S}_2}\leqslant c|\zeta-w|^\alpha,\ \zeta\in\mathbb{D}\},$$

c_α is a constant depending on α, $0<\alpha\leqslant 1$, and on $M(0)$ only.

This theorem implies that CONJECTURES 1 and 2 hold for $I-M(\zeta)\in\mathfrak{S}_1$ and $\omega(\delta)=\delta^\alpha$.

It is possible to prove an analogous proposition for an arbitrary modulus of continuity ω. However, the condition $C(\zeta)\in\mathfrak{S}_1$

seems to be essential [5]. The above Theorem allowed to describe (see [4]) the structure of the singular spectrum of selfadjoint Friedrichs model that is discussed after the statement of Conjecture 2

REFERENCES

4. Н а б о к о С.Н. Теоремы единственности для оператор-функций с положительной мнимой частью и сингулярный спектр в самосопряженной модели Фридрихса. - Докл.АН СССР, 1983 (в печати).

5. Н а б о к о С.Н. Private communication.

4.5. POINT SPECTRUM OF PERTURBATIONS OF UNITARY OPERATORS

Let $\mathcal{U}$ be a unitary operator with purely singular spectrum and let an operator K be of trace class.

QUESTION. Can the point spectrum of the perturbed operator $\mathcal{U}+K$ be uncountable?

If it is not assumed that the spectrum of $\mathcal{U}$ is singular the answer is YES. A necessary and sufficient condition for a subset of $\mathbb{T}$ to be the point spectrum of some trace class perturbation of some (arbitrary) unitary operator was given in [1]: such a subset must be a countable union of Carleson sets (for the definition see, e.g., 9.3 of this "Collection").

A version of the reasoning in [1] allows to reduce our QUESTION to a question of function theory.

PROPOSITION. Let E be a subset of $\mathbb{T}$. The following are equivalent.

1) E is the point spectrum of some trace class perturbation of a unitary operator with singular spectrum.

2) There exist two distinct inner functions θ_1 and θ_2 such that

$$E=\left\{\zeta\in\mathbb{T}:\ \frac{\theta_1(z)-\theta_2(z)}{z-\zeta}\in L^2(\mathbb{T})\right\}.$$

Note that if one of the inner functions is constant then the latter set is countable since it is the point spectrum of a unitary operator, namely of a rank one perturbation of a restricted shift (cf [2]).

REFERENCES

1. Макаров Н.Г. Унитарный точечный спектр почти унитарных операторов. - Зап.научн.семин.ЛОМИ, 1983, 126, 143-149.

2. Clark D. One dimensional perturbations of restricted shifts. - J.Analyse Math., 1972, 25, 169-191.

N.G.MAKAROV
(Н.Г.МАКАРОВ)

СССР, 198904, Ленинград, Петродворец, Ленинградский государственный университет

4.6. RE-EXPANSION OPERATORS AS OBJECTS OF SPECTRAL ANALYSIS

1. Notations. L^2_s and L^2_a are subspaces of even and odd functions in $L^2(\mathbb{R})$; Φ is the Fourier transform; Σ is the multiplication by $sign\, x$ on $L^2(\mathbb{R})$; $K \overset{def}{=} \Phi^* \Sigma \Phi$ is the Hilbert transform; $\mathcal{Y} \overset{def}{=} K\Sigma = \Phi^* \Sigma \Phi \Sigma$. Let $\mathcal{J}_s(\mathcal{J}_a)$ be the following unitary mapping from $L^2_s(L^2_a)$ onto $L^2(\mathbb{R}_+)$:

$$u \mapsto \sqrt{2}\, u | \mathbb{R}_+ .$$

Φ_s and Φ_a are the Fourier cosine and sine transforms on $L^2(\mathbb{R}_+)$; $\Pi \overset{def}{=} \Phi_a^* \Phi_s$, $M \overset{def}{=} i\Pi$. Let σ denote the multiplication by $\sigma_n = sign\,(n+\frac{1}{2})$ on $\ell^2(\mathbb{Z})$. Integrals with singular kernels are understood in the sense of principal value.

2. Re-expansion operators appear quite often in scattering theory Namely, the wave operators for a pair of self-adjoint operators H_0, H

$$W_\pm(H,H_0) = s\text{-}\lim_{t\to\pm\infty} \exp(itH)\exp(-itH_0)$$

can be obtained as follows. A given function is expanded with respect to the eigen-functions of H_0 and then the inverse transform using the eigen-functions of H is taken. Let for example, H_s and H_a be the operators $-\frac{d^2}{dx^2}$ on $L^2(\mathbb{R}_+)$ with the domains defined by $u'(0)=0$ and $u(0)=0$ respectively. Then $W_\pm(H_a,H_s) = \pm M$. Indeed, let

$$g_0 = \Phi_s f_0, \qquad g_1 = \Phi_a f_1,$$

$$\exp(-itH_s) f_0 = u_0(t), \qquad \exp(-itH_a) f_1 = u_1(t).$$

Then

$$u_0(t)(x) = \sqrt{\frac{2}{\pi}} \int_{\mathbb{R}_+} e^{-ik^2t} g_0(k) \cos kx\, dk,$$

$$u_1(t)(x) = \sqrt{\frac{2}{\pi}} \int_{\mathbb{R}_+} e^{-ik^2t} g_1(k) \sin kx\, dk.$$

Simple calculations by the stationary phase method show that $u_0(t) \sim \sim u_1(t)$ when $t \to \pm\infty$ provided $g_1 = \pm i g_0$.

The re-expansion operator Π arises in the polar decomposition of $A = -i\frac{d}{dx}$ on $L^2(\mathbb{R}_+)$ with the boundary condition $u(0)=0$, namely, $A = M^*|A|$ and $A^* = M|A^*|$. Let us verify the first equality. Since $|A| = (A^*A)^{1/2} = H_a^{1/2}$, we have (using the above notation $g_1 = \Phi_a f_1$)

$$(|A|f_1)(x)=\sqrt{\frac{2}{\pi}}\int_{\mathbb{R}_+} k\sin kx\, g_1(k)\,dk,$$

$$(Af_1)(x)=i\sqrt{\frac{2}{\pi}}\int_{\mathbb{R}_+} k\cos kx\, g_1(k)\,dk,$$

hence

$$A=-i\Phi_s^*\Phi_a|A|=M^*|A|.$$

Concrete re-expansion operators are apparently interesting from the analytical point of view and as the objects of spectral analysis. In this connection (see Sect.5) we propose some problems. But at first (in Sect.3,4) we use re-expansion operators as specimens for observations.

The author thanks N.K.Nikol'skii and M.Z.Solomjak, whose remarks are incorporated into the text.

3. Put $\mathcal{Y}=\mathcal{Y}_s\oplus\mathcal{Y}_a$, in accordance with the decomposition $L^2(\mathbb{R})=L^2_s\oplus L^2_a$. It is easy to see that

$$\mathcal{Y}_s=\mathcal{J}_s^* M\mathcal{J}_s\,,\quad \mathcal{Y}_a=\mathcal{J}_a^* M^*\mathcal{J}_a,$$

hence

$$(Mu)(x)=\frac{1}{\pi i}\int_{\mathbb{R}_+}\frac{2xu(t)}{t^2-x^2}\,dt. \qquad (1)$$

The operator V defined by

$$Vu(\sigma)=e^{\sigma/2}u(e^{\sigma})$$

maps isometrically $L^2(\mathbb{R}_+)$ onto $L^2(\mathbb{R})$. Clearly, VMV^* coincides with the convolution operator $f\to f*\varphi$, where

$$\varphi(s)=-\frac{1}{\pi i}\frac{e^{s/2}}{\sinh s}.$$

Taking the Fourier transform, we obtain *)

$$M=V^*\Phi^*E\Phi V, \qquad (2)$$

where E is the multiplication by the function

$$\mathcal{E}(\tau)=\frac{\sinh\pi\tau+i}{\cosh\pi\tau},\ \tau\in\mathbb{R} \qquad (3)$$

on $L^2(\mathbb{R})$. It follows from (2) and (3) that the spectrum of Π (res-

*) The spectral decomposition of M can be also deduced from [1] (see Ch.IX) but the proof given here is more direct and simple.

pectively of $\mathcal{Y}$) is absolutely continuous and fills out the semicircle $\mathbb{T}\cap\{z\in\mathbb{C}:\mathrm{Re}\,z\geqslant 0\}$ (respectively $\mathbb{T}$). These imply that $\mathcal{Y}$ is unitarily equivalent to the shift operator on $\ell^2(\mathbb{Z})$. Note that

$$|\Phi_a-\Phi_s|=|\mathrm{I}-\Pi|=\sqrt{2}$$

and that the equality $\Phi_a u=\Phi_s u$ is impossible for $u\in L^2(\mathbb{R}_+)\setminus\{0\}$, though it holds for $u_0(t)=t^{-1/2}$ and in fact $\Phi_a u_0=\Phi_s u_0=u_0$.

4. A re-expansion operator on $L^2(\Delta)$, $\Delta=(-\pi,\pi)$, with analogous properties appears in connection with the system

$$\left\{\frac{1}{\sqrt{2\pi}}\exp i\left(n+\frac{1}{2}\right)t\right\},\quad n\in\mathbb{Z},$$

(but not with the usual trigonometric system). Let $\widetilde{\Phi}:L^2(\Delta)\to L^2(\mathbb{Z})$ be the Fourier transform corresponding to this system. Let $\Delta_+=(0,\pi)$, $\widetilde{\Phi}_s$ and $\widetilde{\Phi}_a$ be maps of $L^2(\Delta_+)$ onto $\ell^2(\mathbb{Z}_+)$ corresponding to the systems

$$\left\{\sqrt{\frac{2}{\pi}}\sin\left(n+\frac{1}{2}\right)t\right\}\qquad\text{and}\qquad\left\{\sqrt{\frac{2}{\pi}}\cos\left(n+\frac{1}{2}\right)t\right\}.$$

Further put

$$\widetilde{\Pi}=\widetilde{\Phi}_a^*\widetilde{\Phi}_s,\quad \widetilde{M}=i\widetilde{\Pi},\quad \widetilde{K}=\widetilde{\Phi}^*\sigma\widetilde{\Phi}.$$

The sense of the following notations is clear by analogy with Sect.3 The operator $\widetilde{K}$ acts as follows

$$(\widetilde{K}u)(t)=\frac{1}{2\pi i}\int_\Delta\frac{u(s)\,ds}{\sin\frac{s-t}{2}}\,.\tag{4}$$

Changing variables by the formulae

$$f=Gu,\quad f(y)=\sqrt{2}\cos\frac{s}{2}\,u(s),\quad y=\mathrm{tg}\,\frac{s}{2},$$

we reduce $\widetilde{K}$ to the Hilbert transform: $\widetilde{K}=G^*KG$. This implies that $\widetilde{\mathcal{Y}}\overset{\mathrm{def}}{=}\widetilde{K}\widetilde{\Sigma}$ can be written in the form $\widetilde{\mathcal{Y}}=G^*\mathcal{Y}G$. Further, decompose $\widetilde{\mathcal{Y}}$ into two parts (even and odd): $\widetilde{\mathcal{Y}}=\widetilde{\mathcal{Y}}_s\oplus\widetilde{\mathcal{Y}}_a$. Then $\widetilde{\mathcal{Y}}_s=G_s^*\mathcal{Y}_sG_s$ where $G_s=G\,|\,L_s^2$. Since $\widetilde{\mathcal{Y}}_s=\widetilde{\mathcal{J}}_s^*\widetilde{M}\mathcal{J}_s$ and $\mathcal{Y}_s=\mathcal{J}_s^*M\mathcal{J}_s$, we obtain a unitary equivalence of $\widetilde{M}$ and M, namely

$$\widetilde{M}=(\widetilde{\mathcal{J}}_sG_s^*\mathcal{J}_s^*)M(\mathcal{J}_sG_s\widetilde{\mathcal{J}}_s^*)=G_+^*MG_+\,,$$

where $G_+=G\,|\,L^2(\Delta_+)$. Note that $\widetilde{M}$ describes the non-trivial part of the scattering matrix for the diffraction on a semi-infinite screen (this is shown in [2], where a unitary equivalence of $\widetilde{M}$ and of the multiplication by function (3) is presented in an explicit form). Another (and a more elementary) situation where $\tilde{M}$ appears is the following. Let B_0 (resp. B_π) be $-i\frac{d}{dx}$ on $L^2(\Delta_+)$ with the boundary condition $u(0)=0$ (resp. $u(\pi)=0$). Then $B_0=\tilde{M}^*|B_0|$, $B_\pi=\tilde{M}|B_\pi|$. To write $\tilde{\mathcal{Y}}$ in a matrix form we note that the operator $\tilde{\Phi}\tilde{\mathcal{Y}}\tilde{\Phi}^*$ $(=\sigma\tilde{\Phi}\tilde{\Sigma}\tilde{\Phi}^*)$ on $\ell^2(\mathbb{Z})$ has the following bilinear form

$$\frac{2i}{\pi}\sum_{\substack{n,m\in\mathbb{Z}\\ n\neq m\,(\mathrm{mod}\,2)}}\frac{\sigma_n a_n \bar{b}_m}{n-m}\,.$$

5. PROBLEMS.

1) The equality $\Pi=-iM=-i\mathcal{Y}_s\,\mathcal{Y}_s\,\mathcal{Y}_s^*$ implies that Π is bounded on $L^p(\mathbb{R}_+)$, $1<p<\infty$. What is the norm of $I-\Pi$? (If $p=2$ see (4)). It is not excluded that the answer can be extracted from the results of [I], Ch.IX.

2) Multi-dimensional analogues of the operator $\mathcal{Y}$ can be described in the following way Let λ,ν be unimodular functions on $\mathbb{R}^m$ satisfying $\lambda(\tau x)=\lambda(x)$, $\nu(\tau x)=\nu(x)$ for $\tau>0$ Let L, N be multiplications by these functions on $L^2(\mathbb{R}^m)$. If Φ_m is the Fourier transform on $L^2(\mathbb{R}^m)$ then $\mathcal{Y}\overset{def}{=}\Phi_m^* L\Phi_m N$ is a unitary operator. It would be of interest to investigate its spectral properties It might be reasonable to impose some additional conditions on λ and ν (e.g. some symmetry conditions).

3) Let q be an even positive function on $\mathbb{R}$ and $\{p_n\}$ be the orthogonal family of polynomials in $L^2(-a,a;q)$, $0<a\leqslant\infty$. Then an analogue of the operator $\tilde{\Pi}$ appears in $L^2(0,a;q)$, namely, the re-expansion operator from the even polynomials $\{p_{2n}\}$ to the odd ones $\{p_{2n+1}\}$. It would be of interest to investigate its spectral properties.

4) Consider the following systems in $L^2(\Delta_+)$:

$$\left\{\sqrt{\tfrac{2}{\pi}}\sin nt\right\},\quad \left\{\sqrt{\tfrac{2}{\pi}}\cos nt\right\},\quad n\in\mathbb{N}.$$

The second system has defect 1. Let P be the re-expansion operator "from sines to cosines". This is a semi-unitary operator on $L^2(\Delta_+)$ with defect indices (1,0). Is it completely nonunitary? In other words is the orthogonal system $\{P^n\mathbb{1}: n\geqslant 0\}$ complete in $L^2(\Delta_+)$?

5) The operator $\mathcal{Y}$ is connected with the harmonic conjugation. What does the theory of invariant subspaces of the shift mean in terms

of y ? What is the rōle of zeros and poles of the function (3) in this connection?

REFERENCES

I. Г о х б е р г И.Ц., К р у п н и к Н.Я. Введение в теорию одномерных сингулярных интегральных операторов. Кишинев, Штиинца, 1973.

2. И л ь и н Е.М. Характеристики рассеяния для задачи о дифракции на клине и на экране. - Записки науч.семин.ЛОМИ, 1982, 107, 193-197.

M.S.Birman
(М.Ш.Бирман)

СССР, 198904, Петродворец,
Физический факультет
Ленинградский университет

4.7. MAXIMAL NON-NEGATIVE INVARIANT SUBSPACES OF $\mathfrak{J}$-DISSIPATIVE OPERATORS

Let $\mathfrak{H}$ be a $\mathfrak{J}$-space (Krein space) i.e. the Hilbert space with an inner product (x,y) and indefinite $\mathfrak{J}$-form $[x,y]=(\mathfrak{J}x,y)$, $\mathfrak{J}=\mathfrak{J}^*=\mathfrak{J}^{-1}$ (more detailed information see, for instance, in [1] or [2]). A subspace $\mathfrak{L}$ is called non-negative if $[x,x]\geqslant 0$ for $x\in\mathfrak{L}$, and maximal non-negative if it is non-negative and has no proper non-negative extensions.

A linear operator $\mathfrak{A}$ on $\mathfrak{H}$ with a domain $\mathfrak{D}_{\mathfrak{A}}$ is called dissipative ($\mathfrak{J}$-dissipative) if $\mathfrak{Im}(\mathfrak{A}x,x)\geqslant 0$ ($\mathfrak{Im}[\mathfrak{A}x,x]\geqslant 0$) for all $x\in\mathfrak{D}_{\mathfrak{A}}$. Such an $\mathfrak{A}$ is called maximal dissipative (maximal $\mathfrak{J}$-dissipative) if it has no proper dissipative ($\mathfrak{J}$-dissipative) extensions.

PROBLEM 1. <u>Does there exist a maximal non-negative invariant subspace for any bounded $\mathfrak{J}$-dissipative operator $\mathfrak{A}$ with $\mathfrak{D}_{\mathfrak{A}}=\mathfrak{H}$</u> ?

This problem has a positive solution if $\mathfrak{A}$ is a uniformly $\mathfrak{J}$-dissipative operator i.e. there is a constant $\gamma>0$: $\mathfrak{Im}[\mathfrak{A}x,x]\geqslant\gamma\|x\|^2$. In that case $\sigma(\mathfrak{A})\cap\mathbb{R}=\emptyset$ and hence the Riesz projection generated by the set $\sigma(\mathfrak{A})\cap\mathbb{C}^+$ gives us the desired subspace. Note that $\mathfrak{J}$-dissipativity of $\mathfrak{A}$ implies the uniform dissipativity of $\mathfrak{A}_\varepsilon=\mathfrak{A}+i\varepsilon\mathfrak{J}$ $(\varepsilon>0)$. As $\mathfrak{A}_\varepsilon$ posseses a maximal non-negative invariant subspace, it is natural to use the "passage to the limit" for $\varepsilon\to 0$. Such a passage - M.G.Krein's method (see [2]) - leads to a positive solution of Problem 1 if $(I+\mathfrak{J})\mathfrak{A}(I-\mathfrak{J})\in\mathfrak{S}_\infty$. In the general case Problem 1 has not yet been solved and therefore subclasses of operators for which it has a positive solution are being considered and on the other hand attempts are being made to construct counterexamples.

THEOREM. <u>If</u> $\hat{\mathfrak{H}}=\mathfrak{H}\oplus\mathfrak{H}$ <u>is a</u> $\hat{\mathfrak{J}}$<u>-space</u>, $\mathfrak{J}=\begin{pmatrix}0 & I\\ I & 0\end{pmatrix}$, <u>then</u> $\hat{\mathfrak{A}}=\begin{pmatrix}0 & -I\\ \mathfrak{A} & 0\end{pmatrix}$ <u>is a continuous</u> $\hat{\mathfrak{J}}$<u>-dissipative operator if and only if</u> $\mathfrak{A}$ <u>is a continuous dissipative operator in</u> $\mathfrak{H}$; <u>in that case</u> $\hat{\mathfrak{A}}$ <u>has a maximal non-negative (with respect to the</u> $\hat{\mathfrak{J}}$<u>-form) invariant subspace</u>.

PROOF. One verifies immediately that a maximal non-negative subspace $\mathcal{L}$ of $\hat{\mathcal{H}}$ is invariant under $\hat{\mathcal{A}}$ iff it is a graph of an operator $(-i\mathcal{K})$ where $\mathcal{K}$ is dissipative in $\mathcal{H}$ and $\mathcal{K}^2=\mathcal{A}$ $(\mathcal{L}=\bigvee\{\langle x,-i\mathcal{K}x\rangle\}_{x\in\mathcal{H}})$. Such an operator $\mathcal{K}$ does exist and is bounded by the theorem of Matsaev-Palant [3].

Matsaev-Palant's result about the square root of dissipative operator was developed by H.Langer [4]. It was proved there in particular that each maximal dissipative operator posseses a maximal dissipative square root. This result allows to omit the requirement of a continuity of $\mathcal{A}$ in the above Theorem and replace it by the maximal dissipativity condition.

PROBLEM 2. <u>Let</u> $\hat{\mathcal{H}}=\mathcal{H}\oplus\mathcal{H}$, $\hat{\mathcal{J}}=\begin{pmatrix}0 & \mathcal{J}\\ \mathcal{J} & 0\end{pmatrix}$, $\mathcal{J}=\mathcal{J}^*=\mathcal{J}^{-1}$, $\hat{\mathcal{A}}=\begin{pmatrix}0 & -I\\ \mathcal{A} & 0\end{pmatrix}$, <u>where</u> $\mathcal{A}$ <u>is a continuous</u> $\mathcal{J}$<u>-dissipative operator in</u> $\mathcal{H}$. <u>Does there exist a maximal</u> (<u>in</u> $\hat{\mathcal{H}}$) <u>non-negative subspace invariant under the</u> $\hat{\mathcal{J}}$<u>-dissipative operator</u> $\hat{\mathcal{A}}$?

REFERENCES

1. B o g n a r J. Indefinite inner product spaces. - Springer-Verlag, 1974.
2. А з и з о в Т.Я., И о х в и д о в И.С. Линейные операторы в пространствах с индефинитной метрикой и их приложения. "Математический анализ. Том I7 (Итоги науки и техники)", I979, Москва, ВИНИТИ, I05-207.
3. М а ц а е в В.И., П а л а н т Ю.А. О степенях ограниченного диссипативного оператора. - Укр.матем.журнал, I962, I4, 329-337.
4. L a n g e r H. Über die Wurzeln eines maximalen dissipativen Operators. - Acta Math. 1962, XIII, N 3-4, 415-424.

T.Ya.AZIZOV
(Т.Я.АЗИЗОВ)

I.S.IOHVIDOV
(И.С.ИОХВИДОВ)

СССР, 394693, Воронеж,
Университетская пл.I,
Воронежский государственный
университет

4.8. PERTURBATION THEORY AND INVARIANT SUBSPACES

If $\mathcal{C}$ is a given coefficient Hilbert space, let $\mathcal{C}(z)$ be the Hilbert space [1] of square summable power series $f(z)=\sum a_n z^n$ with coefficients in $\mathcal{C}$

$$\|f(z)\|^2_{\mathcal{C}(z)} = \sum |a_n|^2.$$

If $B(z)$ is a power series whose coefficients are operators on $\mathcal{C}$ and which represents a function which is bounded by one in the unit disk, then multiplication by $B(z)$ is contractive in $\mathcal{C}(z)$. Consider the range $\mathcal{M}(B)$ of multiplication by $B(z)$ in $\mathcal{C}(z)$ in the unique norm such that multiplication by $B(z)$ is a partial isometry of $\mathcal{C}(z)$ onto $\mathcal{M}(B)$. Define $\mathcal{H}(B)$ to be the complementary space to $\mathcal{M}(B)$ in $\mathcal{C}(z)$. Then the difference-quotient transformation $f(z)$ into $[f(z)-f(0)]/z$ in $\mathcal{H}(B)$ is a canonical model of contractive transformations in Hilbert space which has been characterized [2] as a conjugate isometric node with transfer function $B(z)$.

If $\varphi(z)$ is a power series whose coefficients are operators on $\mathcal{C}$ and which represents a function with positive real part in the unit disk, then

$$B(z)=[1-\varphi(z)]/[1+\varphi(z)]$$

is a power series which represents a function which is bounded by one in the unit disk. Define $\mathcal{L}(\varphi)$ to be the unique Hilbert space of power series with coefficients in $\mathcal{C}$ such that multiplication by $1+B(z)$ is an isometry of $\mathcal{L}(\varphi)$ onto $\mathcal{H}(B)$. Then the difference-quotient transformation has an isometric adjoint in $\mathcal{L}(\varphi)$.

The overlapping space $\mathcal{L}$ of $\mathcal{H}(B)$ is the set of elements $f(z)$ of $\mathcal{C}(z)$ such that $B(z)f(z)$ belongs to $\mathcal{H}(B)$ in the norm

$$\|f(z)\|^2_{\mathcal{L}} = \|f(z)\|^2_{\mathcal{C}(z)} + \|B(z)f(z)\|^2_{\mathcal{H}(B)}.$$

The overlapping space $\mathcal{L}$ is isometrically equal to a space $\mathcal{L}(\theta)$.

A fundamental theorem of perturbation theory [3] states that a partially isometric transformation exists of $\mathcal{L}(\varphi)$ into $\mathcal{L}(\theta)$ which commutes with the difference-quotient transformation. The transformation is a computation of the wave-limit. The wave-limit is isometric on the square summable elements of $\mathcal{L}(\varphi)$ and annihilates the orthogonal complement of the square summable elements of $\mathcal{L}(\varphi)$. If T denotes the adjoint in $\mathcal{C}(z)$ of multiplication by $B(z)$ as a transformation in $\mathbb{C}(z)$, then the wave limit agrees with $1+T$ on square summable elements of $\mathcal{L}(\varphi)$. A fundamental problem is to determine the range of the wave-limit in $\mathcal{L}(\Theta)$. It is known [4] that the range can be a proper subspace of $\mathcal{L}(\Theta)$. The orthogonal complement of the range of the wave-limit in $\mathcal{L}(\Theta)$ is the overlapping space of a space $\mathcal{H}(C)$ such that $B(z)=A(z)C(z)$ for a space $\mathcal{H}(A)$ which is contained isometrically in $\mathcal{H}(B)$.

CONJECTURE. <u>The range of the wave-limit contains every element of $\mathcal{L}(\Theta)$ if the self-adjoint part of the operator $\varphi(0)$ is of Matsaev class.</u>

The **Matsaev** class seems a reasonable candidate because the existence of invariant subspaces is known for contractive transformations T such that $1-T^{*}T$ is of **Matsaev** class. Invariant subspaces exist which cleave the spectrum of the transformation. An integral representation of the transformation exists in terms of invariant subspaces [5]. For reasons of quasi-analyticity, such results do not hold for any larger class of completely continuous operators.

Some recent improvements in the spectral theory of nonunitary transformations link the **Matsaev** class to the theory of overlapping spaces [6].

REFERENCES

1. De Branges L. Square Summable Power Series, Addison-Wesley, to appear.
2. De Branges L. The model theory for contractive transformations. - In: Proceedings of the Symposium on the Mathematical Theory of Networks and Systems in Beersheva, Springer Verlag, to appear.
3. De Branges L., Shulman L. Perturbations of unitary transformations. - J.Math.Anal.Appl., 1968, 23, 294-326.
4. De Branges L. Perturbation theory. - J.Math Anal.Appl., 1977, 57, 393-415.
5. Гохберг И.Ц., Крейн М.Г. Теория вольтерровых операторов в гильбертовом пространстве и ее приложения, М., Наука, 1967 (Translations of Mathematical Monographs, 24, Amer Math.Soc., 1970).
6. De Branges L. The expansion theorem for Hilbert spaces of analytic functions, Proceedings of the Workshop on Operator

Theory in Rehovot, Birkhäuser Verlag, to appear.

L. DE BRANGES

Department of Mathematics
Purdue University
West Lafayette, Indiana 47907
USA

4.9. old

OPERATORS AND APPROXIMATION

1. What is a "Blaschke product"? As long as we are concerned with scalar-valued analytic functions in the unit disc $\mathbb{D}$, the answer is well-known: this is a function B satisfying one of the following equivalent statements.

(i) B can be represented as a product $B=\prod_{\lambda\in\mathbb{D}} b_\lambda^{k(\lambda)}$ of elementary factors $b_\lambda=\frac{|\lambda|}{\lambda}\cdot\frac{\lambda-z}{1-\bar\lambda z}$ (here k is a function from $\mathbb{D}$ to nonnegative integers with $\sum_\lambda k(\lambda)(1-|\lambda|)<+\infty$)

(ii) B is inner (in Beurling's sense) and the part of the schift operator $\mathbf{Z}^*$ on the invariant subspace K, $K=K_B \overset{def}{=} H^2\ominus \mathbf{B}H^2$ has a complete (in K) family of root subspaces. Here H^2 is the standard Hardy class and $\mathbf{Z}$, $\mathbf{B}$ are operators of multiplication by z and B respectively.

(iii) The same for the operator $T_B \overset{def}{=} P_K\mathbf{Z}|K$ conjugate to $\mathbf{Z}^*|K$ (P_K stands for the orthogonal projection onto K).

(iv) B is inner and

$$\lim_{r\uparrow 1}\int_{\mathbb{T}}\log|B(r\zeta)|\,dm(\zeta)=0.$$

The spectral interpretation of (i) and (iv) is of importance for studying operators in terms of their characteristic functions and the problems discussed in this section are essentially those of a "correct" choice of the notion of Blaschke product in the general case, when operator-valued inner functions are considered (the equality $|B(\zeta)|=1$ a.e. is replaced in this case by the requirement that $B(\zeta)$ be a unitary operator on an auxiliary coefficient space E; H^2 is replaced by $H^2(E)$, and so on). Statements (i)-(iv), appropriately modified, are still equivalent for operators T_B having a determinant (i.e. when $I-T_B^*T_B$ is nuclear). If this is no longer true, (ii) AND (OR) (iii) prove to be the most natural definitions of a Blaschke product.

QUESTION 1. **Is it true in case of an arbitrary operator valued inner function that one of the Conditions (ii), (iii) implies the other one?**

The definition under consideration schould presume a metric criterion for a characteristic function B to belong to the class of "Blaschke products" (that is a criterion for T_B AND (OR) T_B^* to be complete).

QUESTION 2. <u>Do the following conditions give such criteria:</u>

$$\lim_{r\to 1-0}\int_{\mathbb{T}}\log|B(r\zeta)|\,dm(\zeta)=0 \qquad \text{or} \qquad \lim_{r\to 1}\int_{\mathbb{T}}\log B(r\zeta)^*B(r\zeta)\,dm(\zeta)=\mathbb{O}?$$

If we restrict ourselves to the case when $T_B^*=z^*|K$ has a simple spectrum, Question 2 reduces to the following one.

QUESTION 3. <u>How to describe in terms of</u> B <u>the subspaces generated by eigen-functions of</u> z^*, <u>i.e. the subspaces</u>

$$\operatorname{span}\left(\frac{\Delta_\lambda E}{1-\bar{\lambda}z}:\lambda\in\mathbb{D}\right)? \tag{V}$$

Here $\{\Delta_\lambda:\lambda\in\mathbb{D}\}$ is a family of orthogonal projections on E It seems important to know when the space (V) coincides with $H^2(E)$ i.e.

QUESTION 4. <u>For which families</u> $\{\Delta_\lambda:\lambda\in\mathbb{D}\}$ <u>the conditions</u> $f\in H^2(E)$ <u>and</u> $\Delta_\lambda f(\lambda)=\mathbb{O}$, $\lambda\in\mathbb{D}$ <u>imply</u> $f\equiv 0$?

If $\Delta_\lambda=\mathbb{O}$ or I, Question 4 clearly reduces to the scalar uniqueness theorem $\sum_{\Delta_\lambda\neq 0}(1-|\lambda|)=\infty$. The last condition remains necessary in general case. Perhaps the answer to Question 4 is the following: $\sum(1-|\lambda|)\|\Delta_\lambda e\|^2=\infty$ for e belonging to a complete family in E. As to question 3, in case $\dim E=1$ the answer can be expressed in terms of the so called "pseudocontinuation" of functions in (V) (M.M.Djrbashyan, G.C.Tumarkin, R.Douglas, H.Shapiro, A.Shields and others). Possibly the same language fits for $\dim E>1$.

2. <u>Weak generators of the algebra</u> $\mathcal{R}(T_\theta)$. In this section θ is a scalar inner function and $\mathcal{R}(*)$ is the weakly closed algebra of operators generated by the operators $*$ and I. It is known (D.Sarason) that $A\in\mathcal{R}(T_\theta)$ iff $A=\varphi(T_\theta)$ for some φ in H^∞. (The operator $\varphi(T_\theta)$ acts in K_θ by the rule $\varphi(T_\theta)f=P_K\varphi f$, $f\in K$). The description of weak generators

φ of $\mathcal{R}(z) = H^\infty$ is also known (D.Sarason) and can be expressed in a geometrical language, in terms of properties of (necessarily univalent) image $\varphi(\mathbb{D})$. Since the algebras $\mathcal{R}(\mathbf{T}_\theta)$ and $H^\infty/\theta H^\infty$ are isometrically isomorphic, it is plausible that the Sarason theorem should admit "projecting":

QUESTION 5. Is it true that $\mathcal{R}(\varphi(T_\theta)) = \mathcal{R}(\mathbf{T}_\theta)$ if and only if $\varphi + \theta H^\infty$ contains a generator of algebra H^∞?

QUESTION 6. Which operators $\varphi(\mathbf{T}_\theta)$ have simple spectrum? (I.e. for which φ there exists f in K_θ with span $(P_K \varphi^n f : n \geqslant 0) = K_\theta$?)

If φ is a generator of H^∞ then cyclic vectors f from Question 6 do exist and can be easily described. In the particular case $\theta = \exp a \dfrac{z+1}{z-1}$ Question 6 reduces (for some functions φ at least) to the question whether $\varphi(\mathbf{T}_\theta)$ is unicellular (or the same question about the operator $x \mapsto \int_0^t x(s) K(t-s)\, ds$ on $L^2(0,a)$, G.E.Kisilevskii). Related to this matter are a paper of J.Ginsberg and D.Newman (J.Aprox.T., 1970, 24, N 4) and the problem 7.19 of this Collection. Other references, historical comments and more discussion can be found in two papers of N.K.Nikolskii (in books: Итоги науки, Математический анализ, т.I2, I974; Теория операторов в функциональных пространствах, Новосибирск, I977).

N.K.NIKOLSKII
(Н.К.НИКОЛЬСКИЙ)

СССР, I9IOII, Ленинград
Фонтанка 27, ЛОМИ

* * *

COMMENTARY

B.M.Solomyak has answered QUESTION 5 in the negative (oral communication). For the sake of convenience we replace here the unit disc by the upper half-plane $\Pi = \{\zeta : \operatorname{Im}\zeta > 0\}$ and consider corresponding spaces H^2_Π, H^∞_Π. Let $\theta = e^{iz}$, $\varphi = \left(\frac{e^{iz}-1}{iz}\right)^n$, $n > 2$. Clearly $\varphi \in H^\infty_\Pi$ and it is proved in [1] that $\mathcal{R}(\varphi(T_\theta)) = \mathcal{R}(T_\theta)$

(This fact is just equivalent to the unicellularity of $\mathcal{J}^n, (\mathcal{J}f)(x) =$
$= \int_0^x f(t)\,dt$ in $L^2(0,1)$). On the other hand, for any $g \in H^\infty_\Pi$,
$h = \varphi + \Theta g = (i/z)^n + e^{iz} g_1$; $g_1(z + i\eta) \in H^\infty_\Pi$, $\eta > 0$. Since $(i/z)^n$, $n > 2$ is nonunivalent in $\{\operatorname{Im} z > \eta\}$ and e^{iz} tends to zero rapidly as $\operatorname{Im} z \to +\infty$, it can be easily verified that h is also nonunivalent. Thus h cannot be an H^∞_Π-generator. ●

Another counterexample for $\Theta = B$, an interpolation Blaschke product, was constructed by N.G.Makarov.

REFERENCE

1. Frankfurt R., Rovnyak J. Finite convolution operators. - J.Math. Anal.Appl., 1975, 49, 347-374.

4.10. SPECTRAL DECOMPOSITIONS AND THE CARLESON CONDITION
old

Completely nonunitary contractions can be included into the framework of the Szőkefalvi-Nagy-Foiaş model [1]. Especially simple is the case when $\sigma(T)$ does not cover the unit closed disc $\operatorname{clos}\mathbb{D}$ and $d = d_* < \infty$, where $d = \dim(I-T^*T)H$, $d_* = \dim(I-TT^*)H$ are the defect numbers of T. If $s-\lim(T^*)^n = 0$ and the above conditions are satisfied then T is unitarily equivalent to its model, i.e to the operator

$$P\mathfrak{z}|K, \quad K = H^2(E) \ominus \theta H^2(E),$$

where E is an auxiliary Hilbert space with $\dim E = d$, θ is a bounded analytic $(E \to E)$ - operator valued function in $\mathbb{D}$ whose boundary values are unitary almost everywhere on the unit circle $\mathbb{T}$, $H^2(E)$ is the Hardy space of E-valued functions, $\mathfrak{z}$ is the multiplication operator $f \mapsto zf$, P is the orthogonal projection onto K. θ is called the characteristic function of T. It is connected very closely with the resolvent of T, e.g. $\|R(\lambda,T)\| \asymp (1-|\lambda|)^{-1} \times \|\theta(\lambda)^{-1}\|$, $\lambda \in \mathbb{D}$. In terms of θ the operator T can be investigated in details, namely, it is possible to find its spectrum, point spectrum $\sigma_p(T)$, eigenvectors and root vectors, to calculate the angles between maximal spectral subspases etc. (cf.[1-4]). In particular operator T is complete (i.e. the linear hull of its eigenvectors and root vectors is everywhere dense) iff $\det\theta$ is a Blaschke product.

A more detailed spectral analysis should include, however, not only a description of spectral subspaces but also methods of recovering T from its restrictions to spectral subspaces. The strongest method of recovering yields the unconditionally convergent spectral decomposition generated by a given decomposition of the spectrum. For a complete operator T the question is whether its root subspaces $\{K_\lambda : \lambda \in \sigma_p(T)\}$ form an unconditional basis. In the case of a simple point spectrum necessary and sufficient conditions of such "spectrality" (i.e. in the case under consideration for the operator to be similar to a normal one) were found in [2],[3]. These conditions are as follows: the *vectorial Carleson condition*

$$\inf\{\|\tilde{\Delta}_\lambda \theta_\lambda(\lambda)^{-1}\Delta_\lambda\|_E^{-1} : \lambda \in \sigma_p(T)\} > 0 \tag{1}$$

holds and the following **imbedding theorems** are valid:

$$\sum_{\lambda\in\sigma_p(T)} (1-|\lambda|)\,\|\Delta_\lambda f(\lambda)\|_E^2<\infty,\ \sum_{\lambda\in\sigma_p(T)} (1-|\lambda|)\,\|\tilde{\Delta}_\lambda f(\lambda)\|_E^2<\infty,\ \forall f\in H^2(E). \quad (2)$$

Here $\tilde{\Delta}_\lambda$ is the orthoprojection from E onto the subspace $\operatorname{Ker}\theta(\lambda)$ and Δ_λ is the orthoprojection from E onto $\operatorname{Ker}\theta(\lambda)^*$; $\theta=\theta_\lambda\times$ $\times[b_\lambda\tilde{\Delta}_\lambda+(I-\tilde{\Delta}_\lambda)]$ is the factorization of θ corresponding to the eigenspace $K_\lambda=\operatorname{Ker}(T-\lambda I)$, $b_\lambda\overset{def}{=}\frac{|\lambda|}{\lambda}(\lambda-z)(1-\bar{\lambda}z)^{-1}$. From geometrical point of view condition (1) means nothing else as the so called uniform minimality of the family $\{K_\lambda:\lambda\in\sigma_p\}$. Moreover,

$$|\tilde{\Delta}_\lambda\theta_\lambda(\lambda)^{-1}\Delta_\lambda|_E^{-1}=\sin(\widehat{K_\lambda,\ K^\lambda}),$$

where

$$K_\lambda=\operatorname{span}(K_\mu:\mu\in\sigma_p(T)\setminus\{\lambda\}),$$

cf.[2]. In the case $d=d_*=1$ L.Carleson proved that (1) implies (2) (cf.[4]), in the case $d=d_*=\infty$ this is no longer true ([3]).

PROBLEM 1. <u>Prove or disprove the implication</u> (1) $\Longrightarrow$ (2) <u>in the case</u> $1<d=d_*<\infty$.

The case $d=d_*=1$ seems to be an exceptional one, because for an arbitrary family of subspaces the property to be uniformly minimal is very far (in the general case) from the property to form an unconditional basis. However for $d=d_*=1$ these conditions coincide not only for eigenspaces but for root subspaces as well and, moreover, for arbitrary families of spectral subspaces of a contraction T [5]. The proofs of this equivalence we are aware of (cf.[5], [4]) represent some kinds of analytical tricks and depend on the evaluation of the angles between pairs of "complementary" spectral subspaces K_ϑ and $K^\vartheta\overset{def}{=}K_{\vartheta'}$ corresponding to a divisor ϑ of θ. Here ϑ and ϑ' are left divisors of θ corresponding to a given pair of subspaces; if $d=d_*=1$ then $\vartheta'=\frac{\theta}{\vartheta}$. A divisor ϑ is called spectral divisor if K_ϑ is a spectral subspace. So the main part of the above mentioned trick consists in the following implication [5]; let $\dim E=1$ and let $\{\vartheta\}$ be an arbitrary family of spectral divisors of θ , then the condition

$$\inf_{\vartheta} \inf_{\substack{e\in E \\ \|e\|=1}} \inf_{\zeta\in\mathbb{D}} \left\{\|\vartheta(\zeta)e\|_E + \|\vartheta'(\zeta)e\|_E\right\} > 0 \tag{3}$$

implies the following one

$$\inf_{\sigma} \inf_{\substack{e\in E \\ \|e\|=1}} \inf_{\zeta\in\mathbb{D}} \left\{\|\vartheta_\sigma(\zeta)e\|_E + \|\vartheta_\sigma'(\zeta)e\|_E\right\} > 0 \tag{4}$$

where ϑ_σ is the inner function corresponding to the subspace $\operatorname{span}\{K_\vartheta : \vartheta \in \sigma\}$, σ being an arbitrary subset of $\{\vartheta\}$. The proof uses a lower estimate of $\|\theta(\zeta)e\|_E$ depending on $\|\vartheta(\zeta)e\|_E$ and $\|\vartheta'(\zeta)e\|_E$ only. However such an estimate is impossible for $\dim E > 1$ (L.E.Isaev, private communication).

PROBLEM 2. <u>Let</u> $1 < d = d_* < \infty$. <u>Prove or disprove the implication</u> (3) $\Longrightarrow$ (4) <u>for an arbitrary family</u> $\{\vartheta\}$ <u>of spectral divisors of</u> θ .

REFERENCES

1. Szőkefalvi-Nagy B., Foiaş C. Harmonic analysis of operators on Hilbert space, North Holland/Akadémiai Kiadó (Amsterdam/Budapest, 1970).
2. Никольский Н.К., Павлов Б.С. Базисы из собственных векторов вполне неунитарных сжатий и характеристическая функция. - Изв.АН СССР, сер.матем., 1970, 34, № 1, 90-133.
3. Никольский Н.К., Павлов Б.С. Разложения по собственным векторам неунитарных операторов и характеристическая функция. - Зап.научн.семин.ЛОМИ, 1968, 11, 150-203.
4. Никольский Н.К. Лекции об операторе сдвига. II. - Зап. научн.семин.ЛОМИ, 1974, 47, 90-119.
5. Васюнин В.И. Безусловно сходящиеся спектральные разложения и задачи интерполяции. - Труды Матем.ин-та им.В.А.Стеклова АН СССР, 1978, 130, 5-49.

N.K.NIKOL'SKII (Н.К.НИКОЛЬСКИЙ) — СССР, 191011, Ленинград, Фонтанка 27, ЛОМИ

B.S.PAVLOV (Б.С.ПАВЛОВ) — СССР, 198904, Петродворец, Физический факультет Ленинградского университета

V.I.VASYUNIN (В.И.ВАСЮНИН) — СССР, 191011, Ленинград, Фонтанка 27, ЛОМИ

4.11. old SIMILARITY PROBLEM AND THE STRUCTURE OF THE SINGULAR SPECTRUM OF NON-DISSIPATIVE OPERATORS

The similarity problems under cosideration are to find necessary and sufficient conditions for a given operator on Hilbert space to be similar to a selfadjoint (or dissipative) operator. For the first problem an answer was found in terms of the integral growth of the resolvents [3] (see also [2]):

PROPOSITION 1. <u>An operator</u> L <u>is similar to a selfadjoint operator if and only if</u>

$$\sup_{\varepsilon>0} \varepsilon \Big(\int_{\mathbb{R}} \|(L-k+i\varepsilon)^{-1}u\|^2 dk + \int_{\mathbb{R}} \|(L^*-k+i\varepsilon)^{-1}u\|^2 dk \Big) \le C\|u\|^2, \ u\in\mathcal{H}.$$

The second problem is not yet solved. Here we discuss an approach based on the notion of the characteristic function of an operator [1]. For a d i s s i p a t i v e operator $L\ (\mathrm{Im}\,L=(2i)^{-1}(L-L^*)\geqslant \mathbb{O})$ there is a criterion of similarity to a selfadjoint operator (due to B.Sz.-Nagy and C.Foiaş) in terms of its characteristic function S, namely:

$$\sup_{\mathrm{Im}\lambda>0} \| S^{-1}(\lambda)\| < +\infty .$$

The main tool in the proof of this result was the Sz.-Nagy - Foiaş functional model which yields a complete spectral description of a dissipative operator *). For a non-dissipative operator L an analogous condition on its characteristic function Θ $(\sup_{\mathrm{Im}\lambda>0}\|\Theta(\lambda)\|<+\infty,\ \sup_{\mathrm{Im}\lambda>0}\|\Theta^{-1}(\lambda)\|<+\infty)$ is sufficient for L to be similar to a selfadjoint operator (L.A.Sahnovich), but not necessary. It is possible to give counterexamples on finite-dimensional spaces which show that operators whose characteristic function does not satisfy the above condition can be similar to selfadjoint operators (cf [4], where related problems of similarity to a unitary operator and to a contraction are discussed).

To be more precise, consider the characteristic function

$$S \stackrel{\mathrm{def}}{=} I + 2i(|V|)^{1/2}(A-i|V|-\lambda)^{-1}(|V|)^{1/2},\ \ S(\lambda):E\to E, \mathrm{Im}\lambda>0,$$

*) Nevertheless this result can be obtained without using the functional model [3, 5].

of an auxiliary dissipative operator $A+i|V|$, where $A=\operatorname{Re}L$, $V=\operatorname{Im}L$, $E=\operatorname{clos}\operatorname{Range}V$ and let $V=J\cdot|V|$, $J=\operatorname{sign}V$ be the polar decomposition of V. The latter operator for the sake of simplicity is assumed to be bounded. The characteristic function Θ of L and the function S are connected by a triangular factorization

$$\Theta(\lambda)\overset{\text{def}}{=}I+iJ(|V|)^{1/2}(L^*-\lambda)^{-1}(|V|)^{1/2}=(\chi_-+\chi_+S(\lambda))(\chi_++\chi_-S(\lambda))^{-1}$$

where $\chi_\pm=(I\pm J)/2$, [6]. Note that $|\chi_\mp S(\lambda)\chi_\pm|\leq 1$ with $\operatorname{Im}\lambda>0$. Under the additional condition

$$\sup_\lambda \max\{|\chi_-S(\lambda)\chi_+|,\ |\chi_+S(\lambda)\chi_-|\}<1 \tag{1}$$

the above condition of bounded invertibility of Θ is necessary for L to be similar to a selfadjoint operator and the condition $\sup\limits_{\operatorname{Im}\lambda>0}|\Theta(\lambda)|<+\infty$ is necessary and sufficient for the similarity to a dissipative operator [7]. In this case corresponding selfadjoint and dissipative operators can be constructed explicitly in terms of the Sz- Nagy - Foiaş model for $A+i|V|$.

In general case (beyond (1)) serious obstacles appear. The reason is that it is difficult to obtain a complete description of the spectral component of L corresponding to the singular real spectrum The solution of the LATTER PROBLEM would be of independent interest.

Let us dwell upon this question. The operator L is supposed to act on the model space K which can be defined as follows. Let $\mathcal{H}$ be the Hilbert space of pairs $(\tilde{g},g)$ of E-valued functions on $\mathbb{R}$ square summable with respect to the matrix weight $\begin{pmatrix} I & S^* \\ S & I\end{pmatrix}$, $S(K)\overset{\text{def}}{=}S(K+i0)$ being the boundary values (in the strong topology) of the analytic operator-valued function S. Then

$$K=\mathcal{H}\ominus(\mathcal{D}_-\oplus\mathcal{D}_+),\quad \mathcal{D}_+\overset{\text{def}}{=}(H^2(E),\mathbb{O}),\ \mathcal{D}_-=(\mathbb{O},H^2_-(E)),$$

where $H^2_\pm(E)$ are the Hardy classes of E-valued functions in the upper and lower half-planes.

The absolutely continuous subspace N_e in the model representation of L has the following form [6]

$$N_e=N_e(L)=\operatorname{clos}P_K\left(\mathcal{H}\ominus\left(\operatorname{clos}(\chi_-L^2(E),\chi_+L^2(E))\right)\right), \tag{2}$$

where P_K is the orthogonal projection from $\mathcal{H}$ onto K, $L^2(E) = L^2(E,\mathbb{R})_*$. The singular subspace N_i is defined by $N_i = K \ominus N_e^*$, where $N_e^* = N_e(L^*)$. It is natural to distinguish in N_i two subspaces N_i^+ and N_i^-, the first one corresponding to the point spectrum in the upper half-plane and a part of the real singular continuous spectrum, the second one to the point spectrum in the lower half-plane and another (in general) part of the real singular continuous spectrum:

$$N_i^+(L) \overset{def}{=} N_i^+ \overset{def}{=} clos\, P_K\left(H_-^2(E) \ominus (\chi_- + S^*\chi_+)H_-^2(E), \mathbf{0}\right),$$
$$N_i^-(L) \overset{def}{=} N_i^- \overset{def}{=} clos\, P_K\left(\mathbf{0}, H_+^2(E) \ominus (\chi_+ + S\chi_-)H_+^2(E)\right). \qquad (3)$$

The subspace N_i^+ (N_i^-) is analogous to the subspace corresponding to the singular spectrum of a dissipative (adjoint of a dissipative) operator. Nevertheless if (1) fails, N_i does not necessarily coincide with $clos\{N_i^+ + N_i^-\}$. In particular the eigen-vectors and root-vectors of the real isolated spectrum do not belong to $clos\{N_i^+ + N_i^-\}$.

Therefore in general it is necessary to introduce a "complementary spectral component" N_i^0 ($N_i^0 \subset N_i$) which would permit us to take into account the real spectrum of L. Put

$$N_i^0 \overset{def}{=} N_i \ominus clos\left\{N_i^{+*} + N_i^{-*}\right\}$$

where $N_i^{\pm *} \overset{def}{=} N_i^{\pm}(L^*)$.

PROBLEM 1. <u>When</u> $N_i = clos\left\{N_i^+ + N_i^- + N_i^0\right\}$? <u>To estimate angles between</u> N_i^0 <u>and</u> $N_i^{\pm}, N_e$ <u>in terms of the characteristic function</u> Θ. <u>To give an explicit description of</u> N_i^0 (<u>similar to that of</u> $N_i^{\pm}, N_e$), <u>for example as the closure of the projection onto</u> K <u>some linear manifold in</u> $\mathcal{H}$ <u>described in terms of characteristic function</u>.

PROBLEM 2. <u>Find the factor of</u> Θ <u>corresponding to</u> N_i^0. <u>Investigate its further factorization. Describe properties of this factor and connection of its roots with the spectrum of</u> $L|N_i^0$. <u>How to sepa-</u>

rate the spectral singularities of $L|N_e$ from the spectrum of $L|N_i^0$?

PROBLEM 3. Make clear the spectral structure of N_i^0, i.e. in terms of the model space construct spectral projection onto intervals of the real singular spectrum and onto the root-space corresponding to the real point spectrum.

PROBLEM 4. Let L be similar to a dissipative operator. Then does N_i equal to $\operatorname{clos}\{N_i^+ + N_i^- + N_i^0\}$? Does N_i^0 coincide with the subspace of K corresponding to the singular continuous spectrum plus point spectrum of the selfadjoint part of this dissipative operator?

PROBLEM 5. Consider concrete examples (Friedrichs model with rank one perturbation, Schrödinger operator on $\mathbb{R}$ with a powerlike decreasing potential) and describe N_i^0 for such operators

Besides the real discrete spectrum the space N_i^0 apparently can contain one more spectral component. The elements of N_i^0 no longer have the "smoothness" properties as those of $N_i^{\pm}$ (namely for a dense set of vectors $u \in N_i^{\pm}$ we have $(|V|)^{1/2}(L-\lambda I)^{-1}u \in H^2_{\mp}(E)$). Perhaps, the structure of N_i^0 is similar to that of the singular continuous component of a selfadjoint operator (certainly, one should take into account the "non-orthogonalily" caused by the non-self-adjointness of L). It is also important for the similarity problem to know which factor of the characteristic function Θ corresponds to N_i^0 . Note also that all difficulties of the problem appear already in the case when the imaginary part of V is of finite rank $(\dim E \geqslant 2)$.

Let us present here one more assertion closely related to the problem discussed above and especially to the spectral decomposition of L.

PROPOSITION 2. An operator L on a Hilbert space $\mathcal{H}$ is similar to a dissipative operator L_{diss} if and only if there exists an operator M with

$$\|u\|^2 \asymp \lim_{\varepsilon \to 0} \varepsilon \int \|(L-\kappa+i\varepsilon)^{-1}u\|^2 d\kappa + \|M(L-\lambda)^{-1}u\|^2_{H^2_-(\mathcal{H})}, \quad u \in \mathcal{H}.$$

Moreover, if such L_{diss} exists then this M can be chosen satisfying the additional inequality $\operatorname{rank} M \leqslant \operatorname{rank} \operatorname{Im} L_{diss}$.

REFERENCES

1. S z . - N a g y B., F o i a ş C. Harmonic analysis of operators on Hilbert space. North Holland - Akadémiai Kiadó, Amsterdam - Budapest, 1970.
2. S z . - N a g y B. On uniformly bounded linear transformations in Hilbert space. - Acta Sci.Math., 1947, 11, 152-157.
3. Н а б о к о С.Н. Об условиях подобия самосопряженным и унитарным операторам.- Функц.анал. и его прил. (в печати).
4. D a v i s Ch., F o i a ş C. Operators with bounded characteristic functions and their γ-unitary dilation. - Acta Sci.Math , 1971, N 1-2, 127-139.
5. v a n C a s t e r n J. A problem of Sz.-Nagy. - Acta Sci. Math., 1980, 42, N 1-2, 189-194.
6. Н а б о к о С.Н. Абсолютно непрерывный спектр недиссипативного оператора и функциональная модель. II. - Зап.научн.семин.ЛОМИ, 1977, 73, 118-135.
7. Н а б о к о С.Н. О сингулярном спектре несамосопряженного оператора. - Зап.научн.семин.ЛОМИ, 1981, 113, 149-177.

S.N.NABOKO
(С.Н.НАБОКО)

СССР, 198904, Ленинград,
Петродворец,
Физический факультет
Ленинградского университета

4.12. old A PROBLEM ON OPERATOR VALUED BOUNDED ANALYTIC FUNCTIONS

Let $\mathcal{D}$, $\mathcal{D}^*$ be two Hilbert spaces and $\mathcal{B}(\mathcal{D}, \mathcal{D}^*)$ the space of all bounded linear operators mapping $\mathcal{D}$ into $\mathcal{D}^*$. The following was proved in [1].

THEOREM. *Suppose* θ *is a bounded* $\mathcal{B}(\mathcal{D}, \mathcal{D}^*)$ *-valued function analytic in the unit disc* $\mathbb{D}$. *The following assertions are equivalent*:

(a) *there exists a bounded* $\mathcal{B}(\mathcal{D}^*, \mathcal{D})$ *-valued function* Ω *analytic in* $\mathbb{D}$ *and satisfying*

$$\Omega(\lambda)\,\theta(\lambda) = I_{\mathcal{D}} \qquad (\lambda \in \mathbb{D}) \tag{1}$$

(b) *the Kernel function* K_ε :

$$K_\varepsilon(\lambda, \mu) \overset{def}{=\!=} \frac{\theta^*(\mu)\theta(\lambda) - \varepsilon^2 I_{\mathcal{D}}}{1 - \bar{\mu}\lambda} \qquad (\lambda, \mu \in \mathbb{D})$$

is positive definite, i.e.

$$\sum_{k=1}^{n} \sum_{h=1}^{n} (K_\varepsilon(\lambda_k, \lambda_h) d_h, d_k) \geqslant 0 \tag{2}$$

for any finite systems $\{\lambda_1 \ldots \lambda_n\}$, $\{d_1 \ldots d_n\}$, *where* $\lambda_j \in \mathbb{D}$, $d_j \in \mathcal{D}$.

Condition (1) obviously implies that

$$\|\theta(\lambda) d\| \geqslant \varepsilon \|d\| \qquad (|\lambda| < 1) . \tag{3}$$

The QUESTION is *whether (3) implies (2) with the same* ε *or at least with some, possibly different, positive constant.*

In the special case when $\dim \mathcal{D} = 1$ and $\dim \mathcal{D}^* < \infty$ the equivalence of (1) and (3), and thus the equivalence of (2) and (3), follows from the Corona Theorem of L.Carleson, cf. [2]. A proof of the equivalence of (2) and (3) in the general case, and possibly with operator theoretic arguments, would be an important achievement.

REFERENCES

1. S z . - N a g y B., F o i a ş C. On contractions similar to isometries and Toeplitz operators. - Ann.Acad.Scient.Fennicae, Ser.A.I. Mathematica 1976, 2, 553-564.
2. A r v e s o n W. Interpolation problems in nest algebras. - J.Func.Anal., 1975, 20, 208-233.

B.SZŐKEFALVI-NAGY

Bolyai Inst. of Math.
6720 Szeged Aradi Vértanúk tere 1
Hungary

* * *

COMMENTARY

This interesting question has been considered in several publications, but the answers are only partial. Using some refinements of T.Wolff's corona argument, V.A.Tolokonnikov [3] (see also [4], p.101) and M.Rosenblum [7] proved (independently) that (3) ⟹ (1) if $\dim \mathfrak{D} \;(\stackrel{def}{=} d) = 1$. Moreover, Tolokonnikov obtained an estimate of the solution Ω. This estimate (in somewhat simplified form) looks as follows:

$$c_1(\varepsilon) \le \min\left\{ 20\left(\log \frac{e}{\varepsilon}\right)^{3/2} \frac{1}{\varepsilon^2},\ \ (1-46(1-\varepsilon))^{-1} \right\} \qquad (0<\varepsilon<1),$$

where

$$c_d(\varepsilon) \stackrel{def}{=} \sup_{\theta,\ \varepsilon^2 I \le \theta^*\theta \le I} \ \inf_{\Omega,\ \Omega\theta = I} \|\Omega\|_\infty , \qquad d = \dim \mathfrak{D} .$$

For small values of ε a better estimate is due to Uchiyama [6]

$$c_1(\varepsilon) = O(\varepsilon^2 \log \tfrac{1}{\varepsilon});$$

V.I.Vasyunin has shown that

$$c_1(\varepsilon) \leqslant \frac{1}{\varepsilon} + \frac{1}{\varepsilon^2}\left(4\sqrt{e\log\frac{1}{\varepsilon}} + 3e\sqrt{6}\log\frac{1}{\varepsilon}\right).$$

V.I.Vasyunin [5] has proved that (3) $\Longrightarrow$ (1) if $d < +\infty$ (with the estimate $c_d(\varepsilon) \leqslant \sqrt{d}\, c_1(\varepsilon^d)$). In [5] it is also shown that $c_1(\varepsilon) \geqslant \frac{1}{\sqrt{2}\,\varepsilon^2}$, that $c_d(\varepsilon) \geqslant K_d\, \varepsilon^{-d+1}$ $(d = 2, 3, \dots)$, and that if the implication (3)$\Longrightarrow$ (1) were true for $\dim \mathcal{D} = \dim \mathcal{D}^* = \infty$, then the estimate of Ω wouldn't be better than $\exp(a\varepsilon^{-2/3})$ (i.e. that $c_\infty(\varepsilon) \geqslant a \exp(a e^{-2/3})$). Tolokonnikov has noted also that if (3)$\Longrightarrow$ (1) were always true then $c_\infty(\varepsilon)$ would be finite for every $\varepsilon \in (0,1)$ (unpublished).

If $\mathcal{D} \subset \mathcal{D}^*$ and $\dim \mathcal{D} < +\infty$ then assertions (a) and (b) in the Problem are equivalent to the possibility "to enlarge" θ to a SQUARE matrix $\tilde{\theta}$ analytic in $\mathbb{D}$ and satisfying $\theta = \tilde{\theta} \mid \mathcal{D}$, $\sup_{\lambda\in\mathbb{D}} \|\tilde{\theta}(\lambda)\| < \infty$, $\sup_{\lambda\in\mathbb{D}} \|(\tilde{\theta}(\lambda))^{-1}\| < \infty$ (see [5]).

There exists a connection between the corona theorem (for $\dim \mathcal{D} = 1$) and the left invertibility of the vector Toeplitz operator T_{θ^*} ([1], see also [8]).

REFERENCES

3. Толоконников В.А. Оценки в теореме Карлесона о короне и конечнопорожденные идеалы алгебры H^∞ . – Функц.анал. и его прил., 1980, 14, № 4, 85–86.
4. Никольский Н.К. Лекции об операторе сдвига. М., Наука, 1980.
5. Толоконников В.А. Оценки в теореме Карлесона о короне. Идеалы алгебры H^∞ , задача Сёкефальви-Надя. – Зап.научн.семин. ЛОМИ, 1981, 113, 178–198.
6. Uchiyama A. Corona theorems for countably many functions and estimates for their solutions. Preprint, 1981, University of California at Los Angeles.
7. Rosenblum M. A corona theorem for countably many functions. – Integral equat. and operator theory, 1980, 3, N 1, 125–137.
8. Schubert C.F. The corona theorem as an operator theorem. – Proc.Amer.Math.Soc., 1978, 69, N 1, 73–76.

4.13. ON EXISTENCE OF INVARIANT SUBSPACES OF C_{10}-CONTRACTIONS

Let T be a compeletely nonunitary C_{10}-contraction *) with the characteristic function $\Theta \in H^\infty(E, E_*)$. So T can be supposed acting on the space

$$K_\Theta = H^2(E_*) \ominus \Theta H^2(E)$$

as follows

$$Tf = P_\Theta z f, \quad f \in K_\Theta ,$$

where z is multiplication by z (the shift operator) on $H^2(E_*)$ and P_Θ is the orthogonal projection from $H^2(E_*)$onto K_Θ, i.e.

$$P_\Theta = I - \Theta P_+ \Theta^*$$

where P_+ is the Riesz projection from L^2 onto H^2. The adjoint operator acts as follows

$$T^* f = \frac{f - f(0)}{z}, \quad f \in K_\Theta . \tag{1}$$

Recall that $T \in C_{10}$ iff Θ is inner and $*$-outer.

Any C_{11}-contraction is quasi-similar to a unitary operator and this allows us to prove that the lattice of all T-invariant subspaces (Lat T) is non-trivial (see [1]). In our case we have only a quasi-affine transform intertwining T and the *-residual part of its unitary dilation, i.e. multiplication by z on $\operatorname{clos} \Delta_* L^2(E_*)$, where $\Delta_* = (I - \Theta\Theta^*)^{1/2}$. What can we obtain from this for finding a non-trivial invariant subspace?

We can suppose that Ker $\Theta(\lambda) = \{\mathbb{0}\}$ and Ker $\Theta(\lambda)^* = \{\mathbb{0}\}$ for every $|\lambda| < 1$, because otherwise T or T^* has an eigenvector. Hence $H^2_-(E_*) \cap \operatorname{Ker} \Theta^* = \{\mathbb{0}\}$ and therefore we have to investigate only the case $\dim E = \dim E_* = \infty$. Indeed, if $\dim E < \infty$ then there exists an antianalytic solution h_* of the equation $\Theta^* h_* = \mathbb{0}$, but the fact that Θ is inner implies that $\dim E_* \geq \dim E$.

Note that $\operatorname{Ker} \Theta^* = \operatorname{clos} \Delta_* L^2(E_*) = \Delta_* L^2(E_*)$ and

---

*) All used terminology can be found in [1] or [2].

$P_+\Delta_* L^2(E_*)\subset K_\theta$. Here P_+ intertwines T and the $*$-residual part of its unitary dilation, namely $z|\Delta_* L^2(E_*)$.

Let x now be an arbitrary vector in K_θ and $h_*\in \operatorname{Ker}\theta^*$. Then

$$(T^n x, P_+ h_*) = (z^n x, h_*) = \int_{\mathbb{T}} z^n \langle x, h_*\rangle_{E_*} dm ,$$

where m is the normalized Lebesgue measure on the circle $\mathbb{T}$. If

$$m\{\zeta\in\mathbb{T}:\ \operatorname{rank}\Delta_*(\zeta)>1\}>0$$

then the multiplicity of T is greater than 1, i.e. there exists no cyclic vector. Indeed, for every $x\in K_\theta$ we can choose a vector $h_*\in\operatorname{Ker}\theta^*$ such that $\langle x(\zeta), h_*(\zeta)\rangle_{E_*}=0$ a.e. So every nonzero vector x generates a non-trivial invariant subspace.

Further we shall suppose rank $\Delta_*(\zeta)\leqslant 1$ a.e. Putting $\ell=\{\zeta\in\mathbb{T}:\ \operatorname{rank}\Delta_*(\zeta)=1\}$, choose a vector $h_*\in\operatorname{Ker}\theta^*$ such that $\|h_*(\zeta)\|_{E_*}=1$ for a.e. $\zeta\in\ell$ and $h_*(\zeta)=0$ for a.e. $\zeta\in\mathbb{T}\setminus\ell$. Now we have

$$\operatorname{Ker}\theta^* = \Delta_* L^2(E_*) = \{\varphi h_*:\ \varphi\in L^2\}$$

and

$$(T^n x, P_+\varphi h_*) = \int_{\mathbb{T}} z^n\bar{\varphi}\langle x, h_*\rangle_{E_*} dm .$$

That is $P_+\varphi h_*$ is cyclic for T^* iff there exists a nonzero vector $x\in K_\theta$ such that $\langle x,h_*\rangle$ is a cyclic function for the multiplication by z on $L^2(\ell)$, i.e.

$$m\{\zeta\in\ell:\ \langle x(\zeta), h_*(\zeta)\rangle_{E_*}=0\}\neq 0,$$

or $\log|\langle x,h_*\rangle_{E_*}|\in L^1$

Choosing x of special type we can obtain various sufficient conditions for the existence of a non-trivial invariant subspace.

PROPOSITION. <u>Each of the following conditions implies the existence of a non-trivial invariant subspace of</u> T.

1) $\exists\varphi\in L^2$ <u>such that</u> $m\{\zeta\in\ell:\ \langle (P_+\varphi h_*)(\zeta), h_*(\zeta)\rangle_{E_*}=0\}\neq 0;$

2) $\exists \psi \in L^2$ <u>such that</u> $\log|\langle P_+\psi h_*, h_*\rangle| \in L^1$;

3) $\exists \ell_1 \subset \ell$, $m(\ell_1) > 0$ <u>and</u> $\operatorname{span}\{h_*(\zeta) : \zeta \in \ell_1\} \neq E_*$;

4) <u>the antianalytic function</u> $P_-(\|P_+ h_*\|^2_{E_*})$ <u>admits a pseudocontinuation to the unit disc.</u>

CONJECTURE. <u>For every inner</u> $*$<u>-outer function</u> Θ <u>there exists a nonzero noncyclic vector for</u> T^* <u>(defined by (1)) of the form</u> $P_+ h_*$ <u>where</u> $h_* \in \operatorname{Ker} \Theta^*$.

If the CONJECTURE is not true a counter example must have a number of very pathological properties and may be a candidate for an operator without invariant subspace at all.

REFERENCES

1. Sz.-Nagy B., Foiaş C. Harmonic analysis of operators on Hilbert space, North Holland/Akadémiai Kiadó, Amsterdam Budapest, 1970.
2. Никольский Н.К. Лекции об операторе сдвига, М., Наука, 1980.

R.TEODORESCU

Universitatea Braşov
Facultatea de Matematică
B-dul Gh.Gheorghiu - Dej 29,
2200 Braşov, România

V.I.VASYUNIN
(В.И.ВАСЮНИН)

СССР, I9I0II, Ленинград,
Фонтанка 27, ЛОМИ

4.14. old TITCHMARSH'S THEOREM FOR VECTOR FUNCTIONS

In one version (from which others can be derived) Titchmarsh's theorem states: if f and g are functions of $L^2(\mathbb{R}_+)$ such that $f*g$ vanishes on $(0,1)$, and if f vanishes on no interval $(0,\varepsilon)$, then g must vanish on $(0,1)$. Here is a PROOF. Fix f, and denote by M the set of all g such that $f*g$ vanishes on $(0,1)$. M is a closed subspace invariant under shifts to the right. Beurling's theorem states that $\hat{M}$, the space of Fourier transforms of functions in M, is exactly qH^2, where q is inner in the upper half-plane and H^2 is the Hardy space on the half-plane. Since M contains all functions vanishing on $(0,1)$, $(q(z))^{-1}\exp(iz)$ is an inner function too. The known structure of inner functions implies that $q(z)=\exp(isz)$ for some s, $0\le s\le 1$. This means that M contains all functions that vanish on $(0,s)$, and it follows that f vanishes on $(0,1-s)$. Hence $s=1$, so g must vanish on $(0,1)$. ●

Suppose F and G are functions in $L^2(\mathbb{R}_+)$ with values in a Hilbert space H, and suppose the expression

$$\int_0^x (F(t), G(x-t))\,dt$$

vanishes for $0\le x\le 1$. (The integrand is the inner product in H.) What can we say about F and G? More generally, is there a simple characterization of the subspaces M of $L^2(\mathbb{R}_+)$ that are closed, invariant under right shifts, and contain all functions vanishing on $(0,1)$?

By the vectorial version of Beurling's theorem (see [1]) the problem is equivalent to describing the inner functions Q such that $Q(z)^{-1}\exp iz$ is also inner. In the vectorial context, an inner function Q is analytic in the upper half-plane, takes values in the space of operators on H, satisfies $\|Q(z)\|\le 1$, and has boundary values $Q(x)$ that are unitary for almost all real x. In our case $Q(x)$ has spectrum (the support of its Fourier transform) in $[0,1]$ and so is entire.

We obtain inner functions of this kind in the form $\exp izA$, where A is a constant self-adjoint operator satisfying $0\le A\le I$. The corresponding subspace M is easily described. Let (H_t) be the spectral resolution of A ; thus $H_t = \mathbb{O}$ for $t < 0$ and $H_t = H$ for $t > 1$. M is the set of vector functions F such that $F(t)$ lies in H_t for almost every t.

A straightforward extension of Titchmarsh's theorem would assert that the integral above vanishes for $0 \leq x \leq 1$ only if the inner product vanishes identically for such x. This is equivalent to saying the inner function of M necessarily has the form $\exp izA$. This is not true, as shown by an example of Donald Sarason. His example leads to a method for constructing such inner functions. Set $R(z) = (\exp(-iz/2))\,Q(z)$; then the unitary function $R(x)$ has spectrum in $[-\frac{1}{2}, \frac{1}{2}]$. Write $R = S + iT$ with S, T self-adjoint. The fact that R is unitary means that S and T commute at each point, and $S^2 + T^2 = I$.

Suppose H is two-dimensional and $S = rI$, $0 \leq r < 1$. Then on the real axis T must be $\begin{pmatrix} f & g \\ \bar{g} & -f \end{pmatrix}$ where f and g are entire functions of exponential type at most $\frac{1}{2}$, f is real on the real axis, and $f^2 + |g|^2 = 1 - r^2 = s^2$. The choice $f(z) = s\cos bz$, $g(z) = s\sin bz$, $0 < b \leq \frac{1}{2}$, gives

$$Q(z) = r\left(\exp\frac{iz}{2}\right) I + s\left(\exp\frac{iz}{2}\right)\begin{pmatrix} \cos bz & \sin bz \\ \sin bz & -\cos bz \end{pmatrix}.$$

Can the structure of Q be described simply in general, or even when H is two-dimensional?

REFERENCE

1. Helson H. Lectures on invariant subspaces. NY-London, Academic Press, 1964.

HENRY HELSON

Department of Math.
University of California
Berkeley, California 94720
USA

4.15. SOME FUNCTION THEORETIC PROBLEMS CONNECTED WITH THE THEORY OF SPECTRAL MEASURES OF ISOMETRIC OPERATORS

Let V be a completely non-unitary isometric operator in a separable Hilbert space H with the defect spaces N and M :

$$V: H\ominus N \longrightarrow H\ominus M$$

where it is supposed for definiteness that $0<\dim N\leqslant \dim M\leqslant\infty$. Let P_L denote the orthogonal projection of H onto the subspace L, and let $T_V = VP_{H\ominus N}$. The operator V defines in the unit disc $\mathbb{D}$ an operator-valued holomorphic function

$$\chi_V(z) = zP_N(I-zT_V)^{-1}|M$$

which is called the characteristic function of V .

Consider the class $\mathbb{B}(M,N)$ of all operator-valued contractive holomorphic functions in $\mathbb{D}$ taking values in the space of all bounded operators from M to N . Let $\mathbb{B}^\circ(M,N)= =\{\chi\in\mathbb{B}(M,N):\chi(0)=\mathbb{0}\}$. It is known that $\chi_V\in\mathbb{B}^\circ$ and that for every $\chi\in\mathbb{B}^\circ$ there exists an isometric V with given defect spaces N and M such that $\chi_V=\chi$ (see [1]-[3]).

It is also known that in the case $\dim N=\dim M$ all unitary extensions of V not leaving H are described by the formula

$$U_\mathcal{E} = T_V + \mathcal{E}P_N$$

where $\mathcal{E}$ is a unitary "parameter", $\mathcal{E}:N\to M$ ($\mathcal{E}^*\mathcal{E}=I|N$, $\mathcal{E}\mathcal{E}^*=I|M$). The spectral measure $E_\mathcal{E}$ of $U_\mathcal{E}$ can be determined (up to a unitary equivalence) by $\chi=\chi_V$ and $\mathcal{E}$ with a help of the following formula

$$[I+\chi(z)\mathcal{E}][I-\chi(z)\mathcal{E}]^{-1}=\frac{1}{2\pi}\int_{\mathbb{T}}\frac{\zeta+z}{\zeta-z}\,d\sigma_\mathcal{E},\quad z\in\mathbb{D}, \tag{1}$$

where $d\sigma_\mathcal{E}=P_N\,dE_\mathcal{E}|N$

The spectral measures of the minimal unitary extensions of V leaving H (now the case $\dim N\neq\dim M$ is also permitted) can be also determined by (1) where the parameter $\mathcal{E}$ is already an arbitrary function in $\mathbb{B}(N,M)$. The spectral measure of $U_\mathcal{E}$ is absolutely continuous if and only if the measure $\sigma_\mathcal{E}$ in (1) is absolutely continuous with respect to the Lebesgue measure on $\mathbb{T}$.

Consider a subset $\mathbb{B}_a(M,N)$ of $\mathbb{B}^\circ(M,N)$ consisting of functions χ whose measure $\sigma_\mathcal{E}$ in the Riesz-Herglotz representation

$$[I+\chi(z)\mathcal{E}(z)][I-\chi(z)\mathcal{E}(z)]^{-1}=\frac{1}{2\pi}\int_{\mathbb{T}}\frac{\zeta+z}{\zeta-z}\,d\sigma_\mathcal{E}$$

is absolutely continuous for an arbitrary choice of $\mathcal{E}$ in $\mathbb{B}(N,M)$. The inclusion $\chi_V\in\mathbb{B}_a(M,N)$ is clearly equivalent to the condition that all minimal unitary extensions of V have absolutely continuous spectral measures.

PROBLEM. Find criteria for a given χ in $\mathbb{B}^\circ(M,N)$ to belong to $\mathbb{B}_a(M,N)$.

Note that for $\chi\in\mathbb{B}^\circ(M,N)$ the inclusion

$$(1-\|\chi\|)^{-1}\in L^1(\mathbb{T})$$

implies $\chi\in\mathbb{B}_a(M,N)$ being thus a sufficient (but not necessary) condition.

Suppose in the sequel that $\dim M<+\infty$ and let $\mathbb{B}_{Sz}(M,N)$ $(\mathbb{B}^\circ_{Sz}(M,N))$ denote the family of all χ in $\mathbb{B}(M,N)$ $(\mathbb{B}^\circ(M,N))$ with

$$\log\det(I-\chi^*\chi)\in L^1(\mathbb{T}). \tag{2}$$

LEMMA. Given $\chi\in\mathbb{B}(M,N)$ the following are equivalent:

1) there exists an isometric operator $\mathcal{E}:N\to M$ with the corresponding measure in (1) satisfying the Szegö condition

$$\log\det\left(\frac{d\sigma_\mathcal{E}}{dm}\right)\in L^1(\mathbb{T}) \tag{3}$$

2) condition (3) holds for all isometries $\mathcal{E}:N\to M$;

3) $\chi\in\mathbb{B}_{Sz}(M,N)$.

If $\chi\in\mathbb{B}_{Sz}(M,N)$ then the measures $\sigma_\mathcal{E}$ are absolutely continuous for almost all $\mathcal{E}$ (with respect to the invariant measure on the symmetric space of all isometries $\mathcal{E}:N\to M$).

We don't know any example of a function in $\mathbb{B}_a(M,N)$ not satisfying (2).

A more subtle sufficient condition for $\chi \in \mathbb{B}^0_{Sz}(M,N)$ to belong to $\mathbb{B}_a(M,N)$ can be deduced from results of [4]-[6]. Namely, fix χ in $\mathbb{B}^0_{Sz}$ and denote by P_+^{-1} and $(P_-^{-1})^*$ (unique) solutions of the factorization problem

$$I-\chi(\zeta)\chi(\zeta)^* = P_+^{-1}(\zeta)[P_+^{-1}(\zeta)]^*, \quad \zeta\in\mathbb{T},$$

$$I-\chi^*(\zeta)\chi(\zeta) = P_-^{-1}(\zeta)[P_-^{-1}(\zeta)]^*, \quad \zeta\in\mathbb{T},$$

in the classes of outer functions non-negative at the origin and belonging to $\mathbb{B}(N,N)$ and $\mathbb{B}^*(M,M)=\{h(z): h^*(1/\bar{z})\in\mathbb{B}(M,M), |z|>1\}$ respectively. Let

$$q_-(\zeta)=P_-(\zeta)\chi^*(\zeta), \qquad q_+(\zeta)=P_+(\zeta)\chi(\zeta).$$

It follows that the values of $f_0=q_-P_+^{-1}$ are contractions $N\longrightarrow M$ a.e. on $\mathbb{T}$. Consider the Hankel operator Γ with the matrix symbol f_0. The operator Γ maps $\ell^2(N)$ into $\ell^2(M)$ and its matrix in the standard basis is $(\hat{f}_0(-j-k+1))_{j,k\geqslant 1}$, where $\hat{f}_0$ stands for Fourier coefficients of f_0. Consider subspaces

$$N_0=\{e_\xi=(\xi,0,0,\dots):\xi\in N\}, \quad M_0=\{e_\eta=(\eta,0,0,\dots):\eta\in M\}$$

of $\ell^2(N)$ and $\ell^2(M)$ and put

$$P_+(z,\rho)\xi=[(I-\rho^2\Gamma^*\Gamma)^{-1}N^{1/2}(\rho)e_\xi]_+(z), \quad |z|<1,$$

$$q_+(z,\rho)\eta=\rho z[\Gamma^*(I-\rho^2\Gamma\Gamma^*)^{-1}M^{1/2}(\rho)e_\eta]_+(z), \quad |z|<1,$$

where $\rho\in(0,1)$, $\xi\in N$, $\eta\in M$ and $N^{1/2}(\rho)$, $M^{1/2}(\rho)$ denote positive square roots of $[P_{N_0}(I-\rho^2\Gamma^*\Gamma)^{-1}|N_0]^{-1}$ and $[P_{M_0}(I-\rho^2\Gamma\Gamma^*)^{-1}|M_0]^{-1}$ respectively and finally

$$h_+(z)\overset{def}{=\!=}\sum_{j\geqslant 1}\xi_j z^{j-1} \quad (z\in\mathbb{D}) \quad \text{for} \quad h=\{\xi_j\}_1^\infty\in\ell^2(N).$$

It turns out that $\chi \in \mathbb{B}_a(M,N)$ provided

$$\lim_{\rho \to 1-0} P_+^{-1}(z,\rho)\, q_+(z,\rho) = \chi(z) \tag{4}$$

for all z in $\mathbb{D}$.

CONJECTURE. _If_ $\chi \in \mathbb{B}_a(M,N) \cap \mathbb{B}^0_{Sz}(M,N)$ _then_ χ _satisfies_ (4).

REMARK. For $\chi \in \mathbb{B}^0_{Sz}(M,N)$ condition (4) holds iff the following formula

$$f_\mathcal{E}(\zeta) = [q_-(\zeta) + P_-(\zeta)\mathcal{E}(z)][P_+(\zeta) + q_+(\zeta)\mathcal{E}(\zeta)]^{-1}$$

establishes a one-to-one correspondence between the set of operator $(N \longrightarrow M)$- valued contractive functions $f_\mathcal{E}$ with the same principal part as $f_0(\zeta)$ (i.e. $\sum_{k \geq 1} \hat{f}_\mathcal{E}(-k)\zeta^{-k} = \sum_{k \geq 1} \hat{f}_0(-k)\zeta^{-k}$) and all functions $\mathcal{E}$ in $\mathbb{B}(N,M)$.

REFERENCES

I. Л и в ш и ц М.С. Об одном классе линейных операторов в гильбертовом пространстве. - Матем.сб. I946, I9(6I), 236-260.

2. Л и в ш и ц М.С. Изометрические операторы с равными дефектными числами, квазиунитарные операторы. - Матем.сб.,I950,26,247-264.

3. Sz.-N a g y B., F o i a ş C. Harmonic analysis of operators in Hilbert space. Budapest, Akad.Kiadó, 1970.

4. А д а м я н В.М., А р о в Д.З., К р е й н М.Г. Бесконечные ганкелевы матрицы и обобщенные проблемы Каратеодори-Фейера и И.Шура.-Функц.анал.и его прил., I968,2,в.4, I-I7.

5. А д а м я н В.М., А р о в Д.З., К р е й н М.Г. Бесконечные блочно-ганкелевы матрицы и связанные с ними проблемы продолжения.-Изв.АН Арм.ССР, сер.матем.,I97I,6, I8I-206.

6. А д а м я н В.М. Невырожденные унитарные сцепления полуунитарных операторов. - Функц.анал.и его прил., I973 ,7, вып.4, I-I6.

V.M.ADAMYAN (В.М.АДАМЯН) — СССР, 270000, Одесса, Одесский государственный университет

D.Z.AROV (Д.З.АРОВ) — СССР, 270020, Одесса, Одесский педагогический институт

M.G.KREIN (М.Г.КРЕЙН) — СССР, 270057, Одесса, ул.Артёма I4, кв.6

4.16. THREE PROBLEMS ABOUT J-INNER MATRIX-FUNCTIONS

1. Let $J=\begin{pmatrix} I_n & 0 \\ 0 & -I_n \end{pmatrix}$. A matrix-function (m.-f.) W meromorphic in $\mathbb{D}$ is called *J-inner* if

$$W^*(z)JW(z)\leqslant J,\ \forall z\in\mathbb{D},\ W^*(\zeta)JW(\zeta)=J \text{ a.e. on } \mathbb{T}.$$

Let $\mathfrak{D}_+$ denote the class of m.-f. with entry functions representable as a ratio of an H^∞-function and of an outer H^∞-function. A J-inner m.-f. W is called: 1) *singular* if $W^{\pm1}\in\mathfrak{D}_+$ and 2) *regular* if there exists no nonconstant singular J-inner m.-f. W_0 such that WW_0^{-1} is J-inner.

THEOREM 1. <u>An arbitrary J-inner m.-f. W admits a representation $W=W_r W_s$, where W_r and W_s are respectively regular and singular J-inner m.-f.; W_r is uniquely determined by W up to a constant J-unitary right factor</u>

2. The importance of the class of regular J-inner m.-f. is explained in particular by its connection with the generalized interpolation problem of Schur-Nevanlin of finding all m.-f. S such that

$$b_1^{-1}(S-S_0)b_2^{-1}\in H_n^\infty,\quad S\in\mathbb{B}_n, \tag{1}$$

where b_1, b_2, $S_0(\in\mathbb{B}_n)$ are given m.-f. of order n, b_1 and b_2 are inner, $\mathbb{B}_n$ denotes the set of all m.-f. of order n holomorphic and contractive in $\mathbb{D}$; $H_n^p\ (1\leqslant p\leqslant+\infty)$ is the class of m.-f. of order n with entries in H^p.

Fix $z_0\in\mathbb{D}$ with $\det b_k(z_0)\neq 0$ ($k=1,2$). When S ranges over the set of solutions of problem (1), the values $S(z_0)$ fill a matrix ball. If the right and left half-radii of this ball are nondegenerate then problem (1) is called completely indeterminate; this definition does not depend on z_0.

Let $W=[W_{jk}]_1^2$ be an arbitrary J-inner m.-f. It has a meromorphic quasi-continuation to the exterior $\mathbb{D}_e$ of the disc $\mathbb{D}$. We denote $f^\sim(z)=f^*\left(\frac{1}{\bar z}\right)$. We have [2]

$$W_{11}=b_1 p_*, \quad W_{22}=b_2^{-1} p, \qquad s_0 \overset{def}{=\!=} W_{12} W_{22}^{-1} \in \mathbb{B}_n \tag{2}$$

where b_1 and b_2 are inner m.-f., p and $\tilde{p_*}$ are outer m.-f., $p^{-1} \in \mathbb{B}_n$, $(\tilde{p_*})^{-1} \in \mathbb{B}_n$. Singular m.-f. W are characterized by equalities $b_1 = b_2 = I_n$ in (2). The following theorem shows that it is important to establish a criterion of regularity of a J-inner m.f.

THEOREM 2. Let $W=[W_{jk}]_1^2$ be an arbitrary J-inner m.-f. and let b_1, b_2, s_0 be m.-f. defined in (2). Then problem (1) with these data is completely indeterminate and the m.-f. $s = s_\varepsilon$, where

$$s_\varepsilon = (W_{11}\varepsilon + W_{12})(W_{21}\varepsilon + W_{22})^{-1}, \quad \varepsilon \in \mathbb{B}_n \tag{3}$$

are its solutions. The family $\{s_\varepsilon\}$ is the set of all solutions of problem (1) iff the J-inner m.-f. W is regular. For any completely indeterminate problem (1) there exists a regular J-inner m.-f. $W=[W_{jk}]_1^2$, for which formula (3) establishes a one-to-one correspondence between the set of all $\varepsilon \in \mathbb{B}_n$, and the set of all solutions of problem (1). M.-f. W may be chosen so that m.-f. b_1 and b_2 in (2) be the same as in problem (1); in this case W is defined by problem (1) up to a constant J-unitary right factor.

3. J-inner m.-f. $W=[W_{jk}]_1^2$ being arbitrary, let us consider m.-f. b_1 and b_2 corresponding to it by (2) and define a m.-f. f_I,

$$f_I = b_1^{-1}(W_{11}+W_{12})(W_{21}+W_{22})^{-1} b_2^{-1}. \tag{4}$$

It takes unitary values a.e. on $\mathbb{T}$ and $f_I = \varphi_*^{-1}\varphi$, where $\varphi \in H_n^2$, $\tilde{\varphi_*} \in H_n^2$. If a m.-f. f unitary on $\mathbb{T}$ is representable in the form $f = \varphi_-^{-1}\varphi_+$, where $\varphi_+ \in H_n^2$, $\tilde{\varphi_-} \in H_n^2$, $\varphi_\pm$ being determined by f up to the left constant factor x with $\det x \neq 0$, then we write $\operatorname{ind} f = 0$. The following theorem holds (see [1] and Theorem

2).

THEOREM 3. A J-inner m.-f. $W=[W_{jk}]_1^2$ is regular iff for the m.-f. f_I defined in (4) we have $\operatorname{ind} f_I = 0$.

COROLLARY. For a J-inner m.-f. $W=[W_{jk}]_1^2$ to be regular it is sufficient, that

$$(1-\|S_0(\zeta)\|)^{-1} \in L^1, \quad S_0 \overset{\text{def}}{=} W_{12} W_{22}^{-1}. \tag{5}$$

The proof of Theorems 2 and 3 is based on the results about the problem of Nehari [3-5] to which problem (1) is reduced by a substitution $f = b_1^{-1} S b_2^{-1}$.

PROBLEM 1. Find a criterion for a J-inner m.-f. W to be regular without using the notion of the index of a m.-f.

4. It is known [6] that a product of elementary factors of Blaschke-Potapov of the 1st, 2hd and 3rd kind with the poles, respectively in $\mathbb{D}$, in $\mathbb{D}_e$ and on $\mathbb{T}$ (see[7]) is a J-inner m.-f. We have [1]

THEOREM 4. A J-inner m.-f. W is a product of elementary factors of the 1st and the 2nd kind iff it is regular and the m.-f. b_1 and b_2 associated with W by (2) are products of (definite) elementary factors of Blaschke-Potapov.

REMARK. Both m.-f. b_1 and b_2 in (2) are Blaschke-Potapov products iff

$$\lim_{r\uparrow 1} \int_{\mathbb{T}} \log|\det(W_{11}(r^{-1}\zeta) W_{22}(r\zeta))|\, dm(\zeta) = \int_{\mathbb{T}} \log|\det(W_{11}(\zeta) W_{22}(\zeta))|\, dm(\zeta). \tag{6}$$

A corresponding condition also exists for a product of elementary factors of only the 1st (2nd) kind. In this case $b_1 = I_n$ ($b_2 = I_n$) and instead of (6) we have

$$\log|\det W_{11}(\infty)| = \int_{\mathbb{T}} \log|\det W_{11}(\zeta)|\, dm(\zeta), \tag{7}$$

$$\lim_{r\uparrow 1} \int_{\mathbb{T}} \log|\det W_{22}(r\zeta)|\, dm(\zeta) = \int_{\mathbb{T}} \log|\det W_{22}(\zeta)|\, dm(\zeta).$$

$$\left(\log|\det W_{22}(0)| = \int_{\mathbb{T}} \log|\det W_{22}(\zeta)|\,dm(\zeta), \right. \tag{8}$$

$$\left. \lim_{r\uparrow 1} \int_{\mathbb{T}} \log|\det W_{11}(r\zeta)|\,dm(\zeta) = \int_{\mathbb{T}} \log|\det W_{11}(\zeta)|\,dm(\zeta). \right)$$

COROLLARY. Suppose condition (5) holds for a J-inner m.-f. $W=[W_{jk}]_1^2$. Then W is a product of elementary factors of the 1st and 2nd kind (only of the 1st, only of the 2nd) iff condition (6) (respectively (7), (8)) is valid.

PROBLEM 2. Find a criterion for a J-inner m.-f. to be a product of elementary factors of the 1st, 2nd and 3rd kind.

Theorem 4 gives in fact a criterion of completeness of a simple operator in terms of its characteristic m.-f. W in case when its eigenvalues are not on $\mathbb{T}$. The solution of Problem 2 would give a criterion without this restriction.

PROBLEM 3. Find a criterion for a J-inner m.-f. to be a product of elementary factors of the 3rd kind.

Let us point out that such a product is a singular m.-f. A product of elementary factors of the 1st kind arises in the "tangent" problem of Nevanlinna-Pick [8] and products of factors of the 1st and 2nd kind arise in a "bi-tangent" problem in which "tangent" data for $S(z)$ and $S^*(z)$ are given in interpolation knots $z_k (\in \mathbb{D})$. The author's attention was drawn to such a "bi-tangent" problem by B.L.Kogan. Products of elementary factors of the 3-d kind arise in the "tangent" problem which has the interpolation knots on $\mathbb{T}$. The definition and investigation of such problems is much more complicated [9, 10].

REFERENCES

1. А р о в Д.З. Об одной интерполяционной задаче и индефинитном произведении Бляшке-Потапова. Тезисы докладов. Школа по теории операторов в функц.пространствах, Минск, 1982, 14-15.

2. А р о в Д.З. Реализация матриц-функций по Дарлингтону. - Изв. АН СССР, сер.матем., 1973, 37, № 6, 1299-1331.

3. А д а м я н В.М., А р о в Д.З., К р е й н М.Г. Бесконечные ганкелевы матрицы и обобщенные задачи Каратеодори-Фейера

и И.Шура.-Функц.анализ и его прилож., 1968, 2, вып.4, 1-17.

4. А д а м я н В.М., А р о в Д.З., К р е й н М.Г. Бесконечные блочно-ганкелевые матрицы и связанные с ними проблемы продолжения. - Изв.АН Арм.ССР, матем., 1971, 6, № 2-3, 87-112.

5. А д а м я н В.М. Невырожденные унитарные сцепления полуунитарных операторов. - Функц.анализ и его прилож., 1973, 7, вып.4, 1-17.

6. А р о в Д.З., С и м а к о в а Л.А. О граничных значениях сходящейся последовательности J-сжимающих матриц-функций. - Матем.заметки, 1976, 19, № 4, 491-500.

7. П о т а п о в В.П. Мультипликативная структура J-нерастягивающих матриц-функций. - Труды Моск.матем.о-ва, 1955, 4, 125--236.

8. Ф е д ч и н а И.П. Касательная проблема Неванлинны-Пика с кратными точками. - Докл.АН Арм.ССР, 1975, 61, № 4, 214-218.

9. К р е й н М.Г. Общие теоремы о позитивных функционалах. - В кн.: Ахиезер Н.И., Крейн М. О некоторых вопросах теории моментов. Харьков, 121-150. (Ahiezer N.I., Krein M. Some Questions in the Theory of Moments. Trans.Math.Mon. , AMS, 1962, v.2, 124-153.)

10. М е л а м у д Е.Я. Граничная задача Неванлинны-Пика для J-растягивающих матриц-функций. - Известия высших учебных заведений, Матем., 1984 (в печати).

D.Z.AROV
(Д.З.АРОВ)

СССР, 270020, Одесса,
Комсомольская ул., 26,
Одесский государственный
педагогический институт

4.17. EXTREMAL MULTIPLICATIVE REPRESENTATIONS

Let $\mathcal{E}$ be the class of entire functions W of exponential type with values in the space of all bounded operators on a separable Hilbert space and such that

$$W(0)=I,\quad W^*(\lambda)W(\lambda)\leqslant I\quad(\operatorname{Im}\lambda>0),\quad W^*(\lambda)W(\lambda)=I\quad(\operatorname{Im}\lambda=0).$$

For every $W\in\mathcal{E}$ there exists an operator-valued hermitian non-decreasing function E on $[0,\ell]$ $(E(0)=0,\ \operatorname{Var}_{[0,t]} E=t)$ satisfying

$$W(\lambda)=\int_0^{\ell}\!\!\curvearrowleft \exp\{i\lambda dE(t)\}. \tag{1}$$

(see [1, 2]). Let H be the weak derivative of E. Then (1) is equivalent to

$$W(\lambda)=\int_0^{\ell}\!\!\curvearrowleft \exp\{i\lambda H(t)dt\}. \tag{2}$$

The function W determines H uniquely iff $I-W(\lambda)\in\gamma_1$ $(\lambda\in\mathbb{C})$ and W, $\det W$ have the same exponential type [3]. To single out a canonical function from the family of all functions H satisfying (2) in the general case the following definition is introduced.

DEFINITION. Let H be a weakly measurable function on $[a,b]$ $(0\leqslant H,\ |H|\in L^1[a,b])$ and suppose that for every $s\in[\alpha,\beta]\subset[a,b]$ the function

$$W_{\alpha,s}(\lambda H)\overset{\mathrm{def}}{=}\int_{\alpha}^{s}\!\!\curvearrowleft \exp\{i\lambda H(t)dt\}$$

is the greatest divisor (in $\mathcal{E}$) among all divisors of $W_{\alpha,\beta}(\lambda H)$ of type $s-\alpha$. Then H is called an e x t r e m a l f u n c t i o n on $[\alpha,\beta]$.

THEOREM 1. <u>For every W of exponential type σ in $\mathcal{E}$ there exists a unique $\tilde H$ extremal on $[0,\sigma]$ and satisfying $W_{0,\sigma}(\lambda\tilde H)=W(\lambda)$.</u>

This result is a special case of a theorem proved in [2] (compare with [4]).

PROBLEM. Find an intrinsic description of functions $\tilde{H}$ extremal on $[a,b]$ (or, at least, exhibit a large subclass of $\tilde{H}$'s)

The following theorem shows that the description is very likely to be of local character.

THEOREM 2 ([5],[6]). Suppose H is extremal on $[a,b]$. Then it is extremal on any $[\alpha,\beta]\subset[a,b]$. Conversely, if for every $s\in[a,b]$ there exists a segment $[\alpha,\beta]\subset[a,b]$ such that $s\in[\alpha,\beta]$ and H is extremal on $[\alpha,\beta]$ then H is extremal on $[a,b]$.

CONJECTURE. Let H be a continuous (with respect to the norm topology) operator - valued function on $[a,b]$. Then H is extremal iff all values of H are orthogonal projections.

In the particular case $\operatorname{rank} H(t)=1$, $t\in[a,b]$ the conjecture is true by a theorem of G.E.Kisilevskii [7] (in the form given in [8]).

Similar questions in case when $W(\frac{1}{\lambda})$ is the characteristic function of a so called one-block operator have been considered in [9].

REFERENCES

1. П о т а п о в В.П. Мультипликативная структура J-нерастягивающих матриц-функций. - Труды Моск.матем.об-ва, 1955, 4, 125-236.
2. Г и н з б у р г Ю.П. Мультипликативные представления и миноранты ограниченных аналитических оператор-функций. - Функ.анал. и его прил., 1967, 1, № 3, 9-23.
3. Б р о д с к и й М.С. Треугольные и жордановы представления линейных операторов. - Москва, "Наука", 1969.
4. Б р о д с к и й М.С., И с а е в Л.Е. Треугольные представления диссипативных операторов с резольвентой экспоненциального типа. - Докл.АН СССР, 1969, 188, № 5, 971-973.
5. Г и н з б у р г Ю.П. О делителях и минорантах оператор-функций ограниченного вида. - Матем.исследования, Кишинёв, 1967, 2, № 4, 47-72.
6. М о г и л е в с к а я Р.Л. Немонотонные мультипликативные представления ограниченных аналитических оператор-функций. - Матем.исследования, Кишинёв, 4, № 4, 1969, 70-81.

7. К и с и л е в с к и й Г.Э. Инвариантные подпространства вольтерровых диссипативных операторов с ядерными мнимыми компонентами. - Известия АН СССР, сер.матем., 1968, 32, № 1, 3-23.
8. Г о х б е р г И.Ц., К р е й н М.Г. Теория вольтерровых операторов в гильбертовом пространстве и её приложения. Москва, "Наука", 1967.
9. С а х н о в и ч Л.А. О диссипативных вольтерровых операторах.- Матем.сборник, 1968, 76 (118), № 3, 323-343.

YU.P.GINZBURG
(Ю.П.ГИНЗБУРГ)

СССР, 270039, Одесса,
Одесский технологический
институт пищевой промышленности
им.М.В.Ломоносова

4.18. old FACTORIZATION OF OPERATORS ON $L^2(a,b)$

1. A bounded operator S_- (S_+) on $L^2(a,b)$, $-\infty \leqslant a < b \leqslant \infty$, is called *lower triangular* (*upper triangular*) if for every ξ ($a \leqslant \xi \leqslant b$)

$$S_-^* P_\xi = P_\xi S_-^* P_\xi, \quad S_+ P_\xi = P_\xi S_+ P_\xi,$$

where $P_\xi f \overset{\text{def}}{=} \chi_{[a,\xi]} f$.

A bounded operator S on $L^2(a,b)$ is said to *admit the left factorization* if $S = S_- S_+$ where S_- and S_+ are lower and upper triangular bounded operators, with bounded inverses.

I.C.Gohberg and M.G.Krein [1] have studied the problem of factorization under the assumption

$$S - I \in \gamma_\infty . \tag{1}$$

The operators S_+, S_- have been assumed to be of the form

$$S_+ = I + X_+, \quad S_- = I + X_-; \quad X_+, X_- \in \gamma_\infty .$$

(**γ_∞ is the ideal of compact operator**)

Factorization method had played an essential role in a number of problems of the spectral theory. Giving up condition (1) and considering more general triangular operators would essentially widen the scope of applications of this method.

EXAMPLE. Consider [2] the operator

$$S_\beta f = f(x) + \frac{i\beta}{\pi} \, \text{v.p.} \int_0^\omega \frac{f(t)}{x-t} \, dt, \qquad -1 < \beta < 1 .$$

The operator S_β ($\beta \neq 0$) clearly does not satisfy (1). Nevertheless S_β admits a factorization $S_\beta = W_\alpha W_\alpha^*$ with $\alpha = \frac{1}{\pi} \operatorname{arcth} \beta$ and the lower triangular operator W_α defined by the formula:

$$W_\alpha f = \frac{1}{\sqrt{\operatorname{ch} \alpha\pi}} \, \frac{x^{-i\alpha}}{\Gamma(i\alpha+1)} \, \frac{d}{dx} \int_0^x f(t)(x-t)^{i\alpha} \, dt .$$

The following condition is necessary for an operator S to admit the left factorization:

$$S_\xi = P_\xi S P_\xi \quad \underline{\text{is invertible in}}\ L^2(a,\xi)\ \underline{\text{for any}}\ \xi\,(a \leqslant \xi \leqslant b). \quad (*)$$

PROBLEM 1. <u>For what classes of operators condition</u> $(*)$ <u>is sufficient for the existence of the left factorization?</u>

In the general case $(*)$ is not sufficient. Indeed, the operator S defined by

$$Sf = f(x) + \frac{\operatorname{tg} \pi\beta}{\pi}\ \mathrm{v.p.} \int_0^\omega \frac{f(t)}{x-t}\,dt, \quad 0<\beta<1 \qquad (2)$$

satisfies $(*)$ but does not admit the factorization, [2]. Note that $(*)$ follows from $(**)$ defined below:

<u>Operator</u> S <u>is bounded positive and has a bounded inverse</u>. $(**)$

An important particular case of problem 1 is the following

PROBLEM 2. <u>Does</u> $(**)$ <u>imply the existence of a factorization</u>?

If $S-I \in \mathfrak{S}_\omega$ the answer is positive [1] ($\mathfrak{S}_\omega$ **is the Matsaev ideal**).

2. It is interesting to study problems 1-2 for operators of convolution type

$$Sf = \frac{d}{dx}\int_0^\omega f(t)\,s(x-t)\,dt, \qquad 0 \leqslant x \leqslant \omega. \qquad (3)$$

Let $M(x) = s(x)$, $N(x) = -s(-x)$, $0 \leqslant x \leqslant \omega$. Then [3]

$$(A_0 S - S A_0^*) f = i \int_0^\omega f(t)\,[M(x) + N(t)]\,dt, \qquad (4)$$

where $A_0 f = i \int_0^x f(t)\,dt$. If $(*)$ holds, the following matrix-functions of second order make sense:

$$W(\xi,z) = I - i \begin{bmatrix} (S_\xi^{-1} P_\xi (A_0 - zI)^{-1} M, 1), & (S_\xi^{-1} P_\xi (A_0 - zI)^{-1} M, \bar{N}) \\ (S_\xi^{-1} P_\xi (A_0 - zI)^{-1} 1, 1), & (S_\xi^{-1} P_\xi (A_0 - zI)^{-1} 1, \bar{N}) \end{bmatrix}$$

$$B(\xi)=\begin{bmatrix} (S_\xi^{-1}P_\xi M,1) & (S_\xi^{-1}P_\xi M,\bar N) \\ (S_\xi^{-1}P_\xi 1,1) & (S_\xi^{-1}P_\xi 1,\bar N) \end{bmatrix}$$

THEOREM 1. Suppose that the operator S in (3) admits the left factorization. Then the matrix-functions $\xi\mapsto W(\xi,z)$ and $\xi\mapsto B(\xi)$ are absolutely continuous and

$$\frac{dW}{d\xi}=\frac{i}{z}W(\xi,z)H(\xi), \quad H(\xi)=B'(\xi), \tag{5}$$

where the elements $h_{ij}(\xi)$ of the matrix $H(\xi)$ satisfy

$$h_{ij}(\xi)=R_i(\xi)\overline{\Pi_j(\xi)}, \tag{6}$$

and

$$R_1(\xi)\overline{\Pi_1(\xi)}+R_2(\xi)\overline{\Pi_2(\xi)}=1. \tag{7}$$

The functions R_i, Π_i can be expressed in terms of S_-, S_+:

$$\Pi_1(x)=S_+^{*-1}1,\quad \Pi_2(x)=S_+^{*-1}\bar N,\quad R_1(x)=S_-^{-1}M,\quad R_2(x)=S_-^{-1}1. \tag{8}$$

Every operator S satisfying (3) and admitting the factorization defines (via (6)-(8)) a system of differential equations (5). The procedure of this type in the inverse spectral problem have been developed by M.G.Krein [4] provided $S\geq 0$ and $I-S\in\mathfrak{S}_2$. Besides, Theorem 1 means that the "transfer matrix-function" [5] $W(\omega,z)$ admits the multiplicative representation

$$W(\omega,z)=\int_0^{\omega} e^{\frac{i}{z}dB(x)}. \tag{9}$$

If S is positive (4) implies that $W(\omega,z)$ is the characteristic matrix-function of the operator $S^{-1/2}A_0S^{1/2}$. Then formulae (5), (9) are known [6, 7]. The equality

$$\operatorname{rank} B'(x)=1, \qquad 0<x<\omega, \tag{10}$$

is new even in this case.

An immediate consequence of Theorem 1 is the necessity of the following condition for the operator in (3) to admit the factorization.

Operator S in (3) satisfies (*), the matrix-function $B(\xi)$ is absolutely continuous and (10) holds. (***)

Note that all requirements of (***) but (10) are satisfied in example (2).

PROBLEM 3. Does (***) imply the existence of the factorization?

THEOREM 2. If the operator S satisfies both (**) and (***), then it admits the factorization.

REFERENCES

1. Гохберг И.Ц., Крейн М.Г. Теория вольтерровых операторов в гильбертовом пространстве и ее приложения. М., Наука, 1967.
2. Сахнович Л.А. Факторизация операторов в $L^2(a,b)$. - Функц.анал. и его прил., 1979, 13, вып.3, 40-45.
3. Сахнович Л.А. Об интегральном уравнении с ядром, зависящим от разности аргументов. - Матем.исследования, Кишинев, 1973, 8, № 2, 138-146.
4. Крейн М.Г. Континуальные аналоги предложений о многочленах, ортогональных на единичной окружности. - Докл.АН СССР, 1955, 106, № 4, 637-640.
5. Сахнович Л.А. О факторизации передаточной оператор-функции. - Докл.АН СССР, 1976, 226, № 4.
6. Лившиц М.С. Операторы, колебания, волны. Открытые системы. М., Наука, 1966.

7. П о т а п о в В.П. Мультипликативная структура J -нерастягивающих матриц-функций. - Труды Моск.матем.о-ва, I955, 4, I25--I36.

L.A.SAHNOVICH
(Л.А.САХНОВИЧ)

СССР, 27002I, Одесса
Электротехнический институт
связи им.А.С.Попова

4.19. EVALUATION OF AN INFINITE PRODUCT OF SPECIAL MATRICES

An important rôle in studying the integrable models of Field Theory is played by matrix-functions of complex variable of a special form [1]. The simplest example is provided by a rational matrix:

$$L_0(z) = \frac{z+B}{z+\mu}, \tag{1}$$

where B is a matrix of size $n \times n$ and μ is a complex number. It is natural to call it the matrix Weirstrass factor for the complex plane $\mathbb{C}$ (i.e. a meromorphic function on $\mathbb{C}$ with one pole and such that $L(\infty) = 1$).

The next interesting example is given by a matrix Weirstrass factor for a strip. This function L_1 is meromorphic in the strip $\{z \in \mathbb{C} : 0 < \operatorname{Re} z \leqslant 1\}$ has only one pole in it and is regular at infinity, i.e.

$$L_1(z) \longrightarrow \mathcal{D}_\pm, \qquad \operatorname{Im} z \longrightarrow \pm\infty,$$

where $\mathcal{D}_\pm$ are non-degenerate diagonal matrices. The boundary values of $L_1(z)$ satisfy the following relation:

$$L_1(i\lambda+1) = A L_1(i\lambda) A^{-1}, \quad \lambda \in \mathbb{R}, \tag{2}$$

where $A = \operatorname{diag}(1, \varepsilon, \dots, \varepsilon^{n-1})$, $\varepsilon = \exp\left(\frac{2\pi i}{n}\right)$. One can represent such a matrix-function as an infinite product of functions (1). For this purpose introduce the family of matrices

$$L^m(z) = A^m L_0(z+m) A^{-m} \tag{3}$$

and their finite product

$$L_1^N(z) = L^N(z) L^{N-1}(z) \cdots L^{-N+1}(z) L^{-N}(z). \tag{4}$$

It is easy to show that the regularized limit

$$L_1(z) = \lim_{N\to\infty}{}' L_1^N(z) \tag{5}$$

satisfies (2).

For $n=1$ formulae (3)-(5) are nothing but Euler's formulae for $\sin u$, so that

$$L_1(z)=\frac{\sin\pi(z+B)}{\sin\pi(z+\mu)} .$$

We calculated $L_1(z)$ for $n=2$ in [2]. In this case $A=diag(1,-1)$ and

$$B=\begin{pmatrix} S_3 & S_- \\ S_+ & -S_3 \end{pmatrix}$$

so that $trace(B)=0$.

The limit in (5) is defined as follows

$$\lim_{N\to\infty}{}' L_1^N(z)=\lim_{N\to\infty}(A^{-N}D_N L_1^N(z) D_{-N} A^N),$$

where

$$D_N=\begin{pmatrix} N^{S_3} & 0 \\ 0 & N^{-S_3} \end{pmatrix}.$$

The limit matrix $L_1(z)$ has a form:

$$L_1(z)=W^{-1}\tilde{L}_1(z)W,$$

$$W=\begin{pmatrix} h(S_3)^{-1} & 0 \\ 0 & h(S_3) \end{pmatrix}, \quad h(z)=\sqrt{\frac{\Gamma(1+R-z)\Gamma(1-R-z)}{\pi^2(z^2-R^2)}}\, e^{-z},$$

$$\tilde{L}_1(z)=\frac{1}{\sin\pi(z+\mu)}\begin{pmatrix} \sin\pi(z+S_3), & S_-\sqrt{\dfrac{\sin^2\pi S_3-\sin^2\pi R}{\pi^2(S_3^2-R^2)}} \\ S_+\sqrt{\dfrac{\sin^2\pi S_3-\sin^2\pi R}{\pi^2(S_3^2-R^2)}}, & \sin\pi(z-S_3) \end{pmatrix},$$

where $R^2 = S_3^2 + S_+ S_-$.

We pose as a PROBLEM the explicit calculation of the limit in (5) for every n in terms of known special functions.

REFERENCES

1. F a d d e e v L. Integrable models in 1 + 1 dimensional quantum field theory. CEN-SACLAY preprint S.Ph.T./82/76.
2. Р е ш е т и х и н Н.Ю., Ф а д д е е в Л.Д. Гамильтоновы структуры для интегрируемых моделей теории поля. - Теор.Мат.Физ., I983, 57, № I.

L.D.FADDEEV
(Л.Д.ФАДДЕЕВ)
N.Yu.RESHETIHIN
(Н.Ю.РЕШЕТИХИН)

СССР, I9IOII, Ленинград,
Фонтанка 27, ЛОМИ

4.20. old

FACTORIZATION OF OPERATOR FUNCTIONS (CLASSIFICATION OF HOLOMORPHIC HILBERT SPACE BOUNDLES OVER THE RIEMANNIAN SPHERE)

Let H be a Hilbert space, $L=L(H)$ the Banach space of bounded linear operators in H, and $GL=GL(H)$ the group of invertible operators in L. We put $\mathbb{T}_+=\{z\in\mathbb{C}:|z|\leqslant 1\}$ and $\mathbb{T}_-=\{z\in\mathbb{C}\cup\{\infty\}: 1\leqslant|z|\leqslant\infty\}$ and denote by $\mathcal{O}(\mathbb{T},GL)$, $\mathcal{O}(\mathbb{T}_+,GL)$, $\mathcal{O}(\mathbb{T}_-,GL)$ the groups of holomorphic GL-valued functions in a neighborhood of $\mathbb{T},\mathbb{T}_+,\mathbb{T}_-$ respectively. We shall say that two functions S,T ($S,T\in\mathcal{O}(\mathbb{T},GL)$) are *equivalent* if $S=A_-TA_+$ for some $A_\pm, A_\pm\in\mathcal{O}(\mathbb{T}_\pm,GL)$.

PROBLEM. Classify the functions in $\mathcal{O}(\mathbb{T},GL)$ with respect to this notion of equivalence.

REMARK. It is well-known that this problem is equivalent to the classification problem for holomorphic Hilbert space bundles over the Riemannian sphere.

1. What is known about the problem? We shall say that $\mathbb{D}$ is a diagonal function if $\mathbb{D}(z)=\sum_1^n z^{\varkappa_j}P_j$, where $\varkappa_1<\dots<\varkappa_n$ are integers and $P_1,\dots,P_n$ are mutually disjoint projections in $L(H)$ such that $P_1+\dots+P_n=1_H$; the integers $\varkappa_j$ are called the *partial indices* of $\mathbb{D}$ and the dimensions $d_j\overset{def}{=}\dim P_j$ will be called the *dimensions* of the partial indices $\varkappa_j$. It is easily seen that the collection $\varkappa_1,\dots,\varkappa_n$, $d_1,\dots,d_n$ determines a diagonal function up to equivalence. For $\dim H<+\infty$ it is well-known (see, for example, [1], [2]) that every function in $\mathcal{O}(\mathbb{T},GL)$ is equivalent to a diagonal function, a result that is essentially due to G.D.Birkhoff [3]. For $\dim H=\infty$ this is not true. A first counterexample was given in [4]. We present here another counterexample: Let $H=H_1\oplus H_2$ be a decomposition of H and $V\in L(H_1,H_2)$. Then the function defined by the block matrix

$$\begin{pmatrix} z^{-1} & 0 \\ V & z \end{pmatrix} \tag{1}$$

is equivalent to a diagonal function if and only if the operator V has a closed image in H_2, as is easily verified. However there are positive results, too:

THEOREM 1 [5]. Let $A\in\mathcal{O}(\mathbb{T},GL)$. If the values $A(z)-1_H$ are compact for all z, $z\in\mathbb{T}$, then A is equivalent to a diagonal

<u>function whose non-zero partial indices have finite dimensions.</u>

For $A \in \mathcal{O}(\mathbb{T}, GL)$ we denote by W_A the Toeplitz operator defined by $W_A f = P_+(Af)$, where P_+ is the orthogonal projection from $L^2(\mathbb{T}, H)$ onto the subspace $L^2_+(\mathbb{T}, H)$ generated by the holomorphic functions on $\mathbb{T}_+$.

THEOREM 2 [4]. *A function $A \in \mathcal{O}(\mathbb{T}, GL)$ is equivalent to a diagonal function, whose non-zero partial indices $\varkappa_j$ have finite dimension d_j , if and only if W_A is a Fredholm operator in $L^2_+(\mathbb{T}, H)$. If the condition is fulfilled, then* $\dim \operatorname{Ker} W_A = \sum_{\varkappa_j < 0} \varkappa_j d_j$ *and* $\dim \operatorname{CoKer} W_A = \sum_{\varkappa_j > 0} \varkappa_j d_j$.

For further results see [4], [6], and the references in these papers.

2. *A new point of view.* In [6] a new simple proof was given for Theorem 1. The idea of this proof can be used to obtain some new results about general functions in $\mathcal{O}(\mathbb{T}, GL)$, too.

THEOREM 3 (see the proof of Lemma 1 in [6]). *Every function from $\mathcal{O}(\mathbb{T}, GL)$ is equivalent to a rational function of the form*

$$\sum_{j=-s}^{s} z^j T_j \quad , \quad T_j \in L(H).$$

Let $A \in \mathcal{O}(\mathbb{T}, GL)$. A couple $\varphi = (\varphi_-, \varphi_+)$ will be called a $\varkappa$ -section of A if φ_- , φ_+ are holomorphic H-valued functions on $\mathbb{T}_-$, $\mathbb{T}_+$, respectively, and $z^{\varkappa}\varphi_-(z) = A(z)\varphi_+(z)$ for $z \in \mathbb{T}$. Then we put $\varphi(z) = \varphi_+(z)$ for $|z| \leq 1$ and $\varphi(z) = \varphi_-(z)$ for $1 < |z| \leq \infty$. For $0 \neq x \in H$ and $0 \leq |z| \leq \infty$ we denote by $\varkappa(x, z, A)$ the smallest integer $\varkappa$ such that there exists a $\varkappa$ -section φ of A with $\varphi(z) = x$. From Theorem 3 it follows immediately that there are finite numbers m, $m > 1$, depending only on A and A^{-1} , such that $1 \leq \varkappa(x, z, A) \leq m$ for all z , $0 \leq |z| \leq \infty$, and $0 \neq x \in H$.

THEOREM 4. *For every function A , $A \in \mathcal{O}(\mathbb{T}, GL)$, there exist unique integers $\varkappa_1 < \ldots < \varkappa_n$ (the partial indices of A), unique numbers $d_1, \ldots, d_n \in \{1, 2, \ldots, \infty\}$ (the dimensions of the partial indices) and families of (not necessary closed) linear subspaces*

$$0 = M_0(z) \subsetneqq M_1(z) \subsetneqq \cdots \subsetneqq M_n(z) = H \ (0 \leq |z| \leq \infty)$$

such that

(i) $x \in M_j(z) \setminus M_{j-1}(z)$ if and only if $\varkappa(x, z, A) = \varkappa_j$ $(j = 1, \ldots, n;$ $0 \leq |z| \leq \infty)$. If φ is a $\varkappa_j$-section of A and $\varphi(z) \in M_j(z_0) \setminus M_{j-1}(z_0)$ for some point z_0, then $\varphi(z) \in M_j(z) \setminus M_{j-1}(z)$ for all $0 \leq |z| \leq \infty$. If $x_1, \ldots, x_1$ are linearly idependent vectors in H and, for some point z_0, φ_j are $\varkappa(x_j, z_0, A)$-sections of A with $\varphi_j(z_0) = x_j$, then the values $\varphi_1(z), \ldots, \varphi_n(z)$ are linearly independent for all $0 \leq |z| \leq \infty$.

(ii) The function A is equivalent to a diagonal function if and only if the spaces $M_j(z)$ $(0 \leq |z| \leq \infty;\ j = 1, \ldots, n)$ are closed. For this it is sufficient that at least for one point z_0 the spaces $M_j(z_0)$ are closed. Further, it is sufficient that the dimensions d_j are finite with the exception of one of them.

(iii) There are Hilbert spaces $H_1, \ldots, H_{n-1}$ and holomorphic operator functions $S_j^{\pm} : \mathbb{T}_{\pm} \to L(H_j, H)$ such that $M_j(z) = \operatorname{Im} S_j^{+}(z)$ for $|z| \leq 1$, $M_j(z) = \operatorname{Im} S_j^{-}(z)$ for $1 < |z| \leq \infty$ and $z^{\varkappa_j} S_j^{-}(z) = A(z) S_j^{+}(z)$.

(iv) $d_j = \dim(M_j(z)/M_{j-1}(z))$ for all $0 \leq |z| \leq \infty$ $(j = 1, \ldots, n)$ and $\dim \operatorname{Ker} W_A = \sum_{\varkappa_j < 0} \varkappa_j d_j$, $\dim(L^2_+(\mathbb{T}, H)/\operatorname{clos} \operatorname{Im} W_A) = \sum_{\varkappa_j > 0} \varkappa_j d_j$.

The proof of this theorem uses Theorem 3, the method of the proof of Lemma 2 in [6] and the open-mapping-theorem. From Theorem 4 we get a collection of invariants with respect to equivalence, the partial indices and its dimensions. However this collection does not determine the equivalence class uniquely, because, clearly, for every such collection there is a corresponding diagonal function, whereas not every function in $\mathcal{O}(\mathbb{T}, GL)$ is equivalent to a diagonal function.

It is easily seen that, for $\operatorname{Ker} V = \{0\}$ the function (1) has the partial indices $\varkappa_1 = 0$ and $\varkappa_2 = 1$ $(M_1(z) = H_1 \oplus \operatorname{Im} V$ for all z).

PROBLEM. Are all functions in $\mathcal{O}(\mathbb{T}, GL)$ with such partial indices equivalent to a function of the form (1)?

PROBLEM. Can we obtain, in general, a complete classification, adding some special triangular block matrices to the diagonal functions?

REFERENCES

1. P r ö s s d o r f S. Einige Klassen singulärer Gleichungen. - Berlin, 1974.
2. Г о х б е р г И.Ц., Ф е л ь д м а н И.А. Уравнения в свертках и проекционные методы их решения. М., "Наука", 1971.
3. B i r k h o f f G.D. Math.Ann., 1913, 74, 122-138.
4. Г о х б е р г И.Ц., Л а й т е р е р Ю. Общие теоремы о факторизации оператор-функций относительно контура I. Голоморфные функции. - Acta Sci.Math., 1973, 34, 103-120; II. Обобщения. - Acta Sci.Math., 1973, 35, 39-59.
5. Г о х б е р г И.Ц. Задача факторизации оператор-функций. - Изв. АН СССР, сер.матем., 1964, 28, № 5, 1055-1082.
6. Л а й т е р е р Ю. О факторизации матриц и оператор функций. Сообщ.АН Груз.ССР, 1977, 88, № 3, 541-544.

J.LEITERER

Akademie der Wissenschaften der DDR
Institut für Mathematik
DDR, 1030, Berlin
Mohrenstraße 39

4.21. WHEN ARE DIFFERENTIABLE FUNCTIONS DIFFERENTIABLE?

If $f : \mathbb{R} \to \mathbb{R}$ is continuous and A is a C^*-algebra then there is defined by the usual functional calculus a mapping f_A: $x \mapsto f(x)$ from the linear space of hermitian elements of A into itself.

<u>What is a necessary and sufficient condition on</u> f <u>that for all</u> A <u>the function</u> f_A <u>is differentiable everywhere</u>?

Taking $A = \mathbb{C}$ shows that f must be differentiable. In fact:

(1) <u>If</u> f_A <u>is differentiable for all</u> A <u>then</u> $f \in C^1(\mathbb{R})$.

PROOF. Let A be the algebra of bounded functions on an interval $[a, b]$. The differentiability of f_A at a function x asserts that for every ε there is a δ such that for any function h with $\|h\| < \delta$

$$\| f(x+h) - f(x) - f'_A(x) \cdot h \| \leq \varepsilon \cdot \|h\|$$

This shows immediately that $f'_A(x)$ must be the mapping $h \to (f' \circ x)h$. Let $s_0, t_0 \in [a, b]$ satisfy $|s_0 - t_0| < \delta$ and take $x(t)$ to be the identity function t and $h(t)$ the constant function $s_0 - t_0$. Then

$$\|h\| = |s_0 - t_0| < \delta$$

and $f(x+h) - f(x) - f'_A(x)h$ is equal at $t = t_0$ to

$$f(s_0) - f(t_0) - f'(t_0)(s_0 - t_0)$$

Thus $|f(s_0) - f(t_0) - f'(t_0)(s_0 - t_0)| \leq \varepsilon |s_0 - t_0|$. Interchanging s_0 and t_0, adding, and dividing by $|s_0 - t_0|$ give

$$|f'(s_0) - f'(t_0)| \leq 2\varepsilon.$$

It is even easier to show that if $f \in C^1$ and A is commutative than f_A is differentiable. For general A all I know is this:

(2) <u>If in a neighbourhood of each point of</u> $\mathbb{R}$ <u>the function</u> f <u>is equal to a function whose derivative has Fourier transform belonging to</u> $L^1(\mathbb{R})$ <u>then</u> f_A <u>is differentiable for all</u> A.

PROOF. Of course "each point" in the assumption on f can be replaced by "each compact set" and since the differentiability of f_A at x depends on the values of f in an arbitrary neighbourhood of the interval $[-\|x\|, \|x\|]$ we may assume f itself has derivative whose Fourier transform belongs to $L^1(\mathbb{R})$.

Let x and h be hermitian. From the identity

$$\frac{d}{ds} e^{is(x+h)} e^{-isx} = ie^{is(x+h)} h e^{-isx}$$

we obtain upon integrating with respect to s over $[0,t]$ and right multiplying by e^{itx}

$$e^{it(x+h)} = e^{itx} + i\int_0^t e^{is(x+h)} h e^{i(t-s)x} ds .$$

Applying the Fourier inversion formula gives

$$f(x+h) - f(x) = \int_{-\infty}^{+\infty} i\hat{f}(t)\,dt \int_0^t e^{is(x+h)} h e^{i(t-s)x} ds =$$

$$= \int_{-\infty}^{\infty} i\hat{f}(t)\,dt \int_0^t e^{isx} h e^{i(t-s)x} ds + \int_{-\infty}^{+\infty} i\hat{f}(t)\,dt \int_0^t [e^{is(x+h)} - e^{isx}] h e^{i(t-s)x} ds =$$

$$= \mathrm{I} + \mathrm{II},$$

let us say. The inner integral in I has norm at most $|t| \cdot \|h\|$ and so (since $t\hat{f}(t) \in L^1(\mathbb{R})$) the double integral makes sense and represents a continuous linear function of h. In fact it will define $f_A'(x)h$. To show this it suffices to show that the double integral II has norm $o(\|h\|)$ as $\|h\| \to 0$. But the norm of the inner integral in II is $o(\|h\|)$ for each t and is at most $2|t|\,\|h\|$ for all t and so the conclusion follows from the dominated convergence theorem. ●

PROBLEM 1. <u>Fill the gap between</u> (1) <u>and</u> (2). <u>In particular, is</u> $f \in C^1$ <u>a sufficient condition for the differentiability of</u> f_A <u>for all</u> A ?

Here is a concrete example. Let A be the algebra of bounded operators on $L^2(a, b)$. If x is M_t, multiplication by the identity function, and h is the integral operator with kernel

$K(s,t)$, then formally $f'_A(x)h$ is the integral operator with kernel

$$K(s,t)\,\frac{f(s)-f(t)}{s-t} \qquad (*)$$

(This is easily checked by a direct computation if f is a polynomial). Hence we have a concrete analogue of Problem 1:

PROBLEM 2. Find a necessary and sufficient condition on f that whenever $K(s,t)$ is the kernel of a bounded operator on $L^2(a,b)$ then so also is the kernel (*).

HAROLD WIDOM

Natural Sciences Div.
University of California
at Santa Cruz,
Santa Cruz, California, 95064
USA

EDITORS' NOTE

Both problems 1 and 2 were extensively investigated by M.Š.Birman and M.Z.Solomyak within the very general scope of their theory of double operator integrals([1], [2] and references therein; see also previous papers [3], [4]). They obtained a series of sharp sufficient conditions mentioned in Problem 2 and also sharp sufficient conditions for f to be differentiable on the set of all selfadjoint operators (Birman and Solomyak considered the Gâteaux differentiation but their techniques actually gives the existence of the Fréchet differential). Let us cite some results.

Suppose that $[a,b]\subset(0,T)$ and f can be extended from $[a,b]$ as a T-periodic function with Fourier series $\sum_{k=-\infty}^{\infty}\hat f(k)e^{\frac{2\pi i}{T}kt}$. Put $R_n(t)=\sum_{|k|\geq n}\hat f(k)e^{\frac{2\pi i}{T}k\cdot t}$. If there exists a sequence $\{\varepsilon_n\}_{n\geq 1}$ of positive numbers with $\sum\varepsilon_n<+\infty$ such that

$$\sup_{0\leqslant t\leqslant T}\ \sum_{n\geqslant 1}\varepsilon_n^{-1}\,|R_n(t)|^2<+\infty,$$

then the kernel (*) defines a bounded operator on $L^2(a,b)$ whenever $K(s,t)$ does. In particular this is the case if

$$\sum_{n=1}^{\infty} \|R_n\|_{\infty} < +\infty \quad . \tag{1}$$

Condition (1) is satisfied e.g. if f' belongs to the Hölder class Λ_α with a positive (arbitrary small) α or if f' has absolutely convergent Fourier series.

If f is defined on the whole real line and $f|[a,b]$ satisfies the above conditions for any $a,b \in \mathbb{R}$ then f is differentiable on the set of all selfadjoint operators.

The Birman - Solomyak theory encompasses many other related problems (e.g. for unbounded selfadjoint operators and for the differentiation with respect to an operator ideal). In particular they considered Problem 2 in a more general setting, namely replacing the quotient $\frac{f(s)-f(t)}{s-t}$ by a function $h(s,t)$ They reformulated this general problem as follows: for which $h(s,t)$ is $\varphi(s)\psi(t)h(s,t)$ the kernel of a nuclear operator $T_{\varphi,\psi}$ for any $\varphi,\psi \in L^2$ with $\|T_{\varphi,\psi}\|_{\mathfrak{S}_1} \le const \|\varphi\|_{L^2}\|\psi\|_{L^2}$? This equivalence leads (via V.V.Peller's criterion [5] of nuclearity of Hankel operators) to a NECESSARY CONDITION for f to satisfy the requirements of Problems 2 and 1. Indeed, putting $\varphi=\psi\equiv 1$ we see that $\frac{f(s)-f(t)}{s-t}$ should be the kernel of a nuclear operator. It follows from [5], [6] that this is the case iff f belongs to the Besov class $B_1^1[a,b]$ for any $a,b \in \mathbb{R}$. So the condition $f \in C^1$ is not sufficient in both Problems 1 and 2.

Let us mention also an earlier paper by Yu.B.Farforovskaya [7] where explicit examples of selfadjoint operators A_n, B_n with spectra in $[0,1]$ and of functions f_n are constructed such that $\|A_n - B_n\| \to 0$, $|f_n(x)-f_n(y)| \le |x-y|$ and $\frac{\|f_n(A_n)-f_n(B_n)\|}{\|A_n-B_n\|} \to \infty$. Note that the existence of such sequences $\{A_n\},\{B_n\},\{f_n\}$ follows also from the above mentioned Peller's results.

REFERENCES

I. Бирман М.Ш., Соломяк М.З. Замечания о функции спектрального сдвига. – Записки научн.семин. ЛОМИ, I972, 27, 33–46

2. Б и р м а н М.Ш. Двойные операторные интегралы Стилтьеса Ш. Предельный переход под знаком интеграла. – Проблемы мат. физики, изд. ЛГУ, I973, 6, 27–53.
3. К р е й н М.Г. О некоторых новых исследованиях по теории возмущений самосопряженных операторов. В сб.: "Первая летняя математическ. школа" I, Киев, I964, I03–I87.
4. Д а л е ц к и й Ю.Л., К р е й н С.Г. Интегрирование и дифференцирование эрмитовых операторов и приложение к теории возмущений. – Труды семин. по функц.анализу, Воронеж, I956, т.I, 8I–I05.
5. П е л л е р В.В. Операторы Ганкеля класса γ_p и их приложения (рациональная аппроксимация, гауссовские процессы, проблемы мажорации операторов).– Матем. сборник, I980,II3, № 4, 539–58I.
6. P e l l e r V.V. Vectorial Hankel operators, commutators and related operators of the Schatten-von Neumann class γ_p. – Integr. Equat. and Oper.Theory, 1982, 5, N 2, 244-272.
7. Ф а р ф о р о в с к а я Ю.Б. Оценка нормы $\|f(B)-f(A)\|$ для самосопряжённых операторов A и B. – Записки научн.семин.ЛОМИ, I976, 56, I43–I62.

4.22. old ARE MULTIPLICATION AND SHIFT UNIFORMLY ALGEBRAICALLY APPROXIMABLE ?

0. NEW DEFINITION. A family $\mathcal{A} = \{A_\omega : \omega \in \Omega\}$ of bounded operators on Hilbert space H is called *uniformly algebraically approximable* or (briefly) *approximable* if for every positive ε and for every $\omega \in \Omega$ there exists an operator $A_{\omega,\varepsilon}$ such that

a) $\sup_{\omega\in\Omega} \| A_{\omega,\varepsilon} - A_\omega \| < \varepsilon$;

b) the $*$-algebra (i.e. algebra containing B^* together with B) spanned by $\{A_{\omega,\varepsilon}\}_{\omega\in\Omega}$ is finite-dimensional *).

In particular, an operator A is called *approximable* if the family $\{A\}$ is approximable. In this case $\{\mathrm{Re}\, A, \mathrm{Im}\, A\}$ is approximable also. Given an approximable family $\mathcal{A}$ and $\varepsilon > 0$ let $\mathcal{A}^\circ_\varepsilon$ denote the algebra of the least dimension $\dim \mathcal{A}^\circ_\varepsilon$ among algebras satisfying a) and b). The function $\varepsilon \longrightarrow H(\varepsilon, \mathcal{A}) \overset{\mathrm{def}}{=\!=} \log_2 \dim \mathcal{A}^\circ_\varepsilon$ is called *the entropy growth* of $\mathcal{A}$.

1. THE MAIN PROBLEM is to obtain convenient criteria for a family of operators (in particular, for a single non-selfadjoint operator) to be approximable, and to develop functional calculus for approximable families. See concrete analytic problems in section 5.

2. KNOWN APPROXIMABLE FAMILIES. The first is $\{A\}$ with $A = A^*$. Indeed, let $A_\varepsilon = \sum_{i=1}^{n} \lambda_i (P_{\lambda_i} - P_{\lambda_{i-1}})$ where $\{\lambda_i\}_{i=1}^{n}$ forms an ε-net for the spectrum of A and $\{P_\lambda\}$ is the spectral measure of A . In this case $H(\varepsilon, A)$ coincides with the usual ε-entropy of $\mathrm{spec}\, A$ considered as a compact subset in $\mathbb{R}$.

Let $\mathcal{A} = \{A_1, \dots, A_n\}$ be a family of commuting selfadjoint operators. It is clearly approximable with $A_{i,\varepsilon}$ defined analogously. The entropy $H(\varepsilon, \mathcal{A})$ is again the ε-entropy of the joint spectrum in $\mathbb{R}^n$.

The same holds for a finite family of commuting normal operators.

Let now $\mathcal{A}$ be a finite or compact family of compact operators.

---

*) We do not require the identity in A_ε to be the identical operator on H in order to include compact operators into consideration. If the identity of A_ε is the identical operator I on H then $\mathcal{A}_\varepsilon$ does not contain compact operators and defines a decomposition $H = H_1^\varepsilon \otimes H_2^\varepsilon$ with $\dim H_2^\varepsilon < \infty$ and $A_{\omega,\varepsilon} = \{ I_1 \otimes a_{\omega,\varepsilon} : a_{\omega,\varepsilon} \in \mathcal{L}(H_2^\varepsilon),\ \omega\in\Omega \}$. In general it is convenient to consider all algebras in the Calkin algebra. (See [1] for definitions of the theory of C^*-algebras).

Then the operators A_ε can be chosen to have finite rank and therefore $\mathcal{A}$ is approximable.

Given an approximable family $\mathcal{A}'$ and a finite collection $B_1,\dots,B_n$ of compact operators consider the family $\mathcal{A}=\{\mathcal{A}', B_1,\dots,B_n\}$. Then $\mathcal{A}$ is approximable. In particular, any operator with compact imaginary part is approximable.

Let $H=\int_X \oplus H_x\,dx$ and let $\dim H_x = n$. Then $\mathcal{A} = \{A_i=\int_X \oplus A_i^x\,dx : i=1,\dots,n\}$ is approximable.

Consider $\mathcal{A}=\{P_1,P_2\}$, P_i being an orthogonal projection, $i=1,2$. This is a partial case of the previous example because there exists a decomposition $H=\int_X \oplus \mathbb{C}^2\,dx$ such that $P_i=\int_X \oplus P_i^x\,dx$ (see [2] for example).

The unilateral shift U is approximable. If $\mathcal{B}$ denotes the C^*-algebra generated by U then it contains the ideal $\mathcal{LC}(H)$ of all compact operators and $\mathcal{B}/\mathcal{LC}(H)=C(\mathbb{T})$ (cf ,e.g., [3]). It follows that U is approximable in the Calkin algebra

It was actually proved in [4] that any finite family of commuting quasi-nilpotent operators is approximable (see [5]).

3. KNOWN NON-APPROXIMABLE FAMILIES

THEOREM. If a family $\mathcal{A}=\{U_i,\ i=1,\dots,n\}$ of unitary operators is approximable then the C^*-algebra generated by $\{U_i\}$ is amenable; in particular, if $\mathcal{A}$ is a group algebra then the group is amenable

See for example [6] for the proof.

COROLLARY. If $\mathcal{A}=\{P_i\}_{i=1}^n$, $n>2$ is a family of orthogonal projections in general position then $\mathcal{A}$ is not approximable.

Indeed, set $U_i=2P_i-I$. Then $U_i=U_i^*=U_i^{-1}$ and G is a free product of n copies of $\mathbb{Z}_2$ which cannot be amenable for $n>2$.

Therefore a family of two (or more) unitary or selfadjoint operators picked up at random cannot be approximable in general. This implies that the single haphazardly choosen non-selfadjoint operator is not approximable either. The property of approximability imposes some restrictions on the structure of invariant subspaces (see the footnote to Section 1).

Consider a family $\mathcal{A}_n=\{U_1,\dots,U_n\}$ of partial isometries bound by the relation $\sum_{i=1} U_i U_i^* = I$, $n\geq 2$. Then $\mathcal{A}$ is not approximable [7] although the algebra generated by $\mathcal{A}$ is amenable [8].

Any algebra generated by an approximable family, being a subalgebra of an inductive limit of C^*-algebras of type I, is amenable as a C^*-algebra [7]. However, the class of such algebras is narrower than the class of all amenable algebras. If an approximable family generates a factor in H then it is clearly hyperfinite [6]. All that gives necessary conditions of approximability.

4. JUSTIFICATION OF THE PROBLEM. Many families of operators arising in the scope of a single analytic problem turn out to be approximable, apparently because the operators simultaneously considered in applications cannot be "too much non-commutative" (see [9], [10], problems of the perturbation theory, of representations of some non-commutative groups, etc). Besides, approximable families are the simplest non-commutative families after the finite-dimensional ones.

On the other hand an approximable family admits a developed functional calculus based on the usual routine of standard matrix theory. Indeed, functions of non-commutative elements belonging to an approximable family can be defined as the uniform limits of corresponding functions of matrices. Therefore it looks plausible that a well-defined functional calculus as well as symbols, various models and canonical forms can be defined for such a family. This in turn can be applied to the study of lattices of invariant subspaces etc. In particular, if A is an approximable non-selfadjoint operator whose spectrum contains at least two points then, apparently, it can be proved that A has a non-trivial invariant subspace.

It is known that the weak approximation, which holds for any finite family, is not sufficient to develop a substantial functional calculus for non-commuting operators. However, it is possible to consider other intermediate (between the uniform and weak) notions of approximation (see, for instance, the definition of pseudo-finite family in [6]).

5. MORE CONCRETE PROBLEMS. Our topic can be very clearly expressed by the following questions.

a) Let G be a locally compact abelian group with the dual group $\hat{G}$. Given $g \in G$ and $\varphi \in \hat{G}$ consider operators $Uf(x) = f(gx)$ and $Vf = \varphi f$ on $L^2_m(G)$. For example, for $G = \mathbb{Z}$ let

$$(Uf)_n = f_{n+1},\ (Vf)_n = e^{i\alpha n} f_n,\ \alpha \in \mathbb{T},\ f = \{f_n\} \in \ell^2$$

and for $G = \mathbb{T}$

$$Uf(e^{i\theta}) = f(e^{i(\theta+\alpha)}), \quad Vf(e^{i\theta}) = e^{i\theta}f(e^{i\theta}), \quad f \in L^2_m(\mathbb{T})$$

and finally for $G = \mathbb{R}$

$$Uf(x) = f(x+\tau), \quad Vf(x) = e^{itx}f(x), \quad f \in L^2(\mathbb{R}), \; t, \tau \in \mathbb{R}.$$

Is the pair $\{U, V\}$ approximable?

The answer to this question requires a detailed, and useful for its own sake, investigation of the Hilbert space geometry of spectral subspaces of these operators. One of the approaches reduces the problem to the following. Consider a partition of $\mathbb{T} = \bigcup_{i=1}^{n} \ell_i$ by a finite number of arcs ℓ_i. Then $L^2(\mathbb{T}) = \sum_{i=1}^{n} \oplus L^2_{\ell_i}$.

Let $H = H_{\alpha,\varepsilon}$, $\varepsilon > 0$ be the subspace of $L^2(\mathbb{T})$ consisting of functions whose Fourier coefficients may differ from zero only for integers n satisfying $|\{n\alpha\}| < \varepsilon$. Here $\{x\}$ stands for the fractional part of x and α is irrational What is the mutual position of subspaces $H_{\alpha,\varepsilon}$ and $L^2_{\ell_i}$ in $L^2_m(\mathbb{T})$, i.e. what are their stationary angles, the mutual products of the orthogonal projections etc? Since U and V satisfy $VUV^{-1}U^{-1} = e^{i\alpha}I$ (Heisenberg equation), the above question can be reformulated as follows. Is it possible to solve this equation approximately in matrices with any prescribed accuracy in the norm topology?

The shift U can be replaced by a more general dynamical system with invariant measure (X, T, μ). Then $U_T f(x) = f(Tx)$ and $V_\varphi f = \varphi f$, $\varphi \in L^\infty(X)$, $f \in L^2_\mu(X)$ Whether $\{U_T, V_\varphi\}$ is approximable or not depends essentially on properties (and not only spectral ones) of the dynamical system. The author knows no literature on the subject. Note that numerous approximation procedures existing in Ergodic Theory are useless here because it can be easily shown that the restriction of the uniform operator topology to the group of unitary operators generated by the dynamical system induces the discrete topology on the group.

Note also that if the answer is positive, some singular integral operators as well as the operators of Bishop-Halmos type [11] would turn out to be approximable which would lead to the direct proofs of the existence of invariant subspaces (see sec.4).

b) Let A be a contraction on H. Are there convenient criteria for A to be approximable expressed in terms of its unitary dilation or characteristic function?

c) Let

$$Af(x) = \int_X K(x,y)\, f(y)\, d\mu(y), \quad f \in L^2_\mu(X).$$

Find approximability criteria in terms of K.

Non-negative kernels $K \geqslant 0$ are especially interesting.

d) For what countable solvable groups G of rank 2 the regular unitary representation of G in $\ell^2(G)$ generates approximable families? For what general locally compact groups does this hold?

REFERENCES *)

1. D i x m i e r J. Les C^*-algèbres et leurs représentations Paris, Gauthier-Villard, 1969
2. H a l m o s P. Two subspaces. - Trans.Amer.Math.Soc., 1969, 144, 381-389.
3. C o b u r n L. C^*-algebras, generated by semigroups of isometries. - Trans.Amer.Math.Soc., 1969, 137, 211-217.
4. A p o s t o l C. On the norm-closure of nilpotents. III. - Rev. Roum.Math.Pures Appl., 1976, 21, N 2, 143-153.
5. A p o s t o l C., F o i a ş C., V o i c u l e s c u D. On strongly reductive algebras. - ibid., 1976, 21, N 6, 611-633.
6. В е р ш и к А.М. Счетные группы, близкие к конечным. - В кн.: Гринлиф. Инвариантное среднее на топологических группах. М., Мир, 1973 (Revised English version will be published in "Selecta Mathematica Sovietica", 1983. "Amenability and approximation of infinite groups").
7. R o s e n b e r g J. Amenability of cross products of C^*-algebras. - Commun.Math.Phys.,1977, 57, N 2, 187-191.
8. А р з у м а н я н В.А., В е р ш и к А.М. Фактор-представления скрещенного произведения коммутативной C^*- алгебры и полугруппы ее эндоморфизмов. - Докл. АН СССР, 1978, 238, № 3, 511-516.

*) M.I.Zaharevich turned my attention to [4] and A.A.Lodkin to [7].

9. S z - N a g y B , F o i a ş C. Harmonic analysis of operators on Hilbert space Amsterdam - Budapest, 1970

10. Г о х б е р г И.Ц., К р е й н М.Г. Теория вольтерровых операторов в гильбертовом пространстве и ее приложения. М., Наука,1967.

11. D a v i e A. Invariant subspaces for Bishop's operator - Bull London Math Soc , 1974, N 6, 343-348

A.M. VERSHIK
(А.М.ВЕРШИК)

СССР, 198904, Ленинград, Петродворец, Библиотечная пл., 2, Математико-механический факультет Ленинградского университета

* * *

COMMENTARY BY THE AUTHOR

During several recent years a considerable progress in the field discussed in this paper has been made, as well as new problems have arisen. We list the most important facts.

A C^*-algebra will be called an AF-algebra if it is generated by an inductive limit of finite-dimensional C^*-algebras. C^*-subalgebras of AF-algebras will be called AFI-algebras. A family of operators generating an AFI-algebra is called approximable. THE PROBLEM was to find conditions of approximability for a family of operators or one (non-self-adjoint) operator and to give quantitative characteristics of the corresponding AF-algebras, etc.

1. In [12] a positive answer to QUESTION a), Sec.5 was actually given. Namely, the approximability problem is solved for the pair of unitaries U, V.

$$(Uf)(\zeta) = f(\zeta\alpha), \qquad (Vf)(\zeta) = \zeta f(\zeta),$$

$f \in L^2(\mathbb{T})$, ζ, $\alpha \in \mathbb{T}$. This is the simplest of non-trivial cases. In [12] the authors made use of the fact that these operators are the only (up to the equivalence) solutions of the Heisenberg equation: $UVU^{-1}V^{-1} = \alpha I$.

2. In [13], [14] the approximability of an arbitrary dynamical system (i.e. of the pair (U_T, M_φ), where $(U_T f)(x) = f(Tx)$, $f \in L^2(X,\mu)$), T being an automorphism of (X,μ), $M_\varphi f = \varphi f$, $\varphi \in L^\infty$) is proved.

The result is based on a new approximation technique developed for the purposes of ergodic theory ("adic realization" of automorphisms, Markov compacta). Later in [15] conditions on a topological dynamical system (i.e. on a homeomorphism of a compactum) were found under which the skew product $C(X) \tilde{\otimes} \ell^1(\mathbb{Z})$ is an AFI-algebra.

3. These results in turn allowed to describe in some cases an important algebraic invariant of algebras generated by dynamical systems, the K-functor, cf. [12,16,17]. This makes possible to apply K-theory in ergodic theory.

Nevertheless, we still have neither general approximability criteria, nor complete information on approximable non-selfadjoint operators. Since many important families turned out to be approximable, the questions on construction of functional calculus, estimates of the norms of powers, resolvents etc are of great importance. Let us mention SOME CONCRETE QUESTIONS.

A. Is the group algebra of a discrete amenable algebra approximable?

B. Is it true that an arbitrary approximable operator has a non-trivial invariant subspace?

C. How the K-functor (as an ordered group) of an AFI-algebra may look like?

D. How are the properties of a dynamical system related to the entropy growth (i.e. the growth of dimensions of finite-dimensional subalgebras of an AF-algebra which contains the algebra generated by the dynamical system)?

REFERENCES

12. Pimsner M., Voiculescu D. Imbedding the irrational rotation C^*-algebras into an AF-algebra. - J.Oper. Theory, 1980, 4, 201-210.

13. Вершик А.М. Равномерная алгебраическая аппроксимация операторов сдвига и умножения. - Докл. АН СССР, 1981, 259, № 3, 526-529.

14. Вершик А.М. Теорема о марковской периодической аппроксимации в эргодической теории. - Зап.научн.семин.ЛОМИ, 1982, 115, 72-82.

15. P i m s n e r M. Imbedding the compact dynamical system. Preprint N 44, INCREST, 1982.
16. C o n n e s A. An analogue of the Thom isomorphism for crossed products of a C^*-algebra by an action of $\mathbb{R}$. - Adv. in Math., 1981, 39, 31-55.
17. E f f r o s E.G. Dimensions and C^*-algebras. C.B.M.S.Region. Conf.Series, N 46, AMS, Providence, 1981.

4.23. A PROBLEM ON EXTREMAL SIMILARITIES

Let T be a Hilbert space operator which is similar to an isometry S

$$T = L^{-1}SL . \tag{1}$$

Two interesting quantities associated with T are

$$k(T) = \inf\{k : \|p(T)\| \leq K\|p\|_\infty\} ,$$

where $\| \ \|_\infty$ is the sup over $\mathbb{D}$ of the polynomial p , and

$$d(T) = \inf\{\|L\|\,\|L^{-1}\| : (1) \text{ holds}\} .$$

The quantity $k(T)$ is called the k -norm of T and $d(T)$ is related to the distortion coefficient; Holbrook [3].

From (1), it follows that, for any polynomial p,

$$\|p(T)\| \leq \|L\|\,\|L^{-1}\|\,\|p\|_\infty ,$$

and therefore

$$k(T) \leq d(T) . \tag{2}$$

Since estimates on $k(T)$ yield information about functional calculus and spectral sets, this gives one reason why $d(T)$ is interesting. Another reason is the frequent occurence of the quantity $\|L\|\,\|L^{-1}\|$ (see, for example, [2 , p.248]).

One way to try to compute $d(T)$ is to characterize L satisfying (1) and

$$\|L\|\,\|L^{-1}\| = d(T) . \tag{3}$$

<u>Suppose</u> L <u>satisfies (1) and</u>

$$\|Lx\| = \lim_{n\to\infty} \|T^n x\| \tag{4}$$

<u>for all</u> x ; <u>then is</u> L <u>a similarity satisfying (3)</u>?

The answer is "yes" in two (extreme) cases: that in which $\|L\| = 1$ and that in which $\|L^{-1}\| = 1$. In fact, in the latter case $\|L\| = d(T) = k(T)$, by (2) and by

$$k(T)\|x\| \geqslant \|T^n x\| \geqslant \|Lx\| - \varepsilon \geqslant \|L\|\|x\| - 2\varepsilon$$

$$\geqslant \|L\|\|L^{-1}\|\|x\| - 2\varepsilon \geqslant d(T)\|x\| - 2\varepsilon ,$$

if n and x are chosen correctly.

If L satisfies (1), (4) and $\|L\| \leqslant 1$, the proof of (3) is similar and uses the fact that, in this case

$$d(T) = 1/[\inf_x \lim_{n\to\infty} \|T^n x\|]$$

see [1] and [4].

The inequality $\|L\| \leqslant 1$ holds, in particular, when T is a contraction. In that case, strict inequality holds in (2).

In [1], I studied $k(T)$ and $d(T)$ in connection with some recent results on similarity of Toeplitz operators. One result was that, in most cases, the similarity satisfying (1) and (4) could be computed explicitly.

REFERENCES

1. C l a r k D.N. Toeplitz operators and k-spectral sets.–Indiana U.Math.J. (to appear).
2. C o w e n M.J. and D o u g l a s R.G. Complex geometry and operator theory.–Acta Math., 1978, 141, 187-261.
3. H o l b r o o k J.A.R. Distortion coefficients for cryptocontractions.–Linear Algebra Appl., 1977, 18, 229-256.
4. S z . - N a g y B. and F o i a ş C. On contractions similar to isometries and Toeplitz operators.–Ann.Acad.Sci.Fenn.Ser.A.I., 1976, 2, 553-564.

D.N.CLARK

University of Georgia
Athens, Georgia 30602
USA

4.24. ESTIMATES OF FUNCTIONS OF HILBERT SPACE OPERATORS, SIMILARITY TO A CONTRACTION AND RELATED FUNCTION ALGEBRAS

For a given class of operators on Banach spaces we can consider the problem to estimate norms of functions of these operators. Sometimes, dealing with operators on Hilbert space, we can obtain sharper estimates than in the case of an arbitrary Banach space. A remarkable example of such a phenomenon is the following J. von Neumann's inequality:

$$\|\varphi(T)\| \leq \|\varphi\|_\infty = \sup_{|\zeta|<1} |\varphi(\zeta)|$$

for any contraction T (i.e. $\|T\| \leq 1$) on Hilbert space and for any complex polynomial φ (denote by $\mathcal{P}_A$ the set of all complex polynomials).

We consider here some other classes of operators.

<u>Operators with the growth of powers of order</u> $\alpha, \alpha \geq 0$. This class consists of operators satisfying $\|T^n\| \leq c(1+n)^\alpha$, $n \geq 0$. Clearly, for any such operator on a Banach space we have

$$\|\varphi(T)\| \leq c\|\varphi\|_{\mathcal{F}\ell^1_{(\alpha)}} \overset{def}{=} c\sum_{n\geq 0} |\hat{\varphi}(n)|(1+n)^\alpha, \quad \varphi \in \mathcal{P}_A . \tag{1}$$

It is easy to see (cf. [1]) that the fact that inequality (1) cannot be improved for Hilbert space operators is equivalent to the fact that $\mathcal{F}\ell^1_{(\alpha)}$ is an *operator algebra* (with respect to the pointwise multiplication), i.e. it is isomorphic to a subalgebra of the algebra of bounded operators on Hilbert space. It was proved by N.Th.Varopoulos [2] that this is the case if $\alpha > 1/2$ and so for $\alpha > 1/2$ (1) cannot be improved. It follows from [2], [3], [4] that $\mathcal{F}\ell^1_{(\alpha)}$ is not an operator algebra for $\alpha \leq 1/2$ and so (1) can be improved in this case. Now the PROBLEM is to <u>find sharp estimates of</u> $\|\varphi(T)\|$ <u>for</u> T <u>satisfying</u> $\|T^n\| \leq c(1+n)^\alpha$.

Note that some estimates improving (1) are obtained in [1]. This general problem apparently is very difficult. Let us consider the most interesting case $\alpha = 0$.

<u>Power bounded operators</u>. We mean operators on Hilbert space satisfying $\|T^n\| \leq c,\ n \geq 0$. It is well-known (see ref. in [1])

that such operators are not necessarily p o l y n o m i a l l y b o u n d e d (i.e. $|\varphi(T)| \le const\, \|\varphi\|_{L^\infty}, \varphi \in \mathcal{P}_A$). In [1] the following estimates of polynomials of power bounded operators are obtained

$$|\varphi(T)| \le const\, \|\varphi\|_{\mathfrak{X}} \le const\, \|\varphi\|_{VMO_A \hat{\otimes} H^1} \le const\, \|\varphi\|_{\overset{\circ}{B}_{\infty 1}} . \tag{2}$$

Here

$$\|\varphi\| = \inf \{ \|\{\gamma_{mk}\}_{m,k \ge 0}\|_{\ell^1 \check{\otimes} \ell^1} : \sum_{m+k=n} \gamma_{mk} = \hat{\varphi}(n),\ n \ge 0 \} ,$$

where $\ell^1 \check{\otimes} \ell^1$ is the injective tensor product;

$$\|\varphi\|_{VMO_A \hat{\otimes} H^1} = \inf \{ \sum_{m \ge 0} \|f_m\|_{VMO_A} \|g_m\|_{H^1} : \sum_{m \ge 0} f_m * g_m = \varphi \} ,$$

where $VMO_A = \{ f = \sum_{n \ge 0} \hat{f}(n) z^n : \hat{f}(n) = \hat{g}(n), n \ge 0$, for some $g \in C(\mathbb{T})\}$

$$\|\varphi\|_{\overset{\circ}{B}_{\infty 1}} = |\varphi(0)| + \int_0^1 \|\varphi_r'\|_\infty\, dr, \quad \varphi_r(z) \overset{def}{=} \varphi(rz) ,$$

is the norm of φ in the Besov space $\overset{\circ}{B}_{\infty 1}$.

The fact that the inequalities in (2) are precise is equivalent to the fact that the sets $\mathfrak{X}$, $VMO_A \hat{\otimes} H^1$, $\overset{\circ}{B}_{\infty 1}$ form operator algebras with respect to the pointwise multiplication. For $\overset{\circ}{B}_{\infty 1}$ this is not the case [1]. For $\mathfrak{X}$, $VMO_A \hat{\otimes} H^1$ the question is open. It is even unknown whether $\mathfrak{X}$ forms a Banach algebra.

If $VMO_A \hat{\otimes} H^1$ is an operator algebra then the norms $\|\cdot\|_{\mathfrak{X}}$ and $\|\cdot\|_{VMO_A \hat{\otimes} H^1}$ are equivalent. The question of whether this is the case can be reformulated in the following may.

Let $\mathcal{M}H^1$ be the set of all F o u r i e r m u l t i p l i e r s o f H^1 , i.e.

$$\mathcal{M}H^1 = \{ f : g \in H^1 \Longrightarrow f * g \in H^1 \} .$$

Let V^2 be the set of matrices $\{\alpha_{nk}\}_{n,k \ge 0}$ such that

$$\sup_{N > 0} \|\{\alpha_{nk}\}_{0 \le n, k \le n}\|_{\ell^\infty \hat{\otimes} \ell^\infty} < \infty ,$$

where $\ell^\infty \hat{\otimes} \ell^\infty$ is the projective tensor product.

QUESTION 1. Is it true that

$$\psi \in \mathcal{M}H^1 \iff \Gamma_\psi \in V^2 \ ?$$

Recall that $\Gamma_\psi = \{\hat{\psi}(n+k)\}_{n,k\geq 0}$ is the Hankel matrix. It is easy to show (see [1]) that $\Gamma_\psi \in V^2 \Longrightarrow \psi \in \mathcal{M}H^1$.

Similarly we can define the spaces V^M of tensors $\{\alpha_{n_1 \ldots n_M}\}_{n_j \geq 0}$ and the Hankel tensor $\overset{M}{\Gamma}_\psi = \{\hat{\psi}(n_1 + \ldots + n_M)\}_{n_j \geq 0}$.

QUESTION 2. Is it true that

$$\psi \in \mathcal{M}H^1 \Longrightarrow \overset{M}{\Gamma}_\psi \in V^M \qquad \text{and } \log \|\overset{M}{\Gamma}_\psi\|_{V^M} \leq const \cdot M \ ?$$

If Question 2 has a positive answer then $VMO_A \hat{\otimes} H^1$ is an operator algebra (see [1]) and so is $\mathcal{L}$ and the estimates $\|\varphi(T)\| \leq$ $\leq const \|\varphi\|_{\mathcal{L}} \leq const \|\varphi\|_{VMO_A \hat{\otimes} H^1}$ cannot be improved.

QUESTION 3. Is it true that

$$\Gamma_\psi \in V^2 \Longrightarrow \overset{M}{\Gamma}_\psi \in V^M \qquad \text{and } \log \|\overset{M}{\Gamma}_\psi\|_{V^M} \leq const \cdot M \ ?$$

An affirmative answer would imply that $\mathcal{L}$ is an operator algebra and the estimate $\|\varphi(T)\| \leq const \|\varphi\|_{\mathcal{L}}$ is the best possible ([1]). Moreover in this case the estimates is attained on the Davie's example (see [1]) of power bounded non polynomially bounded operator.

Similarity to a contraction. Here we touch the well-known problem (see e.g. [5]) of whether each polynomially bounded operator T on Hilbert space is similar to a contraction (i.e. whether there exists an invertible operator V such that $\|VTV^{-1}\| \leq 1$).

In [1] we considered operators R_f on $\ell^2 \oplus \ell^2$ defined by

$$R_f = \begin{pmatrix} S^* & \Gamma_f \\ \mathbb{O} & S \end{pmatrix},$$

where S is the shift operator on ℓ^2. It was proved in [1] that R_f is power bounded iff f belongs to the Zygmund class Λ_1, i.e. $|f''(\zeta)| \leq const \frac{1}{1-|\zeta|}$, $|\zeta| < 1$. It was also shown in [1] that among

R_f there are many **power bounded operators, non polynomially bounded.** It seems reasonable to try to construct a counterexample to the problem stated above on the class of the operators R_f . It is very easy to calculate **the functions** of R_f. Namely,

$$\varphi(R_f)=\begin{pmatrix} \varphi(S^*) & \Gamma_{\varphi'(S^*)f} \\ \mathbb{O} & \varphi(S) \end{pmatrix}.$$

THEOREM. <u>If</u> $f'\in BMO_A$ <u>then</u> R_f <u>is polynomially bounded</u> (Recall that $BMO_A=\{f=\sum_{n\geq 0}\hat{f}(n)z^n : \hat{f}(n)=\hat{g}(n),\ n\geq 0$, for some g in $L^\infty\}$).

PROOF. By Nehari's theorem (see [6]) $\|\Gamma_{\varphi'(S^*)f}\|\asymp\|\varphi'(S^*)f\|_{BMO_A}$. We have

$$\|\varphi'(S^*)f\|_{BMO_A}=\sup_{\|g\|_{H^1}\leq 1}|(\varphi'(S^*)f,g)|=\sup_{\|g\|_{H^1}\leq 1}|(f,\varphi'g)| .$$

Now $\varphi'g=(\varphi g)'-\varphi g'$. To finish the proof we use the fact that

$$F\in H^1\Longleftrightarrow\int_{-\pi}^{\pi}\Big(\int_0^1|F'(re^{i\theta})|^2(1-r)\,dr\Big)^{1/2}d\theta$$

(see [7]).

It follows that $\int_{-\pi}^{\pi}\Big(\int_0^1|\varphi'g(re^{i\theta})|^2(1-r)\,dr\Big)^{1/2}d\theta<\infty$ and

$$|(f,\varphi'g)|\leq const\,(\|f'\|_{BMO}+|f(0)|)\int_{-\pi}^{\pi}\Big(\int_0^1|\varphi'g(re^{i\theta})|^2(1-r)\,dr\Big)^{1/2}d\theta . \quad\bullet$$

QUESTION 4. <u>Is it true that if</u> R_f <u>is polynomially bounded then</u> $f'\in BMO_A$?

This question is related to a question of R.Rochberg [8] **concerning Hankel operators.**

QUESTION 5. <u>Does there exist</u> f <u>with</u> $f'\in BMO_A$ <u>such that</u> R_f <u>is similar to no contraction</u>?

<u>Operators with the growth of resolvents of order</u> α . We consider here the operators satisfying $\|R(\lambda,T)\|\leq\frac{c}{(|\lambda|-1)^\alpha}$, $|\lambda|>1$, $\alpha\geq 1$. It is not difficult to prove that for any such operator on a Banach space we have

$$|\varphi(T)| \leq const \, \|\varphi\|_{B_1^\alpha} , \qquad (3)$$

where the Besov space B_1^α consists of the functions f satisfying

$$\iint_{\{|z|<1\}} |f^{(n)}(z)|(1-|z|)^{n-\alpha-1}\, dx\, dy < \infty ,$$

n being an integer greater than α.

Inequality (3) is the best possible on the class of all Banach spaces. (It is enough to consider multiplication by z on B_1^α). The fact that (3) is the best possible for Hilbert space operators is equivalent to the fact that the algebra B_1^α is an operator algebra.

Consider the following operators on a commutative Banach algebra A

$$\mathcal{H}_\varphi : A \longrightarrow A^*, \ (\mathcal{H}_\varphi x, y) \stackrel{def}{=} (\varphi, xy), \ \ \varphi \in A^* .$$

It is proved by A.M.Tonge [9] that if all operators $\mathcal{H}_\varphi$ are 2-absolutelly summing (see definition in S.1 of this book) then A is an operator algebra and by P.Charpentier [4] that if A is an operator algebra then all operators $\mathcal{H}_\varphi$ can be factored through Hilbert space.

In the case of B_1^α the operators $\mathcal{H}_\varphi$ look as follows

$$\mathcal{H}_\varphi : B_1^\alpha \longrightarrow B_\infty^{-\alpha} , \quad \mathcal{H}_\varphi x = \mathbb{P}_+ \varphi \tilde{x} , \ \varphi \in B_\infty^{-\alpha} ,$$

where $B_\infty^{-\alpha} = \{ f : |f(\zeta)| \leq const\,(1-|\zeta|)^{-\alpha} , \ |\zeta|<1 \}$,

$\tilde{x}(\zeta) = x(\bar{\zeta}), \ \zeta \in \mathbb{T}, \ \ \mathbb{P}_+ f = \sum_{n \geq 0} \hat{f}(n) z^n .$

The space B_1^α is isomorphic to ℓ^1 (see [10]). It follows from Grothendiek's theorem (see [11]) that operators on ℓ^1 factored through Hilbert space are 2-absolutely summing.

QUESTION 6. <u>Is it true that for any</u> $\varphi \in B_\infty^{-\alpha}$ <u>the operator</u> $\mathcal{H}_\varphi : B_1^\alpha \longrightarrow B_\infty^{-\alpha}$ <u>is</u> 2-<u>absolutely summing</u>?

The answer is positive if and only if B_1^α is an operator algebra which is equivalent to the fact that (3) cannot be improved. If the answer is negative, <u>find estimates sharper than</u> (3). Even the QUESTION <u>of estimating</u> $|T^n|$ <u>is not yet solved</u>. It follows from (3)

that $\|T^n\| \leq c(1+n)^{\alpha}$. This question for $\alpha = 1$ was considered in [12]. The best example I know is due to S.N.Naboko (unpublished), and another one is due to J.A. van Casteren [13]. They constructed for any $\beta < 1/2$ weighted shifts T such that $\|R(\lambda,T)\| \leq \frac{c}{|\lambda|-1}$ and $\|T^n\| \geq c(1+n)^{\beta}$.

REFERENCES

1. P e l l e r V.V. Estimates of functions of power bounded operators on Hilbert spaces. - J.Oper.Theory, 1982, 7, N 2, 341-372.
2. V a r o p o u l o s N.Th. Some remarks on Q-algebras. - Ann. Inst.Fourier (Grenoble), 1972, 22, 1-11.
3. V a r o p o u l o s N.Th. Sur les quotiens des algèbres uniformes. - C.R.Acad.Sci.Paris, 1972, 274, A 1344-1346.
4. C h a r p e n t i e r P. Q-algèbres et produits tensoriels topologiques. Thèse, Orsay, 1973.
5. H a l m o s P. Ten problems in Hilbert space. - Bull.Amer.Math Soc., 1970, 76, N 5, 887-933.
6. S a r a s o n D. Function theory on the unit circle. Notes for lectures at Virginia Polytechnic Inst. and St.Univ., Blacksburg, 1978.
7. F e f f e r m a n Ch., S t e i n E.M. H^p spaces of several variables. - Acta Math., 1972, 129, 137-193.
8. R o c h b e r g R. A Hankel type operator arising in deformation theory. - Proc.Sympos.Pure Math., 1979, 35, N 1, 457-458.
9. T o n g e A.M. Banach algebra and absolutely summing operators. - Math.Proc.Camb.Phil.Soc., 1976, 80, 465-473
10. L i n d e n s t r a u s s J., P e ł c z y n ' s k i A. Contribution to the theory of classical Banach spaces. - J.Funct. Anal., 1971, 8, 225-249.
11. L i n d e n s t r a u s s J., P e ł c z y n ' s k i A. Absolutely summing operators in $\mathcal{L}_p$-spaces and their applications - Studia Math., 1968, 29, N 3, 275-326.
12. S h i e l d s A.L. On Möbius bounded operators. Acta Sci.Math., 1978, 40, N 3-4, 371-374.
13. v a n C a s t e r e n J.A. Operators similar to unitary and self-adjoint ones. - Pacif.J.Math., 1983, 104, N 1, 241-255

V.V. PELLER
(В.В.ПЕЛЛЕР)

СССР, 191011, Ленинград
Фонтанка 27, ЛОМИ

4.25. old ESTIMATES OF OPERATOR POLYNOMIALS ON THE SCHATTEN-VON NEUMANN CLASSES

Precise estimates of functions of operators is an essential part of general spectral theory of operators. In the case of Hilbert space one of the best known and most important inequalities of this type is von Neumann's inequality (cf. [1]):

$$\|\varphi(T)\| \leqslant \max\{|\varphi(\zeta)| : |\zeta| \leqslant 1\}$$

for any contraction T (i.e. $\|T\| \leqslant 1$) on a Hilbert space and for any complex polynomial φ ($\mathcal{P}_A$ is the set of such polynomials). For a Banach space X an explicit calculation of the norm

$$|\!|\!|\varphi|\!|\!|_X \overset{def}{=} \sup\{\|\varphi(T)\| : T \text{ is a contraction on } X\}, \varphi \in \mathcal{P}_A,$$

would be an analogue of von Neumann's inequality.

In 1966 V.I.Matsaev conjectured that for infinite-dimensional L^p -spaces, $|\!|\!|\cdot|\!|\!|_{L^p}$ coincides with the p-multiplier norm. This means that for any contraction T on L^p, $\|\varphi(T)\|_{L^p} \leqslant |\varphi|_p \overset{def}{=} \|\varphi(S)\|_{\ell^p}$, where S is the shift operator on ℓ^p : $S(\lambda_0, \lambda_1, \dots) = (0, \lambda_0, \lambda_1, \dots)$. This inequality was proved in [2] for absolute contractions T (i.e. $\|T\|_{L^1} \leqslant 1, \|T\|_{L^\infty} \leqslant 1$) and in [3], [5] (independently) for operators T having a contractive majorant (for dominated contractions), i.e. for such T that there exists a positive contraction $\tilde{T}$ on L^p satisfying $|Tf| \leqslant \tilde{T}|f|$ a.e. for $f \in L^p$ (See also a survey article [4]).

Let $\mathfrak{S}_p = \mathfrak{S}_p(\mathcal{H})$ be the space of compact operators a on an infinite dimensional Hilbert space $\mathcal{H}$ with $\|a\|_{\mathfrak{S}_p} \overset{def}{=} (\mathrm{trace}(a^*a)^{p/2})^{1/p} < \infty$. The dual of $\mathfrak{S}_p$, $1 < p < \infty$, can be identified with $\mathfrak{S}_{p'}$ with respect to the duality $(a, b) = \mathrm{trace}\, ab$, $a \in \mathfrak{S}_p$, $b \in \mathfrak{S}_{p'}$. We are interested in the $\mathfrak{S}_p$ -version of von Neumann's inequality. Let $\mathcal{K} = \ell^2(\mathcal{H})$ and s be the shift operator on $\mathcal{K}$ $(s(x_0, x_1, \dots) = (\mathbb{0}, x_0, x_1, \dots)$, $x_i \in \mathcal{H}$, and s^* be the adjoint operator. The operator S on $\mathfrak{S}_p(\mathcal{K})$ is defined by $Sa = sas^*$, $a \in \mathfrak{S}_p(\mathcal{K})$. This operator seems to play the rôle similar to that of the shift on ℓ^p. Let us introduce the notation $|\varphi|_{\mathfrak{S}_p} \overset{def}{=} \|\varphi(S)\|_{\mathfrak{S}_p}$.

CONJECTURE 1. $|\!|\!|\varphi|\!|\!|_{\mathfrak{S}_p} = |\varphi|_{\mathfrak{S}_p}$. <u>In other words for any contraction</u> T <u>on</u> $\mathfrak{S}_p$

$$\|\varphi(T)\|_{\mathfrak{S}_p} \leqslant |\varphi|_{\mathfrak{S}_p}, \quad \varphi \in \mathcal{P}_A. \tag{1}$$

The conjecture is true for $p = 1, 2, \infty$.

PROPOSITION 1. <u>Inequality (1) holds for isometries (not necessarily invertible) on the space</u> $\mathcal{X}$.

perhaps it would be useful to apply a known description of the set of $\mathcal{V}_p$ -isometries $\mathcal{U}$ (cf. [6]). It is known that each contraction on a Hilbert space has a unitary dilation on a Hilbert space [1] and can be approximated by unitary operators in the pw-topology (cf. [4]). The set of operators on $L^p[0,1]$, $p \neq 2$, having a unitary dilation on an L^p-space coincides with the closure in the pw-topology of the set of unitary operators and coincides with the set of operators having a contractive majorant (cf. [3], [4]). (Earlier for positive contractions the existence of unitary dilations was established in [7]).

QUESTIONS. Is it true that 1) any $\mathcal{V}_p$ -contraction has an isometric dilation? 2) any absolute $\mathcal{V}_p$ -contraction (i.e. a contraction on $\mathcal{V}_1$ and on $\mathcal{V}_\infty$) has an isometric dilation? 3) $pw - \operatorname{clos} \mathcal{U}$ coincides with the set of all contractions on $\mathcal{V}_p$? 4) $pw-\operatorname{clos} \mathcal{U}$ contains the set of absolute contractions on $\mathcal{V}_p$?

The affirmative answer to 1) or 3) would imply the validity of Conjecture 1. If Conjecture 2 is also valid then this would imply the validity of V.I.Matsaev's conjecture because ℓ^p can be isometrically imbedded into $\mathcal{V}_p$ in such a way that there exists a contractive projection onto its image. In conclusion let us indicate a class of operators satisfying (1).

PROPOSITION 2. Let a, b be contractions on $\mathcal{H}$, $Tc = a c b$. Then the operator T on $\mathcal{V}_p$ has an isometric dilation and can be approximated by isometries.

This follows from the fact that a and b have a unitary dilation on a Hilbert space and can be pw-approximated by unitary operators on $\mathcal{H}$.

REFERENCES

1. S z . - N a g y B., F o i a ş C. Harmonic analysis of operators on Hilbert space. North Holland - Akademiai Kiado, Amsterdam-Budapest, 1970.

2. П е л л е р В.В. Аналог неравенства Дж.фон Неймана для пространства L^p. - Докл.АН СССР, 1976, 231, № 3, 539-542.

3. P e l l e r V.V. L'inégalité de von Neumann, la dilation isométrique et l'approximation par isométries dans L^p . - C.R.Acad. Sci.Paris, 1978, 287, N 5, A 311-314.

4. П е л л е р В.В. Аналог неравенства Дж.фон Неймана, изометрическая дилатация сжатий и аппроксимация изометриями в пространствах измеримых функций. - Труды МИАН, I98I, I55, I03-I50.

5. C o i f m a n R.R., R o c h b e r g R., W e i s s G. Applications of transference: The L^p version of von Neumann's inequality and the Littlewood-Paley-Stein theory. - Proc.Conf. Math.Res.Inst.Oberwolfach, Intern.ser.Numer.Math., v.40, 53-63. Birkhäuser, Basel, 1978.

6. A r a z y J., F r i e d m a n J. The isometries of $C_p^{n,m}$ into C_p. - Isr.J.Math., 1977, 26, N 2, 151-165.

7. A k c o g l u M.A., S u c h e s t o n L. Dilations of positive contractions on L^p spaces. - Canad.Math.Bull., 1977, 20, N 3, 285-292.

V.V.PELLER

(В.В.ПЕЛЛЕР)

СССР, I9I0II, Ленинград,
Фонтанка, 27, ЛОМИ

4.26. A QUESTION IN CONNECTION WITH MATSAEV'S CONJECTURE

This problem is closely related to the preceding one [1], where definitions of all notions used here can be found. I propose to verify Matsaev's conjecture for the operator T on the 2-dimensional space ℓ_2^p, $1 < p < 2$ defined by the matrix

$$\frac{1}{\sqrt[p]{2}} \begin{pmatrix} 1 & -1 \\ 1 & 1 \end{pmatrix}$$

in the standard basis of ℓ_2^p. The contraction T is of interest because it has some resemblance with unitary operators on 2-dimensional Hilbert space and, as is well-known, the abundance of unitary operators plays a decisive role in the proof of von Neumann's inequality. Moreover T has no contractive majorant and so the results of Akcoglu and Peller cannot be applied to it. Thus it seems plausible that the validity of Matsaev's conjecture in general case depends essentially on the answer to the following

QUESTION. Is Matsaev's conjecture true for T ?

REFERENCES

1. Peller V.V. Estimates of operator polynomials on the Schatten - von Neumann classes. - This "Collection", Problem 4.25

A.K.KITOVER
(А.К.КИТОВЕР)

СССР, I9III9, Ленинград,
ул.Константина Заслонова I4, кв.2.

4.27. TO WHAT EXTENT CAN THE SPECTRUM OF AN OPERATOR BE DIMINISHED UNDER AN EXTENSION?

Let X be a Banach space, T be a bounded linear operator on X (i.e. $T\in\mathcal{B}(X)$).

QUESTION 1. <u>Are there a Banach space</u> Y <u>containing</u> X <u>and an operator</u> $S\in\mathcal{B}(Y)$ <u>such that</u> $T=S|X$ <u>and the essential spectrum of</u> T <u>is exactly the spectrum*) of</u> S ?

A stronger version of Question 1 is

QUESTION 2. <u>Given</u> $X, T\in\mathcal{B}(X)$, <u>can one find a Banach space</u> Y <u>containing</u> X <u>and an isometrical algebra homomorphism</u> $\varphi:\mathcal{B}(X)\to\mathcal{B}(Y)$ <u>such that</u> $\varphi(S)|X=S\ \forall S\in\mathcal{B}(X)$ <u>and the spectrum of</u> $\varphi(T)$ <u>is exactly the essential spectrum of</u> T ?

B. BOLLOBÁS

Department of Pure Mathematics
and Mathematical Statistics
University of Cambridge
16 Mill Lane, Cambridge
CB2 1SB, England

EDITORS' NOTE. Related questions are discussed in S.3.

---

*) Apparently, here the essential spectrum of S is the set of $\lambda\in\mathbb{C}$ such that $\|(S-\lambda I)x_n\|\to 0$ for a sequence $\{x_n\}$ in X with $\|x_n\|=1$ - Ed.

4.28. THE DECOMPOSITION OF RIESZ OPERATORS
old

If X is a Banach space, let $B(X)$, $K(X)$ and $Q(X)$ denote the sets of bounded, compact and quasi-nilpotent operators on X (respectively). T, $T \in B(X)$, is a Riesz operator if it has a Riesz spectral theory associated with the compact operators, i.e. the spectrum of T is an at most countable set whose only possible accumulation point is the origin and all of whose non-zero points are poles of the resolvent of finite rank. The set of Riesz operators is denoted by $R(X)$.

Ruston [1] characterised the Riesz operators as $T \in R(X) \Longleftrightarrow$ the coset $T + K(X)$ is a quasi-nilpotent element of the Calkin algebra $B(X)/K(X)$. Clearly $R(X) \supseteq K(X) + Q(X)$. West [2] proved that if X is a Hilbert space then $R(X) = K(X) + Q(X)$. This decomposition has been generalised to a C^*-algebra setting by Smyth [3]. The proof is analogous to the superdiagonalisation of a matrix which is then written as the sum of a diagonal matrix and a superdiagonal matrix with a zero diagonal.

Nothing is known about the decomposition problem in general Banach spaces. It may be that the decomposability of all Riesz operators characterizes Hilbert spaces up to equivalence among Banach spaces.

REFERENCES

1. Ruston A.F. Operators with a Fredholm theory. - J.London Math.Soc., 1954, 29, 318-326.
2. West T.T. The decomposition of Riesz operators. - Proc.London Math.Soc., III Series, 1966, 16, 737-752.
3. Smyth M.R.F. Riesz theory in Banach algebras. - Math.Zeit., 1975, 145, 145-155.

M.R.F.SMYTH
T.T.WEST

39 Trinity college
Dublin 2
Ireland

4.29. OPERATOR ALGEBRAS IN WHICH ALL FREDHOLM OPERATORS ARE INVERTIBLE

Let $\mathcal{L}(X)$ be the algebra of all bounded linear operators in the Banach space X. An operator $A \in \mathcal{L}(X)$ is called a Fredholm operator if $\dim \ker A < \infty$ and $\dim X/\operatorname{Im} A < \infty$.

It is well known that the operator of multiplication by a function (in L^p or C) is invertible if it is a Fredholm operator. The same is true for multidimensional Wiener-Hopf operators in $L^p(\mathbb{R}^n_+)$, for their discrete analogues in $\ell^p(\mathbb{Z}^n_+)$ and for operators in $L^p(\mathbb{T})$ of the form

$$(A\varphi)(t) = a(t)\varphi(t) + b(t)\varphi(\alpha(t)) \qquad (a, b \in L^\infty(\mathbb{T})),$$

where α is a homeomorphism of the circle $\mathbb{T}$ onto itself. This property is valid for the elements of uniformly closed algebras *) generated by the above operators as well (see [1] and the literature cited therein). The usual scheme of the proof consists of two stages. First we prove the invertibility of Fredholm operators in the non-closed algebra $\mathfrak{A}$ generated by the initial operators (using the linear expansion [2] it is reduced to the same operators but with matrix coefficients or kernels [3]). Then we have to extend this statement in some way to the uniform closure $\operatorname{clos} \mathfrak{A}$ of the algebra $\mathfrak{A}$.

QUESTION 1. Let every Fredholm operator from the algebra $\mathfrak{A}$ $(\subset \mathcal{L}(X))$ be invertible. Is every Fredholm operator in the algebra $\operatorname{clos} \mathfrak{A}$ invertible?

In the examples the passage from $\mathfrak{A}$ to $\operatorname{clos} \mathfrak{A}$ becomes easier if $A^{-1} \in \operatorname{clos} \mathfrak{A}$ for each invertible operator $A \in \mathfrak{A}$.

QUESTION 2. Let every Fredholm operator $A \in \mathfrak{A}$ be invertible, and let $A^{-1} \in \operatorname{clos} \mathfrak{A}$. Is every Fredholm operator in the algebra $\operatorname{clos} \mathfrak{A}$ invertible?

We point out two cases, when the answer to Question 1 is positive [1].

---

*) It is supposed that all algebras under consideration contain the identity operator.

1°. Algebra $\mathfrak{A}$ is commutative (or $\mathfrak{A} \subset \mathcal{L}(X^n)$ and $\mathfrak{A}$ consists of operator matrices with elements in some commutative algebra $\mathfrak{A}_0 \subset \mathcal{L}(X)$).

2°. X is a Hilbert space and $\mathfrak{A}$ is a symmetric algebra.

The answer to Question 2 is positive if one of the following conditions is satisfied (see [1]).

3°. The algebra $\operatorname{clos} \mathfrak{A}$ is semi-simple.

4°. The system of minimal invariant subspaces of the algebra $\mathfrak{A}$ is complete in X .

5°. The algebra $\operatorname{clos} \mathfrak{A}$ does not contain nil-ideals consisting of finite-dimensional operators.

We call a non-zero invariant subspace minimal if it doesn't contain other such subspaces. A two-sided ideal is called a nil-ideal if all its elements are nilpotent.

Either of the conditions 3°, 4° implies condition 5°. For 3° this is obvious and for 4° this follows from [4] (comp with [5]).

REFERENCES

I. Крупник Н.Я., Фельдман И.А. Об обратимости некоторых фредгольмовых операторов. - Изв.АН МССР, сер.физ.-техн. и мат.наук, I982, № 2, 8-I4.

2. Гохберг И.Ц., Крупник Н.Я. Сингулярные интегральные операторы с кусочно-непрерывными коэффициентами и их символы. - Изв.АН СССР, сер.матем., I97I, 35, № 4, с.940-964.

3. Крупник Н.Я., Фельдман И.А. О невозможности введения матричного символа на некоторых алгебрах операторов. - В кн.: Линейные операторы и интегральные уравнения. Кишинёв, Штиинца, I98I, 75-85.

4. Ломоносов В.И. Об инвариантных подпространствах семейства операторов, коммутирующих с вполне непрерывным. - Функц.анализ и его прилож., I973, 7, вып.3, 55-56.

5. Маркус А.С., Фельдман И.А. Об алгебрах, порожденных односторонне обратимыми операторами. - В кн.: Исследования по дифференциальным уравнениям. Кишинёв, Штиинца, I983, 42-46.

N.Ya.KRUPNIK (Н.Я.КРУПНИК)
СССР, 277003, Кишинёв, Кишинёвский государственный университет

A.S.MARKUS (А.С.МАРКУС)
I.A.FEL'DMAN (И.А.ФЕЛЬДМАН)
СССР, 277028, Кишинёв, Институт математики АН МССР

4.30. ON THE CONNECTION BETWEEN THE INDICES OF AN OPERATOR MATRIX AND ITS DETERMINANT

Let H be a Hilbert space, and $\mathcal{L}(H)$ be the algebra of all linear bounded operators in H. An operator $A \in \mathcal{L}(H)$ is called a Fredholm operator if $\dim \operatorname{Ker} A < \infty$ and $\dim H/\operatorname{Im} A < \infty$. The number $\operatorname{ind} A = \dim \operatorname{Ker} A - \dim H/\operatorname{Im} A$ is called the index of A.

If H^n is the orthogonal sum of n copies of the space H, then any operator $\mathcal{A} \in \mathcal{L}(H^n)$ can be represented in the form of an operator matrix $\{A_{jk}\}_{j,k=1}^{n}$ $(A_{jk} \in \mathcal{L}(H))$.

Set $[A,B] = AB - BA$. Let $\mathcal{A} = \{A_{jk}\}$ and suppose all commutators $[A_{jk}, A_{j'k'}]$ to be compact. Define the determinant $\det \mathcal{A}$ in the usual way. The order of the factors A_{jk} in each term is of no importance in this connection, since various possible results differ from each other by compact summand. N. Ja. Krupnik showed ([1], see also [2], p. 195) that $\mathcal{A}$ is a Fredholm operator if and only if det $\mathcal{A}$ is. On the other hand it is known that under these conditions the equality

$$\operatorname{ind} \mathcal{A} = \operatorname{ind} \det \mathcal{A} \tag{1}$$

does not hold in general (see an example below).

In [3] it was stated that the equality (1) holds if the condition of compactness of commutators is replaced by the condition of their nuclearity.

The question of preciseness of conditions $[A_{jk}, A_{j'k'}] \in \mathfrak{S}_1$ arises naturally.

CONJECTURE 1. Let $\mathfrak{S}$ be any symmetrically-normed ideal (see [4]) of the algebra $\mathcal{L}(H)$, different from $\mathfrak{S}_1$. There exists a Fredholm operator $\mathcal{A} = \{A_{jk}\}$, such that $[A_{jk}, A_{j'k'}] \in \mathfrak{S}$ but (1) does not hold.

The weaker conjecture given below is also of some interest.

CONJECTURE 2. For any $p > 1$ there exists a Fredholm operator $\mathcal{A} = \{A_{jk}\}$ such that $[A_{jk}, A_{j'k'}] \in \mathfrak{S}_p$ but (1) does not hold.

Note that in the example below Conjecture 2 is confirmed only for $p > 2$.

EXAMPLE. Let $H = L^2(S^2)$ where S^2 is the two-dimensional sphere. There exist singular integral operators $A_{jk} \in \mathcal{L}(H)$ such that

$$\mathcal{A} = \begin{pmatrix} A_{11} & A_{12} \\ A_{21} & A_{22} \end{pmatrix}$$

is a Fredholm operator and $\operatorname{ind} \mathcal{A} = 1$ ([5], Ch.XIV, §4). As $\operatorname{ind} \det \mathcal{A} = 0$ ([5], Ch.XIII, theorem 3.2) equality (1) does not hold. It can be assumed that the symbols of the operators A_{jk} are infinitely smooth (for $\xi \neq 0$) and therefore the commutators $[A_{jk}, A_{j'k'}]$ map $L^2(S^2)$ into $W_1^2(S^2)$ ([6], theorem 3). Hence $s_m([A_{jk}, A_{j'k'}]) = O(m^{-1/2})$ (see e.g. [7]).

We note in conclusion that some conditions sufficient for the validity of (1) have been found in [8-10]. In these papers as well as in [1], [3] the operators in Banach spaces are considered .

REFERENCES

1. К р у п н и к Н.Я. К вопросу о нормальной разрешимости и индексе сингулярных интегральных уравнений. - Уч.зап.Кишиневского университета, 1965, 82, 3-7.
2. Г о х б е р г И.Ц., Ф е л ь д м а н И.А. Уравнения в свертках и проекционные методы их решения. М., Наука, 1971.
3. М а р к у с А.С., Ф е л ь д м а н И.А. Об индексе операторной матрицы. - Функц.анал. и его прил., 1977, II, № 2, 83-84.
4. Г о х б е р г И.Ц., К р е й н М.Г. Введение в теорию линейных несамосопряженных операторов в гильбертовом пространстве. М., Наука, 1965.
5. M i c h l i n S.G. , P r ö s s d o r f S. Singuläre Integraloperatoren. Berlin: Akademie - Verlag, 1980.
6. S e e l e y R.T. Singular integrals on compact manifolds. - Amer.J.Math., 1959, 81, 658-690.
7. П а р а с к а В.И. Об асимптотике собственных и сингулярных чисел линейных операторов, повышающих гладкость. - Матем.сборник, 1965, 68 (110), 623-631.

8. К р у п н и к Н.Я. Некоторые общие вопросы теории одномерных сингулярных операторов с матричными коэффициентами. – В кн.: Несамосопряженные операторы. Кишинев, Штиинца, 1976, 91–112.

9. К р у п н и к Н.Я. Условия существования n-символа и достаточного набора n-мерных представлений банаховой алгебры. – В кн.: Линейные операторы. Кишинев, Штиинца, 1980, 84–97.

10. В а с и л е в с к и й Н.Л., Т р у х и л ь о Р. К теории Ф-операторов в матричных алгебрах операторов. В кн.: Линейные операторы. Кишинев, Штиинца, 1980, 3–15.

I.A.FEL'DMAN
(И.А.ФЕЛЬДМАН)
A.S.MARKUS
(А.С.МАРКУС)

СССР, 277028, Кишинев,
Институт математики АН МССР

4.31. SOME PROBLEMS ON COMPACT OPERATORS WITH POWER-LIKE BEHAVIOUR OF SINGULAR NUMBERS

Classes of compact operators with power-like behaviour of eigenvalues and singular numbers arise quite naturally in studying spectral asymptotics for differential and pseudodifferential operators. Presented are three problems related to the theory of such classes.

Let $\mathcal{B}(H)$ be the algebra of all bounded operators on a Hilbert space H. Given A in the ideal C of all compact operators in $\mathcal{B}$ define $s_n(A)$, $n=1,2,\dots$, the singular numbers of A. For $0<p<\infty$ let

$$\Sigma_p=\{A\in C:\ \|A\|_p \overset{def}{=} \sup_n (n^{1/p}\cdot s_n(A))<\infty\},$$

$$\Sigma_p^\circ=\{A\in\Sigma_p:\ s_n(A)=o(n^{-1/p})\}.$$

See [1-4] for details concerning Σ_p -spaces.

While studying spectral asymptotics the main interest is focused not on the spaces Σ_p, Σ_p° themselves, but on the quotient spaces

$$\sigma_p=\Sigma_p/\Sigma_p^\circ.$$

The spaces σ_p, $0<p<\infty$ (for details see [5]) are complete and non-separable with respect to the quasi-norm $|a|_p = \limsup\{n^{1/p}s_n(A)\}$, $a=A+\Sigma_p^\circ$. The natural limit case of σ_p -spaces is the Calkin algebra $\sigma_\infty=\mathcal{B}/C$.

The multiplication of operators induces the multiplication of elements $a\in\sigma_p$, $b\in\sigma_q$, $0<p,q\leqslant\infty$. The product belongs to the class σ_r, $r^{-1}=p^{-1}+q^{-1}$. Taking adjoints of operators induces the involution $a\mapsto a^*$ in σ_p -spaces. So one can consider commuting classes, self-adjoint classes, normal classes, etc.

PROBLEM 1. <u>Let</u> $a\in\sigma_p$, $a^*a=aa^*$. <u>Is there a normal operator in the class</u> a ?

It is known (see [6]), that the answer is negative if $p=\infty$. It is due to the fact that in the σ_∞ -space there is the Index, i.e. nontrivial homomorphism of the group of invertible elements of the algebra σ_∞ onto the group $\mathbb{Z}$, as well as to the fact that the spectrum of an element $a \in \mathfrak{S}_\infty$ can separate the complex plane $\mathbb{C}$. These two circumstances do not occur if $p < \infty$.

An analogous question on self-adjoint classes has the affirmative answer (and it is trivial): if $a \in \mathfrak{S}_p$, $a^* = a$, then for an arbitrary $A \in a$ the operator $\frac{1}{2}(A + A^*)$ is self-adjoint and belongs to the class a.

Closely related to Problem 1 are the following two problems.

PROBLEM 2. Let $a \in \mathfrak{S}_p$, $b \in \mathfrak{S}_q$, $ab = ba$. Are there commuting operators $A \in a$, $B \in b$?

PROBLEM 3. Let $a = a^* \in \mathfrak{S}_p$, $b = b^* \in \mathfrak{S}_q$, $ab = ba$. Are there self-adjoint commuting operators $A \in a$, $B \in b$?

Problem 1 and Problem 3 in the case $q = p$ are evidently equivalent. To the contrary a positive answer to Problem 2 does not yield automatically the positive answer to Problem 1.

REFERENCES

1. Гохберг И.Ц., Крейн М.Г. Введение в теорию линейных несамосопряженных операторов. М., Наука, 1965.
2. Бирман М.Ш., Соломяк М.З. Оценки сингулярных чисел интегральных операторов. – Успехи матем.наук, 1977, XXXII, № 1(193) 17–84.
3. Simon B. Trace ideals and their applications. – London Math.Soc.Lect.Note Series, 35, Cambridge Univ.Press, 1979.
4. Triebel H. Interpolation theory. Function spaces. Differential operators. Berlin, 1978.
5. Бирман М.Ш., Соломяк М.З. Компактные операторы со степенной асимптотикой сингулярных чисел. – Зап.научн.семин.ЛОМИ, 1983, 126, 21–30.
6. Brown L., Douglas R., Fillmore P Unitary equivalence modulo the compact operators and extensions of C^*-algebras. –Lect.Notes in Math , 1973, 345, 58–128

M.S.BIRMAN
(М.Ш.БИРМАН)

СССР, 198904, Ленинград,Петродворец
Физический факультет
Ленинградского университета

M.Z.SOLOMYAK
(М.З.СОЛОМЯК)

СССР, 198904, Ленинград,Петродворец
Математико-механический факультет
Ленинградского университета

4.32. PERTURBATION OF SPECTRUM OF NORMAL OPERATORS AND OF COMMUTING TUPLES

Recent progress has somewhat clarified the subject of perturbation of spectrum of normal operators and of K-tuples of commuting self-adjoints. This note is a summary. Only the finite-dimensional case is treated here. (The infinite-dimensional case is attacked in [1].)

$\mathcal{H}$ will be a Hilbert space of n dimensions. The spectral resolution of a normal operator A will be written $A=\sum_{j=1}^{n}\alpha_j u_j u_j^*$ here the u_j are orthonormal eigenvectors, with eigenvalues α_j corresponding; and the notation x^*, for any $x\in\mathcal{H}$, denotes the linear functional corresponding to x. Similarly for normal B let us write $B=\sum_{j=1}^{n}\beta_j v_j v_j^*$. As the distance δ between $\sigma(A)$ and $\sigma(B)$ let us use

$$\delta=\min_{\pi}\max_{j}|\alpha_j-\beta_{\pi j}|, \tag{1}$$

the minimum being over all permutations of $\{1,2,\ldots,n\}$.

PROBLEM 1. <u>Find the best constant</u> c <u>such that, for all normal</u> A <u>and</u> B,

$$\delta\leqslant c\|A-B\|. \tag{2}$$

It has long been conjectured that $c=1$ (e.g., [5]). And (2) is known with $c=1$ in some special cases: if A and B are self-adjoint [7], if A is self-adjoint and B skew-self-adjoint [6], if A and B are unitary [3], or if $A-B$ is normal along with A and B [2]. Yet only recently has it even been proved that there exists a universal c for which (2) holds in general [4].

It has been known for many years that if δ were replaced by the Hausdorff distance then (2) would hold with $c=1$. The following stronger assertion is also familiar (e.g., [3]):

PROPOSITION. <u>If</u> K_A <u>is a set of</u> k <u>eigenvalues of</u> A <u>and</u> K_B <u>is a set of</u> $n-k+1$ <u>eigenvalues of</u> B, <u>and if either</u>

(i) <u>the convex hulls of</u> K_A <u>and</u> K_B <u>are at distance</u> $\geqslant d$

<u>or</u>

(ii) <u>for some</u> $\lambda\in\mathbb{C}$ <u>and</u> $p\in\mathbb{R}_+$ <u>we have</u> $K_A\subseteq\{\zeta:|\zeta-\lambda|\leqslant p\}$

while $K_B \subseteq \{\zeta : |\zeta-\lambda| \geq p+d\}$, *then* $d \leq \|A-B\|$.

PROOF. The spectral subspace for A belonging to K_A and the spectral subspace for B belonging to K_B have dimensionalities whose sum is $> n$; hence there is a unit vector x in their intersection. In case (i), $d \leq |x^*Ax - x^*Bx| \leq \|A-B\|$. In case (ii), $\|(A-\lambda)x\| \leq p$ while $\|(B-\lambda)x\| \geq p+d$, so $d \leq \|(A-\lambda)x-(B-\lambda)x\| \leq \|A-B\|$. ●

COROLLARY (R.Bhatia).

$$\min_{\pi} \max_{j} \left| |\alpha_j| - |\beta_{\pi j}| \right| \leq \|A-B\| .$$

(Of course, here too we can translate by any $\lambda \in \mathbb{C}$). Indeed, the left-hand expression is clearly $= \max \left| |\alpha_j| - |\beta_j| \right|$ if both the α_j and the β_j are labelled in order of increasing modulus. Let the maximum be attained for $j = k$, and assume without loss of generality that $|\alpha_k| < |\beta_k|$. Then $K_A = \{\alpha_1, \dots, \alpha_k\}$ and $K_B = \{\beta_k, \dots, \beta_n\}$ satisfy hypothesis (ii) of the Proposition. ●

Let δ' denote the greatest d obtainable for any K_A and K_B satisfying (i) or (ii), or the analogue of (ii) with A and B interchanged. Clearly Hausdorff distance $\leq \delta' \leq \delta$.

EXAMPLE.1. Set $\alpha_1 = -3, \alpha_2 = i\sqrt{3}, \alpha_3 = 1$; and $\beta_j = -\alpha_j$ for all j . The Hausdorff distance is 2 , $\delta' = 3$ (attained for $K_A = \{\alpha_1, \alpha_2\}$ and $K_B = \{\beta_1, \beta_2\}$), but $\delta = 2\sqrt{3}$.

Thus the idea of the Proposition can not be used directly to prove that the constant C above is 1 .

The problem of K-tuples of commuting self-adjoints may be more important, but so far seems less tractable. I will use the following notation. If $A^{(1)}, A^{(2)}, \dots, A^{(K)}$ are self-adjoint and commute, then for orthonormal $u_1, \dots, u_n$ and corresponding real $\alpha_j^{(J)}$ we have $A^{(J)} = \sum_{j=1}^{n} \alpha_j^{(J)} u_j u_j^*$; I will let $\underset{\sim}{A}$ denote the operator-matrix of one column whose J-th entry is $A^{(J)}$, so that $\underset{\sim}{A} = \sum_{j=1}^{n} \underset{\sim}{\alpha}_j u_j u_j^*$, and we may speak of $\underset{\sim}{\alpha}_j \in \mathbb{R}^K$ as the j-th eigenvalue of $\underset{\sim}{A}$. (As an operator from $\mathcal{H}$ to $\mathcal{H}^K$, it does not have eigenvalues in the usual sense.) Similarly $\underset{\sim}{B} = \sum_{j=1}^{n} \underset{\sim}{\beta}_j v_j v_j^*$. As above, the distance will be

$$\delta = \min_{\pi} \max_{j} \| \underset{\sim}{\alpha}_j - \underset{\sim}{\beta}_{\pi j} \| ,$$

the minimum being over all permutations, and the norm being that of $\mathbb{R}^K$.

PROBLEM 2. *Find the best constant* C_K *such that, for all* K -

tuples $\underset{\sim}{A}$ and $\underset{\sim}{B}$, $\delta \leqslant c_K \|\underset{\sim}{A} - \underset{\sim}{B}\|$.

The Proposition and Corollary above have exact analogues for this situation (with $\lambda \in \mathbb{C}$ replaced by $\underset{\sim}{\lambda} \in \mathbb{R}^K$ and so forth); their proofs are almost as brief, and may be left to the reader

To make precise the relationship between Problem 1 and Problem 2 (for the case $K=2$), I recall these elementary facts;

FACT 1. For any self-adjoint H and L on $\mathcal{H}$ (not necessarily commuting), $\sqrt{2}\,\left\|\begin{bmatrix} H \\ L \end{bmatrix}\right\| \geqslant \|H + iL\|$.

PROOF. $H + iL = [1 \;\; i]\begin{bmatrix} H \\ L \end{bmatrix}$ and $\|[1 \;\; i]\| = \sqrt{2}$. ●

FACT 2. For any self-adjoint H and L, $\left\|\begin{bmatrix} H \\ L \end{bmatrix}\right\| \leqslant \|H + iL\|$.

Proof. $\begin{bmatrix} H \\ L \end{bmatrix}^* \begin{bmatrix} H \\ L \end{bmatrix}$ is the average of the operators $(H+iL)^*(H+iL)$ and $(H+iL)(H+iL)^*$. ●

It follows at once that the constants C in Problem 1 and C_2 in Problem 2 are related by $C \leqslant C_2 \leqslant \sqrt{2}\,C$. In particular, C_2 is finite, by virtue of the result of [4] already cited. The existence of a finite C_K for $K>2$ has not been proved

EXAMPLE 2. Adapting Example 1, take $\underset{\sim}{\alpha}_1 = \begin{bmatrix} -3 \\ 0 \end{bmatrix}$, $\underset{\sim}{\alpha}_2 = \begin{bmatrix} 0 \\ \sqrt{3} \end{bmatrix}$, $\underset{\sim}{\alpha}_3 = \begin{bmatrix} 1 \\ 0 \end{bmatrix}$; and $\underset{\sim}{\beta}_j = -\underset{\sim}{\alpha}_j$ for all j. The remarks concerning Example 1 apply to this modification, in particular $\delta = 2\sqrt{3}$. Now choose the eigensystems

$$\mathcal{U}_1 = \frac{1}{4}\begin{bmatrix} 3 \\ -\sqrt{3} \\ 2 \end{bmatrix}, \quad \mathcal{U}_2 = \frac{1}{4}\begin{bmatrix} -\sqrt{3} \\ 1 \\ 2\sqrt{3} \end{bmatrix}, \quad \mathcal{U}_3 = \frac{1}{2}\begin{bmatrix} 1 \\ \sqrt{3} \\ 0 \end{bmatrix};$$

$$V_1 = \frac{1}{4}\begin{bmatrix} -3 \\ -\sqrt{3} \\ 2 \end{bmatrix}, \quad \mathcal{U}_2 = \frac{1}{4}\begin{bmatrix} \sqrt{3} \\ 1 \\ 2\sqrt{3} \end{bmatrix}, \quad V_3 = \frac{1}{2}\begin{bmatrix} -1 \\ \sqrt{3} \\ 0 \end{bmatrix}.$$

It is not hard to compute that $\|\underset{\sim}{A} - \underset{\sim}{B}\|^2 < \frac{23}{2} < 12 = \delta^2$. For a near-by example I have found that $\|\underset{\sim}{A} - \underset{\sim}{B}\|^2/\delta^2 < 11/12$.

The example shows that $C_2 > 1$. However, it remains reasonable to conjecture that $C = 1$, and (as this would imply) that $C_2 \leqslant \sqrt{2}$. I suggest, with little evidence, that in general $C_K \leqslant \sqrt{K}$.

REFERENCES

1. A z o f f E., D a v i s Ch. Perturbation of spectrum of self-adjoint operators. To appear.
2. B h a t i a R. Analysis of spectral variation and some inequalities. - Trans.Amer.Math.Soc. 1982, 272, 323-331.

3. B h a t i a R., D a v i s Ch. A bound for the spectral variation of a unitary operator - Linear Multilinear Alg. To appear.
4. B h a t i a R., D a v i s Ch., M c I n t o s h A. Perturbation of spectral subspaces and solution of linear operator equations. - Linear Alg. and Appl. To apppear.
5. M i r s k y L. Symmetric gauge functions and unitarily invariant norms. - Quarterly J. Math. Oxford Ser. 2, 1960, 11, 50-59.
6. S u n d e r V.S. Distance between normal operators. - Proc.Amer. Math.Soc. 1982, 84, 483-484.
7. W e y l H. Das asymptotische Verteilungsgesetz der Eigenwerte linearer partieller Differentialgleichungen. - Math.Ann., 1912, 71, 441-479.

CHANDLER DAVIS

Department of Mathematics
University of Toronto
Toronto M5S 1A1
Canada

4.33. PERTURBATION OF CONTINUOUS SPECTRUM AND NORMAL OPERATORS

Let T be an invertible bounded operator on Hilbert space $\mathcal{H}$. The continuous spectrum $\sigma_c(T)$ of T is defined as $\sigma(T)\setminus\sigma_0(T)$, where $\sigma_0(T)$ stands for the set of all isolated points of the spectrum $\sigma(T)$ whose spectral subspaces are finite-dimensional. If the origin lies in the unbounded component of $\mathbb{C}\setminus\sigma(T)$ then $0\notin\sigma_c(T+K)$ for any compact operator K. On the other hand, if $\sigma(T)$ separates 0 and ∞, then for any symmetrically-normed ideal $\mathfrak{S}$, $\mathfrak{S}\neq\mathfrak{S}_1$ ($\mathfrak{S}_p$ denotes throughout the Schatten - von Neumann class, $0<p\leq\infty$) there exists $K\in\mathfrak{S}$ such that $0\in\sigma_c(T+K)$ [1]. The question we are interested in concerns the stability of the continuous spectrum under "small" perturbations (e.g., finite rank, nuclear, etc.) For rank one perturbations, the problem can be solved easily in terms of the lattice Lat of invariant subspaces of the operator.

THEOREM 1. *Suppose* 0 *does not belong to the unbounded component of* $\sigma(T)$ *and* $0\notin\sigma(T)$. *Then a rank one operator* K *with* $0\in\sigma_c(T+K)$ *exists if and only if* $\operatorname{Lat}T^{-1}\not\subset\operatorname{Lat}T$.

See [2], for the proof. Given an operator T denote by $\mathcal{R}(T)$ the weakly closed algebra of operators on $\mathcal{H}$ generated by T and the identity I. Suppose N is a normal operator on $\mathcal{H}$. Then $\operatorname{Lat}N^{-1}\subset\operatorname{Lat}N$ iff $N\in\mathcal{R}(N^{-1})$, see [3]. Therefore Theorem 1 together with a theorem of Sarason [4] imply the following criterion for the stability of the continuous spectrum under finite rank one perturbations. Denote by G_M the Sarason hull of the spectral measure of a normal operator M, defined in [4].

THEOREM 2. *Let* N *be an invertible normal operator The following are equivalent:*

1) $0\notin\sigma_c(N+K)$ *for every* K, $\operatorname{rank}(K)=1$;

2) $0\notin G_{N^{-1}}$.

In particular, the continuous spectrum of a unitary operator is stable iff Lebesgue measure on $\mathbb{T}$ is absolutely continuous with respect to its spectral measure. It is shown in [5] that the continuous spectrum of a unitary operator is stable under perturbations of rank one iff it is stable under nuclear perturbations. This result can be extended to normal operators with essential spectra on smooth curves. At the same time it does not hold for arbitrary normal operators [1].

QUESTION 1. Given an invertible operator are the following equivalent?

1) $0 \notin \sigma_c(T+K)$ for every K of rank one.

2) $0 \notin \sigma_c(T+K)$ for every K of finite rank.

QUESTION 2. Are 1) and 2) equivalent for an arbitrary normal operator $N = T$? Is it true that they are equivalent to

3) $0 \notin \sigma_c(N+K)$ for every $K \in \bigcup_{p<1} \mathfrak{S}_p$?

Note that for $T = \mathcal{U} + K$, where $\mathcal{U}$ is unitary and $K \in \mathfrak{S}_1$ the answer to Question 1 is affirmative [2].

QUESTION 3. Is either of the following implications

$0 \notin \sigma_c(T+K), \forall K, \text{rank}(K) < +\infty \Leftrightarrow T \in \mathcal{R}(T^{-1})$ true?

The inclusion $T \in \mathcal{R}(T^{-1})$ being equivalent to the series of inclusions

$$\text{Lat}\, \underbrace{T^{-1} \oplus \cdots \oplus T^{-1}}_{n} \subset \text{Lat}\, \underbrace{T \oplus \cdots \oplus T}_{n},$$

$n = 1, 2, \ldots$ (see [3]), it is natural to ask

QUESTION 4. Are the following statements equivalent for any integer n ?

1) $0 \notin \sigma_c(T+K)$, $\forall K$, $\text{rank}(K) \leq n$.

2) $\text{Lat}\, \underbrace{T^{-1} \oplus \cdots \oplus T^{-1}}_{n} \subset \text{Lat}\, \underbrace{T \oplus \cdots \oplus T}_{n}$.

By the way, we do not know the answer even to the following question.

QUESTION 5. Is it true that

$$T \in \mathcal{R}(T^{-1}) \quad \text{iff} \quad \text{Lat}\, T^{-1} \subset \text{Lat}\, T \, ?$$

Many interesting problems arise when considering **special** perturbations of normal operators. Recall that the problem of stability of continuous spectra in case of normal operators is reduced to the calculation of Sarason hulls.

QUESTION 6. Let N and $N+K$ be normal operators. Is it true that

$$\operatorname{rank} K < +\infty \Rightarrow G_N = G_{N+K} ?$$

$$K \in \mathfrak{S}_p,\ p<1 \Rightarrow G_N = G_{N+K} ?$$

For what normal operators N

$$K \in \mathfrak{S}_1 \Rightarrow G_N = G_{N+K} ? \tag{1}$$

It was noted in [1] that there are normal operators not satisfying (1).

If $\mathcal{U}$ and $\mathcal{U}+K$ are unitary then (1) holds with $\mathcal{U}=N$. This is so because the absolutely continuous parts of $\mathcal{U}$ and $\mathcal{U}+K$ are unitarily equivalent and the Sarason hull of a unitary operator depends on its absolutely continuous part only. Therefore Question 6 may be considered as a question of "scattering theory of normal operators".

Consider a narrower class of perturbations, namely we assume henceforth that N commutes with $N+K$. Then the symmetric difference $G_N \Delta G_{N+K}$ consists of the points of the point spectrum of N or of $N+K$. This reduces the question to the investigation of metric properties of the harmonic measure. L.Carleson has proved in [6] **that the harmonic measure of any simply connected domain is absolutely continuous with respect to Hausdorff measure Λ_β, where $\beta > 1/2$ is an absolute constant. Using this result, it can be proved that $G_N = G_{N+K}$ if $K \in \mathfrak{S}_\beta$ and N commutes with $N+K$.**

QUESTION 7. Let N and $N+K$ be commuting normal operators. Is it true that

$$K \in \mathfrak{S}_p,\ p<1 \Rightarrow G_N = G_{N+K} ?$$

REFERENCES

1. Макаров Н.Г. - Докл.АН СССР (to appear).
2. Makarov N.G., Vasjunin V.I. A model for noncontractions and stability of the continuous spectrum. Complex Analysis and Spectral Theory, Lecture Notes in Math., 1981, 864, 365-412.

3. S a r a s o n D. Invariant subspaces and unstarred operator algebras. - Pacific J.Math., 1966, 17, 511-517.
4. S a r a s o n D. Weak-star density of polynomials. - J.reine und angew.Math., 1972, 252, 1-15.
5. Н и к о л ь с к и й Н.К. О возмущениях спектра унитарных операторов. - Матем.заметки, 1969, 5, 341-349.
6. C a r l e s o n L. On the distortion of sets on a Jordan curve under conformal mapping. - Duke Math.J., 1973, 40, 547-559.

N.G.MAKAROV
(Н.Г.МАКАРОВ)

СССР, 198904, Ленинград, Старый Петергоф, Ленинградский университет, Математико-механический факультет

N.K.NIKOL'SKII
(Н.К.НИКОЛЬСКИЙ)

СССР, 191011, Ленинград, Фонтанка 27, ЛОМИ

4.34. ALMOST-NORMAL OPERATORS MODULO $\mathcal{C}_p$

0. Notations. H = separable complex Hilbert space of infinite dimension; $(\mathcal{L}(H), \|\cdot\|)$ = (bounded operators on H, uniform norm); $(\mathcal{C}_p, |\cdot|_p)$ = (Schatten- von Neumann p-class, p-norm);

$\mathcal{R}_1^+ = \{T \in \mathcal{L}(H) : T$ finite rank, $0 \leqslant T \leqslant I\}$, $\mathcal{P} = \{P \in \mathcal{R}_1^+ : P = P^2\}$.

$AN(H)$ = (almost normal operators on $H) = \{T \in \mathcal{L}(H) : [T^*, T] \in \mathcal{C}_1\}$.

For $T \in AN(H)$ we shall denote by P_T its Helton-Howe measure and by G_T its Pincus G-function (see [6], [10], [5]) so that $P_T = (2\pi)^{-1} G_T \, d\lambda$, where $d\lambda$ is Lebesgue measure on $\mathbb{R}^2$.

I. Basic Analogy. It is known that for $T \in AN(H)$ we have: index $(T - zI) = G_T(z)$ for $z \in \mathbb{C}$ such that $T - zI$ is Fredholm. In [13] we noticed several instances which suggest that this relation is part of a far reaching analogy, in which the G-function plays the same role for $\mathcal{C}_2$-perturbations of almost normal operators as the index for compact perturbations of essentially normal operators.

II. Invariance of P_T under $\mathcal{C}_2$-perturbations. This should correspond to the invariance of the index under compact perturbations. In [13] we proved that if $T, S \in AN(H)$, $T - S \in \mathcal{C}_2$ and T or S has finite multicyclicity then $P_T = P_S$. Our proof in [13] depended on the use of the quasidiagonality relative to $\mathcal{C}_2$. In fact one needs less. Consider

$$k_2(T) = \liminf_{A \in \mathcal{R}_1^+} |[A, T]|_2$$

where the liminf is with respect to the natural order on $\mathcal{R}_1^+$. (see [12]).

PROPOSITION. Let $T, S \in AN(H)$ and suppose $k_2(T) = 0$ and $T - S \in \mathcal{C}_2$. Then we have $P_S = P_T$.

PROOF. As in [13] (Prop.3) the proof reduces to showing that $\operatorname{Tr}[T^*, T] = \operatorname{Tr}[S^*, S]$. Since $k_2(T) = 0$ there are $A_n \in \mathcal{R}_1^+$, $A_n \uparrow I$ such that $|[A_n, T]|_2 \to 0$ as $n \to \infty$ and the same holds also for T replaced by S. Denoting $X = T - S$ we have:

$$|\mathrm{Tr}([T,T^*]-[S,S^*])| =$$

$$=|\mathrm{Tr}([X,T^*]+[X^*,S])| =$$

$$=\lim_{n\to\infty}|\mathrm{Tr}(A_n([X,T^*]+[X^*,S])| \leq$$

$$\leq \limsup_{n\to\infty}|\mathrm{Tr}([A_nX,T^*]+[A_nX^*,S])| +$$

$$+\limsup_{n\to\infty}(|[T^*,A_n]|_2|X|_2+|[S,A_n]|_2|X^*|_2)=0. \qquad \bullet$$

We proved in [13] that $k_2(T)=0$ for $T\in AN(H)$ with finite multiplicity.

CONJECTURE 1. $T\in AN(H)\Longrightarrow k_2(T)=0$.

III. Quasitriangularity. A refinement of Halmos' notion of quasitriangularity [7] was considered in [11]. The corresponding generalization of Apostol's modulus of quasitriangularity is:

$$q_p(T)=\liminf_{P\in\mathcal{P}}|(I-P)TP|_p$$

where the liminf is with respect to the natural order on $\mathcal{P}$. We proved in [13] that for $T\in AN(H)$, $q_2(T)=0\Rightarrow P_T\leq 0$. We conjecture an analogue of the Apostol-Foiaş-Voiculescu theorem on quasitriangular operators [1].

CONJECTURE 2. For $T\in AN(H)$, we have

$$(q_2(T))^2=2\iint_{\mathbb{R}^2}dP_T^+$$

where P_T^+ is the positive part of P_T.

This would imply in particular that $P_T\leq 0\Longrightarrow q_2(T)=0$.

Some results for subnormal and cosubnormal operators supporting the conjecture have been obtained in [13].

IV. Analogue of the BDF theorem. The following conjecture concerns an analogue of the Brown-Douglas-Fillmore theorem [3] on essentially normal operators.

CONJECTURE 3. *Let* $T_1, T_2 \in AN(H)$ *be such that* $P_{T_1} = P_{T_2}$. *Then there is a normal operator* $N \in \mathcal{L}(H)$ *and a unitary* $U \in \mathcal{L}(H \oplus H)$ *such that*

$$U(T_1 \oplus N)U^* - T_2 \oplus N \in \mathcal{C}_2.$$

This conjecture implies the following

CONJECTURE 4. *If* $T \in AN(H)$ *then there is* $S \in AN(H)$ *such that* $T \oplus S$ = normal + Hilbert-Schmidt.

Note that this last statement corresponds to an important part in the proof of the BDF theorem, the existence of inverses in Ext or equivalently the completely positive lifting part in the "Ext is a group" theorem (see [2]). Even for almost normal weighted shifts, there is only a quite restricted class for which Conjecture 4 has been established [8].

Note also that Conjecture 4 implies Conjecture 1 and that by analogy with the proof of the Choi-Effros completely positive lifting theorem one should expect the vanishing of k_2 to be an essential ingredient in establishing Conjecture 4.

REFERENCES

1. Apostol C., Foiaş C., Voiculescu D. Some results on non-quasitriangular operators VI. - Rev.Roum. Math.Pures Appl. 1973, 18, 1473-1494.
2. Arveson W.B. A note on essentially normal operators. - Proc.Royal Irish Acad. 1974, 74, 143-146.
3. Brown L.G., Douglas R.G., Fillmore P.A. Unitary equivalence modulo the compact operators and extensions of C^*-algebras. Lect.Notes in Math., 1973, 345, 58-128.
4. Carey R.W., Pincus J.D. Commutators, symbols and determining functions. - J.Funct.Anal. 1975, 19, 50-80.
5. Clancey K. Seminormal operators. Lect.Notes in Math., 1979, 742.
6. Helton J.W., Howe R. Integral operators, commutator traces, index and homology. Lect.Notes in Math. 1973, 345, 141-209.

7. H a l m o s P.R. Quasitriangular operators. - Acta Sci.Math. (Szeged), 1968, 29, 283-293.
8. P a s n i c u C. Weighted shifts as direct summands mod $\mathcal{C}_2$ of normal operators. INCREST preprint 1982.
9. P e a r c y C. Some recent developments in operator theory. CBMS, Regional Conference Series in Mathematics no.36, Prodidence, Amer.Math.Soc., 1978.
10. P i n c u s J.D. Commutators and systems of integral equations, I. - Acta Math., 1968, 121, 219-249.
11. V o i c u l e s c u D. Some extensions of quasitriangularity. - Rev.Roum.Math.Pures Appl., 1973, 18, 1303-1320.
12. V o i c u l e s c u D. Some results on norm-ideal perturbations of Hilbert space operators. - J.Operator Theory, 1979, 2, 3-37.
13. V o i c u l e s c u D. Remarks on Hilbert-Schmidt perturbations of almost-normal operators. - In: Topics in Modern Operator Theory, Birkhäuser 1981.

D.VOICULESCU

Department of Mathematics
INCREST
Bd.Păcii 220, 79622 Bucharest
ROMANIA

4.35. HYPONORMAL OPERATORS AND SPECTRAL ABSOLUTE CONTINUITY

In the sequel only bounded operators on an infinite dimensional, separable Hilbert space H will be considered. An operator T on H is said to be hyponormal if $T^*T - TT^* \geq 0$. Such an operator is said to be completely hyponormal if, in addition, T has no normal part, that is, if there is no subspace $\neq \{0\}$ reducing T on which T is normal.

If A is selfadjoint with the spectral family $\{E_t\}$ then the set of vectors x in H for which $\|E_t x\|^2$ is an absolutely continuous function of t is a subspace, $H_a(A)$, reducing A (see, e.g., Kato [1], p.516). If $H_a(A) \neq \{0\}$, then $A|H_a(A)$ is called the absolutely continuous part of A, and if $H_a(A) = H$ then A is said to be absolutely continuous. Similar concepts can be defined for a unitary operator.

If T is completely hyponormal then its real and imaginary parts are absolutely continuous. In addition, if T has a polar factorization

$$T = U|T|, \quad U \text{ unitary and } |T| = (T^*T)^{1/2}, \tag{1}$$

then U is also absolutely continuous. (See [2], p.42 and [3], p.193. Incidentally, such a polar factorization (1) exists, and is unique, if and only if 0 is not in the point spectrum of T^*; see [4], p.277.)

In general, if T is completely hyponormal, then its absolute value $|T| = (T^*T)^{1/2}$ need not be absolutely continuous or even have an absolutely continuous part. Probably the simplest example is the simple unilateral shift V for which V^*V is the identity. Of course, V does not have a polar factorization (1), but, nevertheless, there are simple examples of completely hyponormal T satisfying (1) for which $|T|$ has no absolutely continuous part. Moreover, it has recently been shown by K.F.Clancey and the author [5] that if P is selfadjoint on H, then there exists a completely hyponormal T satisfying $|T| = P$ and having the polar factorization (1) if and only if (i) $P \geq 0$ and $\sigma(P)$ contains at least two points, (ii) 0 is not in the point spectrum of P, and (iii) neither $\max \sigma(P)$ nor $\min \sigma(P)$ is in the point spectrum of P with a finite multiplicity.

Let a nonempty compact set of the complex plane be called radially symmetric if whenever z_1 is in the set then so is the entire circle $|z|=z_1$. All examples known to the author of completely hyponormal operators T for which $|T|$ does not have an absolutely continuous part, and whether or not (1) obtains, seem to have radially symmteric spectra. For instance, if $\sigma(|T|)$ has Lebesgue linear measure zero, then $\sigma(T)$ is surely radially symmetric; see [6], p.426, also [7]. At the other extreme, if T is completely hyponormal and if there exists some open wedge $W=\{z: z\neq 0$ and $0<\arg z<\mathrm{const}<2\pi\}$ (or rotated set $e^{i\theta}W$) which does not intersect $\sigma(T)$ then $|T|$ is absolutely continuous. In certain other instances also one can show at least that $H_a(|T|)\neq\{0\}$; see [7] and the references cited there. The following conjecture was made in [7]:

CONJECTURE 1. <u>Let T be completely hyponormal with a polar factorization (1). Suppose that $\sigma(T)$ is not radially symmetric, so that some circle $|z|=r$ intersects both $\sigma(T)$ and its complement in nonempty sets. Then $H_a(|T|)\neq\{0\}$.</u>

The following stronger statement was also indicated in [7] and is set forth here, but with somewhat less conviction than the preceding conjecture, as

CONJECTURE 2. <u>Conjecture 1 remains true without the hypothesis</u> (1).

REFERENCES

1. K a t o T. Perturbation theory for linear operators, Springer-Verlag, New York Inc., 1967.
2. P u t n a m C.R. Commutation properties of Hilbert space operators and related topics, Ergebnisse der Math., 36, Springer-Verlag, New York Inc., 1967.
3. P u t n a m C.R. A polar area inequality for hyponormal specra. - J.Operator Theory, 1980, 4, 191-200.
4. P u t n a m C.R. Absolute continuity of polar factors of hyponormal operators. - Amer.J.Math., Suppl. 1981, 277-283.
5. C l a n c e y K.F., P u t n a m C.R. Nonnegative perturbations of selfadjoint operators. - J.Funct.Anal., 1983, 51 (to appear).
6. P u t n a m C.R. Spectra of polar factors of hyponormal operators. - Trans.Amer.Math.Soc., 1974, 188, 419-428.

7. P u t n a m C.R. Absolute values of hyponormal operators with asymmetric spectra. - Mich.Math.Jour., 1983, 30 (to appear).

C.R.PUTNAM

PURDUE UNIVERSITY
Department of Mathematics
West Lafayette, Indiana 47907
USA

4.36. old OPERATORS, ANALYTIC NEGLIGIBILITY, AND CAPACITIES

Let $\sigma(T)$ and $\sigma_p(T)$ denote the spectrum and point spectrum of a bounded operator T on a Hilbert space H. Such an operator is said to be s u b n o r m a l if it has a normal extension on a Hilbert space K, $K \supset H$. For the basic properties of subnormal operators see [1]. A subnormal T on H is said to be c o m p l e t e l y s u b n o r m a l if there is no nontrivial subspace of H reducing T on which T is normal. If T is completely subnormal then $\sigma_p(T)$ is empty. A necessary and sufficient condition in order that a compact subset of $\mathbb{C}$ be the spectrum of a completely subnormal operator was given in [2].

If X is a compact subset of $\mathbb{C}$, let $R(X)$ denote the functions on X uniformly approximable on X by rational functions with poles off X. A compact subset Q of X is called a p e a k s e t of $R(X)$ if there exists a function f in $R(X)$ such that $f=1$ on Q and $|f|<1$ on $X \setminus Q$; see [3], p.56. The following result was proved in [4].

THEOREM. Let T be subnormal on H with the minimal normal extension $N=\int z\, dE_z$ on K, $K \supset H$. Suppose that Q is a non-trivial proper peak set of $R(\sigma(T))$ and that $E(Q)\neq 0$. Then $E(Q)H \neq \{0\}$ and H, the space $E(Q)H$ reduces T, $T|E(Q)H$ is subnormal with the minimal normal extension $E(Q)N$ on $E(Q)K$, and $\sigma(T|E(Q)H) \subset Q$. Further, if it is also assumed that $R(Q)=C(Q)$, then $T|E(Q)H$ is normal.

Thus, in dealing with reducing subspaces of subnormal operators T, it is of interest to have conditions assuring that a subset of a compact $X (=\sigma(T))$ be a peak set of $R(X)$.

PROBLEM 1. Let X be a compact subset of $\mathbb{C}$ and let C be a rectifiable simple closed curve for which Q =clos ((exterior of C) $\cap$ $\cap X$) is not empty and $C \cap X$ has Lebesgue arc length measure 0. Does it follow that Q must be a peak set of $R(X)$?

In case C is of class C^2 (or piecewise C^2), the answer is affirmative and was first demonstrated by Lautzenheiser [5]. A modified version of his proof can be found in [6], pp.194-195. A crucial step in the argument is an application of a result of Davie and Øksendal [7] which requires that the set $C \cap X$ be analytically negligible.

(A compact set E is said to be analytically negligible if every continuous function on $\mathbb{C}$ which is analytic on an open set V can be approximated uniformly on $V \cup E$ by functions continuous on $\mathbb{C}$ and analytic on $V \cup E$; see [3], p.234.) The C^2 hypothesis is then used to ensure the analytic negligibility of $C \cap X$ as a consequence of a result of Vitushkin [8]. It may be noted that the collection of analytically negligible sets has been extended by Vitushkin to include Liapunov curves (see [9], p.115) and by Davie ([10], section 4) to include "hypo-Liapunov" curves. Thus, for such curves C, the answer to PROBLEM 1 is again yes. The question as to whether a general rectifiable curve, or even one of class C^1, for instance, is necessarily analytically negligible, as well as the corresponding question in PROBLEM 1, apparently remains open however.

As already noted, PROBLEM 1 is related to questions concerning subnormal operators. The problem also arose in connection with a possible generalization of the notion of an "areally disconnected set" as defined in [6] and with a related rational approximation question. Problem 2 below deals with some estimates for the norms of certain operators associated with a bounded operator T on a Hilbert space and with two capacities of the set $\sigma(T)$.

Let $\gamma(E)$ and $\alpha(E)$ denote the analytic capacity and the continuous analytic capacity (or AC capacity) of a set E in $\mathbb{C}$. (For definitions and properties see, for example, [3,11,9]. A brief history of both capacities is contained in [9], pp.142-143, where it is also noted that the concept of continuous analytic capacity was first defined by Dolzhenko [12].) It is known that for any Borel set E,

$$(\pi^{-1}\, meas_2 E)^{\frac{1}{2}} \leq \alpha(E) \leq \gamma(E)$$

see [11], pp.9, 79. The following was proved in [13].

THEOREM. <u>Let</u> T <u>be a bounded operator on a Hilbert space and suppose that</u>

$$(T-z)(T-z)^* \geq D \geq 0 \tag{1}$$

<u>holds for some nonnegative operator</u> D <u>and for all</u> z <u>in the unbounded component of the complement of</u> $\sigma(T)$. <u>Then</u> $\|D\|^{1/2} \leq \gamma(\sigma(T))$. <u>If, in addition, (1) holds for all</u> z <u>in</u> $\mathbb{C}$ <u>and if, for instance,</u> $\sigma_p(T)$ <u>is contained in the interior of</u> $\sigma(T)$ (<u>in particular, if</u>

$\sigma_p(T)$ is empty), then also $\|D\|^{1/2} \leq \alpha(\sigma(T))$.

PROBLEM 2. Does condition (1), if valid for all z in $\mathbb{C}$, but without any rectriction on $\sigma_p(T)$, always imply that $\|D\|^{1/2} \leq \alpha(\sigma(T))$, or possibly even that $\|D\| \leq \pi^{-1}$ $meas_2\, \sigma(T)$?

It may be noted that if T^* is hyponormal, so that $TT^* - T^*T \geq \mathbb{O}$ then (1) holds for all z in $\mathbb{C}$ with $D = TT^* - T^*T$ and, moreover, $\|D\| \leq \pi^{-1}$ $meas_2(\sigma(T))$; see [14].

REFERENCES

1. H a l m o s P.R. A Hilbert space problem book, van Nostrand Co., 1967.
2. C l a n c e y K.F., P u t n a m C.R. The local spectral behavior of completely subnormal operators. - Trans.Amer.Math.Soc., 1972, 163, 239-244.
3. G a m e l i n T.W. Uniform algebras, Prentice-Hall, Inc.,1969.
4. P u t n a m C.R. Peak sets and subnormal operators. - Ill. Jour.Math., 1977, 21, 388-394.
5. L a u t z e n h e i s e r R.G. Spectral sets, reducing subspaces, and function algebras, Thesis, Indiana Univ., 1973.
6. P u t n a m C.R. Rational approximation and Swiss cheeses. - Mich Math.Jour., 1977, 24, 193-196.
7. D a v i e A.M., Ø k s e n d a l B.K. Rational approximation on the union of sets. - Proc.Amer.Math.Soc., 1971, 29, 581-584.
8. В и т у ш к и н А.Г. Аналитическая емкость множеств в задачах теории приближений. - Успехи матем. наук, I967, 22, № 6, I4I-I99.
9. Z a l c m a n L. Analytic capacity and rational approximation. Lecture notes in mathematics, 50, Springer-Verlag, 1968.
10. D a v i e A.M. Analytic capacity and appriximation problems. - Trans.Amer.Math.Soc., 1972, 171, 409-444.
11. G a r n e t t J. Analytic capacity and measure, Lecture notes in mathematics, 297, Springer-Verlag, 1972.
12. Д о л ж е н к о Е.П. О приближении на замкнутых областях и о нуль-множествах. - Докл.АН СССР, I962, I43, № 4, 77I-774.
13. P u t n a m C.R. Spectra and measure inequalities. - Trans. Amer.Math.Soc., 1977, 231, 519-529.

14. P u t n a m C.R. An inequality for the area of hyponormal spectra. - Math.Zeits., 1970, 116, 323-330.

C.R.PUTNAM

PURDUE UNIVERSITY,
Department of Math.,
West Lafayette,
Indiana 47907, USA

EDITORS' NOTE. Two works of Valskii (Р.Э.Вальский, Докл. АН СССР, 1967, 173, N 1, 12-14; Сибирск.матем.ж., 1967, 8, № 6, 1222-1235) contain some results concerning the themes discussed in this section.

4.37. GENERALIZED DERIVATIONS AND SEMIDIAGONALITY

Let A_1, A_2 be bounded linear operators on a Hilbert space $\mathcal{H}$, $\Delta = \Delta_{A_1,A_2}$ be an operator on the space $\mathcal{B}(\mathcal{H})$ of all bounded operators on $\mathcal{H}$ defined by $\Delta(X) = A_1X - XA_2$, $X \in \mathcal{B}(\mathcal{H})$. If $A_1 = A_2$, Δ_{A_1,A_2} is a derivation of $\mathcal{B}(\mathcal{H})$. That is why the operators Δ_{A_1,A_2} are called sometimes *generalized derivations*. Put $\widetilde{\Delta} = \Delta_{A_1^*,A_2^*}$. A question of whether

$$\operatorname{Ker}\widetilde{\Delta} \cap \operatorname{Im}\Delta = \{0\} \tag{1}$$

was raised by various people (see [1], [2], [3]). Equality (1) is true for normal operators A_1, A_2, whence Fuglede - Putnam theorem follows (see [4]). This equality means that $\operatorname{Ker}\widetilde{\Delta}\Delta = \operatorname{Ker}\Delta$ and so $\operatorname{Ker}\widetilde{\Delta} = \operatorname{Ker}\Delta\widetilde{\Delta} = \operatorname{Ker}\widetilde{\Delta}\Delta = \operatorname{Ker}\Delta$. In [3] it is proved that (1) holds when $A_1 = A_2$ is a cyclic subnormal operator or a weighted shift with non-vanishing weights.

Let $p \in [1,\infty]$. An operator B in $\mathcal{B}(\mathcal{H})$ is called *p-semidiagonal* if its modulus of p-quasidiagonality

$$qd_p(A) = \liminf_{P\in\mathcal{P}} \| PA - AP \|$$

($\mathcal{P}$ being the set of all finite rank projections) is finite. Denote by $\mathfrak{M}_p$ the class of $\mathfrak{S}_p$-perturbations of direct sums of p-semidiagonal operators. In [5] it is proved that (1) is true if one of A_j belongs to $\mathfrak{M}_1$. Though that result covers rather extensive class of generalized derivations ($\mathfrak{M}_1$ contains all normal operators with one-dimensional spectrum and their nuclear perturbations, weighted shifts of an arbitrary multiplicity and polynomials of such shifts, Bishop's operators), it is not applicable to many generalized derivations with normal coefficients. Namely, a normal operator belongs in general to $\mathfrak{M}_2$, but not to $\mathfrak{M}_1$. It seems reasonable to try to replace the hypothesis of 1-semidiagonality of one of A_j's by 2-semidiagonality of both.

QUESTION 1. <u>Does</u> (1) <u>hold if</u> $A_1, A_2 \in \mathfrak{M}_2$?

QUESTION 2. <u>Does there exist an operator not in</u> $\mathfrak{M}_2$?

An affirmative answer to the following question would solve QUESTION 2 (see [5]).

QUESTION 3. Do there exist $A\in\mathcal{B}(\mathcal{H})$, $X\in\mathfrak{S}_2$ such that $AX-XA$ is a non-zero projection?

REFERENCES

1. J o h n s o n B., W i l l i a m s J. The range of normal derivations. - Pacif.J.Math., 1975, 58, 105-122.
2. W i l l i a m s J. Derivation ranges: open problems - In: Top Modern Oper.Theory, 5 Int.Conf.Oper.Theory, Timişoara Birkhäuser 1981, 319-328.
3. Y a n g H o . Commutants and derivation ranges. - Tohoku Math J., 1975, 27, 509-514.
4. P u t n a m C. Commutation properties of Hilbert space operators and related topics. Springer-Verlag, Ergebnisse 36, 1967.
5. Ш у л ь м а н В.С. Об операторах умножения и следах коммутаторов. - Записки научн.семин.ЛОМИ, (to appear).

V.S.SHUL'MAN
(В.С.ШУЛЬМАН)

СССР, 160600, Вологда,
ул.Маяковского 6,
Педагогический институт,
Кафедра математики

4.38. WHAT IS A FINITE OPERATOR?

An operator A acting on a complex separable Hilbert space H is called *f i n i t e* if the identity operator 1 is perpendicular to the range of the inner derivation δ_A induced by A, that is,

$$\inf \|\delta_A(X) - 1\| = \|1\| = 1$$

where $\delta_A(X) = AX - XA$ and X runs over the algebra $\mathcal{L}(\mathcal{H})$ of all (bounded linear) operators acting on $\mathcal{H}$.

The notion of finite operator was introduced by J.P. Williams in [7], where he proved the following result.

THEOREM 1. <u>These are equivalent conditions on an operator</u> A :

(i) $\inf(X \in \mathcal{L}(\mathcal{H})) \|\delta_A(X) - 1\| = 1.$

(ii) 0 <u>belongs to the closure of the numerical range of</u> $\delta_A(X)$ <u>for each</u> X <u>in</u> $\mathcal{L}(\mathcal{H})$.

(iii) <u>There exists</u> $f \in \mathcal{L}(\mathcal{H})^*$ <u>such that</u> $f(1) = 1 = \|f\|$ <u>and</u>

$$\ker f \supset \operatorname{ran} \delta_A .$$

The origin of the term *f i n i t e* is this: if $A \subset \mathcal{L}(\mathcal{H})$ has a finite dimensional reducing subspace $\mathcal{M} \neq \{0\}$, $\{e_j\}_{j=1}^{m}$ $(m = \dim \mathcal{M})$ is an orthonormal basis of $\mathcal{M}$, and $f(B) = (1/m)\sum_{j=1}^{m} \langle Be_j, e_j \rangle$, then f satisfies the conditions of Theorem 1 (iii) and therefore A is a finite operator.

Thus, if $R_n = \{T \in \mathcal{L}(\mathcal{H}) : T$ has a reducing subspace of dimension $n\}$ $(n = 1, 2, \ldots)$, then $R = \bigcup_{n=1}^{\infty} R_n$ is a subset of the family (Fin) of all finite operators; furthermore, since (Fin) is closed in $\mathcal{L}(\mathcal{H})$ [7], $R^- \subset$ (Fin).

CONJECTURE 1 (J.P.Williams [7]). (Fin) $= R^-$.

As we have observed above, if $A \in R$, then it is possible to construct f as in Theorem 1 (iii) such that f does not vanish identically on $\mathcal{K}(\mathcal{H})$, the ideal of all compact operators. J.H.Anderson proved in [1] (Theorem 10.10 and its proof) that this fact can actually be used to characterize R : $A \in R$ if and only if there exists $f \in \mathcal{L}(\mathcal{H})^*$ such that $f(1) = 1 = \|f\|$, and ker $f \supset$

$\supset \operatorname{ran}\delta_A$, but ker $f \not\supset \mathcal{K}(\mathcal{H})$; furthermore, if $A \in$ (Fin)$\setminus \mathcal{R}$,then every f as in Theorem 1 (iii) is necessarily a s i n g u l a r functional (i.e., ker $f \supset \mathcal{K}(\mathcal{H})$. This is true, in particular, if $A \in \mathcal{R}^- \setminus \mathcal{R}$.)

Since (Fin) $\supset \mathcal{R}^-$, it is plain that (Fin) contains all q u a s i d i a g o n a l o p e r a t o r s (in the sense of Halmos; see [4]). Moreover, (Fin) contains every operator of the form

$$A = T \oplus B + K, \qquad (*)$$

where $T \in \mathcal{L}(\mathcal{H}_0)$ (for some finite or infinite dimensional subspace $\mathcal{H}_0$ of infinite codimension), $B \simeq \oplus_{n=1}^{\infty} B_n$ for a suitable uniformly bounded sequence $\{B_n\}_{n=1}^{\infty}$ of operators acting on finite dimensional subspoces (i.e , B is a b l o c k - d i a g o n a l operator [4]) and K is compact.

Let $\mathcal{R}_0$ denote the family of all operators of the form (*). It is apparent that

$$\mathcal{R}_0 \subset \mathcal{R}^- \subset (\mathrm{Fin})$$

and

$$\mathcal{R}_0' \overset{\mathrm{def}}{=\!=} \mathcal{R}_0 \setminus \mathcal{R} \subset \mathcal{R}' \overset{\mathrm{def}}{=\!=} \mathcal{R}^- \setminus \mathcal{R} \subset (\mathrm{Fin})' \overset{\mathrm{def}}{=\!=} (\mathrm{Fin}) \setminus \mathcal{R}.$$

CONJECTURE 2 (D.A.Herrero).(Fin)$' = \mathcal{R}_0'$; moreover, if $A \in (\mathrm{Fin})'$ then there exits $K \in \mathcal{K}(\mathcal{H})$ and Q quasidiagonal such that

$$A - K \simeq A \oplus (Q \oplus Q \oplus Q \oplus \ldots). \qquad (**)$$

Several remarks are relevant here:

(1) J.W.Bunce has obtained several other equivalences, in addition to the three given by Theorem 1 (see [2]).

(2) $\mathcal{R}^-$ properly includes $\bigcup_{n \geq 1} (\mathcal{R}_n^-)$ [3, p.262], [5 , Example 11].

(3) In [1, Corollary 10.8], J.H.Anderson proved that every unilateral weighted shift is a finite operator, by exhibiting a functional f satisfying the conditions of Theorem 1 (iii). Recently, D.A. Herrero proved that all the (unilateral or bilateral) weighted shift operators belong to $\mathcal{R}^-$ [5].

(4) Suppose that $A \in \mathcal{R}^-$; then there exists a sequence $\{P_n\}_{n=1}^{\infty}$ of non-zero finite rank orthogonal projections such that $\|AP_n - P_nA\| \to 0$ $(n \to \infty)$. Passing, if necessary, to a subsequence we can directly assume that $P_n \to H$ (weakly, as $n \to \infty$) for some hermitian operator H , $0 \leqslant H \leqslant 1$. It is easily seen that A commutes with H . If H has a non-zero finite rank spectral projection, then $A \in \mathcal{R}$. If either $H=0$ or $H=1$, then it is not difficult to show that $A \in \mathcal{R}_0$.

(5) According to a well-known result of D.Voiculescu, $A-K \simeq A \oplus (Q \oplus Q \oplus Q \oplus \ldots)$ (as in (**)) if and only if the C^*-algebra generated by $\pi(A)$ in the Calkin algebra $\mathcal{L}(\mathcal{H})/\mathcal{K}(\mathcal{H})$ admits a $*$-representation ρ such that $\rho \circ \pi(A) \simeq Q$ [6]. Suppose that $A \in (Fin)'$. Is it possible to use the singular functionals f provided by Theorem 1 (iii) and a Gelfand-Naimark-Segal type construction in order to construct a $*$-representation ρ with the desired properties (i.e., so that $\rho \circ \pi(A)$ is quasidiagonal)?

(6) Is it possible, at least, to show that the existence of such a singular functional implies that, for each $\varepsilon > 0$, A admits a finite rank perturbation F_ε , with $\|F_\varepsilon\| < \varepsilon$, such that $A - F_\varepsilon \simeq A_\varepsilon \oplus A'_\varepsilon$, where A_ε acts on a non-zero finite dimensional subspace? (An affirmative answer to this last question implies that $(\mathrm{Fin})' = \mathcal{R}'_0$).

(7) Any partial answer to the above questions will also shed some light on several interesting problems related to quasidiagonal operators.

REFERENCES

1. A n d e r s o n J.H. Derivations, commutators and essential numerical range. Dissertation, Indiana University, 1971.
2. B u n c e J.W. Finite operators and amenable C^*-algebras. - Proc.Amer. Math.Soc., 1976, 56, 145-151.
3. B u n c e J.W., D e d d e n s J.A. C^*-algebras generated by weighted shifts. - Indiana Univ.Math.J., 1973, 23, 257-271.
4. H a l m o s P.R. Ten problems in Hilbert space.-Bull.Amer.

Math.Soc., 1970, 76, 887-933.
5. H e r r e r o D.A. On quasidiagonal weighted shifts and approximation of operators.-Indiana Univ.Math.J. (To appear).
6. V o i c u l e s c u D. A non-commutative Weyl-von Neumann theorem. - Rev.Roum.Math.Pures et Appl., 1976, 21, 97-113.
7. W i l l i a m s J.P. Finite operators. - Proc.Amer.Math.Soc., 1970, 26, 129-136.

DOMINGO A.HERRERO

Arizona State University
Tempe, Arizona 95287
USA

This research has been partially supported by a Grand of the National Science Foundation

4.39. THE SPECTRUM OF AN ENDOMORPHISM IN A COMMUTATIVE BANACH ALGEBRA

The theorem of H. Kamowitz and S. Scheinberg [1] establishes that the spectrum of a non-periodic automorphism $T: A \to A$ of a semi-simple commutative Banach algebra A (over $\mathbb{C}$) contains the unit circle $\mathbb{T}$. Several rather simple proofs of the theorem have been obtained besides the original one ([2], [3] e.g.) and its various generalizations found (see e.g. [4], [5], [6]...). It is easy to show that under the conditions of the theorem the spectrum is connected. At the same time all "positive" information is exhausted, apparently, by these two properties of the spectrum. There are examples (see [7], [8], [9]) demonstrating the absence of any kind of symmetry structure in the spectrum even if we suppose that the given algebra is regular in the sense of Shilov.

Let for instance K be a compact set in $\mathbb{C}$ lying in the annulus $\{z \in \mathbb{C} : 1/4 \leq |z| \leq 4\}$ and containing $\{z \in \mathbb{C} : 1/2 \leq |z| \leq 2\}$ and equal to the closure of its interior int K. Denote by A the family of all functions continuous on K and holomorphic in int K. Clearly, A equipped with the usual sup-norm on K and with the pointwise operations is a Banach algebra. The spectrum of multiplication by the "independent variable" on A evidently coincides with K. On the other hand the conditions imposed on K imply that A is a Banach algebra (without unit) with respect to the convolution

$$(f * g)(z) = \frac{1}{2\pi i} \int_{|\zeta|=1} f(\zeta)\, g(\bar{\zeta} z) \frac{d\zeta}{\zeta}$$

corresponding to multiplication of the Laurent coefficients. This algebra being semi-simple, its maximal ideal space can be identified with the set of all integers. Adjoining a unit to A thus turns it into a regular algebra. Obviously the above mentioned operator on A is an automorphism.

1. Are there other necessary conditions on the spectrum of a non-periodic automorphism of a semi-simple commutative Banach algebra besides the two mentioned above? In particular, is it obligatory for the spectrum to have interior points when it differs from $\mathbb{T}$? It is known in such cases (see [7], [8]) that the set of interior points

may not be dense in the spectrum and may not be connected either.

Let M_A be the maximal ideal space of a commutative and semi-simple Banach algebra A. An automorphism T of A induces an automorphism of the algebra $C = C(M_A)$. The essential meaning of the Kamowitz - Scheinberg theorem is that $\sigma_C(T) \subset \sigma_A(T)$. It is natural from this point of view to study the inclusion $\sigma_C(L) \subset \sigma_A(L)$ for a more general class of operators L. The case of weighted automorphisms $La \overset{def}{=\!=} u \cdot Ta$ with u an invertible element of A has, for example, been studied in [10]. It turns out that the inclusion does not hold for this class of operators.

2. <u>Does the spectrum of $L = uT$, constructed for a non-periodic automorphism T, contain any circle centred at the origin</u>? If it does then we obtain an instant generalization of the theorem of Kamowitz and Scheinberg.

The spectrum of operators, looking like L, acting on the algebra of all continuous functions on a compact set has a complete description [11]. If A is also a uniform algebra then $\sigma_C(L) = \sigma_A(L)$. Thus $\sigma_A(L) = \sigma_B(L)$ provided that L is a weighted automorphism of two uniform algebras A and B having the same maximal ideal space.

3. Let A be a closed subalgebra of a semi-simple commutative Banach algebra B and let $M_A = M_B$. Let L be a weighted automorphism of A and B simultaneously.

<u>Is it true then that</u> $\sigma_A(L) = \sigma_B(L)$?

We CONJECTURE that this question has a negative answer.

The spectrum of an endomorphism apparently does not have any particular properties even if we suppose that A is a uniform algebra. Given two compacta S_1 and S_2 it is easy to obtain an endomorphism with the spectrum either $S_1 \cup S_2$ or $S_1 \cdot S_2$. The only obvious property of spectra is that λ^n belongs to the spectrum, when λ ranges over its boundary and n over the set of non-negative integers.

4. Let K be a compact subset of $\mathbb{C}$ satisfying $\lambda^n \in K$ for $n = 1, 2, \dots$ and for all points λ in the boundary of K.

<u>Is there an endomorphism of a uniform algebra whose spectrum is equal to</u> K ?

Spectra of endomorphisms of uniform algebras (and even those for weighted endomorphisms) can be described pretty well under the additional assumption that the induced mapping of the maximal ideal space keeps the Shilov boundary invariant (see [12], where one can find references to preceding papers of Kamowitz). Roughly speaking, things, in this case, are going as well as in the case of Banach algebras of functions continuous on a compact set. The situation changes dramatically when the boundary, or only a part of it, penetrates the interior. In such circumstances it is common to begin with the consideration of classical examples. Let $\mathbb{D}$ be the unit disc in $\mathbb{C}$, let $A(\mathbb{D})$ be the algebra (disc-algebra) of all functions continuous on the closure of $\mathbb{D}$ and holomorphic in $\mathbb{D}$, and let $H^\infty(\mathbb{D})$ be the algebra of all functions bounded and holomorphic in $\mathbb{D}$. Both algebras $A(\mathbb{D})$ and $H^\infty(\mathbb{D})$ are equipped with the sup-norm.

5. Every endomorphism of $A(\mathbb{D})$ induces a natural endomorphism of H^∞.

Do the spectra of these endomorphisms coincide?

In this connection it is worth-while to note that the answer to an analogous question concerning the algebra of all continuous functions on a compact set and the algebra of all bounded functions on the same set is in the affirmative [13]. The proof of this result uses, however, a full (though comparatively simple) analysis of the possible spectral pictures depending on the dynamics generated by the endomorphism.

The interesting papers [14], [15], [16] of Kamowitz (see also [6], [9]) deal with spectra of the endomorphisms of $A(\mathbb{D})$ whose induced mappings do not preserve the boundary of $\mathbb{D}$. In the non-degenerate case the spectrum has a tendency to fill out the disc. Discrete and continuous spirals as well as compacta bounded by such spirals may nevertheless appear as the spectrum of an endomorphism. (But only the spirals can appear in the case of Möbius transformations).

6. Is the spectrum of an endomorphism of the disc-algebra a semi-group (with respect to multiplication in $\mathbb{C}$)? What kind of semi-groups can arise as spectra?

7. Is it possible to say something concerning the spectra of endomorphisms of natural multi-dimensional generalizations of the disc algebra?

Note that in the one-dimensional case the theory of Denjoy-Wolff and the interpolation theorem of Carleson-Newman are often involved in the question.

The problem of describing spectra for weighted automorphisms is closely related with an analogous one for the so-called "shift-type" operators which have been studied by A.Lebedev [17] and A.Antonevich [18].

Let A be a uniform algebra of operators on a Banach space X. An invertible operator U on X is called a "shift-type" operator if $UAU^{-1}=A$. Usually X is a Banach space of functions and A is a subalgebra of the algebra of multipliers for X. The transformation $a \to U\cdot aU^{-1}$ determines an automorphism T of A which induces the mapping $\varphi: M_A \to M_A$. It is assumed that:

1) the set of φ-periodic points is of first category in the Shilov boundary ∂A;

2) the spectrum of $U: X \to X$ is contained in ∂A;

3) each invertible operator $a: X \to X$, $a \in A$ is invertible as an element of A;

4) the topological spaces M_A and ∂A have the same stock of clopen (closed and open) φ-invariant subsets.

Then $\sigma_X(aU)=\sigma_A(aT)$ for all a in A [19].

<u>We conjecture that Condition 4 is superfluous.</u>

If this were true it would be possible (in view of [11]) to obtain a complete description of $\sigma_X(aU)$. It is reasonable to ask the same question for other algebras A besides the uniform ones.

REFERENCES

1. K a m o w i t z H., S c h e i n b e r g S. The spectrum of automorphisms of Banach algebras. - J.Funct.An., 1969, 4, N-2, 268-276.
2. J o h n s o n B.E. Automorphisms of commutative Banach algebras. - Proc.Am.Math.Soc., 1973, 40, N 2, 497-499.
3. Л е в и Р.Н. Новое доказательство теоремы об автоморфизмах банаховых алгебр. - Вестн.МГУ, сер.матем., мех., 1972, №4, 71-72.
4. Л е в и Р.Н. Об автоморфизмах банаховых алгебр. - Функц.анализ и его прил., 1972, 6, № 1, 16-18.
5. Л е в и Р.Н. О совместном спектре некоторых коммутирующих операторов. Диссертация, М., 1973.
6. Г о р и н Е.А. Как выглядит спектр эндоморфизма диск-алгебры? - Зап.научн.семин.ЛОМИ, 1983, 126, 55-68.

7. S c h e i n b e r g S. The spectrum of an automorphism. - Bull.Amer.Math.Soc., 1972, 78, N 4, 621-623.

8. S c h e i n b e r g S. Automorphisms of commutative Banach algebras. - Problems in analysis, Princeton Univ.Press., Princeton 1970, 319-323.

9. Г о р и н Е.А. О спектре эндоморфизмов равномерных алгебр. - В кн.: Тезисы Докл.конфер."Теоретические и прикладные вопросы математики" Тарту, I980, I08-II0.

I0. К и т о в е р А.К. О спектре автоморфизмов с весом и теореме Камовица-Шайнберга. - Функц.анализ и его прил., I979, I3,№№ I, 70-7I.

II. К и т о в е р А.К. Спектральные свойства автоморфизмов с весом в равномерных алгебрах. - Зап.научн.семин.ЛОМИ, I979, 92, 288-293.

I2. К и т о в е р А.К. Спектральные свойства гомоморфизмов с весом в алгебрах непрерывных функций и их приложения. - Зап.научн.семин.ЛОМИ, I982, I07, 89-I03.

I3. К и т о в е р А.К. Об операторах в $C^{(1)}$, индуцированных гладкими отображениями. - Функц.анализ и его прил., I982, I6, № 3, 6I-62.

14. K a m o w i t z H. The spectra of endomorphisms of the disk algebra. - Pacif.J.Math., 1973, 46, N 2, 433-440.

15. K a m o w i t z H. The spectra of endomorphisms of algebras of analytic functions. - Pacif. J.Math. 1976, 66, N 2, 433-442.

16. K a m o w i t z H. Compact operators of the form $u C_\varphi$. - Pacif.J.Math., 1979, 80, N 1, 205-211.

I7. Л е б е д е в А.В. Об операторах типа взвешенного сдвига. Диссертация, Минск, I980.

I8. А н т о н е в и ч А.Б. Операторы со сдвигом, порожденным действием компактной группы Ли. - Сибирск.матем.журн.I979, 20, № 3, 467-478.

I9. К и т о в е р А.К. Операторы подстановки с весом в банаховых модулях над равномерными алгебрами (в печати).

E.A.GORIN (Е.А.ГОРИН)

СССР, II7234, Москва Ленинские горы
Механико-математический факультет
Московский государственный университет

A.K.KITOVER (А.К.КИТОВЕР)

СССР, I9III9, Ленинград,
ул Константина Заслонова,
д.I4, кв.2.

4.40. COMPOSITION OF INTEGRATION AND SUBSTITUTION

Consider a continuous function φ on $[0,1]$ satisfying $\varphi(0)=0$, $0\leqslant\varphi(x)\leqslant 1$. The function φ defines a bounded linear operator I_φ on the space $C[0,1]$ of all continuous functions on $[0,1]$:

$$I_\varphi f(x) \stackrel{def}{=\!=} \int_0^{\varphi(x)} f(t)\,dt\,. \tag{1}$$

Recall that a bounded operator T is called quasinilpotent if $\lim\limits_{n\to+\infty}\|T^n\|^{\frac{1}{n}}=0$.

PROBLEM. <u>Describe functions φ corresponding to quasinilpotent operators I_φ</u>.

Clearly I_φ is quasinilpotent provided

$$\varphi(x)\leqslant x\,,\quad 0\leqslant x\leqslant 1\,. \tag{2}$$

<u>Does the inverse conclusion hold</u>? An analogy with the theory of matrices provides arguments in favour of the affirmative answer. Let $\{a_{ij}\}_{i,j=1}^{n}$ be a nilpotent matrix with non-negative elements. It follows from the Perron-Frobenius theorem that it can be transformed to a low-triangular form with zero diagonal by a permutation of the basis. The obtained matrix $\{\tilde a_{ij}\}$ defines an operator on $\mathbb{R}^n$

$$\{\xi_i\}_{1\leqslant i\leqslant n}\mapsto\Big\{\sum_{j=1}^{\varphi(i)}\tilde a_{ij}\,\xi_j\Big\}_{1\leqslant i\leqslant n}$$

with $\varphi(i)<i$.

Consider now a natural generalization of (1):

$$I_{\varphi,K}f(x)=\int_0^{\varphi(x)}K(x,t)\,f(t)\,dt\,, \tag{3}$$

where the kernel $K\geqslant 0$ is continuous. Orientation preserving homeomorphisms of $[0,1]$ replace permutations of the basis in the finite-dimensional case and preserve the inequality $\varphi(x)\leqslant x$.

Consider a counter-argument to the conjecture. If I_φ is not quasinilpotent then by Yentsch's theorem there exist $\rho > 0$ and a non-zero continuous function $f \geqslant 0$ such that

$$\rho f(x) = \int_0^{\varphi(x)} f(t)\,dt\,, \qquad 0 \leqslant x \leqslant 1\,. \tag{4}$$

Suppose $\varphi \in C^\infty([0,1])$. Then evidently $f \in C^\infty([0,1])$ and $f^{(k)}(0) = 0$, $k = 0,1,\dots$, because $\varphi(0) = 0$. If it were possible to prove that f belongs to a quasianalytic class (under some natural restrictions on φ), it would imply clearly that $f \equiv 0$ which contradicts Yentsch's theorem. Are there conditions on φ not demanding $\varphi(x) \leqslant x$ but such that any solution of (4) belongs to a quasianalytic Carleman class? If yes, then there exists φ such that $\varphi(t_0) > t_0$ for a point t_0 in $[0,1]$ but nevertheless I_φ is quasinilpotent.

Yu.I.LYUBIC

(Ю.И.ЛЮБИЧ)

СССР, 310077 Харьков
пл.Дзержинского 4
Харьковский государственный
университет

CHAPTER 5

HANKEL AND TOEPLITZ OPERATORS

A quadratic form is called Hankel (resp. Toeplitz) if entries of its matrix depend on the sum (resp. the difference) of indices only. These forms appeared as objects and tools in works of Jacobi, Stieltjes (and then Hilbert, Plemelj, Schur, Szegö, Toeplitz ...). They play a decisive role in a very wide circle of problems (various kinds of moment problems, interpolation by analytic functions, inverse spectral problems, orthogonal polynomials, Prediction Theory, Wiener-Hopf equations, boundary problems of Function Theory, the extension theory of symmetric operators, singular integral equations, models of statistical physics etc.etc.). It was understood only later that the independent development of this apparatus is a prerequisite for its applications to the above "concrete" fields, and Hankel and Toeplitz operators were singled out as the object of a separate branch of Operator Theory. This branch includes:

- techniques of singular integrals ranging from Hilbert, M.Riesz and Privalov to the Helson-Szegö theorem discovered as a fact of Prediction Theory, and to localization principles of Simonenko and Douglas;

- algebraic schemes originating from the fundamental concept of symbol of a singular integral operator (Mihlin), from the semi-multiplicative dependence of Toeplitz operator on its symbol, (Wiener-Hopf), and culminating in the operator K-theory;

- methods and techniques of extension theory (Krein), which have attracted a new interest to metric properties of Hankel operators and to their numerous connections;

- other important principles and ideas which we have either forgotten or overlooked or had no possibility to mention here.

The inverse influence of Hankel and Toeplitz operators is also considerable. For example, many problems of this chapter fit very well into the context of other chapters: Banach Algebras (Problem 5.6), best approximation (Problem 5.1), singular integrals (Problem 5.14). Problems 2.11, 3.1, 6.6, 10.2 can hardly be severed from spectral aspects of Toeplitz operators, and Problems 3.2, 3.3, 4.15, 4.21, 8.13, S.6, from Hankel operators. Many problems related to the Sz.-Nagy-Foiaş model (4.9-4.14)can be translated into the language of Hankel-Toeplitz (possibly, vectorial) operators, because functions $\varphi(\mathbf{T}_\theta)$ of the model operator $\mathbf{T}_\theta$ coincide essentially with the Hankel operators $H_{\theta^*\varphi}$, and the proximity of model subspaces K_{θ_1} and K_{θ_2} can be expressed in terms of the Toeplitz operator $T_{\theta_1^*\theta_2}$, etc.

Hankel-Toeplitz problems assembled in this book do not exhaust even the most topical problems of this direction*), but contain many interesting questions and suggest some general considerations. Many of the problems are inspired by some other fields and are rooted there so deeply that it is difficult to separate them from the corresponding context. We had to place some Hankel-Toeplitz problems (not without hesitation and disputes) into other chapters. Examples can be found in Chapter 3 (3.1, 3.2, 3.3). Moreover, we believe that

*) To our surprise nobody has asked, for instance, whether every Toeplitz operator has a non-trivial invariant subspace...

Problem 3.3 is one of the most characteristic and essential problems of exactly t h i s Chapter and we hope that the reader looking through this chapter will turn to Problem 3.3 as well.

Problems 5.1-5.3 deal with metric characteristics of Hankel operators (compactness, spectra, s-numbers). In connection with Problems 5.3 and 5.7 concerning operators acting n o t in H^2 we should like to mention recent investigations of S.Janson, J.Peetre and S.Semmes and of V.A.Tolokonnikov (Spring 1983) who have found $(X \to Y)$ -continuity criteria (in terms of symbols) for Hankel and Toeplitz operators in many non-Hilbert function spaces X, Y .

Problems 5.4 and 5.5 treat similarity invariants and some properties of the calculus for Toeplitz operators.

Problems 5.6, 5.13 are related with localization methods, problems 5.8-5.10, 5.15 deal with vectorial and multidimensional variants of Toeplitz operators and with related function-theoretic boundary problems.

Problems 5.11-5.14 treat "limit distributions of spectra" (asymptotics of Szegö determinants, convergence and other properties of projection invertibility methods etc).

The theme of Problem 5.16 may be viewed as a non-commutative analogue of Toeplitz operators arising in the theory of completely integrable systems.

* * *

The field of action and the multitude of connections of Hankel-Toeplitz operators are so impressive that it became fashionable nowadays to find them everywhere - from bases theory to models of Quantum Physics and ... even where they really do not occur...

5.1. old

APPROXIMATION OF BOUNDED FUNCTIONS BY ELEMENTS OF $H^\infty + C$

Every sequence $\{\gamma_n\}_{n\geq 1}$ of complex numbers defines a Hankel matrix $\Gamma = \{\gamma_{j+k-1}\}_{j,k\geq 1}$ which is considered as an operator Γ in the Hilbert space ℓ^2. By Nehari's theorem Γ is bounded if and only if there exists a function f in the algebra L^∞ of all bounded and measurable functions on $\mathbb{T}$ such that $\gamma_n = \hat{f}(-n)$, $n = 1, 2, \ldots$, $\hat{f}(n)$ being the Fourier coefficient $\int_{\mathbb{T}} f \cdot \bar{z}^n dm$ of f. This function is uniquely determined up to a summand from the Hardy algebra H^∞. The norm $\|\Gamma\|$ of Γ coincides with **dist** (f, H^∞). Given $f \in L^\infty$ let $\Gamma(f)$ be the Hankel operator corresponding to the sequence $\{\gamma_n\}_{n\geq 1}$, $\gamma_n = \hat{f}(-n)$, $n = 1, 2, \ldots$ Usual compactness arguments imply that for every $f \in L^\infty$ there exists $h \in H^\infty$ such that

$$\|\Gamma\| = \|f - h\|_{L^\infty}.$$

A criterion of u n i q u e n e s s of the best approximation h as well as a description of all such h's in the non-uniqueness case have been obtained in [2b]. Hartman has shown [3] that Γ is compact iff there exists a function f in the algebra of all continuous functions on $\mathbb{T}$ such that $\Gamma = \Gamma(f)$. Moreover, if Γ is compact then for every $\varepsilon > 0$ there exists $f_\varepsilon \in C$ satisfying $\Gamma(f_\varepsilon) = \Gamma$ and $\|f_\varepsilon\|_C \leq \|\Gamma\| + \varepsilon$.

The results of Nehari and Hartman easily imply the following result discovered independently by Sarason [4] : the algebraic sum $H^\infty + C$ is a closed subalgebra of L^∞ (see also [5] where many properties of this subalgebra are disussed). Hartman's result implies also the following characterization of $H^\infty + C$: an element f in L^∞ belongs to $H^\infty + C$ if and only if $\Gamma(f)$ is a compact operator.

Let $s_n(\Gamma)$, $n = 1, 2, \ldots$ denote s-numbers of Γ counted with multiplicities and let $s_\infty(\Gamma)$ be the least upper bound of the essential spectrum of $(\Gamma^*\Gamma)^{1/2}$ (cf. [6], § 7). Clearly, $s_n(\Gamma) \downarrow s_\infty(\Gamma)$ as $n \to +\infty$.

THEOREM. <u>Let</u> $f \in L^\infty$ <u>and</u> $\Gamma = \Gamma(f)$. <u>Then</u>

$$\operatorname{dist}(f, H^\infty + C) = s_\infty(\Gamma). \qquad (1)$$

Given a function f in $L^\infty \setminus (H^\infty + C)$ consider the following possibilities.

CASE 1. $s_\infty(\Gamma) = \|\Gamma\|$, i.e. Γ does not have s-numbers greater than $s_\infty(\Gamma)$.

CASE 2. There exists only a finite number $\varkappa$, $1 \leqslant \varkappa < \infty$ of s-numbers greater than $s_\infty(\Gamma)$.

CASE 3. The set of s-numbers to the right of $s_\infty(\Gamma)$ is infinite.

Formula (1) is a simple consequence of theorem 3.1 from [2a] but for the purpose of this note it is more convenient to connect it with the investigations of [2c].

For any positive integer k let H_k^∞ denote the set of all sums $f = h + r$, where $h \in H^\infty$ and r is a rational function of degree $\leqslant k$, having all its poles in the open unit disc $\mathbb{D}$ and vanishing at infinity. The set H_k^∞ is neither convex nor linear. This set coincides with the family of symbols f corresponding to the Hankel operators $\Gamma(f)$ of finite rank $\leqslant k$. Clearly $H_0^\infty \ (\overset{def}{=} H^\infty) \subset H_1^\infty \subset \ldots \subset H^\infty + C$ and $\bigcup_{n \geqslant 1} H_n^\infty$ is dense in $H^\infty + C$. Therefore $\operatorname{dist}(f, H^\infty + C) = \lim_{n \to \infty} \operatorname{dist}(f, H_n^\infty)$.

On the other hand it has been shown in [2c] that $\operatorname{dist}(f, H_k^\infty) = s_k(\Gamma)$ provided $s_k(\Gamma) > s_{k+1}(\Gamma)$. Hence (1) holds in case 3. In case 2 we have (see [2c], §5) $s_\infty(\Gamma) = \operatorname{dist}(f, H_\varkappa^\infty)$, which implies (1) also. At last, in case 1 we obviously have $\operatorname{dist}(f, H^\infty) = \operatorname{dist}(f, H^\infty + C)$.

Given $f \in L^\infty$ denote by M_f the set of all $g \in H^\infty + C$ satisfying $\|f - g\|_{L^\infty} = s_\infty(\Gamma(f))$. In case 1 $M_f \cap H^\infty \neq \emptyset$ and in case 2 $M_f \cap H_\varkappa^\infty \neq \emptyset$. A necessary and sufficient condition for $M_f \cap H_\varkappa^\infty$ to be a one-point set as well as a description of when it contains more than one element, have been obtained.

As for case 3 it is UNKNOWN:

a) <u>Can M_f be empty for some $f \in L^\infty$ and if so how to describe such f's</u> ?

b) <u>Can M_f consist of a single element for some f in</u> $L^\infty \setminus (H^\infty + C)$?

c) <u>If $M_f \neq \emptyset$ then is it possible to describe at least a "selected" part of M_f just as in case 2 (with $M_f \cap H^\infty$ as a "selected" part</u>)?

Clearly $M_f \neq \emptyset$ for $f \in L^\infty \setminus (H^\infty + C)$ if and only if there exists $g \in C$ such that for $f_1 = f - g$ case 1 holds, i.e. $\|\Gamma(f_1)\| = S_\infty(\Gamma(f_1))$. Question b) remains interesting for cases 1 and 2 also. The matter is that there are situations when card $M_f = \infty$ but card $(M_f \cap H^\infty) = 1$ (in case 1) and card $(M_f \cap H^\infty_{\varkappa}) = 1$ (in case 2). Indeed, let for example b be an inner function with singularities on an arc $\Delta \subset \mathbb{T}$, $m(\Delta) < 1$, and let φ be any function in C satisfying $\varphi(\zeta) \equiv 1$ on Δ and $|\varphi(\zeta)| < 1$ for $\zeta \in \mathbb{T} \setminus \Delta$. Then $\frac{1-\varphi}{b} \in C$ and setting $f = 1/b$ we have

$$S_\infty(\Gamma(f)) = \mathrm{dist}(f, H^\infty + C) = \mathrm{dist}(f, H^\infty) = 1 = \|f - \frac{1-\varphi}{b}\| .$$

However, $\|f - h\| > 1$ for every $h \in H^\infty$, $\|h\| > 0$.

Almost everything said above can be generalized to the case of matrix-valued functions $F = (f_{jk})$ with entries f_{jk} belonging to L^∞, H^∞ or C. In this case the norm $\|F\|$ of $F \in L^\infty_{m\times n}$ should be defined as $\operatorname{ess\,sup}_{\zeta \in \mathbb{T}} |F(\zeta)|$, where $|A|$ stands for the Hilbert-Schmidt norm of A. In connection with these generalizations we refer to [2d].

REFERENCES

1. N e h a r i A. On bounded bilinear forms. - Ann.Math., 1957, (2), 65, 153-162.

2. А д а м я н В.М., А р о в Д.З., К р е й н М.Г. a) Бесконечные ганкелевы матрицы и обобщенные задачи Каратеодори-Фейера и Ф.Рисса. - Функц.анал. и его прил., 1968, 2, 1, 1-19; b) Бесконечные ганкелевы матрицы и обобщенные задачи Каратеодори-Фейера и И.Шура. - Функц.анал. и его прил., 1968, 2, 4, 1-17; c) Аналитические свойства пар Шмидта ганкелева оператора и обобщенная задача Шура-Такаги. - Матем.сб., 1971, 85 (128), № 19, 33-73; d) Бесконечные блочно-ганкелевы матрицы и связанные с ними проблемы продолжения. - Изв.АН АрмССР, 1971, УI, № 2-3, 87-112.

3. H a r t m a n P. On completely continuous Hankel matrices. - Proc.Amer.Math.Soc., 1958, 9, 862-866.

4. S a r a s o n D. Generalized interpolation in H^∞. - Trans. Amer.Math.Soc., 1967, 127, 179-203.

5. S a r a s o n D. Algebras of functions on the unit circle. - Bull.Amer.Math.Soc., 1973, 79, N 2, 286-299.

6. Г о х б е р г И.Ц., К р е й н М.Г. Введение в теорию линейных несамосопряженных операторов. – М., Наука, 1965.

V.M.ADAMYAN
(В.М.АДАМЯН) СССР, 270000, Одесса, Одесский гос.университет;

D.Z.AROV
(Д.З.АРОВ) СССР, 270020, Одесса, Одесский пед.институт;

M.G.KREIN
(М.Г.КРЕЙН) 270057, Одесса, ул.Артёма 14, кв.6

* * *

COMMENTARY BY THE AUTHORS

Soon after the Collection "99 unsolved problems in linear and complex analysis" LOMI, vol.81 (1978) was published S.Axler, I.D.Berg, N.Jewell and A.Schields wrote an important paper on the theory of approximation of continuous operators in Banach space by compact operators, where they obtained, in particular, the answers to questions a) and b) of **the Problem**. The answers to both questions turned out to be negative. So for every function $f\in L^\infty$ the set M_f is not void and moreover for any $f\in L^\infty\setminus(H^\infty+C)$ the set M_f is infinite.

These results were obtained in [1] as consequences of two remarquable propositions, which we formulate for Hilbert space operators only.

THEOREM. Let $\{T_n\}$ be a sequence of linear compact operators on Hilbert space converging in the strong topology to a bounded operator $T: s\text{-}\lim T_n = T$. Suppose also that $T^* = s\text{-}\lim T_n^*$. Then there exists a sequence $\{a_n\}$ of non-negative numbers such that $\sum a_n = 1$ and

$$\| T-K \| = S_\infty(T)$$

where

$$K = \sum a_n T_n .$$

COROLLARY. Let T and $\{T_n\}$ satisfy the conditions of the theorem. Suppose also that T is not compact. Then there exist two sequences $\{a_n\}$, $\{b_n\}$ of non-negative numbers such that $\sum a_n = \sum b_n = 1$ and

$$\| T-K_a \| = \| T-K_b \| = s_\infty(T)$$

where $K_a = \sum a_n T_n$, $K_b = \sum b_n T_n$, and $K_a \neq K_b$.

The indicated propositions permit also to give answers to questions of type a) and b) for matrix-functions $f \in L^\infty_{m\times n} \setminus (C_{m\times n} + H^\infty_{m\times n})$.

As we got to know from [7] question a) was raised before us by D.Sarason.

Question c) concerning the description of the set M_f or its "selected" part remains open.

REFERENCE

7. Axler S., Berg I.D., Jewell N., Shields A. Approximation by compact operators and the space $H^\infty + C$. - Ann.Math., 1979, 109, 601-612.

* * *

EDITORS' NOTE

Question a) is solved also in a different way by D.Luecking [8]. Let us mention also a recent result of C.Sundberg [9] asserting that the algebra H^∞+ BUC (BUC is the space of bounded uniformly continuous functions on $\mathbb{R}$) does not have the best approximation property, i.e. there exists $f \in L^\infty$ such that there is no g in H^∞+ BUC satisfying $\| f-g \|_\infty = dist_{L^\infty}(f, H^\infty + BUC)$

REFERENCES

8. Luecking D. The compact Hankel operator form an M-ideal in the space of Hankel operators. - Proc.Amer.Math.Soc., 1980, 79, 222-224.

9. Sundberg C. H^∞+ BUC does not have the best approximation property. Preprint, Inst.Mittag-Leffler, 13, 1983.

5.2. QUASINILPOTENT HANKEL OPERATORS

Hankel operators possess little algebraic structure. This fact handicaps attempts to elucidate their spectral theory. The following sample problem is of untested depth and has some interesting function theoretic and operator theoretic connections.

PROBLEM. Does there exist a non-zero quasinilpotent Hankel operator?

A Hankel operator A on H^2 is one whose representing matrix is of the form $(a_{i+j})_{i,j=0}^{\infty}$, with respect to the standard orthonormal basis. A well known theorem of Nehari shows that we may represent A as $A=S_\varphi=P\mathcal{J}M_\varphi|H^2$ where P is the orthogonal projection of L^2 onto H^2, $\mathcal{J}$ is the unitary operator defined by $(\mathcal{J}f)(z)=f(\bar{z})$, for f in L^2, and M_φ denotes multiplication by a function φ in L^∞. The symbol function φ and the defining sequence a_n are connected by $\hat{\varphi}(-n)=a_n$, $n=0,1,2,\ldots$. The following observation appears to be new and provides a little evidence against existence.

PROPOSITION. There does not exist a non zero nilpotent Hankel operator.

PROOF. Suppose $A\neq 0$ and is nilpotent. Then ker A is a non zero invariant subspace for the unilateral shift U, since $AU=U^*A$. By Beurling's theorem this subspace is of the form uH^2 for some non constant inner function u. Thus, with the representation above, we have $S_{\varphi u}=0$ and hence $\widehat{\varphi u}(n)=0$ for $n\geqslant 1$. So the symbol function φ may be written in the factored form $\varphi=z\bar{u}h$ for some h in H^∞, and we may assume (by cancellation) that u and h possess no common inner divisors. The operator $S_{z\bar{u}}$ is a partial isometry with support space $K=H^2\ominus uH^2$ and final space $K^\dagger=H^2\ominus u^\dagger H^2$, where $g^\dagger(z)=\overline{g(\bar{z})}$. By the hypothesised nilpotence of $S_{z\bar{u}h}=S_{z\bar{u}}M_h$ it follows that for some non zero function f in $K^\dagger$, hf belongs to uH^2. Hence u divides f, and $f=uf_1$ with f_1 in H^2. Since f belongs to $K^\dagger$ we have $P(\bar{u}^\dagger uf_1)=0$. This says that the Toeplitz operator $T_{\bar{u}^\dagger u}$ has non trivial kernel. But $\ker T^*_{\bar{u}^\dagger u}=\ker T_{u^\dagger\bar{u}}=\ker T_{(u\bar{u}^\dagger)^\dagger}=$

$=(\ker T_{\bar{u}^\dagger u})^\dagger$, and we have a contradiction of Coburn's alternative: Either the kernel or the co-kernel of a non zero Toeplitz operator is trivial. ●

Function Theory. The evidence for existence is perhaps stronger. There are many compact non self-adjoint Hankel operators, so perhaps a non zero one can be found which has no non zero eigenvalues. A little manipulation reveals that λ is an eigenvalue for S_φ if and only if there is a non zero function f in H^2 (the eigenvector) and a function g in zH^2 such that

$$\varphi(z) = \frac{\lambda f(\bar{z}) + g(z)}{f(z)} . \tag{1}$$

Since continuous functions induce compact Hankel operators it would be sufficient then to find a continuous function φ which fails to be representable in this way for every $\lambda \neq 0$. Whilst the singular numbers of a Hankel operator A (the eigenvalues of $(A^*A)^{1/2}$) have been successfully characterized (see for example [3 , Chapter 5]), less seems to be known about eigenvalues.

Operator theory. It is natural to examine (1) when the symbol can be factored as $\varphi = z\bar{u}h$ (cf. the proof above) with u an interpolating Blaschke product. The corresponding Hankel operators and function theory are tractable in certain senses (see [1], [2, Part 2] and [3, Chapter 4]), partly because the functions $(1-|\alpha_i|^2)^{1/2}(1-\bar{\alpha}_i z)^{-1}$, where $\alpha_1, \alpha_2, \dots$ are the zeros of u , form a Riesz basis for $H^2 \ominus uH^2$. It turns out that S_φ is compact if $h(\alpha_1), h(\alpha_2), \dots$ is a null sequence. A quasinilpotent compact Hankel operator of this kind will exist if and only if the following problem for operators on ℓ^2 can be solved.

PROBLEM. Construct an interpolating sequence α_i and a compact diagonal operator D so that the equation $DXx = \lambda x$ admits no proper solutions x in ℓ^2 when $\lambda \neq 0$. Here X is the bounded(!) operator on ℓ^2 associated with $\alpha_1, \alpha_2, \dots$ determined by representing matrix

$$x_{i,j} = \frac{(1-|\alpha_i|^2)^{1/2}(1-|\alpha_j|^2)^{1/2}}{1-\alpha_i \alpha_j} .$$

REFERENCES

1. Clark D.N. On interpolating sequences and the theory of Hankel and Toeplitz matrices. - J. Functional Anal. 1970, 5, 247-258.

2. H r u š č ë v S.V., N i k o l'skii N.K., P a v l o v B.S. Unconditional bases of exponentials and of reproducing kernels. - Lect.Notes Math. 1981,N864, Springer Verlag.
3. P o w e r S.C. Hankel operators on Hilbert space. - Research Notes in Mathematics. 1982. N 64, Pitman, London.

S.C.POWER

Dept. of Mathematics
Michigan State University
E.Lansing, MI 48824
USA

Usual Address:
Dept.of Mathematics
University of Lancaster
Bailrigg, Lancaster LAi 4YW
England

5.3. HANKEL OPERATORS ON BERGMAN SPACES

Let dA denote the usual area measure on the open unit disk $\mathbb{D}$. The Bergman space $L^2_a(\mathbb{D})$ is the subspace of $L^2(\mathbb{D}, dA)$ consisting of those functions in $L^2(\mathbb{D}, dA)$ which are analytic on $\mathbb{D}$. Let P denote the orthogonal projection of $L^2(\mathbb{D}, dA)$ onto $L^2_a(\mathbb{D})$. For $f \in L^\infty(\mathbb{D}, dA)$, we define the Toeplitz operator

$$T_f : L^2_a(\mathbb{D}) \to L^2_a(\mathbb{D})$$

and the Hankel operator

$$H_f : L^2_a(\mathbb{D}) \to L^2(\mathbb{D}, dA) \ominus L^2_a(\mathbb{D})$$

by $T_f h = P(fh)$ and $H_f h = (I-P)(fh)$.

<u>For which functions</u> $f \in L^\infty(\mathbb{D}, dA)$ <u>is the Hankel operator compact</u>?

If we were dealing with Hankel operators on the circle $\mathbb{T}$ rather than the disk $\mathbb{D}$, the answer would be that the symbol must be in the space $H^\infty + C(\mathbb{T})$. On the disk $\mathbb{D}$, it is easy to see that if $f \in H^\infty + C(\overline{\mathbb{D}})$, then H_f is compact. However, it is not hard to construct an open set $S \subsetneqq \mathbb{D}$ with $\overline{S} \cap \mathbb{T} \neq \emptyset$ such that if f is the characteristic function of S, then H_f is compact. Thus the subset of $L^\infty(\mathbb{D}, dA)$ which gives compact Hankel operators is much bigger (in a nontrivial way) than $H^\infty + C(\overline{\mathbb{D}})$, and it is possible that there is no nice answer to the question as asked above.

A more natural question arises by considering only symbols which are complex conjugates of analytic functions:

<u>For which</u> $f \in H^\infty$ <u>is</u> $H_{\bar f}$ <u>compact</u>?

It is believable that this question has a nice answer. A good candidate is that $\bar f$ must be in $H^\infty + C(\mathbb{D})$.

The importance of this question stems from the identity

$$T_f^* T_f - T_f T_f^* = H_{\bar f}^* H_{\bar f},$$

valid for all $f \in H^\infty$. Thus we are asking which Toeplitz operators on the disk with analytic symbol have compact self-commutator.

Readers familiar with a paper of Coifman, Rochberg, and Weiss

[1] might think that paper answers the question above. Theorem VIII of [1] seems to determine precisely which conjugate analytic functions give rise to compact Hankel operators. However, the Hankel operators used in [1] are (unitarily equivalent to) multiplication followed by projection onto $E=\{\bar{f} : f \in L^2_a(\mathbb{D}) \text{ and } f(0)=0\}$. Since $L^2(\mathbb{D}, dA) \ominus L^2_a(\mathbb{D})$ is far bigger than E, the Hankel operators of [1] are not the same as the Hankel operators defined here.

The Hankel operators as defined in [1] are more natural when dealing with singular integral theory, but the close connection with Toeplitz operators is lost. To determine which analytic Toeplitz operators are essentially normal, the Hankel operators as defined here are the natural objects to study.

REFERENCE

1. Coifman R.R., Rochberg R., Weiss G. Factorization theorems for Hardy spaces in several variables. - Annals of Mathematics 1976, 103, 611-635.

SHELDON AXLER

Michigan State University
East Lansing, MI 48824, USA

5.4. old A SIMILARITY PROBLEM FOR TOEPLITZ OPERATORS

Consider the Toeplitz operator T_F acting on $H^2 = H^2(\mathbb{D})$, where F is a rational function, with $F(\mathbb{T})$ contained in a simple closed curve Γ. Let τ be the conformal map from $\mathbb{D}$ to the interior of Γ, and say that F b a c k s u p at $e^{i\theta}$ if $\arg \tau^{-1} F(e^{it})$ is decreasing in some closed interval $[\theta_1, \theta_2]$, where $\theta_1 < \theta_2$ and $\theta_1 \leqslant \theta \leqslant \theta_2$. Let $\gamma_1, \dots, \gamma_n$ be disjoint arcs on $\mathbb{T}$ such that F is one-to-one on each γ_k and such that $\cup \gamma_k$ is the set of all points where F backs up. Several recent results suggest the following

CONJECTURE. <u>Suppose</u> $F(e^{it})$ <u>has winding number</u> $\nu \geqslant 0$. <u>Then</u> T_F <u>is similar to</u>

$$T_{\tau(z^\nu)} \oplus M_1 \oplus \dots \oplus M_n \tag{1}$$

<u>if</u> $\nu > 0$, <u>and to</u>

$$M_1 \oplus \dots \oplus M_n \tag{2}$$

<u>if</u> $\nu = 0$, <u>where</u> M_k <u>is the operator of multiplication by</u> $F(e^{it})$ <u>on</u> $L^2(\gamma_k)$.

One case of the above conjecture goes back to Duren [1], where it was proved for $F(z) = \alpha z + \frac{\beta}{z}$, $|\alpha| > |\beta|$. In this case $\nu = 1$ and F never backs up, so that $M_1, \dots, M_n$ are not present in (1). Actually, Duren did not obtain similarity, but proved that T_F satisfies

$$L T_F = T_g L,$$

where L is some conjugate-linear operator ($L(\lambda_1 x + \lambda_2 y) = \bar{\lambda}_1 L(x) + \bar{\lambda}_2 L(y)$) and g is the mapping function for the interior of $\operatorname{clos} F(\mathbb{T})$. In [2], the conjecture was proved in case F is ν - to-one in some annulus $s \leqslant |z| \leqslant 1$. Here again F never backs up. In [3], F was assumed to have the form

$$F = \varphi / \psi, \tag{3}$$

where φ and ψ are finite Blaschke products, ψ having only one zero. In this case $\tau(z) = z$ and n can be taken to be 1.

The main tool used in [3] was the Sz.-Nagy-Foiaş characteristic function of T_F , which we computed explicitly and which, as we showed, has a left inverse. A theorem of Sz.-Hagy-Foiaş, [4], Theorem 1.4, was then used to infer similarity of T_F with an isometry. Moreover, the unitary part in the Wold Decomposition of the isometry could be seen to have multiplicity 1, and so the proofs of the representations (1) and (2) were reduced to spectral theory.

If F is of the form (3) where φ and ψ are finite Blaschke products, ψ having m o r e t h a n o n e zero, the computation of the characteristic function of T_F is no longer easy. However, left invertibility can sometimes be proved without explicit computation. This is the case if φ and ψ have the same number of zeros; i.e., when T_F is similar to a unitary operator. This and some other results related to the conjecture are given in [5]. The Sz.-Nagy-Foiaş theory may also be helpful in attempts to formulate and prove a version of the conjecture when (3) holds with φ and ψ arbitrary inner functions. For example, it follows from [3] that if φ is inner and ψ is a Blaschke factor, then T_F is similar to an isometry.

For the case in which Γ is not the unit circle, the only successful techniques so far are those of [2], which do not use model theory. We have not been successful in extending them beyond the case of F satisfying the "annulus hypothesis" described above. There is a model theory which applies to domains other than $\mathbb{D}$ [6], but to our knowledge no results on similarity are a part of this theory. More seriously, to apply the theory, one would NEED TO KNOW that the spectrum of T_F is a spectral set for T_F ; a result which does not seem to be known for rational F at this time.

Finally, it seems hardly necessary to give reasons why the conjecture would be a desirable one to prove. Certainly detailed information on invariant subspaces, commutant, cyclic vectors and functional calculus would follow from this type of result.

REFERENCES

1. D u r e n P.L. Extension of a result of Beurling on invariant subspaces. - Trans.Amer.Math.Soc.1961, 99, 320-324.
2. C l a r k D.N., M o r r e l J.H. On Toeplitz operators and similarity. - Amer.J.Math., 1978, 100, N 5, 973-986.
3. C l a r k D.N., Sz.-Nagy-Foiaş theory and similarity for a class of Toeplitz operators. - Banach Center Publications,v 8,

Spectral Theory, 1982, 221-229

4. Sz.-Nagy B., Foiaş C. On the structure of intertwining operators. - Acta Sci.Math. 1973, 35, 225-254.
5. Clark D.N. Similarity properties of rational Toeplitz operators. In preparation.
6. Sarason D. On spectral sets having connected complement. - Acta Sci.Math. 1965, 26, 289-299.

DOUGLAS N.CLARK

The University of Georgia, Athens, Georgia 30601 USA

* * *

COMMENTARY BY THE AUTHOR

Since my first note on the similarity problem, the following results have been obtained.

THEOREM 1 ([8]). If F is a rational function, mapping $\mathbb{T}$ into a simple closed curve Γ, which is analytic in a neighborhood of $F(\mathbb{T})$; if $\nu \geqslant 0$, where ν is the winding number of $F(\mathbb{T})$ about the points interior to Γ; and if $F(\Sigma) \neq \Gamma$, where Σ is the set (on $\mathbb{T}$) where F backs up; then T_F is similar to $T_\tau^{(\nu)} \oplus V$, where $T_\tau^{(\nu)}$ is the sum of ν copies of the analytic Toeplitz operator associated with the mapping function from $\mathbb{D}$ to the interior of Γ and where V is a normal operator whose spectrum is $F(\Sigma)$. V is absolutely continuous and the spectral multiplicity of a point λ in the spectrum of V is equal to the number of points e^{it} where F backs up and $F(e^{it}) = \lambda$.

THEOREM 2. ([9],[10]). If F is a rational function; if $F(\mathbb{T})$ divides the plane into disjoint regions, from which the ones in which the index of $T_F - \lambda I$ is negative (resp.positive) are labeled ℓ_i (resp. $\mathcal{L}_i$); if the closures of any two of these $(\ell_i, \mathcal{L}_i)$ intersect at only finitely many points (called the multiple points of F); if the boundary of each ℓ_i, $\mathcal{L}_i$ is an analytic curve except at the multiple points, where it is piecewise smooth with inner angle $\neq 0$; if no multiple point is the image under F of a point $z_0 \in T$ where $F'(z_0)=0$; and if F never backs up at a multiple point; then T_F is similar to

$$\Sigma^{\oplus} T_{\tau_i} \oplus \Sigma^{\oplus} T^*_{T_i} \oplus V,$$

where τ (resp. T_i) is the mapping function of $\mathbb{D}$ on l_i (resp. $\mathcal{L}_i$), each summand is included with multiplicity equal to the absolute value of the index of $T_F - \lambda I$ for $\lambda \in l_i$ (resp. $\lambda \in \mathcal{L}_i$), and V is as described in Theorem 1.

THEOREM 3. (Wang, [14]). If $F \in C^4(\mathbb{T})$ and if $F(e^{it})$ is the restriction to $\mathbb{T}$ of a function F analytic in $r<|z|<1$ for some $r<1$, if $F(e^{it})$ is 1-to-1 , $F'(e^{it})$ never vanishes and $t \to F(e^{it})$ is orientation preserving, then T_F is similar to T_τ , where τ is the mapping function from $\mathbb{D}$ to the interior of $F(\mathbb{T})$.

Theorem 1, of course, PROVES THE CONJECTURE posed in my first note on the similarity problem, EXCEPT WHEN the curve $F(\mathbb{T})$ has singularities. The case of a singularity with nonzero inner angle can be settled using the methods of [10], but the case of zero inner angle remains open, owing to our lack of understanding of the behavior of τ and τ' near such a point.

Theorem 2 also excludes zero interior angles in the l_i and $\mathcal{L}_i$. The hypothesis that $F'(z_0)\neq 0$ at the inverse images of the multiple points can be somewhat weakened [10] but not removed. In fact, [10] contains an example of a T_F satisfying Theorem 2 (with such a weakening) and a rational, orientation preserving, homeomorphism φ of $\mathbb{T}$ such that T_F and $T_{F\circ\varphi}$ are not similar.

The case of intersecting loops of $F(\mathbb{T})$ was discussed in [11]. As indicated there, nothing is known unless $F(\mathbb{T})$ is the image of $\mathbb{T}$ under a function analytic is $\mathbb{D}$. Unfortunately, even the proof of Theorem 5 of [11] is incomplete without further hypotheses, as pointed out by Stephenson [13].

Theorem 3 is the first attempt at a systematic similarity theory for non-rational F , and one hopes that it may be generalized to the point of a non-rational version of Theorems 1 and 2. Other examples of similarity for non-rational T_F may be obtained using Cowen's Equivalence Theorem [12], which implies that $T_{F\circ\varphi}$ is unitarily equivalent to a direct sum of countably many copies of T_F , when φ is a non-rational inner function.

An obstruction to similarity between T_F and a "reasonably nice" operator can occur when T_F has a boundary eigenvalue, see Clancey [7, §4.2]. In fact, as Clancey has pointed out to me, the spectrum of T_F is not a k-spectral set, for the example given in [7].

REFERENCES

7. C l a n c e y K.F. Toeplitz models for operators with one dimensional self-commutators (to appear).
8. C l a r k D.N. On a similarity theory for rational Toeplitz operators. - J.Reine Angew.Math. 1980, 320, 6-31.
9. C l a r k D.N. On Toeplitz operators with loops. - J.Operator Theory, 1980, 4, 37-54.
10. C l a r k D.N. On Toeplitz operators with loops, II. - J.Operator Theory 1982, 7, 109-123.
11. C l a r k D.N. On the structure of rational Toeplitz operators. - In: Contributions to Analysis and Geometry, supplement to Amer. J.Math. 1981, 63-72.
12. C o w e n C.C. On equivalence of Toeplitz operators. - J.Operator Theory 1982, 7, 167-172.
13. S t e p h e n s o n K. Analytic functions of finite valence, with applications to Toeplitz operators (to appear).
14. W a n g D. Similarity and boundary eigenvalues for a class of smooth Toeplitz operators (to appear).

5.5. ITERATES OF TOEPLITZ OPERATORS WITH UNIMODULAR SYMBOLS

Each invertible Toeplitz operator T_f on H^2 can be represented as $T_f = T_u T_h$ where h is an outer function with modulus $|f|$ and $u = f/h$ is a unimodular function. The operator T_h, being invertible analytic Toeplitz operator, has simple spectral behaviour. Therefore the Toeplitz operators with unimodular symbols play an especial role (see [1]).

I would like to propose the following questions concerning these operators.

Suppose X is one of the function classes $H^\infty + C$, $QC \overset{def}{=} H^\infty + C \cap \overline{H^\infty + C}$, C, C^k, C^∞.

QUESTION 1. Let u be a unimodular function in X such that $\operatorname{Ker} T_u = \{\mathbb{O}\}$. Is it true that there exists f in H^2 with

$$\inf_{n>0} \| T_u^n f \| > 0 ?$$

If the answer is positive, it is reasonable to ask whether the following stronger conclusion can be done.

QUESTION 2. Is it true that under the hypotheses of Question 1

$$\inf_{n>0} \| T_u^n f \| > 0$$

for every non-zero f in H^2?

It follows from Clark's results [2] that the answer to Question 2 is positive for rational functions u (see also Commentary to 5.4).

In view of T.Wolff's factorization theorem [3] (asserting that each unimodular function u in $H^\infty + C$ can be represented as $u = v\theta$ with $v \in QC$ and θ inner) it seems plausible that if Question 1 (or 2) has a positive answer for $X = QC$ then so is for $X = H^\infty + C$. Note that for general unimodular functions u with $\operatorname{Ker} T_u = \{\mathbb{O}\}$ it may happen that $\lim_{n>0} \| T_u^n f \| = 0$ for any f in H^2. For example, if E is a measurable subset of $\mathbb{T}$, $0 < \operatorname{meas} E < 1$ and $u = 1$ on E and -1 on $\mathbb{T} \setminus E$ then it follows from M.Rosenblum's results [4] that T_u is a selfadjoint operator with absolutely continuous spectrum on $[-1, 1]$.

An affirmative answer to Question 1 for $X = QC$ would imply the existence of a non-trivial invariant subspaces for Toeplitz operators with unimodular symbols in QC (see some results on invariant subspaces of Toeplitz operators in [5]). Indeed, either one of the kernels $\operatorname{Ker} T_u$, $\operatorname{Ker} T_u^*$ is non-trivial or T_u and $T_{\bar u}$ satisfy the hypothesis at Question 1 and so both subspaces

$$\mathcal{L}_1 = \{ f : \lim_n \| T_u^n f \| = 0 \}$$

$$\mathcal{L}_2 = \{ f : \lim_n \| T_u^n f \| = 0 \}^{\perp}$$

are invariant under T_u and $\mathcal{L}_1 \neq H^2, \mathcal{L}_2 \neq \{\mathbb{O}\}$. Therefore either one of these subspaces is non-trivial or $\mathcal{L}_1 = \{\mathbb{O}\}$, $\mathcal{L}_2 = H^2$, i.e. T_u would be a C_{11} contraction (see [6]). But each C_{11} contraction has a non-trivial invariant subspace [6].

REFERENCES

1. S a r a s o n D. Function theory on the unit circle. - Notes for Lect.at a conference at Virginia Polytechnic Inst. and State Univ., 1978.
2. C l a r k D.N. On a similarity theory for rational Toeplitz operators. - J.Reine Angew.Math., 1980, 320, 6-31.
3. W o l f f T. Two algebras of bounded functions. - Duke Math.J., 1982, 49, N 2, 321-328.
4. R o s e n b l u m M. The absolute continuity of Toeplitz's matrices. - Pacif.J.Math., 1960, 10, N 3, 987-996.
5. P e l l e r V.V. Invariant subspaces for Toeplitz operators. - LOMI Preprints, E-7-82, Leningrad, 1982.
6. S z . - N a g y B., F o i a ş C. Harmonic analysis of operators on Hilbert space, North Holland, Amsterdam, 1970.

V.V.PELLER
(В.В.ПЕЛЛЕР)

СССР, I9IOII, Ленинград,
Фонтанка 27 , ЛОМИ

5.6. old LOCALIZATION OF TOEPLITZ OPERATORS

Let H^2 and H^∞ denote the Hardy subspaces of $L^2(\mathbb{T})$ and $L^\infty(\mathbb{T})$, respectively, consisting of the functions with zero negative Fourier coefficients and let P be the orthogonal projection from $L^2(\mathbb{T})$ onto H^2. For φ in $L^2(\mathbb{T})$ the Toeplitz operator with symbol φ is defined on H^2 by $T_\varphi f = P(\varphi f)$. Much of the interest in Toeplitz operators has been directed toward their spectral characteristics either singly or in terms of the algebras of operators which they generate. In particular, one seeks conceptual determinations of why an operator is or is not invertible and more generally Fredholm. One fact which one seeks to explain is the result due to Widom [1] that the spectrum $\sigma(T_\varphi)$ of an arbitrary Toeplitz operator is a connected subset of $\mathbb{C}$ and even [2] the essential spectrum $\sigma_e(T_\varphi)$ is connected. The latter result implies the former in view of Coburn's Lemma.

An important tool introduced in [2], [3] is the algebraic notion of localization. Let $\mathfrak{J}$ denote the closed algebra generated by all Toeplitz operators and QC be the subalgebra

$$(H^\infty + C) \cap \overline{(H^\infty + C)}$$

of L^∞, where C denotes the algebra of continuous functions on $\mathbb{T}$. Each ξ in the maximal ideal space M_{QC} of QC determines a closed subset X_ξ of M_{L^∞} and one can show that the closed ideal ϑ_ξ in $\mathfrak{J}$ generated by

$$\{ T_\varphi : \hat{\varphi} \,|\, X_\xi \equiv 0 \}$$

is proper and that the local Toeplitz operator $T_\varphi + \vartheta_\xi$ in $\mathfrak{J}_\xi = \mathfrak{J}/\vartheta_\xi$ depends only on $\hat{\varphi}\,|\,X_\xi$. Moreover, since $\bigcap_{\xi \in M_{QC}} \vartheta_\xi$ equals the ideal $\mathcal{K}$ of compact operators on H^2, properties which are true modulo $\mathcal{K}$ can be established "locally". For example, T_φ is Fredholm if and only if $T_\varphi + \vartheta_\xi$ is invertible for each ξ in M_{QC}. These localization results are established [4] by identifying QC as the center of $\mathfrak{J}/\mathcal{K}$. One unanswered problem concerning local Toeplitz operators is:

CONJECTURE 1. <u>The spectrum of a local Toeplitz operator is connected</u>.

In [4] it was shown that many of the results known for Toeplitz operators have analogues valid for local Toeplitz operators. Unfortunately a proof of the connectedness would seem to require more refined knowledge of the behavior of H^∞ functions on M_{H^∞} than available and the result would imply the connectedness of $\sigma_e(T_\varphi)$.

A more refined localization has been obtained by Axler replacing X_ξ by the subsets of M_{L^∞} of maximal antisymmetry for $H^\infty + C$ using the fact that the local algebras $\mathfrak{J}_\xi$ have nontrivial centers and iterating this transfinitely.

There is evidence to believe that the ultimate localization should be to the closed support X_η in M_{L^∞} for the representing measure μ_η for a point η in M_{H^∞}. In particular, one would like to show that if $H^2(\mu_\eta)$ denotes the closure in $L^2(\mu_\eta)$ of the functions $\hat{\varphi} | X_\eta$ for φ in H^∞ , P_η the orthogonal projection from $L^2(\mu_\eta)$ onto $H^2(\mu_\eta)$, then the map

$$T_\varphi \longrightarrow T_{\hat{\varphi}|X_\eta}$$

extends to the corresponding algebras, where the local Toeplitz operator is defined by

$$T_{\hat{\varphi}|X_\eta} f = P_\eta(\hat{\varphi}|X_\eta) f$$

for f in $H^2(\mu_\eta)$. If η is a point in M_{L^∞} , then $H^2(\mu_\eta) = \mathbb{C}$ and it is a special case of the result [2] that $\mathfrak{J}$ modulo its commutator ideal is isometrically isomorphic to L^∞ , that the map extends to a character in this case. A generalized spectral inclusion theorem also provides evidence for the existence of this mapping in all cases.

One approach to establishing the existence of this map is to try to exhibit the state on $\mathfrak{J}$ which this "representation" would determine. One property that such a state would have is that it would be multiplicative on the Toeplitz operators with symbols in H^∞ . Call such states *analytically multiplicative*. Two problems connected with such states seem interesting.

CONJECTURE 2 (Generalized Gleason-Whitney). *If σ_1 and σ_2 are analytically multiplicative states on $\mathfrak{J}$ which agree on H^∞ and such that the kernels of the two representations defined by σ_1 and σ_2 are equal, then the representations are equivalent.*

CONJECTURE 3 (Generalized Corona). *In the collection of analytically multiplicative states the ones which correspond to points of*

$\mathbb{D}$ <u>are dense.</u>

One consequence of a localization to X_η when η is an analytic disk $\mathbb{D}_\eta$,would be the following. It is possible for φ in L^∞ that its harmonic extension $\hat{\varphi}|\mathbb{D}_\eta$ agrees with the harmonic extension of a function continuous on $\mathbb{T}$. (Note that this is not the same as saying that $\hat{\varphi}$ is continuous on the boundary of $\mathbb{D}_\eta$ as a subset of M_{H^∞} which is of course always the case.) In that case the invertibility of the local Toeplitz operator would depend on a "winding number" which should yield a subtle necessary condition for T_φ to be Fredholm. Ultimately it may be that there are enough analytic disks in M_{H^∞} on which the harmonic extension $\hat{\varphi}$ is "nice" to determine whether or not T_φ is Fredholm but that would require knowing a lot more about M_{H^∞} than we do now.

REFERENCES

1. W i d o m H. On the spectrum of a Toeplitz operator. - Pacif. J.Math., 1964, 14, 365-375.
2. D o u g l a s R.G. Banach algebra techniques in operator theory. New York, Academic Press, 1972.
3. D o u g l a s R.G. Banach algebra techniques in the theory of Toeplitz operators. CBMS Regional Confer. no.15, Amer.Math.Soc., Providence, R.I., 1973.
4. D o u g l a s R.G. Local Toeplitz operators. - Proc.London Math.Soc., 1978, 3, 36.

R.G.DOUGLAS

State University of New York
Department of Math.
Stony Brook, N.Y. 11794,
USA

5.7. TOEPLITZ OPERATORS ON THE BERGMAN SPACE

Let A^2 denote the Bergman space of analytic functions in $L^2(\mathbb{D})$, and let P be the orthogonal projection of $L^2(\mathbb{D})$ onto A^2. For $\varphi \in L^\infty(\mathbb{D})$ we define the Toeplitz operator with symbol φ by $T_\varphi f = P(\varphi f)$. In general the behaviour of these operators may be quite different from that of the Toeplitz operators on the Hardy space H^2. However it is shown in [1] that Toeplitz operators on A^2 with h a r m o n i c symbols behave quite similarly to those on H^2, and one can prove analogues for this class of many results about Toeplitz operators on H^2.

An important result about Toeplitz operators on H^2 is Widom's Theorem, which states that the spectrum of such an operator is connected ([2]). This suggests our problem.

CONJECTURE. A Toeplitz operator on A^2 with harmonic symbol has a connected spectrum.

In support of this conjecture we mention the following cases for harmonic φ in which the spectra can be explicitly computed.

1) If φ is analytic then $\sigma(T_\varphi) = \operatorname{clos} \varphi(\mathbb{D})$.

2) If φ is real-valued then $\sigma(T_\varphi) = [\inf \varphi, \sup \varphi]$.

3) If φ has piecewise continuous boundary values then $\sigma(T_\varphi)$ consists of the path formed from the boundary values of φ by joining the one-sided limits at discontinuities by straight line segments, together with certain components of the complement of this path.

For proofs of these see [1].

In connection with our conjecture it should be mentioned that there are easy examples of Toeplitz operators on A^2 with disconnected spectra - e.g. $\sigma(T_{1-|z|^2})$ is disconnected since $T_{1-|z|^2}$ is positive and compact. The proof of connectedness in the case of H^2 breaks down almost immediately in the A^2 case. I would expect a solution to the present problem to shed light on Toeplitz operators in general, and perhaps to lead to a different proof and a better understanding of Widom's Theorem.

REFERENCES

1. M c D o n a l d G., S u n d b e r g C. Toeplitz operators on the disc. - Indiana Univ.Math.J., 1979, 28, 595-611.

2. W i d o m H. On the spectrum of Toeplitz operators. - Pacific J.Math. 1964, 14, 365-375.

CARL SUNDBERG

University of Tennessee
Dept. of Math.
Knoxville, TN 37916
USA

and

Institut Mittag-Leffler
Auravägen 17
S-182 62 Djursholm, Sweden

5.8. VECTORIAL TOEPLITZ OPERATORS ON HARDY SPACES

Let $\mathfrak{M}$ be a separable Hilbert space (dim $\mathfrak{M}$ may be finite), $\mathfrak{L}$ be the algebra of all bounded linear operators on $\mathfrak{M}$ and $L^p(\mathfrak{M})$ be the Banach space of weakly measurable $\mathfrak{M}$-valued functions $f:\mathbb{T}\longrightarrow\mathfrak{M}$ with the norm

$$\|f\|=\Big\{\int_{\mathbb{T}}\|f(t)\|^p_{\mathfrak{M}}\,|dt|\Big\}^{1/p}.$$

We denote by $H^p(\mathfrak{M})$ the Hardy space of functions in $L^p(\mathfrak{M})$ with zero negatively indexed Fourier coefficients and by P the Riesz projection onto $H^p(\mathfrak{M})$ $(1<p<\infty)$.

Let $L^\infty(\mathfrak{L})$ be the space of all essentially bounded $\mathfrak{L}$-valued functions and $H^\infty(\mathfrak{L})$ be the Hardy space corresponding to $L^\infty(\mathfrak{L})$. If $\mathcal{A}\in L^\infty(\mathfrak{L})$ then the operator $T_{\mathcal{A}}=P\mathcal{A}|H^p(\mathfrak{M})$ is called a *vectorial Toeplitz operator*.

The following criterion of the invertibility of $T_{\mathcal{A}}$ on $H^2(\mathfrak{M})$ has been obtained by Rabindranathan [1].

THEOREM 1. *Let* $\mathcal{A}\in L^\infty(\mathfrak{L})$. $T_{\mathcal{A}}$ *is invertible on* $H^2(\mathfrak{M})$ *if and only if* $\mathcal{A}=\mathcal{B}_1^*\,\mathcal{U}\,\mathcal{B}_2$ *where*

1) $\mathcal{B}_1^{\pm1},\ \mathcal{B}_2^{\pm1}\in H^\infty(\mathfrak{L})$;

2) $\mathcal{U}$ is a unitary-valued function in $L^\infty(\mathfrak{L})$;

3) there exists an operator-valued function $\mathcal{B}$ with $\mathcal{B}^{\pm1}\in$ $\in H^\infty(\mathfrak{L})$ such that

$$\|\mathcal{U}-\mathcal{B}\|_{L^\infty(\mathfrak{L})}<1. \tag{1}$$

See also [2-3] $(\dim\mathfrak{M}=1)$ and [4] $(\dim\mathfrak{M}<\infty)$.

A sufficient condition for the invertibility of $T_{\mathcal{A}}$ on $H^p(\mathfrak{M})$ has been given by the authors [5] (the case $\dim\mathfrak{M}=1$ had been considered earlier by Simonenko [6]).

THEOREM 2. *Let* $p\in(1,\infty)$, $\mathcal{A}=\mathcal{B}_1^*\,\mathcal{U}\,\mathcal{B}_2$ *and suppose that all conditions of Theorem 1 for* $\mathcal{U},\mathcal{B}_1,\mathcal{B}_2$ *hold except for (1) which has to be replaced by*

$$\|\mathcal{U}-\mathcal{B}\|_{L^\infty(\mathfrak{L})}<\sin\frac{\pi}{\max(p,p')}$$

<u>Then</u> $T_{\mathfrak{A}}$ <u>is invertible on</u> $H^p(\mathfrak{M})$ (<u>and on</u> $H^{p'}(\mathfrak{M})$).

PROBLEM 1. <u>Are the conditions of Theorem 2 necessary for the invertibility of</u> $T_{\mathfrak{A}}$ <u>on both</u> $H^p(\mathfrak{M})$ <u>and</u> $H^{p'}(\mathfrak{M})$?

It is shown in [7] that the answer is affirmative if $\dim \mathfrak{M}=1$.

Let us note that the class of operator-valued functions in Theorem 2 admits an equivalent description [8]: $\mathfrak{A}=\mathfrak{B}_1^* G \mathfrak{B}_2$, $G \in L^\infty(\mathfrak{L})$, $\mathfrak{B}_1^{\pm 1}$, $\mathfrak{B}_2^{\pm 1} \in H^\infty(\mathfrak{L})$ and the numerical range $w(G(t))$ lies in a fixed angle with the vertex at the origin and with the size less than $\pi/\max(p,p')$ a.e. on $\mathbb{T}$.

It is well-known that the problem of invertibility of $T_{\mathfrak{A}}$ in $H^p(\mathfrak{M})$ can be reduced by means of factorization [2-4,7] to the problem of boundedness of P in weighted L^p-spaces. In the case $\dim \mathfrak{M}=1$ a criterion for boundedness of P was given in [9].

PROBLEM 2. <u>Let</u> $\mathfrak{A} \in H^p(\mathfrak{M})$ <u>and</u> $\mathfrak{A}^{-1} \in H^{p'}(\mathfrak{M})$ $(1<p<\infty)$. <u>What are the conditions for</u> $\mathfrak{A} P \mathfrak{A}^{-1}$ <u>to be bounded on</u> $L^p(\mathfrak{M})$?

We don't know the solution of Problems 1 and 2 even in the case of matrix-valued Toeplitz operators $(\dim \mathfrak{M} < \infty)$.

REFERENCES

1. R a b i n d r a n a t h a n M. On the inversion of Toeplitz operators. - J.Math.Mech., 1969/70, 19, 195-206.
2. H e l s o n H., S z e g ö G. A problem in prediction theory. - Ann.Mat.Pura Appl., 1960, 51, 107-138.
3. D e v i n a t z A. Toeplitz operators on H^2 space. - Trans. Amer.Math.Soc. 1964, 112, N 2, 304-317.
4. P o u s s o n H.R. Systems of Toeplitz operators on H^2. - Trans.Amer.Math.Soc., 1968, 133, N 2, 527-536.
5. В е р б и ц к и й И.Э., К р у п н и к Н.Я. Точные константы в теоремах об ограниченности сингулярных операторов в пространствах с весом. - В кн.: Линейные операторы. Кишинев, Штиинца, 1980, 21-35.
6. С и м о н е н к о И.Б. Краевая задача Римана для пар функций с измеримыми коэффициентами и ее применение к исследованию сингулярных интегралов в пространствах с весами. - Изв.АН СССР, сер.мат. 1964, 28, 277-306.

7. К р у п н и к Н.Я. Некоторые следствия из теоремы Ханта-Маккенхаупта-Видена. - В кн.: Операторы в банаховых пространствах. Кишинев, Штиинца, 1978, 64-70.
8. С п и т к о в с к и й И.М. О факторизации матриц-функций, хаусдорфово множество которых расположено внутри угла. - Сообщ. АН Гр.ССР, 1977, 86, с.561-564.
9. H u n t R., M u c k e n h o u p t B., W h e e d e n R. Weighted norm inequalities for conjugate function and Hilbert transform. - Trans.Amer.Math.Soc., 1973, 176, 227-251.

N.Ya.KRUPNIK
(Н.Я.КРУПНИК)

СССР, 277000, Кишинев,
Кишиневский государственный
университет

I.È.VERBITSKY
(И.Э.ВЕРБИЦКИЙ)

СССР, 277000, Кишинев,
Институт геофизики и
геологии АН МССР

5.9. FACTORIZATION PROBLEM FOR ALMOST PERIODIC MATRIX-FUNCTIONS AND FREDHOLM THEORY OF TOEPLITZ OPERATORS WITH SEMI-ALMOST PERIODIC MATRIX SYMBOLS

I. We consider $(n \times n)$ -matrices G defined on $\mathbb{R}$ with elements from the usual algebra AP of almost-periodic functions and Toeplitz operators $T_G = P_- G \mid Im P_-$ generated by these matrices. Here P_- is the Riesz projection onto the Hardy class H^p_- in the lower half-plane, $1 < p < \infty$.

It is well known (this fact holds for an arbitrary $G \in L_\infty$) that condition $G^{-1} \in L_\infty$ is necessary for T_G to be semi-Fredholm. In the case $n = 1$ the converse is true. Moreover, T_G is left-invertible if the almost periodic (a.p.) index ν of the function G is non-positive and T_G is right-invertible if $\nu \geqslant 0$. There exists a certain parallel between Fredholm Toeplitz operators and the factorization problem of their symbols, in accordance with which the formula

$$G(t) = G_+(t) \Lambda(t) G_-^*(t) \tag{1}$$

is valid in the case $n = 1$, $G^{\pm 1} \in AP$. Here $\Lambda(t) = e^{i\nu t}$, the functions $(z+i)^{-1} G_\pm^{\pm 1}$ belong to Hardy class H^r in the upper half-plane and the operator $G_+ P_- G_+^{-1}$ is bounded in all L^r, $1 < r < \infty$.

Formula (1) with $G \in AP$ possessing the above properties (with the natural change of $e^{i\nu t}$ by $diag[e^{i\nu_1 t}, \dots, e^{i\nu_n t}]$, $\nu_j \in \mathbb{R}$) will be called a P-factorization of G. It is easy to check that the partial a.p. indices $\nu_1, \dots, \nu_n$ are uniquely defined by G provided G admits P-factorization; and it is not difficult to describe the freedom of choice of $G_\pm$. However in the case $n > 1$ not each matrix G invertible in AP admits P-factorizations. In this connection the following problems appear.

PROBLEM 1. Obtain a criterion (or at least more or less general sufficient conditions) of existence of P-factorization.

PROBLEM 2. Find out whether the existence of P-factorization of G is a necessary condition for T_G to be (semi-)Fredholm. If not, then is it possible to change the definition of P-factorization

<u>in such a way as to get an equivalent of the semi-Fredholm property for</u> T_G .

Note that if G admits a P-factorization then T_G is left-(right-) Fredholm iff $\nu_j \leqslant 0 \ (\geqslant 0)$, $j=1,\dots,n$. Consequently an affirmative answer to Problem 2 would mean that "Fredholm character" of T_G is the same in all spaces H^{r}_{-} and its fredholmness implies the invertibility. We do not know whether these weaker statements are true if $n>1$.

2. The class SAP (of semi-almost-periodic functions) is a natural extension of AP . This class has been introduced by D. Sarason [1] and may be defined, for example, as $\{g=(0{,}5+u)f+$

$$+(0{,}5-u)h+g_0 : f,h\in AP;\ u,g_0\in C(\dot{\mathbb{R}});\ \lim_{t\to\pm\infty} g_0(t)=0,\ \lim_{t\to\pm\infty} u(t)=\pm 0{,}5\}.$$

The a.p. components f,h (of g) are uniquely determined by g_0 .

A criterion for T_G ($G\in SAP$, $n=1$) to be semi-Fredholm in the space H^2_- was obtained in [1] and was generalized in [2] to the case of an arbitrary H^p_- , $p\in(1,\infty)$. The case $n>1$ is considered in [3,4], where the fredholmness and semi-fredholmness criteria have been established. These results, however, were obtained under the a priori assumption of existence of P_0-factorization of a.p. components F , H of G . The latter means that the factors $F_+^{\pm}$, $(F_-^*)^{\pm 1}$ from the P-factorization $F=F_+\Lambda F_-^*$ belong to the class AP^+ of those matrices from AP whose Fourier exponents are all non-negative; the same holds for H . The following problems arise in connection with the question of removing these a priori assumptions.

PROBLEM 3. <u>Let</u> G <u>be an</u> $(n\times n)$<u>-matrix from</u> SAP, $n>1$, F <u>and</u> H <u>are its a.p. components. Is it true that the semi-fredholmness of</u> T_G <u>implies the semi-fredholmness of</u> T_F <u>and</u> T_H ?

PROBLEM 4. <u>Is the set of matrices admitting a</u> P_0<u>-factorization dense in the set of all matrices admitting a P-factorization? What would the situation be like if we restricted ourselves to matrices with a fixed (non-zero) set of partial a.p. indices?</u>

The positive answer to Problems 3, 4 would allow to extend the criterion for T_G to be (semi-)Fredholm [3, 4] to the case of arbitrary matrices $G\in SAP$.

3. Let us consider a triangular matrix $B \in AP$ of the second order. Under some additional assumptions(e.g. absolute convergence of Fourier series of its elements) the P-factorization property of B is reduced to the corresponding question about

$$B_0(t) = \begin{bmatrix} e^{i\nu t} & 0 \\ r_0(t) & e^{-i\nu t} \end{bmatrix} \tag{2}$$

where $\nu > 0$ and the spectrum Ω of $r_0 \in AP$ is contained in $(-\nu, \nu)$. Assuming that card $(\Omega) < \infty$ define a.p. polynomials r_j $(j = 1, 2, \ldots)$ by the recurrence formula

$$r_{j-2} = q_j r_{j-1} + r_j \quad (r_{-1} = e^{i\nu t}). \tag{3}$$

It is supposed that the sequence μ_j of leading exponents of r_j strictly decreases and $q_j \in AP^+$. Analogously to the case of the usual factorization of continuous triangular matrices [5] there exists an algorithm (say, A) for P-factorization of B_0 which is connected with the **continuous fraction expansion of** r_{-1}/r_0 **or** equivalently with the relations (3). Algorithm A unlike in the continuous case does not necessarily lead to the aim. A sufficient condition to make application possible is $\mu_K + \mu_{K-1} \leq 0$ for some $K \in \mathbb{Z}_+$. In this case the factors $B_\pm$ are a.p. polynomials.

PROBLEM 5. Give conditions for the convergence of the algorithm A to obtain a P_0- (or P-) factorization of matrix (2). These conditions have to be formulated in terms of entries of (2).

Algorithm A **can** be applied to obtain a P_0 -factorization of matrix (2) if, for example, $\Omega \subset (-\nu, 0]$ or the distances between the points of Ω are multiples of a fixed quantity (in particular, if card $(\Omega) \leq 2$). Already in the case card $\Omega = 3$, i.e.

$$r_0(t) = ae^{i\alpha t} + be^{i\beta t} + ce^{i\gamma t} \quad (abc \neq 0, \quad -\nu < \alpha < \beta < \gamma < \nu)$$

there exist situations when algorithm A fails. One of such situations is: $\beta = 0$, $\nu = \gamma - \alpha$ and $\delta\,(= -\alpha/\gamma)$ is irrational.

In this case we found another algorithm based on the successive application of the transformation $B \to A_B B C_B^*$ ($A_B^{\pm 1}, C_B^{\pm 1} \in AP^+$) preserving the structure of B_0 and on the factorization of elements close to the unit matrix. With the help of this algorithm it was established that under the restriction $|c^{\delta} a| \neq |b^{1+\delta}|$ a P_0-factorization exists with $\nu_1 = \nu_2 = 0$, but $B_{\pm}$ are no more a.p. polynomials. In the case $|c^{\delta} a| = |b^{1+\delta}|$ the P-factorization of (2) does not exist. Thus, even in the case card $\Omega = 3$ the following problem is non-trivial.

PROBLEM 6. Describe the cases when matrix (2) admits a P-(or P_0-) factorization, calculate its a.p. partial indices and construct, if possible, corresponding factorizations.

The interest to matrices of the form (2) is motivated by the fact that they naturally arise in connection with convolution equations on a finite interval with kernels K for which $x^{\alpha}(\mathcal{F}K)(x)$ has an a.p. asymptotics at infinity for an $\alpha \in \mathbb{C}$ [3].

REFERENCES

1. Sarason D. Toeplitz operators with semi-almost periodic symbols. - Duke Math.J., 1977, 44, N 2, 357-364.
2. Сагинашвили А.И. Сингулярные интегральные уравнения с коэффициентами, имеющими разрывы полу-почти-периодического типа. - Тр.Тбилис.матем. ин-та, 1980, 64, 84-95.
3. Карлович Ю.И., Спитковский И.М. О нётеровости некоторых сингулярных интегральных операторов с матричными коэффициентами класса SAP и связанных с ними систем уравнений свертки на конечном промежутке. - Докл.АН СССР, 1983, 269, № 3.
4. Карлович Ю.И., Спитковский И.М. О нётеровости, n- и d-нормальности сингулярных интегральных операторов с матричными коэффициентами, допускающими разрывы полу-почти-периодического типа. - Школа по теории операторов в функциональных пространствах (Тезисы докладов), Минск, 1982, 81-82.
5. Чеботарев Г.Н. Частные индексы краевой задачи Римана с треугольной матрицей второго порядка. - Успехи матем.наук, 1956, II, № 3, 199-202.

Yu.I.KARLOVICH
(Ю.И.КАРЛОВИЧ)

I.M.SPITKOVSKII
(И.М.СПИТКОВСКИЙ)

СССР, 270044, Одесса,
Пролетарский бульвар 29,
Морской гидрофизический институт
Отделение экономики и
экологии Мирового океана.

5.10. TOEPLITZ OPERATORS IN SEVERAL VARIABLES

For $\mathbb{C}$ the complex numbers, let Ω be a bounded domain in $\mathbb{C}^m$ with closure Ω^- and with $\partial\Omega$ the Shilov boundary of the uniformly closed algebra $A(\Omega^-)$ generated by all polynomials in the complex variables $z=(z_1,z_2,\ldots,z_m)$ on Ω^-. In general, $\partial\Omega$ is a closed subset of the topological boundary of Ω. When Ω is one of the classical domains of Cartan or in other cases of interest, $\partial\Omega$ is a compact manifold with a "natural" volume element $d\mu$ and the space $L^2(\partial\Omega)$ of μ-square integrable complex valued functions is the setting for our analysis.

The closure of $A(\Omega^-)$ in $L^2(\partial\Omega)$ is denoted by $H^2(\partial\Omega)$ and this (Hardy) space, together with the (unique) orthogonal projection operator P from $L^2(\partial\Omega)$ onto $H^2(\partial\Omega)$, is a basic object in complex analysis on Ω. For φ essentially bounded on $\partial\Omega(\mu)$, the T o e p l i t z o p e r a t o r T_φ is defined for all f in $H^2(\partial\Omega)$ by $T_\varphi f=P(\varphi f)$. The C^*-algebra generated by all T_φ with φ continuous is denoted by $\mathfrak{T}(\partial\Omega)$.

Even for $\Omega=\mathbb{D}\times\mathbb{D}\times\ldots\times\mathbb{D}$ (m times) where $\mathbb{D}$ is the open unit disc in $\mathbb{C}$, many interesting questions about $\mathfrak{T}(\partial\Omega)$ remain open after more than a decade of study. Note that for $\Omega=\mathbb{D}\times\mathbb{D}\times\ldots\times\mathbb{D}$ (m times), $\partial\Omega=\mathbb{T}^m$, the m-torus. The structure of $\mathfrak{T}(\mathbb{T}^m)$ is well-understood for $m=1,2$ [1,2,3].

In particular, necessary and sufficient conditions for A in $\mathfrak{T}(\mathbb{T}^2)$ to be Fredholm of index η are known [2]. It follows from the analysis of [2] that every Fredholm operator of index η in $\mathfrak{T}(\mathbb{T}^2)$ can be joined by an arc of such operators to

$$[T_{z_2}^{u}(I-T_{z_1}T_{z_1}^*)+T_{z_1}T_{z_1}^*][T_{z_1}^{-u-\eta}(I-T_{z_2}T_{z_2}^*)+T_{z_2}T_{z_2}^*].$$

Here, u is an integer and $T_{z_j}^{-|s|}=T_{z_j}^{*\,|s|}$.

PROBLEM 1. <u>Classify the arc-components of Fredholm operators in</u> $\mathfrak{T}(\mathbb{T}^3)$.

This question reduces to:

PROBLEM 2. <u>Classify the arc-components of invertible elements in</u> $\mathfrak{T}(\mathbb{T}^3)$.

REFERENCES

1. C o b u r n L.A. The C^*-algebra generated by an isometry I, II. - Bull.Amer.Math.Soc.,1967, 73, 722-726; Trans.Amer.Math.Soc., 1969, 137, 211-217.
2. C o b u r n L.A., D o u g l a s R.G., S i n g e r I.M. An index theorem for Wiener-Hopf operators on the discrete quarter-plane. - J.Diff.Geom.,1972, 6, 587-593.
3. D o u g l a s R.G., H o w e R. On the C^*-algebra of Toeplitz operators on the quarter-plane. - Trans.Amer.Math.Soc.,1971, 158, 203-217.

L.A.COBURN

State University of New York
Department of Mathematics
Buffalo, N.Y. 14214
USA

5.11. SOME PROBLEMS CONNECTED WITH THE SZEGÖ LIMIT THEOREMS

1. Any sequence of $(r \times r)$ -matrices $\{C_j\}_{j\in\mathbb{Z}}$ determines a sequence of matrix-valued Toeplitz matrices $T_n = \{C_{j-k}\}_{j,k=0}^{n}$. If $\Delta_n \overset{\text{def}}{=\!=} \det T_n \neq 0$ for sufficiently large values of n then the question about the limiting behavior of Δ_{n+1}/Δ_n arises. The analogous question arises about Δ_n/μ^{n+1} provided the non-zero limit $\mu \overset{\text{def}}{=\!=} \lim_n \Delta_{n+1}/\Delta_n$ exists.

It was G. Szegö who studied both questions for the first time. He dealt with the case $r=1$ and supposed that $\{C_j\}_{j\in\mathbb{Z}}$ is a sequence of Fourier coefficients of a positive summable function. See [1] for the precise formulations, for the history of the problem and for its natural generalization

$$C_j = \hat{M}(j), \tag{1}$$

M being a finite non-negative Borel measure on $\mathbb{T}$. By the Riesz-Herglotz theorem the class of sequences satisfying (1) is the class of positive definite sequences.

We consider here the case when $r \geqslant 1$ and $\{C_j\}$ is an α-sectorial sequence for some $\alpha \in [0, \pi/2)$. The latter means that every T_n is α-sectorial i.e. its numerical range (Hausdorff set) lies in the angle $\{z : |\operatorname{Im} z| \leqslant \operatorname{tg}\alpha \cdot \operatorname{Re} z\}$. It is clear that $\{C_j\}$ is an α-sectorial sequence iff there exists a measure M satisfying (1) and taking values in the set of α-sectorial $(r\times r)$-matrices on all arcs of $\mathbb{T}$. The real part M_R of this measure M permits us to construct the Hilbert space $\mathfrak{H}_r = L^2(M_R)$ consisting of r-tuples of functions and equipped with the sesquilinear form $A(f,g) = \int_{\mathbb{T}} f(\zeta)\, dM(\zeta)\, g^*(\zeta)$.

Employing the factorization theorems from [2,3] we have proved in [4] the existence of the limit μ in the case $r \geqslant 1$, $\alpha \geqslant 0$ and have obtained the following formula:

$$\mu = \exp\Big[\int_{\mathbb{T}} \ln \det G(\zeta)\, dm(\zeta)\Big] \tag{2}$$

where $G = dM/dm$ (see [4] for details and the information about earlier results by A. Devinatz and B. Gyires). Formula (2) is valid in the case $\ln \det G \notin L^1$ too; this is the only case when $\mu = 0$.

We propose the following as an UNSOLVED PROBLEM: find an extension of the Szegö second limit theorem to the case of α-sectorial sequences.

We CONJECTURE that the limit $\lambda \overset{def}{=} \lim_{n\to\infty} \Delta_n / \mu^{n+1}$ (finite or not) exists for every α-sectorial sequence satisfying the regularity condition

$$\ln \det G \in L^1(m). \tag{3}$$

We are somewhat encouraged in this conjecture by Theorem 2 of Devinatz [5] related to the case provided $r = 1$, M is an absolutely continuous measure and G satisfies some additional restrictions (including the requirement $G \in L^\infty$).

We have proved in the case $\alpha = 0$, $r > 1$ the existence of λ and have obtained a formula for its calculation using some geometrical considerations from [4]. Before formulating the corresponding result let us remind that under condition (3) there exist two canonical factorizations of the matrix G: the left one $G = G_\ell G_\ell^*$ and the right one $G = G_r^* G_r$ (G_ℓ and G_r are outer matrix functions of the class H^2). Let us denote the Toeplitz operator with the (unitary valued) symbol $F = G_\ell^* G_r^{-1}$ by T_F.

THEOREM 1. Let $\{C_j\}$ be a positive definite sequence of $(r\times r)$-matrices and let M be a measure connected with it by formula (1), $G = dM/dm$. Then under condition (3) there exists a limit $\lambda\ (\leq \infty)$ of the sequence Δ_n / μ^{n+1}. This limit is finite iff M is absolutely continuous and $\sum_1^\infty k \|\hat{F}(-k)\|^2 < \infty$. If these conditions are fulfilled then $\lambda = (\det T_F^* T_F)^{-1}$.

We do not know whether there is any kind of general result in the case $\alpha > 0$, $r > 1$. It is well understood now that the existence of λ (and formulae for its evaluation) may be proved under some additional restrictions with a help of results obtained in another direction (the rejection of the positive-definiteness with simultaneous amplification of restrictions on the smoothness of G) that we do not touch upon here, see [6] and references in it.

2. Considering an α-sectorial matrix measure M concentrated on the line, it is possible to introduce a continuous analogue of the

space $\mathfrak{H}_{r}$ and to establish the following result.

THEOREM 2. The two statements given below are equivalent:

1) $\int_{-\infty}^{+\infty} \ln|\det G(\lambda)| \frac{d\lambda}{1+\lambda^2} > -\infty$, where $G = dM/d\lambda$;

2) the subspace $\mathfrak{N}$ of constant r-tuples has zero intersection with the subspace $\mathcal{L}_s = \bigvee\{e^{it\lambda}\mathfrak{N} : t \geqslant s\}$ for all (at least for one) $s > 0$.

If these conditions are fulfilled and $\alpha = 0$ then the square of the distance from $\xi \in \mathfrak{N}$ to $\mathcal{L}_s$ in $\mathfrak{H}$ metric equals to $\xi \mathfrak{D}_s \xi^*$ where $\mathfrak{D}_s$ is a "distanse matrix" which is calculated by the formula

$$\mathfrak{D}_s = \int_0^s \gamma(t)\gamma(t)^* dt. \tag{4}$$

Here γ is the inverse Fourier transform of the matrix-function G_+ from the left canonical factorization of G ($G = G_+ G_+^*$ a.e. on $\mathbb{R}$, G_+ is outer and belongs to the Hardy class H^2 in the upper half-plane).

For the case $r=1$ Theorem 2 was already proved in [7]; the discrete analogue of (4) was established in [4] in the general case $\alpha \in [0, \pi/2)$, $r \geqslant 1$. We propose a natural

PROBLEM: generalize the second part (concerning formula (4)) of Theorem 2 to the case of distances in the skew A -metric ($\alpha > 0$).

In this case obscure points already appear after first attempts to interprete the right-hand side of the formula of type (4). The fact is that the inverse Fourier transform of the factors $G_\pm$ from canonical factorizations of α -sectorial matrix-functions [2,3] in general are not elements of L^2 .

The problem to find continuous generalizations for the Szegö second limit theorem admits different formulations and even in the definite case corresponding investigations form an "unordered set" (see [8] and the papers cited there). There are still more unsolved questions in the case $\alpha > 0$ but we shall not go into this matter here.

REFERENCES

I. Голинский Б.Л., Ибрагимов И.А. О предельной теореме Г.Сегё. - Изв.АН СССР, серия матем., 1971, 35, вып.2, 408–427.

2. Крейн М.Г., Спитковский И.М. О факторизации матриц-функций на единичной окружности. - Докл.АН СССР, 1977, 234, № 2, 287-290.

3. Крейн М.Г., Спитковский И.М. О факторизации α -секториальных матриц-функций на единичной окружности. - Матем.исследования, 1978, 47, 41-63.

4. Крейн М.Г., Спитковский И.М. О некоторых обобщениях первой предельной теоремы Сеге. - Anal.Math., 1983, 9, N 1.

5. Devinatz A. The strong Szegö limit theorem. - Illinois J.Math., 1967, II, 160-175.

6. Basor E., Helton J.W. A new proof of the Szegö limit theorem and new results for Toeplitz operators with discontinuous symbol. - J.Oper.Theory, 1980, 3, N 1, 23-39.

7. Крейн М.Г. Об одной экстраполяционной проблеме А.Н.Колмогорова. - Докл.АН СССР, 1945, 46, № 8, 306-309.

8. Микаелян Л.В. Матричные континуальные аналоги теорем Г.Сегё о тёплицевых детерминантах. - Изв.АН АрмССР, 1982, 17, № 4, 239-263.

M.G.KREIN
(М.Г.КРЕЙН)

СССР, 270057, Одесса,
ул.Артёма 14, кв.6

I.M.SPITKOVSKII
(И.М.СПИТКОВСКИЙ)

СССР, 270044, Одесса,
Пролетарский бульвар 29,
Морской гидрофизический институт
Отделение экономики и
экологии Мирового океана

5.12. THE DIOPHANTINE MOMENT PROBLEM, ORTHOGONAL POLYNOMIALS AND SOME MODELS OF STATISTICAL PHYSICS

I. In [1], [2] it was shown that in investigations of the Ising model in the presence of a magnetic field the following one-parametric Diophantine trigonometrical moment problem (DTMP) appears.

PROBLEM. Describe all non-negative measures $d\sigma(\theta,\xi)$ on the circle $\mathbb{T}=\{\zeta\in\mathbb{C}:\ \zeta=e^{i\theta},\ \theta\in[-\pi,\pi]\}$ even in θ, depending on a parameter ξ, $0\leqslant\xi\leqslant 1$, and such that

$$M_0(\xi)\overset{def}{=}\frac{1}{\pi}\int_0^{\pi} d\sigma(\theta,\xi)\equiv 1$$

and the moments

$$M_k(\xi)\overset{def}{=}\frac{1}{\pi}\int_0^{\pi}\cos k\theta\, d\sigma(\theta,\xi),\quad k=1,2,\dots$$

are polynomials (in ξ) of degree $\varkappa k$ with integer coefficients; the parity of $M_k(\xi)$ coincides with the parity of $\varkappa k$. Here $\varkappa$ is an integer $\geqslant 2$ ($\varkappa$ is the number of the nearest neighbours in the lattice).

It is known that the description of such measures can be reduced to the description of the corresponding generating functions

$$I(z,\xi)=1+2\sum_{k=1}^{\infty}M_k(\xi)z^k,\quad |z|<1.$$

EXAMPLES. 1. $M_k(\xi)=\xi^{\varkappa k}$, $I(z,\xi)=\dfrac{1+z\xi^{\varkappa}}{1-z\xi^{\varkappa}}$

2. $M_k(\xi)=T_k(\xi^{\varkappa})$, where T_k are Tchebyshёv polynomials, $I(z,\xi)=\dfrac{1-z^2}{1-2z\xi^2+z^2}$.

3. $M_k(\xi)=T_{k\varkappa}(\xi)$, $I(z,\xi)=\dfrac{1-z^2}{1-2zT_{\varkappa}(\xi)+z^2}$.

4. $M_k(\xi)=\frac{1}{2}\left[P_k(1-2\xi^2)-P_{k-1}(1-2\xi^2)\right]$, where P_k are Legendre polynomials,

$$I(z,\xi)=\frac{1-z}{\sqrt{(1-z)^2+4z\xi^2}}.$$

We note that in examples 1-3 the generating function is **rational**, whereas in example 4 it is algebraic (this case corresponds to the one-dimensional Ising model).

QUESTION. Has the generating function corresponding to a DTMP, to be algebraic?

Fixing a rational value of the parameter ξ , $\xi = \frac{p}{q}$, p, q integers, $0 \leq p \leq q$, $q \geq 1$, we see that our DTMP implies the following "quasi-DTMP":

$$M_k \overset{\text{def}}{=} \frac{1}{\pi} \int_0^{\pi} \cos k\theta \, d\sigma(\theta) = \frac{c_k}{q^{\text{æ}k}} , \quad c_k \qquad \text{being integers,}$$

In particular for $\xi = 0$ or $1 (q=1)$ we obtain the following moment PROBLEM:

Describe non-negative even measures whose trigonometrical moments are integers.

This problem is solved by the known Helson theorem [3]:

$$d\sigma(\theta) = \sum_{s=0}^{N-1} a_s \, \delta\left(\theta - \frac{2\pi s}{N}\right) + \sum_{s=0}^{M-1} b_s \cos s\theta \, d\theta$$

under some additional conditions on a_s, b_s [2].

II. It was shown in [4] that the theory of Toeplitz forms and orthogonal polynomials is closely connected with some problems of statistical physics and in particular with the Gauss model on the semi-axis. In this connection some mathematical problems appear whose solution would be useful for the further investigation of such models.

1. Let $f(\theta)$ be an even non-negative summable function on $\mathbb{T}$ satisfying the Szegö condition

$$\int_0^{\pi} \ln f(\theta) \, d\theta > -\infty .$$

We define the function

$$\pi(z) = \exp\left[-\frac{1}{2\pi} \int_0^{\pi} \frac{1-z^2}{1-2z\cos\theta + z^2} \ln f(\theta) \, d\theta\right] =$$

$$= \sum_{k=0}^{\infty} g_k z^k, \quad |z| < 1 .$$

PROBLEM. <u>Find necessary and sufficient conditions for</u> $g_k \to 0$ <u>as</u> $k \to \infty$ (<u>physically the last condition means the absence of a long-order parameter</u>).

It was shown in [4] that the condition $\frac{1}{\sqrt{f}} \in L^1$ implies $g_k \to 0$ as $k \to \infty$ and

$$\sum_{k=0}^{\infty} \frac{|g_k|}{1+k} < \infty .$$

2. Let $d\sigma(\theta)$ be an even non-negative measure on $\mathbb{T}$ and

$$J_{|k-j|} = \frac{1}{\pi} \int_0^{\pi} \cos(k-j)\theta \, d\sigma(\theta) ; \quad k, j = 0, 1, \ldots$$

is the corresponding Toeplitz matrix. We denote by $\left(J_{kj}^{(N)}\right)^{-1}$ the inverse matrix for $J_{|k-j|}^{(N)} = \left(J_{|k-j|}\right)^N{}_{k,j=0}$.

PROBLEM. <u>Find an asymptotics for</u>

$$\sum_{j,k=0}^{N} \left(J_{j,k}^{(N)}\right)^{-1} \qquad \underline{\text{as}} \quad N \to \infty .$$

This problem appears in the study of the free energy in the Gaussian model on the semi-axis with an external field (see [4]). For example, when $d\sigma(\theta) = d\theta + 2\pi a \delta(\theta)$, $a > 0$, this expression tends to $1/a$.

The numbers $\left(J_{0k}^{(N)}\right)^{-1}$, $k = 0, 1, \ldots, N$, are proportional to the coefficients of the orthogonal polynomials $\varphi_N(z)$ (see [4], [5]). This leads to a PROBLEM of a more detailed investigation of the asymptotics of $\varphi_N(e^{i\theta})$ as $N \to \infty$ in the presence of non-zero singular part of the measure $d\sigma$. As we know only the case of an absolutely continuous measure was considered in detail (see, e.g. [6], [7]).

3. The multidimensional Gaussian model.

<u>Calculate the free energy and correlation functions under less restrictive conditions than in</u> [4], [8].

REFERENCES

1. Barnsley M., Bessis D., Moussa P. The Diophantine moment problem and the analytic structure in the activity of the ferromagnetic Ising model. - J.Math.Phys., 1979, N 4, 20, 535-552.
2. Владимиров B.C., Волович И.В. Модель Изинга с магнитным полем и диофантова проблема моментов. - Теор.Матем. Физ., 1982, 53, № 1, 3-15.
3. Helson H. Note on harmonic functions. - Proc.Amer.Math.Soc., 1953, 4, N 5, 686-691.
4. Владимиров B.C., Волович И.В. Об одной модели статистической физики. - Теор. Матем. Физ., 1983, 54, № 1, 8-22.
5. Владимиров B.C., Волович И.В. Уравнение Винера - Хопфа, задача Римана - Гильберта и ортогональные многочлены. - Докл.АН СССР, 1982, 266, № 4, 788-792.
6. Szegö G. Orthogonal polynomials. AMS Coll.Publ., 23, 2 ed., 1959.
7. Голинский Б.Л. Асимптотическое представление ортогональных многочленов. - Успехи матем.наук, 1980, 35, № 2, 145-196.
8. Линник И.Ю. Многомерный аналог теоремы Сегё. - Изв.АН СССР, сер.матем., 1975, 39, № 6, 1393-1403.

V.S.VLADIMIROV
(В.С.ВЛАДИМИРОВ)

I.V.VOLOVICH
(И.В.Волович)

СССР, 117966, Москва
ул.Вавилова, 42
Математический институт
АН СССР

5.13. THE BANACH ALGEBRA APPROACH TO THE REDUCTION METHOD FOR TOEPLITZ OPERATORS

Let H^2 denote the Hardy subspace of $L^2 = L^2(\mathbb{T})$, consisting of the functions f with $\hat{f}(n) = 0$, $n < 0$, and let P be the orthogonal projection from L^2 onto H^2. For $a \in L^\infty = L^\infty(\mathbb{T})$ the Toeplitz operator with symbol a is defined on H^2 by $T(a)\varphi = P(a\varphi)$.

Let $\mathcal{B}(H^2)$ be the Banach algebra of linear and bounded operators on H^2. Given a closed subalgebra B of L^∞ denote by $alg\,T(B)$ the smallest closed subalgebra of $\mathcal{B}(H^2)$ containing all operators $T(a)$ with $a \in B$. Furthermore, let $Q(B)$ denote the so-called *quasicommutator ideal* of $alg\,T(B)$, i.e. the smallest closed twosided ideal in $alg\,T(B)$ containing all operators of the form $T(ab) - T(a)T(b)$ $(a, b \in B)$. It is a rather surprising fact that this ideal plays an important role not only in the Fredholm theory of Toeplitz operators, but also in the theory of the reduction method for operators $A \in alg\,T(L^\infty)$ (with respect to the projections P_n defined by $P_n \sum_{k=0}^{\infty} \hat{f}(k)\zeta^k = \sum_{k=0}^{n} \hat{f}(k)\zeta^k$).

For $A \in \mathcal{B}(H^2)$ write $A \in \Pi\{P_n\}$ if the reduction method is applicable to A (see [3] for a precise definition). Finally, put $Q_n = I - P_n$ and denote by $G\mathfrak{A}$ the group of invertible elements of a Banach algebra $\mathfrak{A}$ with identity.

For $A \in \mathcal{B}(H^2)$, the following statements are easily seen to be equivalent:

(i) $A \in \Pi\{P_n\}$; (ii) $A^* \in \Pi\{P_n\}$;

(iii) $A \in G\mathcal{B}(H^2)$, $P_nAP_n + Q_n \in G\mathcal{B}(H^2)$ $(n \geq n_0)$, and

$$\sup_{n \geq n_0} \|(P_nAP_n + Q_n)^{-1}\| < \infty;$$

(iv) $A \in G\mathcal{B}(H^2)$, $Q_nA^{-1}Q_n + P_n \in G\mathcal{B}(H^2)$ $(n \geq n_0)$,

$$\sup_{n \geq n_0} \|(Q_nA^{-1}Q_n + P_n)^{-1}\| < \infty;$$

(v) $A \in G\mathcal{B}(H^2)$, $V_{-n}A^{-1}V_n \in G\mathcal{B}(H^2)$ $(n \geq n_0)$ and

$$\sup_{n \geq n_0} \|(V_{-n}A^{-1}V_n)^{-1}\| < \infty, \quad \text{where} \quad V_n = T(\zeta^n),$$
$$V_{-n} = T(\zeta^{-n}) \ (n \in \mathbb{N}).$$

There is an important estimate closely related to (v) (see [1]*)):

$$\Big\|\sum_{i=1}^{k} \sum_{j=1}^{\ell} T(a_{ij})\Big\| \geq \Big\|T\Big(\sum_{i=1}^{k} \prod_{j=1}^{\ell} a_{ij}\Big)\Big\|, \qquad (1)$$

which holds for every finite collection of functions $a_{ij} \in L^\infty$. Now, given a closed subalgebra B of L^∞ it follows from (1) that

$$S\Big(\sum_{i=1}^{k} \prod_{j=1}^{\ell} T(a_{ij})\Big) \overset{\text{def}}{=\!=} T\Big(\sum_{i=1}^{k} \prod_{j=1}^{\ell} a_{ij}\Big) \quad (a_{ij} \in B)$$

defines a bounded projection S on $\operatorname{alg} T(B)$. One can show that

$$\operatorname{im} S = \{T(a) : a \in B\}, \quad \ker S = Q(B), \quad S(A) = s\text{-}\lim_{n \to \infty} V_{-n} A V_n .$$

If $A \in \operatorname{alg} T(L^\infty) \cap G\mathcal{B}(H^2)$, then $A \in G \operatorname{alg} T(L^\infty)$, since $\operatorname{alg} T(L^\infty)$ is a C^*-algebra. Thus $S(A^{-1})$ makes sense and belongs to $\operatorname{alg} T(L^\infty)$. Moreover, if $A \in \Pi\{P_n\}$ then (ii) and (v) imply the invertibility of $S(A^{-1})$.

CONJECTURE 1. <u>Let</u> $A \in \operatorname{alg} T(L^\infty)$. <u>Then</u> $A \in \Pi\{P_n\}$ <u>if and only if</u> A <u>and</u> $S(A^{-1})$ <u>are in</u> $G\mathcal{B}(H^2)$.

The following special cases are of particular interest;

(a) $A = T(a)$;

---

*) See also Н.К.Никольский, Операторы Ганкеля и Тёплица. Спектральная теория. – Препринт ЛОМИ Р-I-82, Ленинград, 1982. – Ed.

(b) $A=\sum_{i=1}^{k}\prod_{j=1}^{l} T^{\varepsilon_{ij}}(a_{ij})$, where $\varepsilon_{i,j}\in\{-1,1\}$ and,

of course, for $\varepsilon_{ij}=-1$ the invertibility of $T(a_{ij})$ is part of the hypotheses.

For $a_{ij}\in PC$ (PC is the algebra of piecewise continuous functions on $\mathbb{T}$ with only finitely many jumps) the case (b) is of importance in connection with the asymptotic behavior of Toeplitz determinants generated by singular functions (cf. [1]). In the case (a) the conjecture 1 is confirmed for $a\in C+H^\infty$ or $a\in \mathrm{clos}_{L^\infty} PC$ (see [3], [6]), and in the case (b) for $A=(T^{\varepsilon}(a))^n$ ($\varepsilon=\pm 1$, $n\in\mathbb{N}$, $a\in \mathrm{clos}_{L^\infty}PC$) (see [4], [7]).

One possible way to attack these problems concerned with the reduction method is to formulate them in the language of Banach algebras and then to use localization techniques (cf. [6]). Define $W_n: H^2\longrightarrow H^2$ by

$$W_n\sum_{k=0}^{\infty}\hat{f}(k)\zeta^k=\sum_{k=0}^{n}\hat{f}(k)\zeta^{n-k}$$

and denote by $\mathfrak{A}$ the collection of all sequences $\{A_n\}_{n=0}^{\infty}$, $A_n: \mathrm{im}\,P_n\longrightarrow \mathrm{im}\,P_n$ having the following property: there exist two operators $A,\tilde{A}\in\mathcal{B}(H^2)$ such that

$$A_nP_n\longrightarrow A,\quad A_n^*P_n\longrightarrow A^*,\quad \tilde{A}_nP_n\overset{def}{=}W_nA_nW_n\longrightarrow\tilde{A},\quad \tilde{A}_n^*P_n\rightarrow\tilde{A}^*$$

(strong convergence). By definition $\{A_n\}+\{B_n\}=\{A_n+B_n\}$, $\{A_n\}\{B_n\}=\{A_nB_n\}$, $\|\{A_n\}\|=\sup_n|A_nP_n|$, the set $\mathfrak{A}$ becomes a Banach algebra with identity. If $A\in\mathfrak{T}_\infty$ ($\mathfrak{T}_\infty$ is the ideal of compact operators on H^2) or even if $A\in \mathrm{alg}\,T(L^\infty)$, then $\{P_nAP_n\}\in\mathfrak{A}$ (see [5]). Notice that this is obvious for $A=T(a)$, $a\in L^\infty$, since $W_nT(a)W_n=P_nT(\tilde{a})P_n$, where $\tilde{a}(\zeta)=a(1/\zeta)$. It can be proved that the set

$$J=\{\{A_n\}: A_n=P_nTP_n+W_n\tilde{T}W_n+C_n,\, T,\tilde{T}\in\mathcal{T}_\infty,\ |C_nP_n|\to 0\}$$

actually forms a closed two-sided ideal in $\mathcal{A}$, and that the problem of the applicability of the reduction method to $A\in alg\,T(L^\infty)$ admits the following reformulation (see [6]):

$A\in\Pi\{P_n\}\Leftrightarrow A,\tilde{A}\in G\mathcal{B}(H^2)$ and the coset of $\mathcal{A}/J$ containing $\{P_nAP_n\}$ is invertible in $\mathcal{A}/J$.

Note that now localization techniques can very advantageously be applied to study invertibility in the algebra A/J .

There is a construction which is perhaps of interest in connection with the case (a). Denote by $alg_{\{P_n\}}T(B)$ the smallest closed subalgebra of $\mathcal{A}$ containing all elements of the form $\{P_nT(a)P_n\}$, where $a\in B$. If $C(\mathbb{T})\subset B$ then $\mathcal{T}_\infty\subset alg\,T(B)$ and $J\subset alg_{\{P_n\}}T(B)$ (cf.[2]). Assign to each sequence $\{A_n\}\in alg_{\{P_n\}}T(B)$ the coset $[A]\in alg\,T(B)/\mathcal{T}_\infty$ containing $A\overset{def}{=}s\text{-}\lim A_n$. In this way a continuous homomorphism $\varphi: alg_{\{P_n\}}T(B)\longrightarrow alg\,T(B)/\mathcal{T}_\infty$ is produced, and one has $\ker\varphi\supset J$.

CONJECTURE 2. <u>If</u> $B=L^\infty$ <u>then</u> $\ker\varphi=J$.

A confirmation of this conjecture would imply that

$$alg_{\{P_n\}}T(B)/J\cong alg\,T(B)/\mathcal{T}_\infty, \tag{2}$$

which, on its hand, would verify conjecture 1 for $A=T(a),\ a\in L^\infty$. It is already of interest to find sufficient conditions for the validity of (2) in the case $B\neq L^\infty$. Note that (2) was proved for $B=C+H^\infty$ or $B=clos_{L^\infty}PC$ in [2].

REFERENCES

1. B ö t t c h e r A., S i l b e r m a n n B. Invertibility and Asymptotics of Toeplitz Matrices. Berlin, Akademie-Verlag, (to appear).
2. B ö t t c h e r A., S i l b e r m a n n B. The finite section method for Toeplitz operators on the quarter-plane with

piecewise continuous symbols. - Math.Nachr. (to appear).

3. Г о х б е р г И.Ц., Ф е л ь д м а н И.А. Уравнения в свертках и проекционные методы их решения. Москва, Наука, 1971. (Transl. Math.Monogr., Vol.41, AMS, Providence, R.I., 1974).

4. R o c h S., S i l b e r m a n n B. Das Reduktionsverfahren für Potenzen von Toeplitzoperatoren mit unstetigem Symbol. - Wiss. Z. d. TH Karl-Marx-Stadt 1982, 24, Heft 3, 289-294.

5. R o c h S., S i l b e r m a n n B. Toeplitz-like Operators, Quasicommutator Ideals, Numerical Analysis. - Math.Nachr. (to appear).

6. S i l b e r m a n n B. Lokale Theorie des Reduktionsverfahrens für Toeplitzoperatoren. - Math.Nachr. 1981, 104, 137-146.

7. В е р б и ц к и й И.Э. О методе редукции для степеней тёплицевых матриц. - Математические исследования, 1978, вып.47, 3-11.

B.SILBERMANN

Technische Hochschule
Karl-Marx-Stadt
Sektion Mathematik
DDR-9010 Karl-Marx-Stadt
PSF 964

5.14. STARKE ELLIPTIZITÄT SINGULÄRER INTEGRALOPERATOREN UND SPLINE-APPROXIMATION

Sei Γ ein Kurvensystem in $\mathbb{C}$, das aus endlich vielen einfachen geschlossenen oder offenen Ljapunowkurven besteht, die keine gemeinsamen Punkte haben. Des weiteren seien $t_1,\dots,t_m \in \Gamma$ paarweise verschiedene Punkte, $-1<\alpha_k<1$ $(k=1,\dots,m)$ und $\rho(t) \stackrel{def}{=} \prod_{k=1}^{m} |t-t_k|^{\alpha_k}$. Mit $L^2(\Gamma,\rho)$ bezeichnen wir den Hilbertraum aller auf Γ meßbaren Funktionen f mit $\rho^{1/2} f \in L^2(\Gamma)$. Wir betrachten die singulären Integraloperatoren der Gestalt

$$A_\Gamma \stackrel{def}{=} aI + bS_\Gamma, \quad (S_\Gamma f) \stackrel{def}{=} \frac{1}{\pi i}\int_\Gamma \frac{f(\tau)}{\tau - t}\, d\tau \quad (t\in\Gamma)$$

mit stückweise stetigen Koeffizienten $a,b \in PC(\Gamma)$. Bekanntlich gilt $A_\Gamma \in \mathcal{L}(L^2(\Gamma,\rho))$. Mit $\mathfrak{A}(PC)$ bezeichnen wir die kleinste abgeschlossene Teilalgebra von $\mathcal{L}(L^2(\Gamma,\rho))$, die alle Operatoren A_Γ sowie das Ideal $\mathcal{T}_\infty$ der kompakten Operatoren in $L^2(\Gamma,\rho)$ enthält, und mit $\mathrm{Sym}\, A$ das Symbol eines Operators $A \in \mathfrak{A}(PC)$ (vgl. [6] oder [10]).

Für stetige Koeffizienten $a,b \in C(\Gamma)$ gilt

$$\mathrm{Sym}\, A = a(t) + b(t)\, \mathrm{Sym}\, S_\Gamma(t,z) \qquad ((t,z)\in\Lambda),$$

und $\mathrm{Sym}\, S_\Gamma$ ist eine stetige Funktion auf einer gewissen Raumkurve $\Lambda = \Lambda(\Gamma)$ [7]. Im Falle des Intervalls $\Gamma = [a,b]$ ist Λ der Rand des Rechtecks $[a,b]\times[-1,1]$. Wenn Γ nur aus geschlossenen Kurven besteht und $\rho \equiv 1$ ist, dann gilt

$$\mathrm{Sym}\, A_\Gamma(t,z) = a(t) + b(t)z; \quad t\in\Gamma, \; z=\pm 1.$$

Die Abbildung Sym ist ein isometrischer Isomorphismus der symmetrischen Algebra $\mathfrak{A}(C)/\mathcal{T}_\infty$ auf $C(\Lambda)$ mit $\mathrm{Sym}\, A^* = \overline{\mathrm{Sym}\, A}$ $(A\in\mathfrak{A}(C))$ [5], [3].

Der Operator $A \in \mathcal{L}(L^2(\Gamma,\rho))$ heißt *stark elliptisch*, wenn gilt $\mathrm{Re}\, A \stackrel{def}{=} \frac{1}{2}(A+A^*) = D + T$, wobei D positiv definit und $T \in \mathcal{T}_\infty$ ist. Wir nennen A *θ-stark elliptisch*, wenn eine Funktion $\theta \in C(\Gamma)$ $\theta(t) \neq 0$ $(\forall t\in\Gamma)$, existiert derart, daß θA stark elliptisch ist.

Für Operatoren der Algebra $\mathfrak{A}(\mathcal{C})$ gelten folgende Kriterien:

1°. $A \in \mathfrak{A}(\mathcal{C})$ ist genau dann stark elliptisch, wenn $\operatorname{Re} \operatorname{Sym} A > 0$.

2°. $A_\Gamma = aI + bS_\Gamma \quad (a, b \in C(\Gamma))$ ist genau dann θ -stark elliptisch, wenn

$$a(t) + b(t)\lambda \neq 0, \quad \forall t \in \Gamma, \quad \forall \lambda \in K_\rho(t), \tag{1}$$

wobei $K_\rho(t)$ die konvexe Hülle der Menge $\{\operatorname{Sym} S_\Gamma(t,z)\}_{(t,z)\in\Lambda}$ bei festem $t \in \Gamma$ bezeichnet. Im Falle $\rho \equiv 1$ ist $K_1(t) = [-1, 1]$

Die Hinlänglichket der Bedingung $\operatorname{Re} \operatorname{Sym} A > 0$ folgt leicht aus den obengenannten Eigenschaften der Abbildung $\operatorname{Sym} A$ (vgl. [9]); ihre Notwendigkeit ergibt sich aus der Hinlänglichkeit und der Eigenschaft, daß der Operator $A \in \mathfrak{A}(\mathcal{C})$ genau dann ein Fredholmoperator ist, wenn $\operatorname{Sym} A \neq 0$ [6], [10]. Die Notwendigkeit von (1) ist eine direkte Folgerung der Eigenschaft 1°; ihre Hinlänglichkeit kann man mit Hilte einer Einheitszerlegung der Kurve Γ beweisen [3]. Wegen $\sigma_{ess}(S_\Gamma) = \{\operatorname{Sym} S_\Gamma(t,z)\}_{(t,z)\in\Lambda}$ (vgl. [7]) zieht die Bedingung

$$a(t) + b(t)\lambda \neq 0, \quad \forall t \in \Gamma, \ \forall \lambda \in \operatorname{conv} \sigma_{ess}(S_\Gamma), \tag{2}$$

die Bedingung (1) nach sich; für konstante Koeffizienten a, b sind beide Bedingungen (1) und (2) äquivalent.

Die starke Elliptizität ist eine notwendige und hinreichende Bedingung dafür, daß für den invertierbaren Operator A die Reduktionsmethode bezüglich einer beliebigen Orthonormalbasis konvergiert [4]. Wenn $\Gamma = \mathbb{T}$ der Einheitskreis ist, so konvergieren für den singulären Operator $A_\mathbb{T}$ in $L^2(\mathbb{T})$ gewisse Projektionsmethoden mit Spline-Basisfunktionen genau dann, wenn $A_\mathbb{T}$ θ -stark elliptisch ist [12], [13].

Wir betrachten auf $\mathbb{T}$ die stückweise linearen Splines

$$\varphi_k^n(t) \overset{def}{=} \begin{cases} (t-t_{k-1})(t_k-t_{k-1})^{-1} & \text{für } t \in t_{k-1}t_k \\ (t_{k+1}-t)(t_{k+1}-t_k)^{-1} & \text{für } t \in t_k t_{k+1} \\ 0 & \text{sonst,} \end{cases}$$

wobei $t_k = t_k^{(n)} = e^{2\pi i k/n} \quad (k = 0, \ldots, n-1)$. Mit P_n bezeichnen wir den Orthoprojektor in $L^2(\mathbb{T})$ auf die lineare Hülle

$H^n \overset{def}{=} \operatorname{span}\{\varphi_0^{(n)}, \dots, \varphi_{n-1}^{(n)}\}$ und mit Q_n den Interpolationsprojektor, der jeder beschränkten Funktion f den Polygonzug

$$(Q_n f)(t) \overset{def}{=} \sum_{k=0}^{n-1} f(t_k)\varphi_k^{(n)}(t)$$

zuordnet. Wenn die Operatoren $A_n \overset{def}{=} Q_n A_{\mathbb{T}} P_n$ im Unterraum $H^n \subset L^2(\mathbb{T})$ (für alle $n > n_0$) invertierbar sind und $\sup \|A_n^{-1}\| < \infty$ ist, dann schreiben wir $A_{\mathbb{T}} \in \Pi\{P_n, Q_n\}^n$; in diesem Falle gilt $A_n^{-1} Q_n f \xrightarrow{L^2} A_{\mathbb{T}}^{-1} f$, $n \to \infty$, für alle f mit $Q_n f \to f$, insbesondere für alle Riemannintegrierbaren Funktionen [12]. Das soeben beschriebene Projektionsverfahren heißt Kollokationsmethode (auch Polygonmethode). Analoge Bedeutung hat $\Pi\{P_n, P_n\}$ (Reduktionsmethode). Es gilt folgender

SATZ ([12],[13]). <u>Sei</u> $A_{\mathbb{T}} = aI + bS_{\mathbb{T}}$ $(A_{\mathbb{T}} = aI + S_{\mathbb{T}} bI)$ <u>mit</u> $a, b \in PC(\Gamma)$. <u>Dann gilt</u> $A_{\mathbb{T}} \in \Pi\{P_n, Q_n\}$ <u>genau dann, wenn folgende Bedingung erfüllt ist:</u>

$$[c(t+0)d(t-0)\mu + c(t-0)d(t+0)(1-\mu)]\nu + \tag{3}$$
$$+ d(t+0)d(t-0)(1-\nu) \neq 0, \quad \forall t \in \Gamma, \ \forall \mu, \nu \in [0,1]$$

<u>wobei</u> $c \overset{def}{=} a+b, \quad d \overset{def}{=} a-b.$

Wir bemerken, daß aus (3) die Invertierbarkeit von $A_{\mathbb{T}}$ in $L^2(\mathbb{T})$ folgt [7],[10]. Im Falle $a, b \in C(\mathbb{T})$ bedeutet (3) gerade die θ-starke Elliptizität von $A_{\mathbb{T}}$.

HYPOTHESE 1. <u>Bedingung (3) ist äquivalent der θ-starken Elliptizität des Operators $A_{\mathbb{T}}$ im Raum $L^2(\mathbb{T})$</u> $(\theta \in PC(\mathbb{T}))$.

Aus der Gültigkeit dieser Hypothese würde insbesondere $A_{\mathbb{T}} \in \Pi\{P_n, P_n\}$ folgen. Die Schwierigkeiten beim Überprüfen der Hypothese 1 bestehen darin, daß $\operatorname{Sym} A_{\mathbb{T}}$ eine Matrixfunktion und $\mathfrak{A}(PC)$ eine nichtsymmetrische Algebra ist.

Von großem theoretischen und praktischen Interesse sind Bedingungen, die die Konvergenz entsprechender Kollokationsmethoden mit gewichteten Splines auf offenen Kurven garantieren (s.z.B. [2],[8]). Sei der Einfachheit halber $\Gamma = [0,1]$, $t_j^{(n)} = j/n$, $(j = 0,1,\dots,n)$ $\varphi_j^{(n)}$ die entsprechenden stückweise linearen Splines und $\psi_j^{(n)} =$

$= \rho^{-1/2} \varphi_j^{(n)}$. Mit P_n^1 bezeichnen wir den Orthoprojektor in $L^2(\Gamma,\rho)$ auf $\mathrm{span}\{\psi_0^{(n)},\ldots,\psi_n^{(n)}\}$ $(\subset L^2(\Gamma,\rho))$ und mit Q_n^1 den entsprechenden Interpolationsprojektor.

HYPOTHESE 2. Sei $A_\Gamma = aI + bS_\Gamma$ $(a,b) \in C(\Gamma)$ ein θ-stark elliptischer Operator. Dann gilt $A_\Gamma \in \Pi\{P_n^1, Q_n^1\}$.

Im Falle $\rho \equiv 1$ ergibt sich die Richtigkeit der Hypothese 2 aus dem obengenannten Satz durch Abbildung von Γ auf eine Hälfte von $\mathbb{T}$ und anschließende Fortsetzung der Koeffizienten auf ganz $\mathbb{T}$ (vgl. [10], Seite 86).

HYPOTHESE 3. Hypothese 2 gilt für beliebige polynomiale Splines $\varphi_j^{(n)}$ ungeraden Grades.

Für geschlossene Kurven Γ wurde die Konvergenz der Kollokations - und Reduktionsmethoden mit Splines beliebigen Grades in nichtgewichteten Sobolewräumen in [1], [11], [14] untersucht.

LITERATUR

1. A r n o l d D.N., W e n d l a n d W.L. On the asymptotic convergence of collocation methods. - Math. of Comput., 1983.
2. D a n g D.Q., N o r r i e D.H. A finite element method for the solution of singular integral equations. - Comp. Math. with Appl., 1978, 4, 219-224.
3. E l s c h n e r I., P r ö s s d o r f S. Über die starke Elliptizität singulärer Integraloperatoren. - Math. Nachr. (im Druck).
4. Г о х б е р г И.Ц., Ф е л ь д м а н И.А. Уравнения в свертках и проекционные методы их решения. М., Наука, 1971.
5. Г о х б е р г И.Ц., К р у п н и к Н.Я. Об алгебре, порожденной одномерными сингулярными интегральными операторами с кусочно-непрерывными коэффициентами. - Функц. анал. и его прил., 1970, 4, № 3, 26-36.
6. Г о х б е р г И.Ц., К р у п н и к Н.Я. Сингулярные интегральные операторы с кусочно-непрерывными коэффициентами и их символы. - Изв. АН СССР, сер. матем., 1971, 35, № 4, 940-961.
7. Г о х б е р г И.Ц., К р у п н и к Н.Я. Введение в теорию одномерных сингулярных интегральных операторов. Кишинев, Штиинца, 1973.

8. I e n E., S r i v a s t a v R.P. Cubic splines and approximate solution of singular integral equations. - Math. of Comput., 1981, 37, N 156, 417-423.

9. K o h n I.I., N i r e n b e r g L.I. An algebra of pseudodifferential operators. - Comm.Pure and Appl.Math., 1965, 18, N 112, 269-205.

10. M i c h l i n S.G., P r ö s s d o r f S., Singuläre Integraloperatoren. - Akademie-Verlag, Berlin, 1980.

11. P r ö s s d o r f S. Zur Splinekollokation für lineare Operatoren in Sobolewräumen. - Teubner - Texte zur Math. "Recent Trends in Math.", 1983, Bd.50, 251-262.

12. P r ö s s d o r f S., S c h m i d t G., A finite element collocation method for singular integral equations. - Math.Nachr., 1981, 100, 33-60.

13. P r ö s s d o r f S., R a t h s f e l d A. Finite-Elemente Methoden für singuläre Integralgleichungen mit stückweise stetigen Koeffizienten. - Math.Nachr. (im Druck).

14. S c h m i d t G. On spline collocation for singular integral equations. - Preprint P-Math.-13/82, Akademie der Wissenschaften der DDR, Inst.f.Math., 1982.

S.PRÖSSDORF

Institut für Mathematik AdW
Mohrenstraße 39,
1086 Berlin,
DDR

5.15. HOW TO CALCULATE THE DEFECT NUMBERS OF THE GENERALIZED RIEMANN BOUNDARY VALUE PROBLEM?

The question concerns the problem of finding functions $\varphi, \psi \in H^p$ satisfying the boundary condition

$$\varphi(\alpha(t)) = a(t)\overline{\psi(t)} + b(t)\psi(t) + h(t), \quad t \in \mathbb{T}. \tag{1}$$

Here $a, b \in L^\infty$, $h \in L^p$, $1<p<\infty$, α is a non-singular orientation preserving diffeomorphism of $\mathbb{T}$ ("the shift") with the derivative in $\mathrm{Lip}\,\mu$, $0<\mu<1$. The case of orientation-changing shift α comes to this by an evident replacement of $\alpha(t)$ by $\overline{\alpha(t)}$; a by $\bar{b}$; b by $\bar{a}$.

The investigation of (1) and of its generalizations is connected with a number of questions of elasticity theory ([1], Ch 7), the rigidity problem for piecewise-regular surfaces [2], etc., and has already a rather long history, starting with A.I.Markushevitch's work of 1946 (see [3] and a detailed bibliography contained therein).

Fredholmness conditions and the index of the operator corresponding to the problem (1) are known and don't depend upon "the shift". If $a^{-1}b$ is sufficiently small, then under certain additional conditions on a (e.g. $\|a^{-1}b\| < \sin\frac{\pi}{\max(p,p')}$, $a \in \mathfrak{M}$, $\mathfrak{M}$ being the class (introduced in [5]) of multipliers not affecting the factorizability) one of the defect numbers of (1) is equal to zero, and therefore defect numbers don't depend upon "the shift". I.H.Sabitov's example (see [3], p.272) shows that this is not the case in general.

PROBLEM. <u>Calculate the defect numbers of the problem (1). Find the conditions on the coefficients</u> a, b, <u>under which the defect numbers do not depend upon "the shift"</u>.

The defect numbers l and l' of the problem (1) without "shift" ($\alpha(t) = t$, $t \in \mathbb{T}$) are connected [3,4] by formulas $l = \max(\varkappa_1, 0) + \max(\varkappa_2, 0) - 1$; $l' = \varkappa_1 + \varkappa_2 - l - 2$ with partial indices $\varkappa_1, \varkappa_2$ of the matrix

$$\bar{a}^{-1}\begin{pmatrix} |b|^2 - |a|^2 & b \\ \bar{b} & 1 \end{pmatrix}$$

(here the defect numbers are calculated over $\mathbb{R}$).

The problem to calculate partial indices of matrix-valued functions even of this special kind, however, is far from final solution.

Under assumptions

$$p=2,\ a\in \mathfrak{M};\quad dist\,(\omega, H^{\infty}+C)<1 \tag{2}$$

(where $\omega = a_+^{-1} b$, a_+ is an outer function with $|a_+|=|a|$ a.e.), as it is shown in [6], $\varkappa_1$ and $\varkappa_2$ are expressed in terms of the multiplicity of s-number 1 of the Hankel operator $Q\omega P$ (P is the Riesz projection of L^p onto H^p, $Q=I-P$). Using this fact and the results of V.M.Adamjan, D.Z.Arov and M.G.Krein [7,8,9] the defect numbers of problem (1) are expressed in [6] in terms of approximation characteristics of its coefficients. The elimination of restrictions (2), and the generalization of the above-mentioned results to the weighted spaces seems to be of interest.

In the case $\alpha(t)\not\equiv t$ the calculation of defect numbers of (1) may reduce to the problem of calculation of the dimension of the kernel of the block operator

$$\begin{pmatrix} QW_\alpha^{-1} a Q & QW_\alpha^{-1} b P \\ PW_\alpha^{-1} \bar{b} Q & PW_\alpha^{-1} \bar{a} P \end{pmatrix},$$

composed by "shifted" Hankel and Toeplitz operators (here $W_\alpha f = f\circ\alpha$ is the so-called "shift" (translation) operator). It is interesting to remark, that these operators also appear while investigating the so-called one-sided boundary-value problems, studied in [10]. Thus, the investigation of the problem $\varphi(\alpha)=a\bar{\varphi}+h$, $\varphi\in H^p$, $h\in L^p$ with an involutory orientation-changing "shift" α under ordinary conditions [3] $a\cdot\overline{a(\alpha)}=1$; $h+a\overline{h(\alpha)}=0$ can reduce to the study of the operator $QW_\alpha a Q$: both have the same defect numbers, their images are closed simultaneously and so on. Thus, the new information about "shifted" Toeplitz and Hankel operators may be employed in the study of the boundary value problems with the shift.

In conclusion it should be remarked, that by an analytical continuation of $\bar{\varphi}$ into the domain $\{z\in\mathbb{C}:|z|>1\}$ and by a conformal mapping, the problem (1) can be reduced to the problem of finding the pair of functions in Smirnov classes $E_\pm^p$, satisfying a "nonshifted" boundary condition on a certain contour. We note, by the way, that the related question of the change of partial indices of matrix-valued functions under a conformal mapping (i.e., practically, the question of calculating defect numbers of vectorial "shifted" Riemann boundary value problem) put by B.V.Bojarskiĭ [11], has received no satisfactory solution so far.

The authors are grateful to I.M.Spitkovskiĭ for useful discussions.

REFERENCES

I. В е к у а Н.П. Системы сингулярных интегральных уравнений и некоторые граничные задачи, М., Наука, I970.

2. В е к у а И.Н. Обобщенные аналитические функции, М., ФМ, I959.

3. Л и т в и н ч у к Г.С. Краевые задачи и сингулярные интегральные уравнения со сдвигом, М., Наука, I977.

4. С п и т к о в с к и й И.М. К теории обобщенной краевой задачи Римана в классах L^p. - Укр.матем.журн., I979, 3I, № I, 63-73.

5. С п и т к о в с к и й И.М. О множителях, не влияющих на факторизуемость. - Докл.АН СССР, I976, 23I, № 6, I300-I303.

6. Л и т в и н ч у к Г.С., С п и т к о в с к и й И.М. Точные оценки дефектных чисел обобщенной краевой задачи Римана, факторизация эрмитовых матриц-функций и некоторые проблемы приближения мероморфными функциями. - Матем.сборн., I982, II7, № 2, I96-2I4.

7. А д а м я н В.М., А р о в Д.З., К р е й н М.Г. О бесконечных ганкелевых матрицах и обобщенных задачах Каратеодори-Фейера и Ф.Рисса. - Функц.анал. и его прил., I968, 2, № I, I-I9.

8. А д а м я н В.М., А р о в Д.З., К р е й н М.Г. Бесконечные ганкелевы матрицы и обобщенные задачи Каратеодори-Фейера и И.Шура. - Функц.анал. и его прил., I968, 2, № 4, I-I7.

9. А д а м я н В.М., А р о в Д.З., К р е й н М.Г. Аналитические свойства пар Шмидта ганкелева оператора и обобщенная задача Шура - Такаги. - Матем.сборн., I97I, 86, № I, 33-73.

I0. З в е р о в и ч Э.И., Л и т в и н ч у к Г.С. Односторонние краевые задачи теории аналитических функций. - Изв.АН СССР, сер. матем., I964, 28, № 5, I003-I036.

II. Б о я р с к и й Б.В. Анализ разрешимости граничных задач теории функций. - В кн.: Исследования по совр.проблемам теории функций комплексного переменного, М., ФМ, I96I, 57-79.

Yu.D.LATUSHKIN
(Ю.Д.ЛАТУШКИН)

G.S.LITVINCHUK
(Г.С.ЛИТВИНЧУК)

СССР, 220000, Одесса,
ул.Петра Великого 2,
Одесский государственный
университет

5.16. POINCARÉ-BERTRAND OPERATORS IN BANACH ALGEBRAS

Let A be an associative algebra over $\mathbb{C}$. A linear operator $R \in \operatorname{End} A$ is said to satisfy the Poincaré-Bertrand identity if for all $X, Y \in A$

$$R(X \cdot RY + RX \cdot Y) = RX \cdot RY + XY . \tag{1}$$

THEOREM. <u>Suppose R satisfies (1). Then</u>

(i) <u>the formula</u>

$$X \times_R Y = RX \cdot Y + X \cdot RY$$

<u>defines an associative product in</u> A (We denote the corresponding <u>algebra by</u> A_R);

(ii) <u>The mappings</u> $R \pm 1$ <u>are homomorphisms from</u> A_R <u>into</u> A. <u>Let</u> $A_\pm = \operatorname{Im}(R \pm 1)$, $N_\pm = \operatorname{Ker}(R \mp 1)$. <u>Then</u> $A_\pm \subset A$ <u>is a subalgebra and</u> $N_\pm \subset A_\pm$ <u>is a two-sided ideal. Also</u>, $A = A_+ + A_-$, $N_+ \cap N_- = 0$;

(iii) <u>The mapping of the quotient algebras</u>

$$\theta : A_+ / N_+ \longrightarrow A_- / N_-$$

<u>given by</u> $\theta : (R+1)X \longrightarrow (R-1)X$ <u>is an algebra isomorphism</u>;

(iv) <u>each</u> $X \in A$ <u>can be uniquely decomposed as</u>

$$X = X_+ - X_- , \; X_\pm \in A_\pm , \; \theta(\overline{X}_+) = \overline{X}_-$$

(we denote by $\overline{X}_\pm$ the residue class of $X_\pm$ modulo $N_\pm$).

EXAMPLE. Let $A = W$ be the Wiener algebra. Define $R \in \operatorname{End} W$ by

$$RX = \begin{cases} X, & \text{if } X \text{ is analytic in } \mathcal{D} , \\ -X, & \text{if } X \text{ is antianalytic in } \mathcal{D} . \end{cases}$$

Then R satisfies (1).

PROBLEM. <u>For</u> $A = W \otimes \operatorname{Mat}(n, \mathbb{C})$, <u>or</u> $A = L^\infty \otimes \operatorname{Mat}(n, \mathbb{C})$ <u>describe all linear operators satisfying (1)</u>.

NOTES. 1. This problem arises as a byproduct from the studies of completely integrable systems. This connection is fully explained in [1]. For partial results in the classification problem cf. [1], [2]. In most papers the problem is considered in the Lie algebraic setting In that case equation (1) is replaced by

$$R([X,RY]+[RX,Y])=[RX,RY]+[X,Y]$$

which is commonly known as the (classical) Yang-Baxter equation.

2. Given a solution of (1) an operator $X \longmapsto X \times_R Y$ can be regarded as an analogue of the Toeplitz operator with the symbol Y. It seems interesting to study the corresponding operator calculus in detail.

REFERENCES

I. Семёнов-Тян-Шанский М.А. Что такое классическая r-матрица. - Функц.анал. и его прил., I983. I7, № 4.

2. Белавин А.А., Дринфельд В.Г. О решениях классического уравнения Янга-Бакстера для простых алгебр Ли. - Функц.анал. и его прил., I982, I6, № 3, I-29.

M.A.SEMËNOV-TIAN-SHANSKY (М.А.СЕМЁНОВ-ТЯН-ШАНСКИЙ)

СССР, I9I0II, Ленинград, Фонтанка 27, ЛОМИ

CHAPTER 6

SINGULAR INTEGRALS, BMO, H^p

This chapter is a natural continuation of the preceding one: eight problems opening the chapter deal with singular integrals. The two first are "old" (and are essentially influenced by Calderón's 1977 breakthrough in L^2-estimates of Cauchy-type integrals on Lipschitz curves (see S.5 and Commentrary therein). Others cover various aspects of the theory of singular integrals (continuity, two-sided estimates as in 6.8 and even the exact values of their norms as in 6.6).

Unlike the preface of the "old" Chapter 9 we dispense here with emotions caused at that time by the very appearance of BMO and the real H^p-spaces. The H^p-BMO ideology has shared the destiny of all significant theories (see Introduction to Chapter 1) being now - together with L^p's or C - a necessary prerequisite for analytical activity, as though they (i.e. BMO and H^p)"existed always," but passed unnoticed for a period of time.

BMO is ubiquitous as is seen, e.g., from the items of this (and not only of this) Chapter. In Problem 6.10 BMO is intertwined with famous coefficient problems for univalent functions. In some problems its presence or influence is not so explicit (as e.g. in 6.11, 6.12, 6.14 or in Problem 6.13, dealing with a quantitative variant of the John - Nirenberg inequality) but nevertheless undeniable The same can be said about VMO-setting of the "old" Problem by Sarason (S.6). Various aspects of the H^p -theory (real or complex) are discussed in 6.4, 6.9, 6.13, 6.15-6.19. Other interesting connections are represented by items 6.9 and 6.10. These problems are of importance for Toeplitz operators (see Chapter 5). The solution of S.6 found by T.Wolff (see Commentary in S.6) yields a useful factorization of unimodular functions in $H^\infty + C$ leading to a factorization of Toeplitz operators with $(H^\infty + C)$ -symbols. The prediction contained in the last phrase of Section 2 in 6.9 was more than justified: the negative solution obtained by T.Wolff (see Commentary to 6.9) also has an important application, namely, the existence of a non-invertible Toeplitz operator whose symbol has a Poisson extension bounded away from zero This disproves the famous conjecture of Douglas.

We conclude by the indication of Problem S.11 inspired by the abstract H^p-theory of Coifman - Weiss. (We first included the problem into this Chapter, but people became aware of it before the volume was ready and got so interested that we had - at the last moment - to remove it to "Solutions".)

6.1. old ON THE CAUCHY INTEGRAL AND RELATED INTEGRAL OPERATORS

Let Γ be a rectifiable curve in $\mathbb{C}$. The Cauchy integral of a function defined on Γ and integrable relative to arc length is defined as:

$$C(f)(z) = \frac{1}{2\pi i} \int_\Gamma \frac{f(\zeta)}{\zeta - z}\, d\zeta .$$

Recently A.P.Calderón [1] has proved the existence almost everywhere of nontangential boundary values for the function $C(f)$ (denoted as $C(f)(\zeta)$). This pointwise existence theorem follows from the following estimate proved in [1] by an ingenious complex variable method.

THEOREM 1 (A.P.Calderón). <u>There is a constant</u> η_0, $\eta_0 > 0$, <u>such that for all functions</u> φ <u>with</u> $\|\varphi'\|_\infty < \eta$ <u>there exists a constant</u> C_φ <u>for which</u>:

$$\int \left| \int_{-\infty}^{+\infty} \frac{f(t)\, dt}{(s-t) + i(\varphi(s) - \varphi(t))} \right|^2 ds \leq C_\varphi \int |f(s)|^2\, ds .$$

It is not hard(by using singular integral techniques) to reduce the existence a.e. result mentioned above to theorem 1. SEVERAL IMPORTANT QUESTIONS remain open.

I. <u>Is the restriction</u> $\|\varphi'\|_\infty < \eta_0$ <u>necessary to obtain the estimate of theorem 1</u>? Calderón's method as well as other techniques are unable to eliminate this restriction.

II. Since the operator $C(f)(\zeta)$ exists almost everywhere for all functions in $L^2(\Gamma, |d\zeta|)$, it is natural <u>to conjecture the existence of a weight</u> $\omega_\Gamma(\zeta)$ (> 0 a.e.) <u>for which</u>

$$\int |C(f)(\zeta)|^2\, \omega_\Gamma(\zeta)\, |d\zeta| \leq C_\Gamma \int |f|^2 |d\zeta| .$$

(The existence of such a weight for a weak L^2 estimate is guaranteed by general considerations related to the Nikishin-Stein theorem.)

III. The integral operator appearing in Theorem 1 is related to a general class of operators like Hilbert-transforms, of which the following are typical examples.

a) The so called commutators of order n

$$A_n(f) = \int_{-\infty}^{+\infty} \left(\frac{A(x) - A(y)}{x - y} \right)^n \frac{f(y)}{x-y}\, dy$$

b) $$\int_0^\infty \frac{A(x+t) - 2A(x) + A(x-t)}{t^2} f(x-t)\, dt$$

c) $$\int \frac{A(x) - A(y)}{|x-y|^{2+i\gamma}} f(y)\, dy\,, \quad \gamma > 0 \qquad \text{(Here } A' \in L^\infty \text{)}.$$

It is easily seen that Theorem 1 is equivalent to the following estimates on the operators A_n :

$$\| A_n(f) \|_2 \leq C^n \| A' \|_\infty^n \| f \|_2$$

for some constant C . The boundedness in L^2 of the operators in a) b) c) has been proved in [2], [3] by using Fourier analysis and real variable techniques (which extend to $\mathbb{R}^n$). Unfortunately the estimate obtained (by these methods) on the growth of the constant in (*) is of the order of $n!$ (and not C^n). It will be highly desirable to obtain a proof of Calderón's result which does not depend on special tricks or complex variables. Any such technique will extend to higher dimensions and is bound to imply various sharp estimates for operators arising in partial differential equations.

REFERENCES

1. C a l d e r ó n A.P. On the Cauchy integral on Lipschitz curves and related operators. - Proc.N.Ac.Sc.1977, 4, 1324-27.
2. C o i f m a n R.R., M e y e r Y. Commutateurs d'integrales singulières et opérateurs multilinéaires.- Ann. Inst. Fourier (Grenoble), 1978, 28, N 3, xi, 177-202.
3. C o i f m a n R.R., M e y e r Y. Multilinear pseudo-differential operators and commutators, to appear.

R.R.COIFMAN
Department of Mathematics
Washington University
Box 1146, St.Louis, MO.63130
USA

YVES MEYER
Faculté des Sciences d'Orsay
Université de Paris-Sud
France

* * *

COMMENTARY

The solution of Problem I is discussed in Commentary to S.5.

6.2. old

SOME PROBLEMS CONCERNING CLASSES OF DOMAINS DETERMINED BY PROPERTIES OF CAUCHY TYPE INTEGRALS

Investigation of boundary properties of analytic functions representable by Cauchy-Stieltjes type integrals in a given planar domain G (i.e. functions of the form $z \longmapsto \int_{\partial G} (\zeta - z)^{-1} d\mu(\zeta)$; if $d\mu = \omega d\zeta$ we denote this function by $\mathcal{K}^{\omega}$), as well as some other problems of function theory (approximation by polynomials and rational fractions, boundary value problems, etc.) have led to introduction of some classes of domains. These classes are defined by conditions that the boundary singular integral $S_\Gamma \omega \, (\Gamma = \partial\Omega)$ should exist and belong to a given class of functions on Γ or (which is in many cases equivalent) that analytic functions representable by Cauchy type integrals should belong to a given class of analytic functions in G. See [1] for a good survey on solutions of boundary value problems. An important role is played by the class of curves (denoted in [1] by R_p) for which the singular integral operator is continuous on $L^p(\Gamma)$: $\Gamma \in R_p$ if and only if

$$\forall \omega \in L^p(\Gamma) \quad \|S_\Gamma(\omega)\| \leq C_p \|\omega\|_p \; . \tag{1}$$

This means that the M.Riesz theorem (well-known for the circle) holds for Γ. Some sufficient conditions for (1) were given by B.V.Hvedelidze, A.G.Džvarsheishvily, G.A.Huskivadze and others. I.I.Danilyuk and V.Yu.Shelepov (a detailed exposition can be found in the monograph [2]) have shown that (1) is true for all $p>1$ for simple rectifiable Jordan curves Γ with bounded rotation and without cusps. Some general properties of the class R_p were described by V.P.Havin, V.A.Paatashvily, V.M.Kokilashvily and others. It was shown, e.g., that (1) is equivalent to the following condition:

$$\forall \; \omega \in L^p(\Gamma) \qquad f \stackrel{def}{=} K^{\omega} \in E_p(G) \, ,$$

$E_p(G)$ being the well-known V.I.Smirnov class (cf. e.g. [3]) of functions f analytic in G and such that integrals of $|f|^p$ over some system of closed curves $\{\gamma_i\}$ (with $\gamma_i \subset G$, $\gamma_i \to \Gamma$) are bounded. That is, R_p can be characterized by the property

$$\forall\ \omega \in L^p(\Gamma)\ (f\circ\varphi)\sqrt[p]{\varphi'} \in H_p\ \ (f = K^{\omega}), \tag{2}$$

where φ is a conformal mapping of $\mathbb{D}$ onto G.

Another class of domains (denoted by K) has been introduced and investigated earlier by the author (cf. [4] and references to other author's papers therein). We quote a definition of K that is closely connected with definition (2) of R_p : $G \in K$ if for any function f in G, analytic and representable as a Cauchy type integral, the function $(f\circ\varphi)\varphi'$ (φ being a conformal mapping of $\mathbb{D}$ onto G) is also representable as a Cauchy type integral:*)

$$f = \mathcal{K}^{\omega},\ \ \omega \in L^1(\Gamma) \Rightarrow (f\circ\varphi)\varphi' = \mathcal{K}^{\Omega},\quad \Omega \in L^1(\mathbb{T}).$$

Note that by Riesz theorem it is sufficient for $\Gamma \in R_p$ that the function $(f\circ\varphi)\sqrt[p]{\varphi'}$ in (2) be representable in the form $\mathcal{K}^{\Omega}$ with $\Omega \in L^p(\mathbb{T})$. This allows to consider K as a counterpart of R_p for $p=1$ (it is well-known that to use (2) directly is impossible for $p=1$ even for $\Gamma = \mathbb{T}$). It is established in [5] (see also [1]), using Cotlar's approach, that the classes R_p coincide for $p>1$. Thus the following problem arises naturally.

PROBLEM 1. <u>Do the classes</u> $R_p\ (p>1)$ <u>and</u> R_1 <u>coincide</u>? <u>If not, what geometric conditions guarantee</u> $\Gamma \in R_p \cap R_1$?

Note that for $G \in K$ it becomes easier to transfer many theorems, known for the disc, on approximation by polynomials or by rational fractions in various metrics (cf. references in [4]). It is possible to obtain for such domains conditions that guarantee convergence of boundary values of Cauchy type integrals [4]. As I have proved, K is a rather wide class containing in particular all domains G bounded by curves with finite rotation (cusps are allowed) [4]. At the same time, it follows from characterizations of K proved by me earlier that K coincides with the class of Faber domains, introduced and used later by Dyn'kin (cf. e.g. [6], [7]) to investigate uniform approximations by polynomials and by Anderson and

---

*) In virtue of a well-known V.I. Smirnov theorem, the analog of this property for Cauchy integrals is always true.

Ganelius [8] to investigate uniform approximation by rational fractions with fixed poles. This fact seems to have stayed unnoticed by the authors of these papers, because they reprove for the class of Faber domains some facts established earlier by me (the fact that domains with bounded rotation and without cusps belong to this class, conditions on the distribution of poles guaranteeing completeness, etc.). The following question is of interest.

PROBLEM 2. Suppose that the interior domain G^+ of a curve Γ belongs to $K(=R_1)$. Is it true that the exterior domain G^- also belongs to K? (Of course, we use here a conformal mapping of G^- onto $\{|w|>1\}$).

For R_p with $p>1$ the positive answer to the analogous question is evident. At the same time the similar problem formulated in [9] for the class S of Smirnov domains remains still open. At last it is of interest to study the relationship between the classes S of Smirnov domains and A_0 of Ahlfors domains (bounded by quasicircles [10]), on the one hand, and K and R_p (considered here) on the other. See [9] for more details on S and A_0. It is known that $R_p \subset S$, $K \subset S$ ([4],[11]). At the same time there exist domains with a rectifiable boundary in A_0 which do not belong to S (cf. [3],[9]). Simple examples of domains bounded by piecewise differentiable curves with cusp points show that $K \setminus A_0 \neq \emptyset$.

PROBLEM 3. Find geometric conditions guaranteeing

$$G \in K \cap R_p \cap A_0 .$$

Once these conditions are satisfied, it follows from the papers cited above and [12], [13] that many results known for the unit disc can be generalized.

One of such conditions is that Γ should be of bounded rotation and without cusps.

REFERENCES

1. Хведелидзе Б.В. Метод интегралов типа Коши в разрывных граничных задачах теории голоморфных функций одной комплексной переменной. "Современные проблемы математики", т.7, Москва, 1975, 5-162.
2. Данилюк И.И. Нерегулярные граничные задачи на плоскости, Москва, Наука, 1975.

3. D u r e n P.L., S h a p i r o H.S., S h i e l d s A.L. Singular measures and domains not of Smirnov type. - Duke Math. J., 1966, v.33, N 2, 247-254.

4. Т у м а р к и н Г.Ц. Граничные свойства аналитических функций, представимых интегралами типа Коши. - Матем.сб., I97I, 84 (I26), № 3, 425-439.

5. П а а т а ш в и л и В.А. О сингулярных интегралах Коши. - Сообщ. АН Груз.ССР, I969, 53, № 3, 529-532.

6. Д ы н ь к и н Е.М. О равномерном приближении многочленами в комплексной плоскости. - Зап.научн.семин.ЛОМИ, I975, 56, I64-I65.

7. Д ы н ь к и н Е.М. О равномерном приближении функции в жордановых областях. - Сиб.мат.ж. I977, I8, № 4, 775-786.

8. A n d e r s s o n Jan-Erik, G a n e l i u s Tord. The degree of approximation by rational function with fixed poles. - Math.Z., 1977, 153, N 2, 161-166.

9. Т у м а р к и н Г.Ц. Граничные свойства конформных отображений некоторых классов областей.-сб."Некоторые вопросы современной теории функций", Новосибирск, I976, I49-I60.

I0. А л ь ф о р с Л. Лекции по квазиконформным отображениям. Москва, Мир, I969.

II. Х а в и н В.П. Граничные свойства интегралов типа Коши и гармонически сопряженных функций в областях со спрямляемой границей. - Матем.сб., I965, 68 (II0), 499-5I7.

I2. Б е л ы й В.И., М и к л ю к о в В.М. Некоторые свойства конформных и квазиконформных отображений и прямые теоремы конструктивной теории функций. - Изв.АН СССР, серия матем.,I974, № 6, I343-I36I.

I3. Б е л ы й В.И. Конформные отображения и приближение аналитических функций в областях с квазиконформной границей. - Мат.сб., I977, I02, № 3, 33I-36I.

G.C.TUMARKIN
(Г.Ц.ТУМАРКИН)

СССР, I039I2, Москва,
просп.Маркса I8,
Московский геолого-
разведочный институт

* * *

COMMENTARY

A complete geometric description of the class R_p, $1<p<\infty$, has been obtained by Guy David. See Commentary to S.5 for more information.

6.3. BILINEAR SINGULAR INTEGRALS AND MAXIMAL FUNCTIONS

While the boundedness of Cauchy integrals on curves is now fairly well understood [1], there remain some difficult one dimensional problems in this area. One such example is the operator

$$T_1(f,g)(x) = \text{p.v.} \int_{-\infty}^{\infty} f(x+t)\, g(x-t)\, \frac{dt}{t} .$$

Is T_1 a bounded operator from $L^2 \times L^2$ to L^1? A.P.Calderón first considered these operators during the 1960's, when he noticed (unpublished) that the boundedness of T_1 implies the boundedness of the first commutator (with kernel $\frac{A(x)-A(y)}{(x-y)^2}$, $A' \in L^\infty$) as an operator from L^2 to L^2. In order to make sense out of T_1, it seems that one must first study the related maximal operator

$$T_2(f,g)(x) = \sup_{h>0} \frac{1}{2h} \int_{-h}^{h} f(x+t)\, g(x-t)\, dt ,$$

and see whether T_2 is a bounded operator from $L^2 \times L^2$ to L^1. It is easy to see that T_2 maps to weak L^1.

REFERENCE

1. Coifman R.R., McIntosh A., Meyer Y. L'intégrale de Cauchy définit un opérateur borné sur L^2 pour les courbes Lipschitziennes. - Ann.Math.,1982, 116, 361-387.

PETER W.JONES

Institut Mittag-Leffler
Auravägen 17
S-182 62 Djursholm
Sweden

Usual Address:

Dept.of Mathematics
University of Chicago
Chicago, Illinois 60637
USA

6.4. WEIGHTED NORM INEQUALITIES

The problems to be discussed here are of the following type. GIVEN p SATISFYING $1<p<\infty$ AND TWO OPERATORS T AND S, DETERMINE ALL PAIRS OF NONNEGATIVE FUNCTIONS U, V SUCH THAT

$$\int_{\mathbb{R}^n} |Tf(x)|^p U(x)\,dx \leq C \int_{\mathbb{R}^n} |Sf(x)|^p V(x)\,dx,$$

throughout this paper C denotes a constant independent of f but not necessarily the same at each occurrence. There is a question of what constitutes a solution to this sort of problem; it is to be hoped that the conditions are simple and that it is possible to decide easily whether a given pair U, V satisfies the conditions. In some cases, particularly with the restriction $U=V$, this problem has been solved; for a survey of such results and references to some of the literature see [3]. Some of the most interesting unsolved and partially solved problems of this type are as follows.

1. <u>For</u> $1<p<\infty$ <u>find all nonnegative pairs</u> U <u>and</u> V <u>such that</u>

$$\int_0^\infty\int_0^\infty \left| \int_0^x\int_0^y f(t,u)\,du\,dt \right|^p U(x,y)\,dx\,dy \leq C \int_0^\infty\int_0^\infty |f(x,y)|^p V(x,y)\,dx\,dy. \tag{1}$$

This two dimensional version of Hardy's inequality appears easy because f can be assumed nonnegative and no cancellation occurs on the left. The solution of the one dimensional case is known; the obvious two dimensional version of the one dimensional characterization is

$$\left[\int_s^\infty\int_t^\infty U(x,y)\,dy\,dx\right]\left[\int_0^s\int_0^t V(x,y)^{-\frac{1}{p-1}}\,dy\,dx\right]^{p-1} \leq C \tag{2}$$

for $0<s,t<\infty$. This condition is necessary for (1) but not sufficient except for $p=1$. See [7] for a proof that (2) is not sufficient for (1) and for additional conditions under which (2) does imply (1).

2. <u>For</u> $1<p<\infty$ <u>find a simple characterization of all non-</u>

negative pairs U , V such that

$$\int_{-\infty}^{+\infty}[Mf(x)]^p U(x)\,dx \leq C\int_{-\infty}^{+\infty}|f(x)|^p V(x)\,dx \ , \tag{3}$$

where $Mf(x) = \sup_{y\neq x}(y-x)^{-1}\int_x^y |f(t)|\,dt$ is the Hardy-Littlewood maximal function. This problem was solved by Sawyer in [5]; his condition is that for every interval I

$$\int_I [M(\chi_I(x)V(x)^{-\frac{1}{p-1}})]^p U(x)\,dx \leq C\int_I V(x)^{-\frac{1}{p-1}}\,dx \tag{4}$$

with C independent of I. It seems that there should be a characterization that does not use the operator M . One CONJECTURE is that (3) holds if and only if for every interval I and every subset E of I with $|E| = |I|/2$ we have

$$\left[\int_I U(x)\,dx\right]\left[\frac{1}{|I|}\int_I V(x)^{-\frac{1}{p-1}}\,dx\right]^p \leq C\int_E V(x)^{-\frac{1}{p-1}}\,dx \tag{5}$$

with C independent of E and I. Condition (5) does give the right pairs for some of the usual troublesome functions and is not satisfied by the counter example in [5] to an earlier conjecture.

3. For $1 < p < \infty$ find a simple characterization of all non-negative pairs U , V such that

$$\int_{-\infty}^{+\infty}|Hf(x)|^p U(x)\,dx \leq C\int_{-\infty}^{\infty}|f(x)|^p V(x)\,dx \ , \tag{6}$$

where $Hf(x) = \lim_{\varepsilon\to 0+}\int_{|y|>\varepsilon} f(x-y)/y\,dy$ is the Hilbert transform. There is a complicated solution to the periodie version of this by Cotlar and Sadosky in [1]. One CONJECTURE here is that a pair U, V satisfies (6) if and only if U, V satisfy (3) and

$$\int_{-\infty}^{+\infty} [Mf(x)]^{p'} V(x)^{-\frac{1}{p-1}} dx \leq C \int_{-\infty}^{+\infty} |f(x)|^{p'} U(x)^{-\frac{1}{p-1}} dx, \tag{7}$$

where $p' = p/(p-1)$.

4. For $1 \leq p < +\infty$ find a simple characterization of all non-negative pairs U, V for which the weak type inequality

$$\int_{|Hf(x)|>a} U(x) dx \leq C a^{-p} \int_{-\infty}^{+\infty} |f(x)| V(x) dx \tag{8}$$

is valid for $a > 0$. A CONJECTURED SOLUTION is that (7) is a necessary and sufficient condition for (8).

5. For $1 < p < \infty$ find a characterization of all nonnegative functions U such that

$$\int_{-\infty}^{+\infty} |Hf(x)|^p U(x) dx \leq C \int_{-\infty}^{+\infty} [Mf(x)]^p U(x) dx . \tag{9}$$

A necessary condition for (8) is the existence of positive constants C and ε such that for all intervals I and subsets E of I

$$\int_E U(x) dx \leq C \left[\frac{|E|}{|I|}\right]^{\varepsilon} \int_{-\infty}^{\infty} \frac{|I|^q U(x) dx}{|I|^q + |x - x_I|^q}, \tag{10}$$

where x_I denotes the center of I and $q = p$. In [6] it is shown that if (10) holds for some $q > p$, then (9) holds. It is CONJECTURED that (10) with $q = p$ is also sufficient for (9).

6. For $1 < p \leq q < \infty$ find all nonnegative pairs U, V for which

$$\left[\int_{\mathbb{R}^n} |\hat{f}(x)|^q U(x) dx\right]^{1/q} \leq C \left[\int_{\mathbb{R}^n} |f(x)|^p V(x) dx\right]^{1/p} \tag{11}$$

It was shown by Jurkat and Sampson in [2] that if for $r>0$

$$\left[\int_0^r U^*(x)\,dx\right]^{1/q}\left[\int_0^{1/r}\left(V(x)^{-\frac{1}{p-1}}\right)^*dx\right]^{1/p'}\leqslant C, \tag{12}$$

where $*$ indicates the nonincreasing rearrangement and $p'=p/(p-1)$, then (11) holds. Furthermore, if (11) holds for all rearrangements of U and V, then (12) is true. However, (12) is not a necessary condition for (11) as shown in [4]. This problem is probably difficult since if $p=q=2$ and $V(x)=|x|^a$, $0<a<1$, then the necessary and sufficient condition on U is a capacity condition. Its difficulty is also suggested by the fact that a solution would probably solve the restriction problem for the Fourier transform.

REFERENCES

1. Cotlar M., Sadosky C. On some L^p versions of the Helson-Szegö theorem. - In: Conference on Harmonic Analysis in Honor of Antoni Zygmund, Wadsworth, Belmont, California, 1983, 306-317.
2. Jurkat W.B., Sampson G. On rearrangement and weight inequalities for the Fourier transform, to appear.
3. Muckenhoupt B. Weighted norm inequalities for classical operators. - Proc.Symp. in Pure Math 35 (1), 1979, 69-83.
4. Muckenhoupt B. Weighted norm inequalities for the Fourier transform. - Trans.Amer.Math.Soc., to appear.
5. Sawyer E. Two weight norm inequalities for certain maximal and integral operators. In: Harmonic Analysis, Lecture Notes Math. 908, Springer, Berlin 1982, 102-127.
6. Sawyer E. Norm inequalities relating singular integrals and the maximal function, to appear.
7. Sawyer E. Weighted norm inequalities for the n-dimensional Hardy operator, to appear.

BENJAMIN MUCKENHOUPT

Math.Dept.
Rutgers University
New Brunswick
N.J. 08903, USA

6.5. A SUBSTITUTE FOR THE WEAK TYPE (1,1) INEQUALITY FOR MULTIPLE RIESZ PROJECTIONS

Let $C_A(\mathbb{T}^n)$ denote the polydisc algebra, i.e. the subspace of $C(\mathbb{T}^n)$ consisting of the restrictions to the n-dimensional torus $\mathbb{T}^n$ of functions analytic in the open polydisc $\mathbb{D}^n$ and continuous in $\operatorname{clos}\mathbb{D}^n$. By $H^2(\mathbb{T}^n)$ we denote the closure of $C_A(\mathbb{T}^n)$ in $L^2(\mathbb{T}^n)$ and by $i: C_A(\mathbb{T}^n) \to H^2(\mathbb{T}^n)$ the identity operator. The space $H^2(\mathbb{T}^n)^*$ will be identified with $\overline{H^2(\mathbb{T}^n)}$, the bar standing for the complex conjugation (we use throughout the duality established by the pairing $\langle f,g\rangle = \int_{\mathbb{T}^n} f(\theta)\,g(\theta)\,d\theta$).

PROBLEM 1. <u>Does there exist a positive function φ on $(0,1]$ with $\varphi(\varepsilon)\searrow 0$ as $\varepsilon\searrow 0$ such that for each $g\in\overline{H^2(\mathbb{T}^n)}$ with $\|g\|_2=1$ the following inequality holds</u>:

$$\|g\|_1 \leqslant \varphi\left(\|i^*g\|_{C_A(\mathbb{T}^n)^*}\right)? \tag{1}$$

(throughout $\|\cdot\|_p$ denotes the L^p -norm).

If $n=1$ the answer is evidently "yes".

Indeed, in this case the Riesz projection P_- (i.e. the orthogonal projection of $L^2(\mathbb{T}^1)$ onto $\overline{H^2(\mathbb{T}^1)}$) is of weak type (1,1) and so the above function g satisfies the estimate

$$m\{|g|>t\} \leqslant \frac{const}{t}\,\|i^*g\|_{C_A(\mathbb{T}^1)^*}\,.$$

Using this and $\|g\|_2=1$ it can be shown by means of a simple calculation that (1) holds with $\varphi(\varepsilon)=const\,\varepsilon\,(1+\log\varepsilon^{-1})$ (and moreover, for all p, $1<p\leqslant 2$ we have $\|g\|_p \leqslant c\,(p-1)^{-1}\|i^*g\|^{\theta}$ with θ given by $p^{-1}=\theta+(1-\theta)/2$).

For $n\geqslant 2$ the orthogonal projection of $L^2(\mathbb{T}^n)$ onto $\overline{H^2(\mathbb{T}^n)}$ (which is nothing else as the n-fold tensor product $P_-\otimes\ldots\otimes P_-$) is no longer of weak type (1,1) and the above argument fails. Nevertheless for $n=2$ PROBLEM 1 also has a positive solution. This was proved by the author [1] with φ a power function. Using the same idea as in [1] but more careful calculations it can be shown that for $n=2$ and $1<p<2$ we have

$$\|g\|_p \leqslant const\,(p-1)^{-2}\,\|i^*g\|^{\theta}$$

(θ is the same as for $n=1$) provided $g\in\overline{H^2(\mathbb{T}^2)}$ and $\|g\|_2=1$. Consequently, (1) holds for $n=2$ with $\varphi(\varepsilon)=const\,\varepsilon\,(1+\log\varepsilon^{-1})^2$ (to see this set $p=1+(\log\|i^*g\|^{-1})^{-1}$ in the preceding inequality).

Nothing is known for $n>2$. It seems plausible that for such n

PROBLEM 1 should also have a positive solution. Moreover, I think that for $1<p\leq 2$ the inequality $\|g\|_p \leq C_n (p-1)^n \|i^* g\|^\theta$ should be true (and so (1) should hold with $\varphi(\varepsilon)=C_n\varepsilon(1+(\log\varepsilon^{-1})^n)$).

The estimate (1) for $n=2$ was used in [1] to carry over from $C_A(\mathbb{T})$ to $C_A(\mathbb{T}^2)$ some results whose standard proofs for $C_A(\mathbb{T})$ use the weak type (1,1) inequality for $\mathbb{P}_-$. (For example, it was established in [1] that, given a Λ_2-subset E of $(\mathbb{Z}_+)^2$, the operator S, $Sf = \{\hat{f}(n)\}_{n\in E}$ maps $C_A(\mathbb{T}^2)$ onto $\ell^2(E)$. It is still unknown if the same is true with $\mathbb{T}$ and $(\mathbb{Z}_+)^2$ replaced by $\mathbb{T}^n$ and $(\mathbb{Z}_+)^n$, $n\geq 2$.) So (1) is really a substitute for the weak type inequality. Profound generalizations of inequality (1) for $n=2$ with very interesting applications can be found in [2] (some of these applications are quoted in COMMENTARY to S.1).

The proof of (1) for $n=2$ in [1] is based on the weak type (1,1) inequality for $\mathbb{P}_-$ and a complex variable trick, and essentially the same trick appears in [2]. Investigating the case $n>2$ one may seek a more complicated trick that also involves analyticity. But to seek a purely real variable proof is probably more promising from different viewpoints. The solution of the following problem might be the first step in this direction.

PROBLEM 2. Find a real-variable proof of (1) for $n=2$.

In connection with PROBLEM 2 we formulate another problem which is also rather vague but probably clarifies what is meant in the former. The inequality (1) is clearly equivalent to the following one:

$$\|(\mathbb{P}_-\otimes\ldots\otimes\mathbb{P}_-)h\|_1 \leq \varphi(\|h\|_1)$$

provided $h\in L^1(\mathbb{T}^n)$ and $\|(\mathbb{P}_-\otimes\ldots\otimes\mathbb{P}_-)h\|_2=1$.

PROBLEM 3. At least for $n=2$, find and prove a "right" analog of the above inequality involving n-fold tensor products of operators of the form $f\mapsto f*(\mu_i+K_i)$, μ_i being a measure on a multidimensional torus and K_i being a Calderon - Zygmund kernel on the same torus, rather than tensor products of Riesz projections.

REFERENCES

1. К и с л я к о в С.В. Коэффициенты Фурье граничных значений функций, аналитических в круге и в бидиске. - Труды Матем.ин-та им. В.А.Стеклова, 1981, 155, 77-94.

2. B o u r g a i n J. Extensions of H^{∞} -valued functions and bounded bianalytic functions. Preprint, 1982.

S.V.KISLIAKOV
(С.В.КИСЛЯКОВ)

СССР, 191011, Ленинград
Фонтанка 27,
ЛОМИ АН СССР

6.6. THE NORM OF THE RIESZ PROJECTION

The operator of the harmonic conjugation S and the Riesz projection P (i.e. the orthogonal projection onto H^2 in $L^2(\mathbb{T})$) are connected by the simple formula $S = 2P - I$. It has been proved in [1] that

$$\|S\|_{L^p} = \operatorname{ctg}\frac{\pi}{2p} \ (p=2^n), \quad \|S\|_{L^p} \geq \operatorname{ctg}\frac{\pi}{2r}, \quad \|P\|_{L^p} \geq \sin^{-1}\frac{\pi}{p} \ (r=\max(p,p')). \tag{1}$$

In [1] it has been also conjectured that the inequalities in (1) can be replaced by equalities. In the case of operator S this conjecture has been proved in [2,3], but for P the question remains open. The following refinement of the main inequality of [2] has been obtained in [4]:

$$\frac{1}{\cos \pi/2r}\|\operatorname{Im} h\|_{L^p} \leq \|h\|_{H^p} \leq \frac{1}{\sin \pi/2r}\|\operatorname{Re} h\|_{L^p}, \tag{2}$$

where $h \in H^p$, $\operatorname{Im} h\,(0) = 0$. The right-hand side of (2) gives the norm of the restriction of P onto the space of all real-valued functions in L^p satisfying $\hat{f}(0) = 0$.

The same situation occurs for the weighted L^p spaces

$$L^p(w) = \{f : \int_{\mathbb{T}} |f|^p w\,dm < +\infty\}, \quad w(t) = |t - t_0|^\beta,$$

where $t_0 \in \mathbb{T}$, $-1 < \beta < p-1$. The formula for the norm of S in $L^p(w)$ has been obtained in [5]. For P it is known only that (see [6])

$$\|P\| \geq \sin^{-1}\pi/r, \quad r = \max(p, p', p/1+\beta, p/(p-1-\beta)). \tag{3}$$

CONJECTURE. $\|P\| = \dfrac{1}{\sin\frac{\pi}{r}}$.

The conjecture holds for $p=2$ because in this case the problem can be reduced to the calculation of the norm of the Hilbert matrix $\left\{\frac{1}{j+k+\lambda}\right\}_{j,k\geq 0}$. Here is a SKETCH OF THE PROOF. Let

$$a(t)=t^{-\beta/2}, \quad a_+(t)=(1-t)^{\beta/2} \quad (t\in\mathbb{T}, \ -1<\beta<1)$$

and $T_a = Pa\,|\,H^2$. It is known [6] that the Toeplitz operator T_a is invertible in H^2 and $T_a^{-1} = a_+ P \bar{a}_+^{-1}$. Consequently

$$|T_a^{-1}| = |a_+ P \bar{a}_+^{-1}|_{L^2} = |P|_{L^2(\rho)}.$$

The operator T_a is invertible and $|a|=1$, therefore (see [7]) $|T_a^{-1}|^{-2} = 1-|PaQ|^2$. Here $Q=1-P$ and $PaQ = (\hat{a}(j+k+1))_{j,k\geq 0}$ is a Hankel operator. Let us note that

$$\hat{a}(k+1)=\frac{\exp(\pi i\beta/2)}{\pi}\,\frac{\sin \pi\beta/2}{k+1+\beta/2}.$$

It is known [8] that the norm of matrix $\left(\frac{1}{j+k+\lambda}\right)_{j,k\geq 0}$ equals π if $\lambda > 1/2$. We have $|PaQ| = |\sin \pi\beta/2|$, $|P|_{L^2(\rho)} = \cos^{-1}\pi\beta/2 = \sin^{-1}\pi/r$. ●

Let Γ be a simple closed oriented Lyapunov curve; $t_1,\dots,t_n$ be points on Γ, $|P|_{ess}$ ($|P|_{ess}^{(K)}$) be the essential norm of P in the space $L^p(\Gamma,\varrho)$ ($L^p(\Gamma,\varrho_k)$) on Γ with the weight $\varrho(T)=\prod|t-t_k|^{\beta_k}$ ($\varrho_k(t)=|t-t_k|^{\beta_k}$); here $-1<\beta_k<p-1$, $1<p<\infty$. In [6] it was proved that $|P|_{ess}\geq \max\gamma(p,\beta_k)$ ($\gamma(p,\beta)\overset{def}{=}\sin^{-1}\pi/r$, r being defined by (3)). Then in [5] it was proved that $|P|_{ess}^{(k)}=\max\limits_k |P|_{ess}^{(k)}$. If our Conjecture is true then $|P|_{ess}=\max\limits_k \gamma(p,\beta_k)$.

In conclusion we note that in the space L^p on the circle $\mathbb{T}$ (without weight) $|P|_{ess}=|P|$ ([3]). But in general the norm $|P|$ depends on the weight and on the contour Γ ([3],[5]).

REFERENCES

1. Г о х б е р г И.Ц., К р у п н и к Н.Я. О норме преобразования Гильберта в пространстве L^p. – Функц.анал. и его прил., I968, 2, № 2, 9I–92.
2. P i c h o r i d e s S.K. On the best values of the constants in in the theorems of M.Riesz, Zygmund and Kolmogorov. – Studia Math., 1972, 44, N 2, 165–179.
3. К р у п н и к Н.Я., П о л о н с к и й Е.П. О норме оператора сингулярного интегрирования. – Функц.анал. и его прил., I975, 9, № 4, 73–74.
4. В е р б и ц к и й И.Э. Оценка нормы функции из пространства Харди через норму ее вещественной и мнимой части. – В сб."Матем. исследования", Кишинев, Штиинца, I980, № 54, I6–20.
5. В е р б и ц к и й И.Э., К р у п н и к Н.Я. Точные константы в теоремах К.И.Бабенко и Б.В.Хведелидзе об ограниченности сингулярного оператора. – Сообщ.АН Груз.ССР, I977, 85, № I, 2I–24.
6. Г о х б е р г И.Ц., К р у п н и к Н.Я. Введение в теорию сингулярных интегральных операторов. – Кишинев, Штиинца, I973.
7. Н и к о л ь с к и й Н.К. Лекции об операторе сдвига. М.: Наука, I980.
8. H a r d y G.H., L i t t l e w o o d J.E., P ó l y a G. Inequalities. 2nd ed. Cambridge Univ.Press, London and New York, 1952.

I.È.VERBITSKY
(И.Э.ВЕРБИЦКИЙ)

СССР, 277028, Кишинев,
Институт геофизики и геологии
АН МССР

N.Ya.KRUPNIK
(Н.Я.КРУПНИК)

СССР, 277003, Кишинев,
Кишиневский государственный
Университет

6.7. IS THIS OPERATOR INVERTIBLE?

Let G denote the group of increasing locally absolutely continuous homeomorphisms h of $\mathbb{R}$ onto itself such that h' lies in the Muckenhoupt class A^∞ of weights. Let V_h denote the operator defined by $V_h(f)=f\circ h$, so that V_h is bounded on $BMO(\mathbb{R})$ if and only if $h\in G$ (Jones [3]). Suppose that P is the usual projection of BMO onto BMOA. For which $h\in G$ is it true that there exists a $c>0$ such that $\|PV_h(f)\|_{BMO}\geqslant c\|f\|_{BMO}$ for all $f\in$ BMOA? Is this true for all $h\in G$?

This questions asks about a quantitative version of the notion that a direction-preserving homeomorphism cannot take a function of analytic type to one of antianalytic type. For nice functions and homeomorphisms this can be proved using the argument principle, but there are examples where it fails; see Garnett-O'Farrell [2].

We should point out that the natural predual formulation of this problem takes place not on H^1 but on $H^1_1=\{f: f'\in H^1\}$ because V_h is self-adjoint with respect to the pairing $\langle f,g\rangle=\int fg'\,dx$. This also has the advantage of working with analytic functions whose boundary values trace a rectifiable curve.

An equivalent reformulation of the problem is to ask when $H+V_h^{-1}HV_h$ is invertible on BMO, if H denotes the Hilbert transform. This question is related to certain conformal mapping estimates; see the proof of Theorem 2 in [1]. In particular, it is shown there that this operator is invertible if $\|\log h'\|_{BMO}$ is small enough.

REFERENCES

1. D a v i d G. Courbes corde-arc et espaces de Hardy généralisés. - Ann.Inst.Fourier (Grenoble), 1982, 32, 227-239.
2. G a r n e t t J., O ' F a r r e l l A. Sobolev approximation by a sum of subalgebras on the circle. - Pacific J.Math. 1976, 65, 55-63.
3. J o n e s P. Homeomorphisms of the line which preserve BMO, to appear in Arkiv för Matematik.

STEPHEN SEMMES Dept. of Mathematics Yale University
New Haven, Connecticut
06520 USA

6.8. AN ESTIMATE OF BMO NORM IN TERMS OF AN OPERATOR NORM

Let b be a function in BMO $(\mathbb{R}^n)$ with norm $\|b\|_*$ and let K be a Calderon-Zygmund singular integral operator acting on $L^2(\mathbb{R}^n)$. Define K_b by $K_b(f)=e^{-b}K(e^b f)$. The theory of weighted norm inequalities insures that K_b is bounded on L^2 if $\|b\|_*$ is small. In fact the map of b to K_b is an analytic map of a neighborhood of the origin in BMO into the space of bounded operators (for instance, by the argument on p.611 of [3]). Much less is known in the opposite direction.

QUESTION: <u>Given</u> b, K ; <u>if</u> $|K-K_b|$ <u>is small, must</u> $\|b\|_*$ <u>be small</u>?

The hypothesis is enough to insure that $\|b\|_*$ is finite but the naive estimates are in terms of $|K|+|K_b|$.

If $n=1$ and K is the Hilbert transform then the answer is yes. This follows from the careful analysis of the Helson-Szegö theorem given by Cotlar, Sadosky, and Arocena (see, e.g. Corollary (III.d) of [1]).

A similar question can be asked in more general contexts, for instance with the weighted projections of [2]. In that context one would hope to estimate the operator norm of the commutator $[M_b,P]$ (defined by $[M_b,P](f)=bPf-P(bf)$) in terms of the operator norm of $P-P_b$.

REFERENCES

1. A r o c e n a R. A refinement of the Helson-Szegö theorem and the determination of the extremal measures. - Studia Math, 1981, LXXI, 203-221.
2. C o i f m a n R., R o c h b e r g R. Projections in weighted spaces, skew projections, and inversion of Toeplitz operators. - Integral Equations and Operator Theory, 1982, 5, 145-159.
3. C o i f m a n R., R o c h b e r g R., W e i s s G. Factorization Theorems for Hardy Spaces in Several Variables. - Ann. Math. 1976, 103, 611-635.

RICHARD ROCHBERG

Washington University
Box 1146
St.Louis, MO 63130
USA

6.9. old SOME OPEN PROBLEMS CONCERNING H^∞ AND BMO

1. An interpolating Blaschke product is a Blaschke product having distinct zeros which lie on an H^∞ interpolating sequence. Is H^∞ the uniformly closed linear span of the interpolating Blaschke products? See [1], [2]. It is known that the interpolating Blaschke products separate the points of the maximal ideal space (Peter Jones, thesis, University of California, Los Angeles 1978).

2. Let φ be a real locally integrable function on $\mathbb{R}$. Assume that for every interval

$$m(\{x\in I : |\varphi(x) - \varphi_I| > \lambda\}) \leqslant C e^{-\lambda} |I| ,$$

where φ_I is the mean value of φ over I, and where C is a constant. Does it follow that $\varphi = u + Hv$, where $u \in L^\infty$ and $\|v\|_\infty \leqslant \frac{\pi}{2}$? (H denotes the Hilbert transform). This is the limiting case of the equivalence of the Muckenhoupt (A_2) condition with the condition of Helson and Szegö. See [3] and [4]. This question is due to Peter Jones. A positive solution should have several applications.

3. Let f be a function of bounded mean oscillation on $\mathbb{R}$. Construct L^∞ functions u and v so that $f = u + Hv$, $\|u\|_\infty + \|v\|_\infty \leqslant C\|f\|_{BMO}$ with C a constant not depending on f. See [5] and [6].

4. Let $T_1, T_2, \ldots, T_n$ be singular integral operators on $\mathbb{R}^n$. See [7]. Find necessary and sufficient conditions on $\{T_1, T_2, \ldots, T_n\}$ such that $f \in H^1(\mathbb{R}^n)$ if and only if $|f| + \sum_{j=1}^{n} |T_j f| \in L^1(\mathbb{R}^n)$ See [5] and [8].

REFERENCES

1. Marshall D. Blaschke products generate H^∞. - Bull.Amer. Math.Soc., 1976, 82, 494-496.
2. Marshall D. Subalgebras of L^∞ containing H^∞. - Acta Math., 1976, 137, 91-98.
3. Hunt R.A., Muckenhoupt B., Wheeden R.L. Weighted norm inequalities for the conjugate function and Hilbert transform. - Trans.Amer.Math.Soc., 1973, 176, 227-251.

4. H e l s o n H., S z e g ö G. A problem in prediction theory. - Ann.Math.Pure Appl.,1960, 51, 107-138.
5. F e f f e r m a n C., S t e i n E.M. H^p spaces of several variables. - Acta Math.,1972, 129, 137-193.
6. C a r l e s o n L. Two remarks on H^1 and BMO. - Advances in Math.,1976, 22, 269-277.
7. S t e i n E.M. Singular integrals and differentiability properties of functions. Princeton N.J.,1970.
8. J a n s o n S. Characterization of H^1 by singular integral transforms on martingales and R^n . - Math.Scand.,1977, 41,140--152.

JOHN GARNETT
University of California
Los Angeles, California
90024 USA

* * *

COMMENTARY

QUESTION 2 has been answered in the negative by T.Wolff [9].

QUESTION 3 has been solved by P.Jones [10]. Other constructive (and more explicit) decompositions were given later in [11], [12] and [13]. One more constructive decomposition of BMO functions can be obtained from a remarkable paper [14]. See also [15], [17].

QUESTION 4 has the following answer found by A.Uchiyama in [12] (he obtained a more general result). Let $T_j f = K_j * f$, $1 \le j \le m$, M_j be the Fourier transform of K_j . Suppose M_j are homogeneous of degree zero and C^∞ on the unit sphere S^{n-1} of $\mathbb{R}^n$. Then

$$f \in H^1(\mathbb{R}^n) \iff \sum_{j=1}^{m} |T_j f| \in L^1(\mathbb{R}^n)$$

if and only if the matrix

$$\begin{pmatrix} M_1(\xi), \dots, M_m(\xi) \\ M_1(-\xi), \dots, M_m(-\xi) \end{pmatrix}$$

is of rank 2 everywhere on S^{n-1}. The "only if" part is essentially due to S.Janson [8]. In particular

$f \in H^1(\mathbb{R}^n) \Longleftrightarrow |f| + \sum_{j=1}^{m} |T_j f| \in L^1(\mathbb{R}^n)$ iff for any $\xi \in S^{n-1}$ there exists j such that $M_j(\xi) \neq M_j(-\xi)$.

In connection with this result see also PROBLEM 6.16.

REFERENCES

9. W o l f f T. Counterexamples to two variants of the Helson - Szegö theorem. Preprint, 1983, Institut Mittag-Leffler, 11.
10. J o n e s P. Carleson measures and the Fefferman - Stein decomposition of BMO($\mathbb{R}$). - Ann. of Math., 1980, 111, 197-208.
11. J o n e s P. L^∞ -estimates for the $\bar{\partial}$ -problem. To appear in Acta Math.
12. U c h i y a m a A. A constructive proof of the Fefferman - Stein decomposition of BMO($\mathbb{R}^n$). - Acta Math., 1982, 148, 215-241.
13. S t r a y A. Two applications of the Schur - Nevanlinna algorithm. - Pacif. J. of Math., 1980, 91, N 1, 223-232.
14. R u b i o d e F r a n c i a J.L. Factorization and extrapolation by weights. - Bull.Amer.Math.Soc., 1982, 7, N 2, 393-395.
15. A m a r E. Représentation des fonctions de BMO et solutions de l'équation $\bar{\partial}_b$. Preprint, 1978, Univ. Paris XI Orsay.
16. C o i f m a n R., J o n e s P.W., R u b i o d e F r a n c i a J.L. Constructive decomposition of BMO functions and factorization of A_p weights. - Proc.Amer.Math.Soc., 1983, 87, N 4, 675-680.

6.10. old TWO CONJECTURES BY ALBERT BAERNSTEIN

In [1] I proved a factorization theorem for zero-free univalent functions in the unit disk $\mathbb{D}$. Let S_0 denote the set of all functions F analytic and $1-1$ in $\mathbb{D}$ with $0 \notin F(\mathbb{D})$, $F(0)=1$.

THEOREM 1. <u>If</u> $F \in S_0$, <u>then, for each</u> λ, $\lambda \in (0,1)$, <u>there exist functions</u> B <u>and</u> Q <u>analytic in</u> $\mathbb{D}$ <u>such that</u>

$$F(z)^\lambda = B(z)Q(z), \quad z \in \mathbb{D},$$

<u>where</u> $B \in H^\infty$, $1/B \in H^\infty$, <u>and</u> $|\arg Q| < \pi$.

The "Koebe function" for the class S_0 is $K(z)=\left(\frac{1+z}{1-z}\right)^2$ which maps $\mathbb{D}$ onto the slit plane $\{w \in \mathbb{C} : |\arg w| < \pi\}$. This suggests that it might be possible to let $\lambda \to 1$ in Theorem 1.

CONJECTURE 1. <u>If</u> $F \in S_0$, <u>then there exist functions</u> B <u>and</u> Q <u>analytic in</u> $\mathbb{D}$ <u>such that</u>

$$F(z) = B(z)Q(z), \quad z \in \mathbb{D},$$

<u>where</u> $B \in H^\infty$, $1/B \in H^\infty$, <u>and</u> $|\arg Q| < \pi$.

We do not insist that B or Q be univalent, nor that $Q(0)=1$. However, when the functions are adjusted so that $|Q(0)|=1$, then $\|B\|_\infty$ and $\|B^{-1}\|_\infty$ should be bounded independently of F.

Using the fact that $Q^{1/2}$ has positive real part, it is easy to show that the power series coefficients $\{a_n\}$ of Q satisfy $|a_n| \leq 4n$, $n \geq 1$, with equality when $Q(z) = K(z)$. L i t t l e w o o d ' s C o n j e c t u r e asserts that this inequality is true for coefficients of functions in S_0. A proof of CONJECTURE 1 could possibly tell us something new about how to attempt Littlewood's conjecture, and this in turn might lead to fresh ideas about how to prove (the stronger) Bieberbach's conjecture.

THEOREM 1 is easily deduced from a decomposition theorem obtained by combining results of Helson and Szegö [2] and Hunt, Muckenhoupt, and Wheeden [3]. Suppose $f \in L^1(\mathbb{T})$ and f real valued. Consider the zero-free analytic function F defined by $F(z)=\exp(f(z)+i\tilde{f}(z))$, $z \in \mathbb{D}$, where $f(z)$ denotes the harmonic extension of $f(e^{i\theta})$ and $\tilde{f}$ the conjugate of f. Also, let $S(F)$ denote the set of all functions obtained by "hyperbolically translating" F and then normalizing,

$$S(F) = \left\{ F\left(\frac{z+a}{1+\bar{a}z}\right) F(a)^{-1} : a \in D \right\}$$

and let H^p denote the usual Hardy space. Part of Theorem 1 of [3] can be phrased in the following way.

THEOREM 2. <u>For</u> $f \in L^1(\mathbb{T})$ <u>the following are equivalent.</u>

(1) $f = u_1 + \tilde{u}_2$ <u>where</u> $u_1, u_2 \in L^\infty(\mathbb{T})$ <u>and</u> $\|u_2\|_\infty < \frac{\pi}{2}$

(2) $S(F) \cup S(1/F)$ <u>is a bounded subset of</u> H^1.

THEOREM 1 follows, since $F^{\lambda/2}$ satisfies (2) when $F \in S_0$ and $0<\lambda<1$. ●

THEOREM 2 may be regarded is a sharpened form of the theorem of Fefferman and Stein [4], which asserts that $f = u_1 + \tilde{u}_2$ for s o m e pair of bounded functions if and only if f is of bounded mean oscillation.

To obtain CONJECTURE 1 in the same fashion as THEOREM 1, we need a result like THEOREM 2 in which the $<\pi/2$ of (1) is replaced by $\leq \pi/2$. Consideration of $F(z) = \frac{1+z}{1-z}$ leads to the following guess.

CONJECTURE 2. <u>For</u> $f \in L^1(\mathbb{T})$ <u>the following are equivalent.</u>

(1′) $f = u_1 + \tilde{u}_2$, <u>where</u> $u_1, u_2 \in L^\infty(\mathbb{T})$ <u>and</u> $\|u_2\|_\infty \leq \pi/2$.

(2′) $S(F) \cup S(1/F)$ <u>is a bounded subset of weak</u> H^1.

Statement (2′) means the following: There is a constant C such that for every t, $t \in R_+$, and every G, $G \in S(F) \cup S(1/F)$

$$m\{\theta : |G(e^{i\theta})| > t\} \leq Ct^{-1}.$$

It is not hard to prove, using subordination, that (1′) implies (2′). If the implication (2′) $\rightarrow$ (1′) is true, then so is CONJECTURE 1.

Condition (2′) can be restated in a number of equivalent ways. We mention one which is closely related to the subharmonic maximal type function used by the author in [5] and elsewhere.

(2$''$) <u>There is a constant</u> C <u>such that</u>

$$\int_E \left[f\left(\frac{e^{i\theta}+a}{1+\bar{a}e^{i\theta}}\right) - f(a)\right] d\theta \leq \int_{-\frac{1}{2}mE}^{\frac{1}{2}mE} \log\left|\frac{1+e^{i\theta}}{1-e^{i\theta}}\right| d\theta + C|E|$$

<u>for every measurable set</u> $E,\ E\subset\mathbb{T}$, <u>and every</u> $a,\ a\in\mathbb{D}$.

For $F\in S_0$, Theorem 6 of [5] asserts that (2$''$) holds with $C=0$.

In both the Fefferman-Stein and Helson-Szegö theorems the splitting $f=u_1+\tilde{u}_2$ is accomplished via duality and pure existence proofs from functional analysis. It would be of considerable interest if, given f , $f\in BMO$, one could show how to actually c o n s t r u c t the bounded functions u_1 and u_2 . We remark that if $f\in BMO$ then some constant multiple of f satisfies (2$''$).

I can prove that (2$''$) $\longrightarrow$ (1$'$) provided we assume also that f is m o n o t o n e on $\mathbb{T}$, i.e., there exist $\theta_1<\theta_2<\theta_1+2\pi$ such that

$$f(e^{i\theta})\downarrow \text{ as } \theta\uparrow \text{ on } (\theta_1,\theta_2) \text{ and } f(e^{i\theta})\uparrow \text{ as } \theta\uparrow \text{ on } (\theta_2,\theta_1+2\pi).$$

By composing with a suitable Möbius transformation, we may assume $\theta_1=0$, $\theta_2=\pi$. Then, when $C=0$, u_2 can be constructed as follows. Let $\theta\in(0,\pi)$ and $x\in(-1,1)$ be related by $(1+x)(1-x)^{-1}= |1+e^{i\theta}||1-e^{i\theta}|^{-1}$. Let V be the harmonic function in $\mathbb{D}$ with boundary values $V(e^{i\theta})=f(x)$, $0<\theta<\pi$, and $V(e^{-i\theta})=V(e^{i\theta})$. Then it turns out that $|\tilde{V}|\leq\pi/2$ and $f-V=O(1)$, so that $u_2=-\tilde{V}$ gives us (1$'$).

It follows that CONJECTURE 1 is true for functions $F,\ F\in S_0$, which map $\mathbb{D}$ onto the complement of a "monotone slit".

REFERENCES

1. B a e r n s t e i n A. II. Univalence and bounded mean oscillation. - Mich.Math.J., 1976, 23, 217-223.

2. H e l s o n H., S z e g ö G. A problem in prediction theory. - Ann.Mat.Pura Appl., 1960, 51, (4), 107-138.

3. H u n t R., M u c k e n h o u p t B., W h e e d e n R. Weighted norm inequalities for the conjugate function and Hilbert transform. - Trans.Amer.Math.Soc., 1973, 176, 227-251.
4. F e f f e r m a n C., S t e i n E.M. H^p spaces of several variables. - Acta Math., 1972, 129, 137-193.
5. B a e r n s t e i n A. II. Integral means, univalent functions and circular symmetrization. - Acta Math., 1974, 133, 139-169.

ALBERT BAERNSTEIN

Washington University
St.Louis, Missouri 63130
USA

* * *

COMMENTARY

Conjecture 2 has been disproved by T.Wolff (see ref. [9] after the Commentary to Problem 6.9)

6.11. old BLASCHKE PRODUCTS IN $\mathcal{B}_0$.

The class $\mathcal{B}_0$ consists of those functions f that are holomorphic in $\mathbb{D}$ and satisfy $\lim_{|z|\to 1}(1-|z|)|f'(z)|=0$. It can be described alternatively as the class of functions in $\mathbb{D}$ that are derivatives of holomorphic functions having boundary values in the Zygmund class (the class of uniformly smooth functions) [1, p.263]. It is a subclass of the class $\mathcal{B}$ of Bloch functions (those holomorphic f in $\mathbb{D}$ satisfying $\sup_{|z|<1}(1-|z|)|f'(z)|<\infty$) ; see, for example, [2]. It contains VMOA, the class of holomorphic functions in $\mathbb{D}$ whose boundary values have vanishing mean oscillation [3]. The class $\mathcal{B}_0\cap H^\infty$ has an interesting interpretation: it consists of those functions in H^∞ that are constant on each Gleason part of H^∞.

It is not too hard to come up with an example to show that the inclusion VMOA $\subset \mathcal{B}_0$ is proper. Indeed, it is known that λ_* contains functions that are not of bounded variation [1, p.48]. If u is the Poisson integral of such a function and v is its harmonic conjugate, then the derivative of $u+iv$ will be such an example. In connection with a problem in prediction theory mentioned in [4], I was interested in having an example of a b o u n d e d function in $\mathcal{B}_0$ which is not in VMOA, and that seems somewhat more difficult to obtain. Eventually I realized one can produce such an example on the basis of a result of H.S.Shapiro [5] and J.-P.Kahane [6]. They showed, by rather complicated constructions, that there exist positive singular measures on $\mathbb{T}$ whose indefinite integrals are in λ_*. It is easy to check that the singular inner function associated with such a measure is in $\mathcal{B}_0$. That does it, because the only inner functions in VMOA are the finite Blaschke products.

If f is an inner function in $\mathcal{B}_0$ and $|c|<1$, then $\frac{f-c}{1-\bar{c}f}$ is also an inner function in $\mathcal{B}_0$, and it is a Blaschke product for "most" values of c. Thus, $\mathcal{B}_0$ contains infinite Blaschke products. I should like to propose THE PROBLEM of <u>characterizing the Blaschke products in $\mathcal{B}_0$ by means of the distribution of their zeros</u>. One has the feeling that the zeros of a Blaschke product in $\mathcal{B}_0$ must, in some sense, be "spread smoothly" in $\mathbb{D}$. A natural first step in trying to find the correct condition would be to try to give a direct construction of an infinite Blaschke product in $\mathcal{B}_0$. The only information I can offer on the problem is very meagre: A Blaschke product in $\mathcal{B}_0$ cannot have an isolated singularity on $\mathbb{T}$. The proof, unfortunately, is too involved to indicate here. As A TEST QUESTION one might ask whether a Blaschke product in $\mathcal{B}_0$ can have a singular set which meets some subarc of $\mathbb{T}$ in a nonempty set of measure zero.

ANOTHER QUESTION, admittedly vague, concerns the abundance of Blaschke products in $\mathcal{B}_0$. For instance, a Blaschke product should be in $\mathcal{B}_0$ if its zeros are evenly spread throughout $\mathbb{D}$. One is led to suspect that, in some sense, a Blaschke product with random zeros will be almost surely in $\mathcal{B}_0$.

REFERENCES

1. Zygmund A. Trigonometric series, vol.I. Cambridge, Cambridge Univ.Press. 1959.
2. Anderson J.M., Clunie J., Pommerenke Ch. On Bloch functions and normal functions. - J.Reine Angew.Math. 1974, 270, 12-37.
3. Pommerenke Ch. On univalent functions, Bloch functions and VMOA. - Math.Ann., 1978, 236, N 3, 199-208.
4. Sarason D. Functions of vanishing mean oscillation. - Trans.Amer.Math.Soc. 1975, 207, 391-405.
5. Shapiro H.S. Monotonic singular functions of high smoothness. - Michigan Math.J. 1968, 15, 265-275.
6. Kahane J.-P. Trois notes sur les ensembles parfaits linéaires. - Enseignement Math. 1969, (2), 15, 185-192.

DONALD SARASON University of California, Dept.Math., Berkeley, California, 94720, USA

* * *

COMMENTARY BY THE AUTHOR

The problem is still open. T.H.Wolff has pointed out that the measures constructed by Kahane and Shapiro can be taken with supports of Lebesgue measure 0 , so there do exist infinite Blaschke products in $\mathcal{B}_0$ whose singularities form a set of measure 0 . (The author was remiss in failing to notice this.) Wolff (unpublished) has shown that the set of singularities on the unit circle of an inner function in $\mathcal{B}_0$ meets each open subarc either in the empty set or in a set of positive logarithmic capacity. He conjectures that "positive logarithmic capacity" can be replaced by "Hausdorff dimension 1."

6.12. old ALGEBRAS CONTAINED WITHIN H^∞.

Let $A=\{f: f$ analytic in $\mathbb{D}$, f continuous in $\operatorname{clos}\mathbb{D}=\mathbb{D}\cup\mathbb{T}\}$. Then A is an algebra contained within H^∞, but there are two intermediate algebras that present some interest. First we require some notation.

Let B denote the Banach space of functions f, analytic in $\mathbb{D}$ for which the norm

$$\|f\|_B = |f(0)| + \sup_{|z|<1} (1-|z|^2)|f'(z)|$$

is finite. This is called the B l o c h space. We also define

$$B_0=\{f: f\in B, f'(z)=o(1-|z|^2)^{-1}, |z|\to 1\}.$$

For a survey of these spaces see [1]. The following facts are easily established:

a) $H^\infty\subset B$, b) $H^\infty\not\subset B_0$, c) $H^\infty\cap B_0 \overset{def}{=\!=} X$ is a subalgebra of H^∞.

Similarly we define BMOA (a n a l y t i c f u n c t i o n s o f b o u n d e d m e a n o s c i l l a t i o n) to be the space of those functions f, analytic in $\mathbb{D}$ for which the norm

$$\|f\| = |f(0)| + \sup_{|\zeta|<1} \|f_\zeta\|_2$$

is finite. Here $\|\cdot\|_2$ is the ordinary H^2 norm and

$$f_\zeta(z) = f\left(\frac{z+\zeta}{1+\bar\zeta z}\right) - f(\zeta), \quad z\in\mathbb{D}.$$

Similarly

$$\mathrm{VMOA}=\{f: f\in \mathrm{BMOA}, \|f_\zeta\|_2=o(1), |\zeta|\to 1\}.$$

The space VMOA consists of those analytic functions in $\mathbb{D}$ whose boundary values on $\mathbb{T}$ have vanishing mean oscillation (see [2], p.591). It is also easy to see that

d) $H^\infty\subset \mathrm{BMOA}$, e) $H^\infty\not\subset \mathrm{VMOA}$, f) $H^\infty\cap \mathrm{VMOA} \overset{def}{=\!=} Y$ is a subalgebra of H^∞.

It is not difficult to establish the following relation (see e.g. [3])

$$A \subsetneq Y \subsetneq X \subsetneq H^{\infty}.$$

The algebra X has already been studied. It was shown by Behrens, unpublished, that X consists precisely of those f , $f \in H^{\infty}$, whose Gelfand transform $\hat{f}$ is constant on all the non-trivial Gleason parts of the maximal ideal space of H^{∞} . It is also known [3] that X does not possess the f-property or K-property in the sense of Havin [4].

IT WOULD BE NICE TO HAVE A SIMILAR STUDY MADE OF Y . The space Y cannot contain any inner functions [3], other than finite Blaschke products, in contrast to X . But Y does, of course, contain functions having an inner factor - for example the function of [5], p.29 belongs to A .

REFERENCES

1. Anderson J.M., Clunie J., Pommerenke Ch. On Bloch functions and normal functions. - J.Reine Angew. Math., 1974, 270, 12-37.
2. Pommerenke Ch. Schlichte Funktionen und analytische Funktionen von beschränkter mittlerer Oszillation. - Comment.Math. Helv., 1977, 52, 591-602.
3. Anderson J.M. On division by Inner Factors. - Comment. Math.Helv., 1979, 54, N 2, 309-317.
4. Хавин В.П. О факторизации аналитических функций, гладких вплоть до границы. - Зап.научн.семин.ЛОМИ, 1971, 22, 202-205.
5. Гурарий В.П. О факторизации абсолютно сходящихся рядов Тэйлора и интегралов Фурье. - Зап.научн.семин.ЛОМИ, 1972, 30, 15-32.

J.M.ANDERSON

Department of Mathematics,
University College, London
London WC1E 6BT
England

6.13. ANALYTIC FUNCTIONS WITH FINITE DIRICHLET INTEGRAL

If f is an analytic function defined on $\mathbb{D}$, let $\mathcal{D}(f)=$ $=\left(\iint_{\mathbb{D}}|f'(z)|^2\frac{dx\,dy}{\pi}\right)^{1/2}$ be the Dirichlet integral of f. In [1], the following theorem is proved.

THEOREM. There is a constant $C_0<\infty$, such that if f is analytic on $\mathbb{D}$, $f(0)=0$ and $\mathcal{D}(f)\leq 1$ then

$$\int_0^{2\pi} e^{\alpha|f(e^{i\theta})|^2}\,d\theta\leq C_0 \qquad \text{for all } \alpha\leq 1 .$$

It would be interesting to know the size of C_0 and also the extremal functions (if exist) which correspond to the sharp constant C_0. Actually, the above theorem is only a part of results similar to Moser's sharp form of the Trudinger inequality (see [2]). It would actually be interesting to see if there is a general form of extremal functions which correspond to Moser's sharp inequalities.

REFERENCES

1. Chang S.-Y. A., Marshall D. A sharp inequality concerning the Dirichlet integral. 1982, preprint.
2. Moser J. A sharp form of an inequality by N.Trudinger. - Ind. Univ.Math.J., 1971, 20, 1077-1092.

SUN-YUNG A.CHANG

University of Maryland
Math.Dept., College Park,
Maryland 20742
USA

6.14. SUBALGEBRAS OF $L^\infty(\mathbb{T}^2)$ CONTAINING $H^\infty(\mathbb{T}^2)$.

Let $H^\infty(\mathbb{T})$ denote the Hardy space of boundary values of bounded analytic functions defined on $\mathbb{D}$. There has been a systematic study of the subalgebras (called the Douglas algebras) between $L^\infty(\mathbb{T})$ and $H^\infty(\mathbb{T})$ in the past 10 years. (For a survey article, see [1]). In particular, it has been noticed there is a parallel relationship between subalgebras of $L^\infty(\mathbb{T})$ containing $H^\infty(\mathbb{T})$ to subspaces of B.M.O. (functions of bounded mean oscillations) which contain V.M.O. (functions of vanishing mean oscillations). For example, based on the fact that on $\mathbb{T}$, B.M.O. $= L^\infty + H(L^\infty)$, where H denotes the Hilbert transform, one can deduce that each Douglas algebra can be written as $H^\infty(\mathbb{T})$+some C^*-algebra. There are some indications that relations of this type may still hold on the bi-disc $\mathbb{D}^2$ (with distinguished boundary $\mathbb{T}^2$). For example, if one views B.M.O. $(\mathbb{T}^2)$ as $L^\infty(\mathbb{T}^2) + H_1(L^\infty) + H_2(L^\infty) + H_1H_2(L^\infty)$ where the H_i $\mathfrak{s}, i=1,2$ are Hilbert transforms acting on z_i variables independently with $(z_1, z_2) \in \mathbb{D}^2$ and H_1H_2 is the composition of H_1 with H_2, one can ask the question whether each subalgebra of $L^\infty(\mathbb{T}^2)$ containing $H^\infty(\mathbb{T}^2)$ has the structure of $H^\infty(\mathbb{T}^2)$+ some other three C^*-algebras. It seems this problem can be studied independently of the maximal ideal structure of $H^\infty(\mathbb{T}^2)$. So far the only case which has been worked out is the subalgebra of $L^\infty(\mathbb{T}^2)$ generated by $H^\infty(\mathbb{T}^2)$ and $C(\mathbb{T}^2)$ (see [2]).

REFERENCES

1. S a r a s o n D. Algebras between L^∞ and H^∞. - Lect.Notes in Math.Springer-Verlag, 1976, 512, 117-129.
2. C h a n g S.-Y. A. Structure of some subalgebra of L^∞ of the torus. - Proc.Symposia in Pure Math., 1979, 35, Part 1, 421-426.

SUN-YUNG A. CHANG

University of Maryland
Math.Dept., College Park,
Maryland 20742
USA

6.15. INNER FUNCTIONS WITH DERIVATIVE IN H^p, $0<p<1$.

Let φ be an inner function defined in the unit disc $\mathbb{D}$. For $\alpha\in\mathbb{D}$ let $\varphi_\alpha(z)=(\varphi(z)-\alpha)/(1-\bar\alpha\varphi(z))$. Let $\{z_n(\alpha)\}_{n=1}^{\infty}$ denote the zero set of φ_α. From [1], theorem 6.2, we have:

THEOREM. Suppose that $\varphi(z)=\sum_{n=0}^{\infty}a_n z^n$ is an inner function and that $1/2<p<1$. Then the following are equivalent:

1. $\varphi'\in H^p$

2. $\sum_{n=0}^{\infty}|a_n|^2 n^p<\infty$

3. $\sum_{n=1}^{\infty}(1-|z_n(\alpha)|)^{1-p}<\infty$ for all $\alpha\in\mathbb{D}$ with the exception of a set of capacity zero.

For $0<p<1/2$ the situation is quite different. It is still true that 1 implies 2 and 3. However, as is pointed out in [1], page 342, there is a Blaschke product $\varphi(z)=\sum_{n=0}^{\infty}a_n z^n$ such that 2 and 3 hold for φ for all p, $0<p<1/2$, but φ' is not a function of bounded characteristic.

PROBLEM. Find a condition on the Taylor coefficients or on the distribution of values of an inner function φ that is equivalent to the condition $\varphi'\in H^p$, $0<p<1/2$.

REFERENCE

1. A h e r n P. The mean modulus and the derivative of an inner function. - Indiana Univ.Math.J., 1979, 28, 2, 311-347.

PATRICK AHERN

University of Wisconsin
Madison, Wisconsin, USA

* * *

EDITORS' NOTE

I.È.Verbitskii has informed us about his result pertaining to the Problem.

THEOREM. <u>If</u> $\frac{1}{p}-1<s<\frac{1}{p}$ <u>then the following are equivalent:</u>

1) $\varphi^{(s)}\in H^p$;

2) $\sum|a_n|^2 n^{sp}<\infty$;

3) $\sum(1-|z_n(\alpha)|)^{1-sp}<\infty$

<u>for all</u> $\alpha\in\mathbb{D}$ <u>with the exception of a set of capacity zero;</u>

4) $\varphi\in B_p^s$.

Here $\varphi^{(s)}$ denotes the fractional derivative of φ of order s, B_p^s is the Besov class, i.e.

$$B_p^s \overset{\text{def}}{=\!=} \left\{ f \text{ analytic in } \mathbb{D} : \iint_{\mathbb{D}} |f^{(n)}(z)|^p (1-|z|)^{(n-s)p-1}\, dx\, dy<\infty \right\},$$

n being any integer $>s$, $z=x+iy$.

This theorem is implied by results of [1] when $p\leqslant 2$, $0<s\leqslant\leqslant 1$, $sp>\frac{1}{2}$. It is not valid when $s\leqslant\frac{1}{p}-1$ and no analogous result seems to be known in that case.

6.16. EQUIVALENT NORMS IN H^p

Let H^p denote the real variables Hardy space on $\mathbb{R}^n$. Let K_j be a Fourier multiplier operator whose symbol θ_j is $C^\infty(\mathbb{R}^n \setminus \{0\})$ and homogeneous of degree zero. <u>For which families</u> $\{K_j\}_{j=1}^m$ <u>is it true that</u>

$$\|f\|_{H^p} \sim \sum_{j=1}^m \|K_j f\|_{L^p}$$

<u>for all</u> $f \in H^p \cap L^2$? This problem was solved for $p=1$ in [1] and the results were extended in [2] to the case where p is only slightly less than one. A subproblem is to decide whether the above equivalence holds for all $p<1$ when the family consists of the identity operator and the first order Riesz kernels. See [3] for related results.

REFERENCES

1. U c h i y a m a A. A constructive proof of the Fefferman - Stein decomposition of BMO ($\mathbb{R}^n$). - Acta Math.,1982, 148, 215-241.
2. U c h i y a m a A. The Fefferman - Stein decomposition of smooth functions and its application to $H^p(\mathbb{R}^n)$.- University of Chicago, Ph.D.thesis, 1982.
3. C a l d e r ó n A.P., Z y g m u n d A. On higher gradients of harmonic functions.- Studia Math.,1964, 24, 211-226.

PETER W.JONES

Institut Mittag-Leffler
Auravägen 17
S-182 62 Djursholm
Sweden

Usual Address:

Dept.of Mathematics
University of Chicago
Chicago, Illinois 60637
USA

6.17. ON THE DEFINITION OF $H^p(\mathbb{R}^n)$.

Suppose T is a distribution on $\mathbb{R}^n$, φ a compactly supported C^∞-function on $\mathbb{R}^n$, $\int_{\mathbb{R}^n}\varphi=1$, $p>0$. Put

$$\varphi_{\varepsilon,x}(y)=\frac{1}{\varepsilon^n}\varphi\left(\frac{x-y}{\varepsilon}\right)\qquad (x,y\in\mathbb{R}^n,\ \varepsilon>0),$$

$$T_\varphi^+(x)=\sup\left\{|T(\varphi_{\varepsilon,x})|:\ \varepsilon>0\right\}$$

(the r a d i a l maximal function of T corresponding to the mollifier φ).

QUESTION. <u>Does the inclusion</u> $T_\varphi^+\in L^p(\mathbb{R}^n)$ <u>imply</u> $T\in H^p(\mathbb{R}^n)$?

The answer is YES if $p\geq 1$ or under the supplementary assumption $T\in S'(\mathbb{R}^n)$ [1] or if we replace T_φ^+ by T_φ^*, the a n g u l a r maximal function (because then the inclusion $T\in S'(\mathbb{R}^n)$ is easy to prove). We were unable to answer the question following the patterns of [1].

REFERENCE

1. F e f f e r m a n C., S t e i n E.M. H^p spaces of several variables. - Acta Math., 1972, 129, 137-193.

A.B.ALEKSANDROV
(А.Б.АЛЕКСАНДРОВ)
V.P.HAVIN
(В.П.ХАВИН)

СССР, 198904, Ленинград, Петродворец, Библиотечная 2, Ленинградский государственный университет, математико-механический факультет

6.18. HARDY CLASSES AND RIEMANN SURFACES OF PARREAU-WIDOM TYPE
old

The theory of Hardy classes on the unit disk and its abstract generalization have received considerable attention in recent years (cf. Hoffman [1], Helson [2], Gamelin [3]). The case of compact bordered surfaces has also been studied in detail. It is thus natural that we should try to increase our knowledge concerning the theory of Hardy classes on infinitely connected Riemann surfaces. OUR BASIC QUESTION is this: For which class of Riemann surfaces can one get a fruitful extension of the Hardy class theory on the disk? A candidate we believe most promising is the class of Riemann surfaces of Parreau -Widom type, which is defined as follows:

DIFINITION. Let R be a hyperbolic Riemann surface, $G(a,z)$ the Green function for R with pole at a point $a \in R$ and $B(a,\alpha)$ the first Betti number of the region $R(a,\alpha)=\{z\in R: G(a,z)>\alpha\}$ with $\alpha>0$. We say that R is of Parreau-Widom type if $\int_0^\infty B(a,\alpha)d\alpha<\infty$.

We first sketch some relevant results showing that such surfaces are nice. In the following, R denotes a surface of Parreau-Widom type, unless stated otherwise.

(1) PARREAU [4]: (a) Every positive harmonic function on R has a limit along almost every Green line issuing from any fixed point in R. (b) The Dirichlet problem on Green lines on R for any bounded measurable boundary function has a unique solution, which converges to the boundary data along almost all Green lines.

(2) WIDOM [5]: For a hyperbolic Riemann surface R, it is of Parreau-Widom type if and only if the set $H^\infty(R,\xi)$ of all bounded holomorphic sections of any given complex flat unitary line bundle ξ over R has nonzero elements.

(3) HASUMI [6]: (a) Every surface of Parreau-Widom type is obtained by deleting a discrete subset from a surface of Parreau-Widom type, R, which is regular in the sense that $\{z\in R: G(a,z)\geqslant\varepsilon\}$ is compact for any $\varepsilon>0$. (b) Brelot-Choquet's problem (cf. [7]) concerning the relation between Green lines and Martin's boundary has a completely affirmative solution for any surface of Parreau-Widom type. (c) The inverse Cauchy theorem holds for R.

In view of (3)-(a), we assume in what follows that R is a regular surface of Parreau-Widom type. The Parreau-Widom condition stated in the definition above is then equivalent to the inequality

$\sum\{G(a,w): w\in Z(a)\}<\infty$, where $Z(a)$ denotes the set of critical points, repeated according to multiplicity, of the function $z\mapsto$ $\mapsto G(a,z)$. We set $g^{(a)}(z) = \exp[\sum\{G(a,w): w\in Z(a)\}]$.

Moreover, let Δ_1 be Martin's minimal boundary of R and $d\chi_a$ the harmonic measure, carried by Δ_1, at the point a. Look at the following STATEMENT (DCT): Let h be a meromorphic function on R such that $|h|g^{(a)}$ has a harmonic majorant on R. Then $h(a)=$ $=\int_{\Delta_1}\hat{h}(b)\,d\chi_a(b)$, where $\hat{h}$ denotes the fine boundary function for h. (Note: DCT stands for Direct Cauchy Theorem).

(4) HAYASHI [8]: (a) (DCT) is valid for all points a in R if it is valid for some a. (b) (DCT) is valid if and only if each β-closed ideal of $H^\infty(R)$ is generated by some (multiple-valued) inner function on R. (c) There exist surfaces of Parreau-Widom type for which (DCT) fails.

We now mention SOME PROBLEMS related to surfaces of Parreau-Widom type. (i) Find simple sufficient conditions for a surface of Parreau-Widom type to satisfy (DCT). Hayashi [8] has found a couple of conditions equivalent to (DCT) including (4)-(b) above. But none of them are easy enough to be used as practical tests. (ii) Is there any criterion for a surface of Parreau-Widom type to satisfy the Corona Theorem? Known results: there exist surfaces of Parreau-Widom type for which the Corona Theorem is false; there exist surfaces of Parreau-Widom type with infinite genus for which the Corona Theorem is valid. Hayashi asks the following: (iii) Does $H^\infty(R,\xi)$ for any ξ have only constant common inner factors? (iv) Is a generalized F. and M.Riesz theorem true for measures on Wiener's harmonic boundary, which are orthogonal to H^∞? Another problem: (v) Characterize those surfaces R for which $H^\infty(R,\xi)$ for every ξ has an element without zero. This was once communicated from Widom and seems to be still open. On the other hand, plane domains of Parreau-Widom type are not very well known: (vi) Characterize closed subsets E of the Riemann sphere S for which $S\setminus E$ is of Parreau-Widom type (cf. Voichick [9], Neville [10]).

Finally we note that interesting observations may be found in work of Pommerenke [11], Stanton [12], Pranger [13] and others.

REFERENCES

1. H o f f m a n K. Banach Spaces of Analytic Functions. Prentice -Hall, Englewood Cliffs, N.J., 1962.
2. H e l s o n H. Lectures on Invariant Subspaces. Academic Press, New York, 1964 .
3. G a m e l i n T. Uniform Algebras, Pretice-Hall. Englewood Cliffs, N.J., 1969.
4. P a r r e a u M. Théorème de Fatou et problème de Dirichlet pour les lignes de Green de certaines surfaces de Riemann. - Ann.Acad.Sci.Fenn.Ser.A. I, 1958, no.250/25, 8 pp.
5. W i d o m H. H_p sections of vector bundles over Riemann surfaces. - Ann. of Math., 1971, 94, 304-324.
6. H a s u m i M. Invariant subspaces on open Riemann surfaces. - Ann.Inst.Fourier, Grenoble,1974, 24, 4, 241-286; II, ibid. 1976, 26, 2, 273-299.
7. B r e l o t M. Topology of R.S. Martin and Green lines. Lectures on Functions of a Complex Variable, pp.105-121. Univ. of Michigan Press, Ann Arbor, 1955.
8. H a y a s h i M. Invariant subspaces on Riemann surfaces of Parreau-Widom type. Preprint (1980).
9. V o i c h i c k M. Extreme points of bounded analytic functions on infinitely connected regions. - Proc.Amer.Math.Soc., 1966, 17, 1366-1369.
10. N e v i l l e C. Invariant subspaces of Hardy classes on infinitely connected open surfaces. - Memoirs of the Amer.Math.Soc., 1975, N 160.
11. P o m m e r e n k e Ch. On the Green's function of Fuchsian groups. - Ann.Acad.Sci.Fenn. Ser. A. I, 1976, 2, 408-427.
12. S t a n t o n C. Bounded analytic functions on a class of open Riemann surfaces. - Pacific J.Math., 1975, 59, 557-565.
13. P r a n g e r W. Riemann surfaces and bounded holomorphic functions. - Trans.Amer.Math.Soc., 1980, 259, 393-400.

MORISUKE HASUMI

Ibaraki University,
Department of Mathematics,
Mito, Ibaraki, 310, Japan

EDITORS' NOTE. A Parreau-Widom surface with a corona has been constructed in the paper

N a k a i   M i t s u r u , Corona problem for Riemann surfaces of Parreau-Widom type. - Pacif.J.Math., 1982, 103, N 1, 103-109.

6.19. INTERPOLATING BLASCHKE PRODUCTS

If $B=\prod_{n=1}^{\infty}\frac{\bar{\alpha}_n}{|\alpha_n|}\frac{\alpha_n-z}{1-\bar{\alpha}_n z}$ is a Blaschke product, the interpolation constant of B, denoted $\delta(B)$ is $\inf_m \prod_{\substack{n=1\\ n\neq m}}^{\infty}\left|\frac{\alpha_n-\alpha_m}{1-\bar{\alpha}_n\alpha_m}\right|$.

A well known result of L. Carleson asserts that B is an interpolating Blaschke product if and only if $\delta(B)>0$. It is also well known that the following open problems are equivalent:

PROBLEM 1. <u>Can every inner function be uniformly approximated by interpolating Blaschke products?</u> i.e., <u>Given any Blaschke product</u> B <u>and an</u> $\varepsilon>0$ <u>is there an interpolating Blaschke product</u> B_1 <u>such that</u> $\|B-B_1\|<\varepsilon$?

PROBLEM 2. <u>Is there a function</u> $f(\varepsilon)$ <u>so that for any finite Blaschke product</u> B <u>and any</u> $\varepsilon>0$, <u>there is a (finite) Blaschke product</u> B_1 <u>such that</u> $\|B-B_1\|<\varepsilon$ <u>and</u> $\delta(B_1)\geqslant f(\varepsilon)$?

These problems are stronger than Problem 1 posed by John Garnett in "Some open problems concerning H^∞ and BMO" in this problem book, Problem 6. 9.

If these problems are eventually answered in the negative, then the obvious question is to classify those inner functions which can be so approximated. T. Trant and P. Casazza have observed (and this may already be known) that changing convergence in norm to convergence uniform on compacta produces satisfactory classifications. For example,

PROPOSITION 3. <u>The following are equivalent for a function</u> $F\in H^\infty$:

(1) <u>There is a sequence</u> $\{B_n\}$ <u>of finite Blaschke products which converge to</u> F <u>uniformly on compacta for which</u> $\inf_n \delta(B_n)>0$,

(2) $F=BG$ <u>where</u> B <u>is an interpolating Blaschke product and</u> G <u>is an outer function satisfying</u>

$$0 < \inf_{z \in \partial D} |G(z)| \leq \sup_{z \in \partial D} |G(z)| \leq 1.$$

The proof that (2) $\Longrightarrow$ (1) follows by calculating the interpolation constants of the approximating Blaschke products given in the proof of Frostman's Theorem. By using some techniques developed in [1], it is easily shown that (1) $\Longrightarrow$ (2).

I am particularly interested in the form of the function $f(\varepsilon)$ given in problem 2. A variation of this relates to a problem stated in [1]. If K is a compact subset of the unit circle with Lebesgue measure zero, let A_K denote the ideal in the disk algebra A consisting of the functions which vanish on K. The most general closed ideals in A have the well known form $J_F = \{g \cdot F : g \in A_K\}$ where F is an inner function continuous on the complement of K in the closed disk. A sequence $\{z_n\}$ in the open disk is called a Carleson sequence if $M = \sup\{\sum_{n=1}^{\infty}(1-|z_n|^2)|f(z_n)| : f \in H^1, \|f\| \leq 1\} < \infty$.

In [1], the following problem appeared:

PROBLEM 4. <u>If</u> $\{z_n\}$ <u>is a Carleson sequence and</u> B <u>the Blaschke product with zeroes</u> $\{z_n\}$ <u>continuous off</u> K, <u>does there exist absolute constants</u> a <u>and</u> A <u>so that</u>

$$a \cdot \log M \leq \inf\{\|Q\| : Q : A_K \longrightarrow J_B \text{ is a projection onto}\} \leq A \log M ?$$

I have since discovered that the left hand inequality is true (there does exist a universal constant a) but the right hand inequality is false (there does not exist a universal constant A). A new conjecture for the norm of the best projection onto an ideal in A is needed. The calculations involved in computing this seem to be related to those needed for problems (1) and (2) above.

REFERENCE

1. Casazza P.G., Pengra R. and Sundberg C. Complemented ideals in the Disk Algebra. Israel J. Math., vol. 37. No1-2, (1980), p.76-83.

PETER G. CASAZZA
USA

Department of Mathematics University of Missouri-Columbia, Columbia, Missouri 65211

CHAPTER 7

SPECTRAL ANALYSIS AND SYNTHESIS

Problems of Spectral Analysis - Synthesis arose long before they were stated in a precise form. They stimulated, e.g., the development of Linear Algebra ("The Fundamental Theorem of Algebra", Jordan Theory) and of basic ideas of Fourier Analysis. The success and the universal character of the last are the reasons why the present theme was confined for a very long time to the sphere of influence of Harmonic Analysis. The well developed theory of trigonometrical series and integrals, group representations, Abstract Harmonic Analysis - all these disciplines are directed at the same twofold problem: what are "the elementary harmonics" of an object (= a function, an operator, ...) which is undergoing the action of a semi-group of transformations; what are the ways of reconstructing the object, once its spectrum, i.e. the intensity of every harmonic, is known? Another apparently different, but essentially identical aspect stimulating the development of the theme has roots in Differential Equations. The ritual of writing down the general solution of $p(\frac{d}{dx})f = 0$ using the zeros of the symbol p generated numerous investigations of differential-difference and more general convolution operators. The results always reflect the same routine: the general solution is the limit of linear combinations of elementary solutions $z^k e^{\lambda z}$ corresponding to the zeros λ of the symbol (Ritt,

Valiron, Delsarte). It was L.Schwartz who formulated the circle of ideas in its real meaning and appropriate generality (in his classical paper in Ann Math , 1947, 48, N 4, 857-927) Now the Problem of Spectral Analysis - Synthesis can be stated as follows: given a linear topological space X and a semi-group of its endomorphisms,

describe τ-invariant closed subspaces, containing non-trivial τ-invariant finite-dimensional parts ("Analysis"), and then

describe subspaces spanned topologically by the above parts ("Synthesis").

If τ has a single generator then our problem actually deals with eigen- and root-subspaces of the generator and with the subsequent recovery of all its invariant subspaces via these "elementary" ones. Systems of differential and general convolution equations lead to finitely-generated τ-invariant subspaces, τ being the corresponding group (or semi-group) of translations (in $\mathbb{R}^n$, $\mathbb{C}^n$, $\mathbb{T}^n$ etc). Annihilators of such subspaces become (via Fourier transform) modules over the ring of trigonometrical(resp. "analytic" trigonometrical) polynomials; the Analysis-Synthesis Problem converts into the well-known problem of "localization of ideals". Roughly speaking the principal role is played in this context by the concept of the divisor of an analytic function, and the Problem reduces to the description of divisorial ideals (or submodules). After this reduction is accomplished, we may forget the origin of our problem and confine ourselves exclusively to Function Theory. Namely, we are led to one of its key questions, the interplay of local and global properties of analytic functions. Thus, starting with Analysis - Synthesis, we come to the multiplicative structure of analytic functions (Weierstrass products and their generalizations), the factorization theory of Nevanlinna-Smirnov, uniqueness theorems characterizing non-trivial divisors and to many other accoutrements of Complex Analysis.

The problems of this chapter treat the above ideas in various

ays. Localization of ideals (submodules) in spaces of analytic func-
ions determined by growth conditions is discussed in Problems 7.1-
.6, andin more special spaces in 7.7-7.11. These Problems overlap
ssentially. We add to the references given in the text of Problems
ooks of L.Schwartz (Théorie des distributions, Paris, Hermann, 1966),
.Ehrenpreis (Fourier analysis in several complex variables, N.Y.
970) and J.-P.Ferrier (Spectral Theory and Complex Analysis, N.Y.,
973) (see also the bibliography in the survey [1] cited in Problem
.7). Analyzing spaces of holomorphic functions defined by a family
f majorants requires a study of the intrinsic properties of majorants
see e.g. Problem 11.8 and §7.3 of Ferrier's book).

Problem 7.13 deals with an interesting question concerning fi-
itely generated (algebraically) ideals in H^∞ , a generalization
f Corona Problem. And we mention once more Ferrier's book in connec-
ion with analogues for "Hörmander algebras" of that problem (inc-
uding multidimensional settings).

The more "rigid" is the topology of a space, the more profound
is the concept of divisor (and, as a rule, the more difficult it is
to prove that z-invariant subspaces are divisorial). The series of
Problems 7.4-7.16 is very instructive in this respect. Another feature they have in common is that they aim at the well-known "secon-
dary" approximaton problem of Analysis - Synthesis: to prove or to
disprove that any subspace with a trivial divisor is dense (cf.
Wiener's Tauberian Theorem). This problem is implicit in arguments
of items 7.7-7.11, 7.14, concerning weakly invertible (cyclic) functions in corresponding subspaces.

Classical Harmonic Analysis has led to very delicate and difficult theorems in Spectral Synthesis and to a vast variety of problems-
from numerous generalizations of periodicity (which corresponds to
the simplest convolution equation $(\delta_0-\delta_a)*f=0$) to the theory of
resolvent sets of Malliavin - Varopoulos. This direction is repre-

sented by Problem 7.17-7.23 (see also Problem S.4).

Other problems related to Analysis - Synthesis are 4.9, 4.14, 6.11, 6.12, 8.1, 8.8, 9.1, 9.3, 9.13, 10.1,10.3, 10.6, 10.8.

We conclude by some articles connected with 7.1: В.П.Паламодов, Доклады АН СССР, 1966, 168, N 6, 1251-1253; R.Narasimhan, Proc.Conf. Univ.Maryland, 1970, Berlin, Springer, 1970, 141-150; H.Skoda, Ann. Inst.Fourier, 1971, 21, 11-23. The references in 7.7 contain several items concerning the localization of ideals (for $n=1$) in the spirit of 7.2. Many problems in 7.9 are discussed in the book [3] cited in 7.7.

7.1. ABOUT HOLOMORPHIC FUNCTIONS WITH LIMITED GROWTH

old

Can one develop a theory of holomorphic functions satisfying growth conditions analogous to the theory of holomorphic functions on Stein manifolds?

Let δ be a continuous non-negative function on $\mathbb{C}^n$ which tends to zero at infinity; $\mathcal{O}(\delta)$ will be the set of all holomorphic functions u on the set $S_\delta : \delta > 0$ such that $\delta^N u$ is bounded for N large enough.

Research about the holomorphic functional calculus [1] led the author to the consideration of the algebras $\mathcal{O}(\delta)$. The only relevant algebras however were the algebras $\mathcal{O}(\delta)$ where δ is Lipschitz and $|s|\,\delta(s)$ is bounded.

L.Hörmander [2-4], has obtained results concerning algebras that he called $A(\varphi)$, but $A(\varphi) = \mathcal{O}(e^{-\varphi})$. His proofs used assumptions about φ which imply (up to equivalence) that $e^{-\varphi}$ is Lipschitz and $|s|\,e^{-\varphi(s)}$ is a bounded function of s.

He also assumed that φ, i.e. $-\log\delta$, is a plurisubharmonic function. This is an expected hypothesis, it means that $\mathcal{O}(\delta)$ behaves like the algebra of holomorphic functions on a domain of holomorphy. From the point of view of the holomorphic functional calculus, the condition "$-\log\delta$ is p.s.h." is also significant, as I.Cnop [5] showed (using Hörmander's results).

The reason why L.Hörmander and the author looked more specially at the algebras $\mathcal{O}(\delta)$, δ Lipschitz, were quite different. For Hörmander it appears that better estimates can be obtained when δ is Lipschitz. For the author, the only algebras relevant to a significant application of the theory were the algebras $\mathcal{O}(\delta)$, with δ Lipschitz. This coincidence suggests that the Lipschitz property is an important property δ has to possess if we want $\mathcal{O}(\delta)$ to behave somewhat like holomorphic functions on an open set.

Unfortunately, it is not clear what should take the place of this Lipschitz property when we investigate holomorphic functions on manifolds. The Lipschitz property is expressed in global coordinates. Manifolds only have local coordinate systems. An auxiliary Riemann or Kähler metric could be defined on the manifold. Or one may notice that $\mathcal{O}(\delta)$ is nuclear when δ is Lipschitz.

The plurisubharmonicity of $-\log\delta$ involves the structure of the complex manifold only. It generalizes the holomorphic convexity of Stein manifolds.

L.Hörmander has proved an analogue of Cartan's theorem B for holomorphic functions satisfying growth conditions. The full force of the Oka-Cartan theory of ideals and modules of holomorphic functions does not follow, until an analogue of Cartan's theorem on invertible matrices has been proved, with bounds, and bounds have been inserted in Oka's theorem on the coherence of the sheaf of relations.

We shall call $B(s,\varepsilon\delta(s))$ the open ball with center s and radius $\varepsilon\delta(s)$, and shall assume that ε is small. This ensures that $B(s,\varepsilon\delta(s))\subset S_\delta$, also that $\delta(z)/\delta(s)$ is bounded from above and bounded away from below when $z\in B(s,\varepsilon\delta(s))$, and that $\delta(s)/\delta(t)$ is bounded from above and bounded away from below when $B(s,\varepsilon\delta(s))$ and $B(t,\varepsilon\delta(t))$ have a non empty intersection (because δ is lipschitzian).

The following results should be a part of the theory.

CONJECTURE. <u>Let</u> $f_1,\dots,f_K$ <u>be elements of</u> $\mathcal{O}(\delta)^q$. <u>Let</u> $g\in\mathcal{O}(\delta)^q$ <u>be such that</u> $u_{1,s},\dots,u_{K,s}$ <u>can be found for each</u> s, <u>holomorphic on</u> $B(s,\varepsilon\delta(s))$, <u>with</u>

$$g=\sum_{1\le i\le K} u_{i,s}f_i$$

<u>on</u> $B(s,\varepsilon\delta(s))$, <u>and</u> $|u_{i,s}(z)|<M\delta(s)^{-N}$ <u>for some</u> M, $M\in\mathbb{R}_+$, <u>and</u> N, $N\in\mathbb{N}$. <u>Then</u> g <u>is in the submodule of</u> $\mathcal{O}(\delta)^q$ <u>generated by</u> $f_1,\dots,f_K$.

CONJECTURE. <u>With the same conventions, assume that</u> g_s <u>is given for each</u> s, $s\in S_\delta$, <u>such that</u> $\|g_s(z)\|<M\delta(s)^{-N}$, <u>when</u> $z\in B(s,\varepsilon\delta(s))$, <u>if</u> M <u>and</u> N <u>are large enough. Assume also that</u>

$$g_s-g_t=\sum u_{i,st}f_i$$

<u>on</u> $B(s,\varepsilon\delta(s))\cap B(t,\varepsilon\delta(t))$ <u>with</u> g_s-g_t <u>holomorphic on this open set, and less than</u> $M\delta(s)^{-N}$. <u>Then it is possible to find</u> g, M', N', ε' <u>such that</u> $g\in\mathcal{O}(\delta)$, <u>and</u>

$$g-g_s=\sum V_{i,s}f_i$$

n $B(s,\varepsilon'\delta(s))$, with $V_{i,s}$ holomorphic on $B(s,\varepsilon'\delta(s))$ and $|V_{i,s}(z)| < M'\delta(s)^{-N'}$ when $z \in B(s,\varepsilon'\delta(s))$.

A local description of the submodules of $\mathcal{O}(\delta)^q$ would also be welcome. Let M be a submodule of $\mathcal{O}(\delta)^q$. Then, for each s , M generates a submodule M_s of $\mathcal{O}(B(s,\varepsilon\delta(s)))^q$. When $B(s,\varepsilon\delta(s))$ and $B(t,\varepsilon\delta(t))$ intersect, M_s and M_t generate the same submodule of $\mathcal{O}(B(s,\varepsilon\delta(s)) \cap B(t,\varepsilon\delta(t)))^q$. Is it possible to find conditions from functional analysis which ensure that a family of modules M_s , which agree in the manner described, would be generated by a submodule M of $\mathcal{O}(\delta)^q$? J.-P. Ferrier [6], [7] considers Runge's theorem in the above context. Assuming $\delta \gg \delta'$ to be two Lipschitz functions he shows that the set of limits in $\mathcal{O}(\delta')$ of restrictions of elements of $\mathcal{O}(\delta)$ is - or can be identified with - some $\mathcal{O}(\delta_1)$, and δ_1 has some analogy with an " $\mathcal{O}(\delta)$ -convex hull" of δ' .

However the limits that Ferrier handles are bornological, not topological. Ferrier cannot show that $\mathcal{O}(\delta_1)$ is a closed subspace of $\mathcal{O}(\delta)$. It might very well be that the limits of elements of $\mathcal{O}(\delta_1)$ would be elements of $\mathcal{O}(\delta_2)$ with $\delta_1 \gg \delta_2 \gg \delta'$, etc. This specific problem is therefore open. So is the generalization of Ferrier's results to algebras of holomorphic functions satisfying growth conditions on a Stein manifold ... once we know what is a good analogue to the condition " δ is Lipschitz".

The general problem described in this note is vaguer than the editors of the series wish. It intrigued the author eighteen years ago, when [1] was published, but the author could not make any headway and went on to other things. Hörmander's breakthrough came later. The author has not taken the time to investigate all of the consequences of Hörmander's results. Results have been obtained by several authors, after Hörmander. They do not solve the problem as it is put. But they indicate that significant progress at the boundary of complex and functional analysis would follow from a good understanding of the question.

REFERENCES

1. Waelbroeck L. Étude spectrale des algèbres complètes. - Acad. Royale Belg. Mém. Cl. Sci., 1960, (2) 31.
2. Hörmander L. L^2-estimates and existence theorems for the $\bar\partial$-operator. - Acta Math., 1965, 113, p. 85-152.
3. Hörmander L. An introduction to complex analysis in

several variables. New York, Van Nostrand. 1966.

4. H ö r m a n d e r L. Generators for some rings of analytic functions. - Bull.Amer.Math.Soc., 1967, 73, 943-949.

5. C n o p I. Spectral study of holomorphic functions with bounded growth. - Ann.Inst.Fourier, 1972, 22, 293-309.

6. F e r r i e r J.-P. Approximation des fonctions holomorphes de plusieurs variables avec croissance. - Ann.Inst.Fourier, 1972, 22, 67-87.

7. F e r r i e r J.-P. Spectral theory and complex analysis. North Holland Math.Stud. 4. Amsterdam. North Holland. 1973.

L.WAELBROECK Univ.Libre de Bruxelles, Dép. de Math.
Campus Plaine. C.P.214 BRUXELLES
BELGIQUE

.2. LOCALIZATION OF POLYNOMIAL SUBMODULES IN SOME SPACES OF
ld HOLOMORPHIC FUNCTIONS AND SOLVABILITY OF THE $\bar\partial$-EQUATION

Let K be a compact in $\mathbb{R}^n$. Consider its support function

$$m_K(y) = max\{(y,\xi) : \xi \in K\},\ y \in (\mathbb{R}^n)'.$$

'or every positive integer q define a norm $\|\cdot\|_{q,K}$ on the space ·f complex-valued functions in $\mathbb{C}^n$ by

$$\|f\|_{q,K} = sup\{|f(z)|(|z|+1)^q exp(-m_K(y)) : z = x+iy \in \mathbb{C}^n\}.$$

.et S_K be the space of all entire functions f in $\mathbb{C}^n$ with $\|f\|_{q,K} < \infty$ for every q. This space can be considered as module over the algebra $\mathbb{C}[z]$ of polynomials in $\mathbb{C}^n$ with res- ›ect to the pointwise multiplication. Therefore each ideal I of $\mathbb{C}[z]$ generates a submodule $I \cdot S_K$ of S_K.

DEFINITION. A submodule $I \cdot S_K$ is called l o c a l if it :ontains all functions $f \in S_K$ satisfying the following condition: for every $w \in \mathbb{C}^n$ the Taylor series of f in w

$$\sum \frac{f^{(j)}(w)}{j!}(z-w)^j,\quad j=(j_1,\dots,j_n),\quad j!=j_1!\dots j_n!$$

belongs to the submodule $I \cdot T_w$, where T_w is the $\mathbb{C}[z]$-module of all formal power series in $z-w=(z_1-w_1,\dots,z_n-w_n)$.

CONJECTURE 1. <u>For any compact set</u> K <u>and for any ideal</u> I <u>in</u> $\mathbb{C}[z]$ <u>the submodule</u> $I \cdot S_K$ <u>is local</u>.

The CONJECTURE can be generalized to the case where the ideal of $\mathbb{C}[z]$ is replaced by an arbitrary submodule I of $\overset{\ell}{\oplus}\mathbb{C}[z]$ (the direct sum of ℓ copies of $\mathbb{C}[z]$). This more general CONJECTURE is easily reduced to the case of the ideal I.

Since the support function of a compact set coincides with that of its convex hull, we can suppose K to be a convex compact set. In this case the space of the Fourier transforms of S_K coincides with the space $\mathcal{D}_K$ of infinitely differentiable functions in $\mathbb{R}^n$ supported on K. The validity of CONJECTURE 1 would lead, in view of this connection, to some interesting consequences in the theory of differential equations with constant coefficients. Let us mention one of them.

COROLLARY. Let P be a $(t \times s)$ matrix of differential operators in $\mathbb{R}^n$ with constant coefficients. Then the system of equations $Pu = f$, $u = (u_1, \ldots, u_s)$ has a solution in the class $\overset{s}{\oplus} \mathcal{D}'_K$ of distributions on K for any $f \in \overset{t}{\oplus} \mathcal{D}'_K$ satisfying the formal compatibility condition (i.e. $Qf = 0$ for any matrix Q of operators with constant coefficients such that $QP = 0$).

Conjecture 1 is induced by the following result.

THEOREM OF MALGRANGE AND PALAMODOV ([1],[2]). Let Ω be a convex domain in $\mathbb{R}^n$, S_Ω be the union of S_K over all compact subsets K of Ω. Then for any ideal I of $\mathbb{C}[z]$ the submodule $I \cdot S_\Omega$ is local.

The proof of this Theorem depends on the triviality of the Čech cohomologies for holomorphic cochains in $\mathbb{C}^n$ with an estimation of the growth at infinity or on the equivalent theorem on the solvability of the $\bar{\partial}$-equation in $\mathbb{C}^n$ with the estimation at infinity as well. To use this way for the proof of CONJECTURE 1 one needs the following assertion.

Let S_K^* be the space of $\bar{\partial}$-differential forms

$$f = \sum_{j_1,\ldots,j_s} f_{j_1,\ldots,j_s} \, d\bar{z}_{j_1} \wedge \ldots \wedge d\bar{z}_{j_s}$$

such that all derivatives $f^{(\ell)}_{j_1,\ldots,j_s}$ have finite norms $\|\cdot\|_{q,K}$ for every q.

CONJECTURE 2. For every $K \subset \mathbb{R}^n$ and every α in S_K^* such that $\bar{\partial}\alpha = 0$ there exists β in S_K^* satisfying $\bar{\partial}\beta = \alpha$.

In this CONJECTURE the essential point is not the local properties of the coeffitients but their growth at infinity. We can assume them to be locally square summable or even to be distributions. The operator $\bar{\partial}$ being elliptic the complexes corresponding to the different local conditions are homotopic and therefore can satisfy CONJECTURE 2 only simultaneously.

The following result obtained for another purpose can be considered as an approach to CONJECTURE 2.

LEMMA ([3]). *Let G be a ball in $\mathbb{R}^n$ centered at the origin, G^+ be the intersection of G and a half-space of $\mathbb{R}^n$. Then for every q and for every $\bar\partial$-closed form $\alpha \in S_K^*$ there exists a $\bar\partial$-form β such that $\partial\beta = \alpha$ and $\|\beta^{(\ell)}\|_{q, K+G^+} < \infty$ with $\ell < q$.*

The following result was obtained recently.

THEOREM (Dufresnoy [4]). *Conjecture 2 is valid for any convex compact set with C^2 boundary.*

The proof is based on a well-known Hörmander's theorem on solvability of the $\bar\partial$-equation. A non-trivial point is the choice of an appropriate weight e^α with plurisubharmonic α. It is here where the smoothness of boundary is used.

REFERENCES

1. M a l g r a n g e B. Sur les systèmes differentiels à coefficients constants. Paris, Coll.Int. CNRS, 1963.
2. П а л а м о д о в В.П. Линейные дифференциальные операторы с постоянными коэффициентами. М., "Наука", 1967.
3. П а л а м о д о в В.П. Комплекс голоморфных волн. - В кн.: Труды семинара им.И.Г.Петровского, 1975, № I, 177-210.
4. D u f r e s n o y A. Un résultat de d''-cohomologie; applications aux systèmes differentiels à coefficients constants. - Ann. Inst.Fourier 1977, 27, N 2, 125-143.
5. H ö r m a n d e r L. Linear partial differential operators. Springer-Verlag, Berlin-Göttingen - Heidelberg, 1963.

V.P.PALAMODOV
(В.П.ПАЛАМОДОВ)

СССР, 117234, Москва
Ленинские Горы,
Московский государственный
университет, мех.-мат.факультет

7.3. old INVARIANT SUBSPACES AND THE SOLVABILITY OF DIFFERENTIAL EQUATIONS

1. Let Ω be a convex domain in $\mathbb{C}^n$ and let $H(\Omega)$ be the space of all functions analytic in Ω supplied with the natural topology. L.Schwarz posed and solved (for $\Omega = \mathbb{C}$) [1] the following PROBLEM.

Does any closed subspace $W \subset H(\Omega)$ invariant under the operator of differentiation contain exponential monomials, and if it does then do such monomials span W ?

This problem is completely explored in [2] for $n=1$. In case $n>1$ the problem has not been solved so far even for principal subspaces such as

$$W = \{ f \in H(\Omega) : \int f(\zeta + z)\, d\mu(\zeta) = 0, \ \operatorname{supp}(\mu) \subset \Omega \}$$

for example.

The positive answer to the question of L.Schwarz has been obtained only for special domains in $\mathbb{C}^n$, namely for $\Omega = \mathbb{C}^n$ [3], [4], for half-spaces in $\mathbb{C}^n$ [5]; for tube domains [6] and for domains in $\mathbb{C}^n$ satisfying $\Omega + \Omega \subset \Omega$ [7]. The proof in all listed cases, besides the tube domains, exploits essentially the fact that W is invariant under the translations. The condition $\Omega + \Omega \subset \Omega$ embraces a general class of domains with required invariance property. As in the one-dimensional case the proof of the following conjecture could be the key to the solution of the whole problem. Let $T \to \langle e^{(z,\zeta)}, T_\zeta \rangle$ be the generalized Laplace transform and let E_Ω be the space of entire functions coinciding with the Laplace transform of continuous linear functionals on $H'(\Omega)$. The space E_Ω is endowed with the natural topology borrowed from $H'(\Omega)$.

CONJECTURE. Given $\varphi, \psi \in E_\Omega$ such that ψ/φ is an entire function there exists a sequence $\{P_n\}_{n \geq 0}$ of polynomials satisfying $E_\Omega - \lim\limits_n \varphi \cdot P_n = \psi$.

The proof of this statement in case $n=1$ hinges on the employment of canonical products and therefore cannot be directly transferred to the case of several variables.

2. It is well known in the theory of differential equations that

$PC^\infty(\Omega) = C^\infty(\Omega)$ for every differential operator P with constant coefficients if and only if Ω is a convex domain. A natural complex analog of this statement can be formulated as follows.

CONJECTURE. Let Ω be a pseudo-convex domain in $\mathbb{C}^n$. Then $PH(\Omega) = H(\Omega)$ for every differential operator $P = P\left(\frac{\partial}{\partial z_1}, \dots, \frac{\partial}{\partial z_n}\right)$ with constant coefficients if and only if Ω is strongly linearly convex (see 1.13 of this volume for the definition).

The following facts are in favour of the conjecture. The property of strong linear convexity is a sufficient condition [8]. Conversely if Ω is a pseudo-convex domain and $PH(\Omega) = H(\Omega)$ then all slices of Ω by one-dimensional complex planes are simply connected (the proof follows the lines of [9]). It is known ([10],[11] in 1.13) that this implies that Ω is strongly linearly convex provided all slices of Ω are connected.

REFERENCES

1. Schwarz L. Théorie générale des fonctions moyenne-périodiques.- Ann.Math., 1947, 48, N 4, 857-925.
2. Красичков-Терновский И.Ф. Инвариантные подпространства аналитических функций. I. Спектральный синтез на выпуклых областях.- Матем.сб., 1972, 87, № 4, 459-489; II - Матем. сб.., 1972, 88, № 1, 3-30
3. Malgrange B. Existence et approximation des solution des équations aux derivées partielles et des équations de convolution.- Ann.Inst.Fourier, 1955, 6, 271-354.
4. Ehrenpreis L. Mean periodic functions. - Amer.J.Math., 1955, 77, N 2, 293-328.
5. Напалков В.В. О подпространствах аналитических функций, инвариантных относительно сдвига. - Изв.АН СССР, сер.матем., 1972, 36, 1269-1281.
6. Напалков В.В. Уравнение типа свертки в трубчатых областях $\mathbb{C}^2$. - Изв.АН СССР, Сер.матем., 1974, 38, 446-456.
7. Трутнев В.М. Об уравнении в свертках в выпуклых областях пространства $\mathbb{C}^n$. - В кн.: Вопросы математики. Сб.научн.трудов № 510, Ташкент, ТГУ, 1976, 148-150.
8. Martineau A. Sur la notion d'ensemble fortement liné-

ellement convexe. - Ann. Acad.Brasil., Ciens., 1968, 40, N 4, 427-435.

9. П и н ч у к С.И. О существовании голоморфных первообразных. - Докл.АН СССР, I972, 204, № 2, 292-294.

V.M.TRUTNEV
(В.М.ТРУТНЕВ)

СССР, 660075, Красноярск,
Красноярский государственный
университет

* * *

COMMENTARY

D.I.Gurevich proved in [10] that in the space $H(\mathbb{C}^2)$ there exist closed non-trivial translation invariant subspaces without exponential polynomials. The same holds in $C(\mathbb{R}^3)$, $C^\infty(\mathbb{R}^3)$, $\mathcal{D}'(\mathbb{R}^3)$ too.

REFERENCES

10. Г у р е в и ч Д.И. Контрпримеры к проблеме Л.Шварца. - Функц. анализ и его прил., 1975, 9, 2, 29-35.

7.4. LOCAL DESCRIPTION OF CLOSED SUBMODULES AND THE PROBLEM OF OVER-SATURATION

old

The space $\mathcal{H}^q$ of all $\mathbb{C}^q$-valued functions analytic in a domain G of the complex plane $\mathbb{C}$ becomes a module over the ring of all polynomials $\mathbb{C}[z]$ under pointwise algebraic operations. Consider a submodule P of $\mathcal{H}^q$ endowed with the structure of a Hausdorff locally convex space such that the multiplication operators by polynomials are continuous. A great many problems in Analysis, such as the problem of polynomial approximation [1], convolution equations [2], mean periodic functions [3], the problem of spectral synthesis [4],[5] etc., is connected with the problem of local description of closed submodules $I \subset P$. Such a submodule I defines a *divisor* $div(I)$. The *divisor* is a mapping which transforms any point $\lambda \in G$ into a submodule I_λ of the module $\mathcal{O}_\lambda^q$ of all germs at λ of $\mathbb{C}^q$-valued analytic functions. The mapping $\varphi_\lambda : \mathcal{H}^q \to \mathcal{O}_\lambda^q$ transforms every function in $\mathcal{H}^q$ into its germ at λ. The module I_λ is the smallest $\mathcal{O}_\lambda$-submodule of $\mathcal{O}_\lambda^q$ containing $\varphi_\lambda(I)$.

A submodule I is called a *divisorial submodule* if

$$I = I(div) \stackrel{def}{=} \{ f \in P : \varphi_\lambda(f) \in I_\lambda, \ \forall \lambda \in G \} .$$

The module $P = \mathcal{H}^q$ equipped with the topology of uniform convergence on compact subsets of G provides an example of a module whose all closed submodules are divisorial [6]. Many antipodal examples can be found in [1], [4], [7].

The PROBLEM of localization consists in <u>the characterization of those conditions which ensure that every submodule of a given module is divisorial</u>.

The following concepts are useful for the solution of the problem of localization. Namely, these are the concepts of stability and saturation, which separate the algebraic and analytic difficulties of the problem. Define $I_\lambda = \mathcal{H}^q$ if $\lambda \notin G$.

DEFINITION 1. <u>A submodule</u> I <u>is called stable if for every</u> $\lambda \in \mathbb{C}$

$$f \in I, \quad f/(z-\lambda) \in I_\lambda \Rightarrow f/(z-\lambda) \in I$$

It is natural to consider stable submodules for modules P possessing the property of the uniform stability. This property ensures a certain kind of "softness" of the topology in P.

DEFINITION 2. A module P is called uniformly stable if for every neighbourhood $V\subset P$ of zero there exists a neighbourhood $\mathcal{U}\subset P$ of zero satisfying

$$f\in\mathcal{U},\ f/(z-\lambda)\in\mathcal{H}^{q}\Rightarrow f/(z-\lambda)\in V.$$

The following theorem explains the importance of the concept of saturation which will be defined later.

THEOREM 1. Let P be a uniformly stable module. Then the submodule $I\subset P$ is divisorial iff it is stable and saturated.

The saturated submodules for $q=1$ can be described as follows. Let V be a neighbourhood in P and let $f\in I(div)$ Set

$$C_{f,V}(z)=1+\inf\{|\frac{f(z)}{g(z)}|: g\in I\cap V\}.$$

Suppose that for each $f\in I(div)$ and each $\Phi\in\mathcal{H}^{1}$

$$|\Phi(z)|\leq C_{f,V}(z),\ z\in G\Rightarrow|\Phi|\leq const \tag{1}$$

Then I is called s a t u r a t e d. Note that (1) automatically holds for $f\in I$.

In general we proceed as follows. The dimension $\dim I_\lambda$ of I_λ over $\mathcal{O}_\lambda\ (\lambda\in G)$ is clearly not greater than q. Put $k\overset{def}{=}\max_{\lambda\in G}\dim I_\lambda$. Then it is easy to show, using standard arguments with determinants, that $\dim I_\lambda\equiv k$ in G. Moreover there exists a family $u^{(1)},\dots,u^{(k)}$ in I such that $\varphi_\lambda(u^{(1)}),\dots,\varphi_\lambda(u^{(k)})$ forms a basis of I_λ for every $\lambda\in G$. Set $\dim I\overset{def}{=}k$ (the local rank of I). If $\varphi_\lambda(f)\in I_\lambda$ then

$$f=c_1u^{(1)}+\dots+c_ku^{(k)},\ c_j\in\mathcal{O}_\lambda \tag{2}$$

and the germs c_j can be found as follows. Consider in $\mathbb{C}^q$ the orthogonal projection P_J onto the subspace spanned by $e_{j_1}, \ldots, e_{j_k}$. Here $J=(j_1,\ldots,j_k)$ and $\{e_j\}_{j=1}^q$ is the standard basis in $\mathbb{C}^q$. The system of linear equations (with a $(k\times k)$-matrix) $P_J f = c_1 P_J \overset{(1)}{u} + \ldots + c_k P_J \overset{(k)}{u}$ can be solved which leads to the formulae

$$c_s = \frac{\det_s(f, J)}{\det(u, J)}, \quad s = 1, \ldots, k,$$

where the determinants are defined in accordance with Kramer's rule.

DEFINITION 3. <u>A submodule</u> I, $\dim I = k$ <u>is called saturated with respect to</u> $f \in P$ <u>if for every neighbourhood of zero</u> $V \subset P$ <u>the following holds</u>

$$|\Phi| \leq 1 + \inf\left\{\sum_{s=1}^{k}\left|\frac{\det_s(f,J)}{\det(u,J)}\right| : \overset{(1)}{u}, \ldots, \overset{(k)}{u} \in I \cap V, \forall J\right\}$$

$$\Longrightarrow |\Phi| \leq const.$$

A submodule I is called s a t u r a t e d if it is saturated with respect to every $f \in I(div)$ and I is called o v e r - s a t u r a t e d provided it is saturated with respect to every $f \in P$.

The existence of suitable estimates for holomorphic ratios f/g (see [1],[8],[9]) in many cases permits to prove that a given submodule I is saturated. In particular the local description of ideals in algebras can be obtained in this way [10–12]. If P is an algebra then every ideal of P is stable (as a rule). But if P is only a module, as for example in [4], then the role of stability may turn out to be dominant.

THEOREM 2. <u>Suppose that for every collection</u> $\overset{(1)}{f}, \ldots, \overset{(n)}{f}$ <u>of elements of</u> P <u>the set</u>

$$\{f \in \mathcal{H}^q : |f_i(z)| \leq |\overset{(1)}{f_i}(z)| + \ldots + |\overset{(n)}{f_i}(z)|, \quad z \in G, \quad i = 1, \ldots, q\} \qquad (3)$$

<u>is contained in</u> ρ <u>and bounded. Then every divisorial submodule</u> $I \subset \rho$, $\dim I = 1$ <u>is over-saturated</u>.

The proofs of theorems 1 and 2 are to appear in Izvestia Acad. Nauk SSSR *).

It follows from Theorems 1 and 2 that for a uniformly stable module ρ satisfying (3) every submodule of local rank 1, containing a submodule with the same properties, is divisorial. This shows how important is it to extend Theorem 2 to submodules of an arbitrary local rank.

THE PROBLEM OF SATURATION. <u>Let</u> ρ <u>be a uniformly stable submodule satisfying (3). Is it true that every divisorial submodule</u> $I \subset \rho$ <u>is over-saturated? If not, what are general conditions ensuring that</u> I <u>is over-saturated</u>?

The solution of the problem would clarify obscure points in the theory of the local description and in its own turn would lead to solutions of some problems of real and complex analysis.

REFERENCES

1. Никольский Н.К. Избранные задачи весовой аппроксимации и спектрального анализа. - Тр.Мат.ин-та АН СССР, 1974, 120.
2. Красичков-Терновский И.Ф. Однородное уравнение типа свертки на выпуклых областях. - Докл.АН СССР, 1971, 197, № 1, 29-31.
3. Schwartz L. Théorie générale des fonctions moyenne-périodiqué. - Ann.Math., 1947, 48, N 4, 857-929.
4. Красичков-Терновский И.Ф. Инвариантные подпространства аналитических функций I. Спектральный синтез на выпуклых областях. - Матем.сб., 1972, 87, № 4, 459-488.
5. Красичков-Терновский И.Ф. Инвариантные подпространства аналитических функций II. Спектральный синтез на выпуклых областях. - Матем.сб., 1972, 88, № 1, 3-30.
6. Cartan H. Idéaux et modules de fonctions analytiques de variables complexes.-Bull.Soc.Math.France,1950,78,N1,29-64.
7. Kelleher J.J., Taylor B.A. Closed ideals in locally convex algebras of analytic functions. - J.reine und angew.Math., 1972, 225, 190-209.

*) Cf. Изв.АН СССР, сер. мат., 1979, 43, N 1, 44-46 and N 2, 309-341 - Ed.

8. К р а с и ч к о в - Т е р н о в с к и й И.Ф. Оценка субгармонической разности субгармонических функций I. - Матем.сб.,1977, 102, № 2, 216-247.
9. К р а с и ч к о в - Т е р н о в с к и й И.Ф. Оценка субгармонической разности субгармонических функций П. - Матем.сб.,1977, 103, № 1, 69-111.
10. Р а ш е в с к и й П.К. О замкнутых идеалах в одной счетно-нормированной алгебре целых аналитических функций. - Докл.АН СССР, 1965, 162, № 3, 513-515.
11. К р а с и ч к о в И.Ф. О замкнутых идеалах в локально-выпуклых алгебрах целых функций I, П. - Изв.АН СССР,сер.мат. 1967, 31, 37-60; 1968, 32, 1024-1032.
12. М а ц а е в В.И., М о г у л ь с к и й Е.З. Теорема деления для аналитических функций с заданной мажорантой и некоторые ее приложения. - Зап.научн.семин.ЛОМИ, 1976, 56.

I.F.KRASICHKOV-TERNOVSKII
(И.Ф.КРАСИЧКОВ-ТЕРНОВСКИЙ)

СССР, 450057, Уфа
ул.Тукаева 50
отдел физики и математики
Башкирский филиал АН СССР

7.5. ON THE SPECTRAL SYNTHESIS IN SPACES OF ENTIRE FUNCTIONS OF NORMAL TYPE

Let ρ be a positive real number and let H be a 2π-periodic lower semi-continuous trigonometrically ρ-convex function with values in $(-\infty, \infty]$. Denote by $\mathfrak{M}(H)$ the set of all trigonometrically ρ-convex functions h satisfying $h(\theta) < H(\theta)$ for every $\theta \in [0, 2\pi]$. For $h \in \mathfrak{M}(H)$ let $F(h)$ be the Banach space of all entire functions φ with the norm $\|\varphi\|_h = \sup_{r,\theta} |\varphi(re^{i\theta})| \exp(-h(\theta)r^\rho)$. The family $\{F(h)\}_{h \in \mathfrak{M}(H)}$ is inductive with respect to natural imbeddings and its inductive limit $[\rho, H(\theta))$ is a space of $(L\mathcal{N}^*)$-type in the sense of J.Sebastião-e-Silva.

Multiplication by the independent variable is a continuous mapping of $[\rho, H(\theta))$ into itself, so that $[\rho, H(\theta))$ is a topological module over the ring of polynomials and one may consider the lattice of the closed (invariant) submodules. The submodule $I_{a,k} = \{\varphi \in [\rho, H(\theta)) : \varphi(a) = \cdots = \varphi^{(k)}(a) = 0\}$ defined by $a \in \mathbb{C}$ and $k \in \mathbb{Z}^+$ is of the simplest structure.

By a commonly accepted definition (cf. [1]) a submodule $I \subset [\rho, H(\theta))$ *admits the spectral synthesis* (or is *localized*) if it coincides with the intersection of all submodules $I_{a,k}$ containing it.

PROBLEM. Find necessary and sufficient conditions (on $H(\theta)$) for every closed submodule $I \subset [\rho, H(\theta))$ to admit spectral synthesis.

In 1947 L.Schwartz [2] proved that every ideal is localized in the algebra of all entire functions of exponential type ($\rho = 1$, $H(\theta) \equiv +\infty$). The progress in the localization theory for spaces invariant with respect to multiplication by the independent variable in the weighted algebras and modules of entire functions is described in the survey by N.K.Nikol'skii [1]; there is an extensive list of references there.

For $\rho = 1$ the discussed problem was solved by I.F.Krasičkov: every closed submodule $I \subset [1, H(\theta))$ is localized iff H is unbounded [3].

For an arbitrary $\rho > 0$ it was proved in [4] that if the length of every interval, where H is finite, does not exceed π/ρ then every closed submodule $I \subset [\rho, H(\theta))$ admits localization.

There is an indirect evidence that the last condition is not only sufficient but necessary. Unfortunately, all my attempts to prove its necessity failed.

REFERENCES

1. Никольский Н.К. Инвариантные подпространства в теории операторов и в теории функций. – В кн.: Итоги науки и техники. М.: ВИНИТИ, Матем.анализ, I974, I2, I99–4I2.
2. Schwartz L. Théorie générale des fonctions moyennes périodiques. – Ann.Math., 1947, 48, N 4, 857–929.
3. Красичков-Терновский И.Ф. Инвариантные подпространства аналитических функций.I. – Матем.сборн.,I972, 87, № 4, 459–489; II. – Матем.сборн., 88, № I, 3–30.
4. Ткаченко В.А. О спектральном синтезе в пространствах аналитических функционалов. – Докл.АН СССР, I975, 223, № 2, 307–309.

V.A.TKACHENKO
(В.А.ТКАЧЕНКО)

СССР, 3I0I64, Харьков
пр.Ленина 47,
Физико-технический институт
низких температур АН УССР

7.6. old A PROBLEM IN SPECTRAL THEORY OF ORDINARY DIFFERENTIAL OPERATORS IN THE COMPLEX DOMAIN

Let Ω be a domain in $\mathbb{C}$ and $A(\Omega)$ be the space of functions analytic in Ω supplied with the topology of uniform convergence on compact subsets of Ω. Let $a_k \in A(\Omega)$, $k=0,\ldots,n-2$. Consider a differential polynomial ℓ, $\ell=(d/dz)^n+a_{n-2}(d/dz)^{n-2}+\ldots+a_0$. Choose n linear functionals $\varphi_1,\ldots,\varphi_n$ on $A(\Omega)$ and set $\mathcal{D}_1 = \{f\in A(\Omega): \langle f,\varphi_k\rangle=0,\ k=1,\ldots,n\}$. The differential polynomial ℓ defines a linear operator $\mathcal{L}$ on $\mathcal{D}_1$ that maps $\mathcal{D}_1$ into $A(\Omega)$ by the rule $\mathcal{L}f=\ell f$. If $\{y_k(z,\lambda)\}_1^n$ is a fundamental system of solutions of the equation

$$\ell y-\lambda y=0 \tag{1}$$

normalized by equalities $y_k^{(p)}(0,\lambda)=\delta_{k,p+1}$, $k=1,\ldots,n$; $p=0,\ldots,n-1$, the spectrum of $\mathcal{L}$ coincides with the set of solutions of the characteristic equation

$$\Delta(\lambda)=0\ , \text{ where } \Delta(\lambda)=\det(\langle y_k(z,\lambda),\varphi_j\rangle)_{k,j=1}^{n}.$$

Since Δ is an entire function of λ, the spectrum is, unless $\Delta\equiv 0$, a discrete set with possibly a unique limit point at infinity. In this case a root subspace of finite dimension corresponds to each point of the spcetrum.

The PROBLEM mentioned in the title consists in obtaining a description of the $A(\Omega)$-closure of the linear span of root vectors of $\mathcal{L}$. This problem is closely connected with completeness questions for the system $\{y(z,\lambda_j)\}$ of solutions of equation (1) in $A(\Omega)$, with the construction of general solutions of differential equations of infinite order with respect to ℓ, with the theory of convolution equations and of mean periodic analytic functions. An analogous problem for differential operators on the real line is well-known.

CONJECTURE. <u>If Ω is convex and</u> $\Delta\not\equiv 0$ <u>then the closure</u> $C\ell$ <u>of the linear span of root vectors of</u> $\mathcal{L}$ <u>coincides with the domain of all its powers, i.e. with the subspace</u>

$$\mathcal{D}_\infty = \{f \in A(\Omega) : \langle \ell^q f, \varphi_k \rangle = 0,\ k=1,\dots,n;\ q=0,1,\dots\}.$$

The inclusion $\mathcal{O} \subset \mathcal{D}_\infty$ follows immediately from the $A(\Omega)$-continuity of ℓ and $\varphi_1, \dots, \varphi_n$. The inverse inclusion is non-trivial and has been proved only in some particular cases: by A.F.Leont'ev [1] in the problem of completeness of the system $\{y(z,\lambda_j)\}$; by Yu.N.Frolov [2] in the problem of constructing a general solution of equations of infinite order under some additional restrictions on $\Delta(\lambda)$, see also the papers of the same authors cited in [1] and [2]; by V.I.Matsaev [3] for a general system $\{\varphi_k\}_{k=1}^n$ but with $\Omega = \mathbb{C}$; by the author [4] in some weighted spaces of entire functions. In the case of an arbitrary convex domain Ω and $n=1$ the question under consideration is equivalent to that of the possibility of spectral synthesis in the space of solutions of a homogeneous convolution equation; spectral synthesis is really possible in this situation, this has been proved by I.F.Krasickov-Ternovskii [5],[6], further generalizations can be found in [7]. The results of [5],[6] imply that the convexity condition imposed on Ω cannot be dropped. The question whether the above conjecture is true for an arbitrary convex domain Ω and an arbitrary n remains open.

REFERENCES

I. Л е о н т ь е в А.Ф. К вопросу о последовательностях линейных агрегатов, образованных из решений дифференциальных уравнений. - Матем.сб., I959, 48, № 2, I29-I36.

2. Ф р о л о в Ю.Н. Об одном методе решения операторного уравнения бесконечного порядка. -Матем.сб., I972, 89, № 3, 46I-474.

3. М а ц а е в В.И. О разложении целых функций по собственным и присоединенным функциям обобщенной краевой задачи. - Теор.функц., функц.анализ и их прил., I972, I6, I98-206.

4. Т к а ч е н к о В.А. О разложении целой функции конечного порядка по корневым функциям одного дифференциального оператора. - Матем.сб., I972, 89, № 4, 558-568.

5. К р а с и ч к о в - Т е р н о в с к и й И.Ф. Однородные уравнения типа свертки на выпуклых областях. - Докл.АН СССР, I97I, I97, № I, 29-3I.

6. Красичков-Терновский И.Ф. Инвариантные подпространства аналитических функций. II. Спектральный синтез на выпуклых областях. - Матем.сб., 1972, 88, № 1, 3-30.
7. Ткаченко В.А. О спектральном синтезе в пространствах аналитических функционалов. - Докл.АН СССР, 1975, 223, № 2, 307--309.

V.A.TKACHENKO
(В.А.ТКАЧЕНКО)

СССР, 310164, Харьков
проспект Ленина, 47,
Физико-технический институт
низких температур АН УССР

* * *

COMMENTARY BY THE AUTHOR

S.G.Merzlyakov has discovered that my CONJECTURE IS FALSE. Namely, pick up two entire even functions φ and ψ of exponential type and of completely regular growth such that the zero-set of $\varphi\psi$ is an $\mathcal{R}$-set and all zeros are simple. For example

$$\varphi(\lambda)=\prod_{k\geq 1}\left(1-\frac{\lambda^4}{k^4}\right), \qquad \psi(\lambda)=\prod_{k\geq 1}\left(1+\frac{\lambda^4}{k^4}\right)$$

fit. Let $\{\lambda_k\}$ and $\{\mu_k\}$ denote the zero-sets of φ and ψ. Then the functions

$$\beta(\lambda)=\sum_{k\geq 1}\frac{\varphi(\lambda)}{\psi(\lambda_k)\varphi'(\lambda_k)(\lambda-\lambda_k)}, \qquad \alpha(\lambda)=\frac{1-\psi(\lambda)\beta(\lambda)}{\varphi(\lambda)}$$

are entire functions of exponential type and $\beta(\lambda)+\lambda\varphi(\lambda)$, $\alpha(\lambda)-\lambda\psi(\lambda)$ define continuous linear functionals on $A(\Omega)$, Ω being the interior of the indicator diagram of $\varphi\psi$. The operator $\mathcal{L}$ defined by $(d/dz)^2$ and these functionals is an operator with the void spectrum because $\Delta(\lambda)=\alpha(\lambda)\varphi(\lambda)+\beta(\lambda)\psi(\lambda)\equiv 1$. However, the domain $\mathcal{D}_\infty$ of $\mathcal{L}$ contains a non-zero element, namely, the holomorphic function defined in Ω by

$$\sum_{k\geq 1} \frac{e^{\lambda_k z}}{\varphi'(\lambda_k)\varphi(\lambda_k)} = -\sum_{k=1}^{\infty} \frac{e^{\mu_k z}}{\varphi(\mu_k)\psi'(\mu_k)} .$$

;.G.Merzlyakov has communicated that ANALOGOUS COUNTER-EXAMPLES EXIST FOR UNBOUNDED DOMAINS AS WELL.

Nevertheless, to my knowledge THE GENERAL PROBLEM of describing the closure of the family of roots vectors for an arbitrary operator $\mathcal{L}$ REMAINS UNSOLVED.

7.7. old TWO PROBLEMS ON THE SPECTRAL SYNTHESIS

1. Synthesis is impossible. We are concerned with the synthesis of (closed) invariant subspaces of z^*, the adjoint of the operator z of multiplication by the independent variable z on some space of analytic functions. More precisely, let X be a Banach space of functions defined in the unit disc $\mathbb{D}$ and analytic there, and suppose that $zX \subset X$ and the natural embedding $X \longrightarrow Hol(\mathbb{D})$ is continuous, $Hol\, \mathbb{D}$ being the space of all functions holomorphic in $\mathbb{D}$. If $f \in X$ then $k_f(\zeta)$ denotes the multiplicity of zero of f at a point ζ in $\mathbb{D}$, and for any function k from $\mathbb{D}$ to nonnegative integers let

$$X_k \overset{def}{=\!=} \{ f : f \in X ,\ \ k_f \geqslant k \} .$$

A closed z-invariant subspace E of X is said to be DIVISORIAL (or to have THE d-PROPERTY) if $E = X_k$ for some k (necessarily $k(\zeta) = k_E(\zeta) \overset{def}{=\!=} \min_{f \in E} k_f(\zeta)$, $\zeta \in \mathbb{D}$).

CONJECTURE 1. In every space X as above there exist non-divisorial z-invariant subspaces.

The dualized d-property means that the spectral synthesis is possible. To be more precise, let Y be the space dual (or predual) to X equipped with the weak topology $\sigma(Y, X)$ (the duality of X and Y is determined by the Cauchy pairing, i.e. $\langle f, g \rangle = \sum_{n \geqslant 0} \hat{f}(n)\hat{g}(n)$ for polynomials f, g). A z^*-invariant subspace E of Y is said to be SYNTHESABLE (or simply s-SPACE) if

$$E = span((\mathbb{1} - \lambda z)^{-n} z^{n-1} : 1 \leqslant n \leqslant k(\lambda)) \qquad (1)$$

with $k = k_{E^\perp}$. In other words E is an s-space if it can be recovered by the root vectors of z^* it contains.

All known results on z-invariant subspaces (cf. [1]) support Conjecture 1. The main hypothesis on X here is that X should be a Banach space. The problem becomes non-trivial if, e.g. the set of polynomials $\mathcal{P}_A$ is contained and dense in X and $\{ \zeta : \zeta \in \mathbb{C} : \sup_{p \in \mathcal{P}_A} | p(\zeta) | \, \| p \|_X^{-1} < \infty \} = \mathbb{D}$. The existence of a single norm defining the topology should lead to some limit stable peculiarities of the boundary behaviour of elements of X, and it is hese peculiarities that should be responsible for the presence of non-divisorial $\mathbb{Z}$-invariant subspaces. Spaces topologically contained in the Nevanlinna class provide leading examples. The aforementioned boundary effect consists here in the presence of a non-trivial inner factor (i.e. other than a Blaschke product) in the canonical factorization. Analogues of inner functions are discovered in classes of functions defined by growth restrictions ([2], [3], [4]); these classes are even not necessarily Banach spaces but their topology is still "sufficiently rigid" (i.e. the seminorms defining the topology are of "comparable strength"). On the contrary, in spaces X with a "soft" topology the invariant subspaces are usually divisorial. Sometimes the "softness" of the topology can be expressed in purely quantitative terms (for example, under some regularity restrictions on λ, all ideals in the algebra $\{ f : f \in Hol(\mathbb{D}), |(f(\zeta)| = O(\lambda^{c}(\zeta)), \quad c = c_f \}$ are divisorial if and only if $\int_0^1 \left(\frac{\log \lambda(r)}{1-r}\right)^{1/2} dr = +\infty$, [5], [10]). This viewpoint can be given a metric character; it can be connected with the multiplicative structure of analytic functions, with some problems of weighted polynomial approximation, with generalizations of the corona theorem, etc. (cf. [1,3,6]).

2. Approximative synthesis is possible. Let us read formula (1) in the following manner: there is an increasing sequence $\{E_n\}$ of $\mathbb{Z}^*$-invariant subspaces of finite dimension that approximates E:

$$E = \lim_n E_n \stackrel{def}{=} \{ f : f \in X, \ \lim_n dist(f, E_n) = 0 \} .$$

Removing one word from this sentence seems to lead to a universal description of $\mathbb{Z}^*$-invariant subspaces.

CONJECTURE 2. Let Y be a space from section 1 and E be a $\mathbb{Z}^*$-invariant subspace of Y. Then there exist subspaces E_n with $\mathbb{Z}^* E_n \subset E_n$, $\dim E_n < \infty \ (n \in \mathbb{N})$ so that $E = \lim_n E_n$.

There is a further extension of this Conjecture that still could look probable. Namely, let T be a continuous linear operator on a linear space Y and suppose that the system of root vectors of T is complete in Y. Is it true that $TE \subset E \Rightarrow E = \lim_n E_n$ for some sequence E_n with $TE_n \subset E_n$, $\dim E_n < \infty \ (n \in \mathbb{N})$?

But it is easy to see that without additional restrictions on T the answer to the last question is "no". A counterexample is provided by the left shift (i.e. still $\mathbb{Z}^*$!) $(a_0, a_1, \dots) \longmapsto (a_1, a_2, \dots)$ on $\ell^p(w_n)$ with an appropriate weight $\{w_n\}_{n\geq 0}$ (decreasing rapidly and irregularly). This operator posesses invariant subspaces that can not be approximated by root subspaces, [3]. In examples of such kind it is essential that the spectrum of the operator reduces to the single point 0.

A plenty of classical theorems on $\mathbb{Z}$-invariant subspaces (as e.g., Beurling's theorem) not only support Conjecture 2, but also allow to describe $\mathbb{Z}^*$-cyclic vectors (that is, functions f with the property span $(\mathbb{Z}^{*n} f : n \geq 0) = Y$) in terms of the approximation by rational functions with bounded " X-capacities". If r is a rational function with poles in $\mathbb{C} \setminus \operatorname{clos} \mathbb{D}$, $r(\infty) = 0$ then $\operatorname{cap}_X r \overset{def}{=} \inf\{\|g\|_X : \langle z^n g, r \rangle = 0,\ n \geq 1;\ g(0) = 1\}$. The capacity of an arbitrary $\mathbb{Z}^*$-invariant subspace is defined similarly. If $f = (Y) - \lim\limits_n r_n$ and $\sup\limits_n \operatorname{cap}_X r_n < \infty$ then f is not cyclic for $\mathbb{Z}^*$; analogously, $\sup\limits_n \operatorname{cap}_X E_n < \infty \Longrightarrow \underline{\lim}\, E_n \neq Y$.

The last assertion can be converted, after a slight modification of the notion of "capacity" [7,8]. Probably techniques of rational approximation should allow to prove Conjecture 2 avoiding estimates of " X-capacities" of rational functions (that appears to be a more difficult question; it is worth mentioning that this question is a quantitative form of the uniqueness theorem for X). The results on this matter known up to now use, on the contrary, not only classical uniqueness theorems but also the explicit description of $\mathbb{Z}$-invariant subspaces in terms of the inner-outer factorization.

REFERENCES

1. Никольский Н.К. Инвариантные подпространства в теории операторов и теории функций. - В кн.: Итоги науки и техники. Математический анализ, т.12, М., ВИНИТИ, 1974, 199-412.
2. Красичков-Терновский И.Ф. Инвариантные подпространства аналитических функций. II. Спектральный синтез на выпуклых областях. - Матем. сб., 1972, 88, № 1, 3-30.
3. Никольский Н.К. Избранные задачи весовой аппроксимации и спектрального анализа. - Труды МИАН, 120, М.-Л., Наука, 1974.
4. Korenbljum B. A Beurling-type theorem. - Acta Math., 1975, 135, 187-219.
5. Апресян С.А. Описание алгебр аналитических функций, допускающих локализацию идеалов. - Зап.научн.семин.ЛОМИ, 1977, 70, 267-

269.

6. Никольский Н.К. Опыт использования фактор-оператора для локализации z - инвариантных подпространств. - Докл.АН СССР, 1978 240, № 1, 24-27.

7. Грибов М.Б., Никольский Н.К. Инвариантные подпространства и рациональная аппроксимация. - Зап.научн.семин.ЛОМИ, 1979, 92, 103-114.

8. Никольский Н.К. Лекции об операторе сдвига I. - Зап.научн.семин.ЛОМИ, 1974, 39, 59-93.

9. Hilden H.M., Wallen L.J. Some cyclic and non-cyclic vectors of certain operators. - Indiana Univ.Math.J., 1974, 23, N 7, 557-565.

10. Шамоян Ф.А. Теоремы деления и замкнутые идеалы в алгебрах аналитических функций с мажорантой конечного роста. - Изв. АН Арм. ССР, Математика, 1980, 15, № 4, 323-331.

N.K.NIKOL'SKII
(Н.К.НИКОЛЬСКИЙ)

СССР, 191011, Ленинград
Фонтанка 27, ЛОМИ

7.8. old CYCLIC VECTORS IN SPACES OF ANALYTIC FUNCTIONS

Let X be a Banach space of analytic functions in $\mathbb{D}$ satisfying the following two conditions: (i) for each ζ, $\zeta \in \mathbb{D}$, the map $f \to f(\zeta)$ is a bounded linear functional on X, (ii) $zX \subset X$. It follows from (ii), by means of the closed graph theorem, that multiplication by z is a bounded linear transformation (more briefly, an operator) on X. Finally, $f \in X$ is said to be a *cyclic vector* for the operator of multiplication by z if the finite linear combinations of the vectors f, zf, $z^2 f, \ldots$ are dense in X (when the constant function $\mathbb{1}$ is in X, one also says that f is *weakly invertible* in X; this terminology was first used in [1]).

QUESTION 1. Does strong invertibility imply weak invertibility? (That is, if $\mathbb{1}, \frac{1}{f}, f$ are all in X, is f cyclic?)

Consider the special case when X is the Bergman space, that is the set of square-integrable analytic functions: $\|f\|^2 = \int_{\mathbb{D}} |f|^2 < \infty$.

CONJECTURE 1. If f is in the Bergman space and if $|f(z)| > c(1-|z|)^a$ for some $c, a > 0$, then f is cyclic.

If correct this would imply an affirmative answer to QUESTION 1 when X is the Bergman space. The conjecture is known to be correct under mild additional assumptions (see [2], [3], [4]). In particular it is correct when f is a singular inner function. In this case the condition in the hypothesis of the conjecture is equivalent to the condition that the singular measure associated with f has modulus of continuity $O(\delta \log 1/\delta)$ (see [1]).

CONJECTURE 2. A singular inner function is cyclic in the Bergman space if and only if its associated singular measure puts no mass on any Carleson set. (For the definition of Carleson set see [5], pp. 326-327.)[*]

For more discussion of the cyclicity of inner functions see §6 of [6], pages 54-58, where the possibility of an "inner-outer" factorization for inner functions is considered.

---

[*] or else 9.3 - Ed.

QUESTION 2. Does there exist a Banach space of analytic functions, satisfying (i) and (ii), in which a function f is cyclic if and only if it has no zeros in $\mathbb{D}$?

N.K.Nikolskii has shown [7] that no weighted sup-norm space of a certain type has this property. If such a space X existed then the operator of multiplication by z on X would have the property that its set of cyclic vectors is non-empty, and is a closed subset of the space $X \setminus \{0\}$ (this follows since the limit of non-vanishing analytic functions is either non-vanishing or identically zero). No example of an operator with this property is known. (This may no longer be correct; Per Enflo has announced an example of an operator on a Banach space with no invariant subspaces; that is, every non-zero vector is cyclic. The construction is apparently exceedingly difficult.) H.S.Shapiro has shown that for any operator the set of cyclic vectors is always a G_δ set (see [8], §11, Proposition 40, p.110). For a discussion of some of these questions from the point of view of weighted shift operators, see [8], §§11 , 12.

QUESTION 3. Let X be as before, and let $f, g \in X$ with g cyclic. If $|f(z)| \geqslant |g(z)|$ in $\mathbb{D}$, is f cyclic?

This question has a trivial affirmative answer in spaces like the Bergman space, since bounded analytic functions multiply the space into itself. It is unknown for the Dirichlet space (that is, the space of functions with $\int_{\mathbb{D}} |f'|^2 < \infty$); the special case $g =$ constant is established in [9].

REFERENCES

1. Shapiro Harold S. Weakly invertible elements in certain function spaces, and generators in ℓ^1 . - Mich.Math.J., 1964, 11, 161-165.
2. Shapiro Harold S. Weighted polynomial approximation and boundary behaviour of holomorphic functions. - В кн.: Современные проблемы теории аналитических функций, М., Наука, 1966, 326-335.
3. Шапиро Г. Некоторые замечания о весовой полиномиальной аппроксимации голоморфных функций. - Матем.сб., 1967, 73, 320--330.
4. Aharonov D., Shapiro H.S., Shields A.L. Weakly invertible elements in the space of square-summable holo-

morphic functions. - J.London Math.Soc.,1974, 9, 183-192.

5. C a r l e s o n L. Sets of uniqueness for functions regular in the unit circle. - Acta Math.,1952, 87, 325-345.

6. D u r e n P.L., R o m b e r g B.W., S h i e l d s A.L. Linear functionals on H^p spaces with $0<p<1$. - J.für reine und angew.Math.,1969, 238, 32-60.

7. Н и к о л ь с к и й Н.К. Спектральный синтез и задача весовой аппроксимации в пространствах аналитических функций. - Изв.АН Арм. ССР. Сер.матем., 1971, 6, № 5, 345-367.

8. S h i e l d s A l l e n L. Weighted shift operators and analytic function theory. - In: Topics in operator theory, Math.Surveys N 13, 49-128; Providence, Amer.Math.Soc., 1974.

9. S h i e l d s A l l e n L. Cyclic vectors in some spaces of analytic functions. - Proc.Royal Irish Acad.,1974, 74, Section A, 293-296.

ALLEN L.SHIELDS

Department of Mathematics
University of Michigan
Ann Arbor, Michigan 48109
U.S.A.

* * *

COMMENTARY

QUESTION 1 has been answered in the negative by Shamoyan [10].

THEOREM ([10]). Let $\varphi(\zeta)=\left|\frac{1+\zeta}{1-\zeta}\right|^2+\left|\frac{1+\zeta}{1-\zeta}\right|^\alpha$, $\zeta\in\mathbb{D}$, $0<\alpha<1$ and let X_φ denote the space of all functions f analytic in the unit disc $\mathbb{D}$, continuous in $(\operatorname{clos}\mathbb{D})\setminus\{1\}$ and such that $|f(\zeta)|=O(e^{\varphi(\zeta)})$ for $|\zeta|<1$, $\zeta\to 1$. Then polynomials are dense in X_φ and for $f\overset{def}{=}\exp\left(-\left(\frac{1+z}{1-z}\right)^2\right)$ we have: $f, f^{-1}\in X_\varphi$ but f is not weakly invertible in X_φ.

CONJECTURE 2. The "only if" part can be found in [2] of 7.10, the "if" part is proved in [11]. The same criterion of weak invertibility of inner functions holds in all Bergman spaces $\mathcal{H}^p$, $1\leqslant p<\infty$, in spaces $A^p\overset{def}{=}\{f:\ f$ is analytic in $\mathbb{D}$ and $|f(\zeta)|=O((1-|\zeta|)^{-p})$, $|\zeta|\to 1\}$ and in $\bigcup_{p>0}\mathcal{H}^p=\bigcup_{p>0}A_p$.

Note, by the way, that 7.7 contains a conjecture in the spirit of QUESTION 2, and that both QUESTION 2 and 3 (together with some others) are discussed in ref. [3] of 7.7.

REFERENCES

0. Шамоян Ф.А. О слабой обратимости в некоторых пространствах аналитических функций. - Докл.АН Арм.ССР, 1982, 74, № 4, 157-161.

1. Korenblum B. Cyclic elements in some spaces of analytic functions. - Bull.Amer.Math.Soc., 1981, 5, N 3, 317-318.

7.9. old WEAK INVERTIBILITY AND FACTORIZATION IN CERTAIN SPACES OF ANALYTIC FUNCTIONS

A measure μ on $\mathbb{D}$ is called a *symmetric measure* if μ has the form $d\mu(r,\theta)=(2\pi)^{-1}d\nu(r)\,d\theta$, where ν is a finite, positive Borel measure on $[0,1]$, having no mass at 0, and such that $\nu([r,1])>0$ for all $0\leqslant r<1$. For any function f analytic in $\mathbb{D}$ and any p, $0<p<\infty$, we define the generalized mean

$$M_p(r;f;\mu)=\Big(\int_{\operatorname{clos}\mathbb{D}}|f(rw)|^p d\mu(w)\Big)^{\frac{1}{p}} \tag{1}$$

$0\leqslant r<1$. The class $E^p(\mu)$ consists of all functions f analytic in $\mathbb{D}$ such that

$$\|f\|_{p,\mu}=\sup_{r<1} M_p(r;f;\mu)<\infty . \tag{2}$$

In the special case where ν is a single unit point mass at 1, the means (1) reduce to the classical H^p means, and the $E^p(\mu)$ classes to the standard Hardy classes on $\mathbb{D}$. In all cases, $E^p(\mu)$ is isometrically isomorphic to the $L^p(\mu)$ -closure of the polynomials. General properties of these classes are outlined in [1,2,3]. Numerous investigations of special cases (e.g., the Bergman classes, μ=area measure) are scattered throughout the literature. A complete bibliography would be quite extensive, and so references here are restricted to those which have had the most direct influence upon the author's work.

A function f, $f\in E^p(\mu)$ is said to be *weakly invertible* if there is a sequence of polynomials $\{p_n\}$ such that $p_n f\to \mathbb{1}$ in the metric of $E^p(\mu)$. From an operator-theoretic point of view, such functions are significant in that an element of $E^p(\mu)$ is weakly invertible if and only if it is a *cyclic vector* for the operator of multiplication by z on $E^p(\mu)$. (When $p=2$, this operator is unitarily equivalent to a subnormal weighted shift.) In the special case of the Hardy classes, Beurling [4] showed that a function is weakly invertible if and only if it is outer. In the more general context of the $E^p(\mu)$ classes, a complete characterization of the weakly invertible functions awaits discovery. At this juncture, however, it is not even lear what general shape such a characterization might take. We know f only a handful of scattered results which are applicable to these pacial classes. The earliest of these can be found in three papers y Shapiro [5, 6, 7] and in the survey article by Mergelyan [8]. ore recent contributions have been made by the author [1, 2], Aha-onov, Shapiro and Shields [9]; and Hedberg (see Shields [10,p.112]).

Many of the known results on weakly invertible functions in the $E^p(\mu)$ classes are essentially either multiplication or factoriza-tion theorems. It is well known that the product of two outer func-tions is outer, and that any factor of an outer function is outer. Do these properties carry over to weakly invertible functions in the $E^p(\mu)$ classes? We list a number of specific questions along these lines.

(a) Suppose $f, g, h \in E^p(\mu)$ and $f = gh$. If g and h are weakly invertible, is f weakly invertible? Conversely, if f is weakly invertible, are g and h weakly invertible?

(b) If $f \in E^p(\mu)$ is nonvanishing and $\frac{1}{f} \in E^q(\mu)$ for some q, $q > 0$, is f weakly invertible?

(c) If $f \in E^p(\mu)$ is weakly invertible and $\alpha > 0$, is f^α weakly invertible in $E^{p/\alpha}(\mu)$?

(d) If $f \in E^p(\mu)$ and f is weakly invertible in $E^q(\mu)$ for some q, $q < p$, is f weakly invertible in $E^p(\mu)$?

(e) Let $f \in E^p(\mu)$, $g \in E^q(\mu)$ and $h \in E^s(\mu)$, and let $f = gh$. If g and h are weakly invertible in $E^q(\mu)$ and $E^s(\mu)$, respectively, is f weakly invertible in $E^p(\mu)$? What about the converse?

Of course, all these things are trivially true in the special case of the Hardy classes. Question (e) is the most general of the list. The reader can easily convince himself that affirmative ans-wers to (e) would imply affirmative asnwers to all the others. Con-versely, affirmative answers to (a) and (d) together would yield affirmative answers to (e).

The answers to question (a) are known to be affirmative if $h \in H^\infty$ or if $g \in E^{p'}(\mu)$ and $h \in E^{q'}(\mu)$ with $\frac{1}{p'} + \frac{1}{q'} = \frac{1}{p}$ (see [2])*). These results are inspired by an earlier result of Shapiro

*) See also [3] of 7.7 - Ed.

[5 , Lemma 2]. The question remains unanswered for unrestricted g and h .

Question (b) has a long history, and versions of it appear in numerous sources. An affirmative answer may be obtained by imposing the additional condition $f \in E^{p+\delta}(\mu)$ for some δ , $\delta > 0$. The legacy of results of this type seems to begin with the paper of Shapiro [6], and has been carried forth into a variety of different settings in the separate researches of Brennan [11], Hedberg [12], and the author [2]. A similar result with a different kind of side condition is to be found in the work of Aharonov, Shapiro and Shields [9]. In its full generality, however, the question remains unanswered.

Question (d) seems in some sense to be the crucial question, certainly in moving from the setting of question (a) to that of question (e), but perhaps also in removing the side conditions from the results cited above. Presently, however, there seems to be little evidence either for or against an affirmative answer, nor can we offer any tangible ideas on how to attack the problem. The key to its solution in the special case of the Hardy classes rests upon the fact that, there, weak invertibility can be accounted for in terms of behavior within the larger Nevanlinna class. Unfortunately, in the more general setting of the $E^p(\mu)$ classes, none of the several different generalizations of the Nevanlinna theory discovered to date seems to shed any light upon the matter. It may very well be that the answer to the question is negative. Clearly, a negative answer would introduce complications which have no parallel in the Hardy classes. However, in view of the negative results of Horowitz [13] concerning the zero sets of functions in the Bergman classes, such complications would not be too surprising, and perhaps not altogether unwelcome.

REFERENCES

1. F r a n k f u r t R. Subnormal weighted shifts and related function spaces. - J.Math.Anal.Appl., 1975, 52, 471-489.
2. F r a n k f u r t R. Subnormal weighted shifts and related function spaces. II. - J.Math.Anal.Appl., 1976, 55, 1-17.
3. F r a n k f u r t R. Function spaces associated with radially symmetric measures. - J.Math.Anal.Appl., 1977, 60, 502-541.
4. B e u r l i n g A. On two problems concerning linear transformations in Hilbert space. - Acta Math., 1949, 81, 239-255.
5. S h a p i r o H.S. Weakly invertible elements in certain function spaces, and generators of ℓ_1 . - Mich.Math.J., 1964, 11,

161-165.

6. Shapiro H.S. Weighted polynomial approximation and boundary behaviour of analytic functions. - В кн.: Современные проблемы теории аналитических функций. - М., "Наука", 1966,326-335.

7. Шапиро Г. Некоторые замечания о весовой полиномиальной аппроксимации голоморфных функций. - Матем.сб., 1967, 73, № 3, 320-330.

8. Мергелян С.Н. О полноте систем аналитических функций. - Успехи матем.наук, 1953, 8, № 4, 3-63.

9. Aharonov D., Shapiro H.S., Shields A.L. Weakly invertible elements in the space of square-summable holomorphic functions. - J.London Math.Soc.,1974, 9, 183-192.

0. Shields A.L. Weighted shift operators and analytic function theory. - In: Topics in Operator Theory. Providence, R.I., Amer.Math.Soc., 1974, pp.49-128.

1. Brennan J. Invariant subspaces and weighted polynomial approximation. - Ark.Mat.,1973, 11, 167-189.

2. Hedberg L.I. Weighted mean approximation in Caratheodory regions. - Math.Scand.,1968, 23, 113-122.

3. Horowitz C. Zeros of functions in the Bergman spaces. - Duke Math.J.,1974, 41, 693-710.

RICHARD FRANKFURT

Dept.of Math., College of Arts and Sciences.
University of Kentucky,
Lexington 40506, USA

EDITORS' NOTE. See also 7.8 and Commentary to 7.8.

7.10. old WEAKLY INVERTIBLE ELEMENTS IN BERGMAN SPACES

DEFINITION 1. $\mathcal{H}^2$ is the Hilbert space of analytic functions f in $\mathbb{D}$ with the norm

$$\|f\| = \left(\frac{1}{\pi}\int_{\mathbb{D}} |f(x+iy)|^2\,dx\,dy\right)^{\frac{1}{2}}. \tag{1}$$

DEFINITION 2 [1]. Let $\mathcal{K}$ be the set of all open, closed and half-closed arcs $I, I\subset\mathbb{T}$, including all single points, $\mathbb{T}$ and $\emptyset$. A function $\mu:\mathcal{K}\to\mathbb{R}$ is called a p r e m e a s u r e iff

(i) $\mu(I_1\cup I_2) = \mu(I_1)+\mu(I_2)$, where $I_1, I_2\in\mathcal{K}$,

$$I_1\cup I_2\in\mathcal{K}, \quad I_1\cap I_2=\emptyset;$$

(ii) $\lim_n \mu(I_n)=0$, where $I_n\in\mathcal{K}$, $I_1\supset I_2\supset\ldots$ and $\bigcap_n I_n=\emptyset$.

DEFINITION 3. A closed set F, $F\subset\mathbb{T}$ is called B e u r - l i n g - C a r l e s o n (B.-C. set) iff

(i) $|F|=0$;

(ii) $\sum_n |I_n| \log\frac{2\pi}{|I_n|} < +\infty$,

where I_n are the components of $\mathbb{T}\setminus F$ and $|\cdot|$ denotes the linear Lebesgue measure.

PROPOSITION 1 [1;2]. Let $f\in\mathcal{H}^2$ and $f(z)\neq 0$ $(z\in\mathbb{D})$. Then the following properties hold:

(i) The limit

$$\mu^*(I) = \lim_{r\to 1-0}\int_I \log|f(r\zeta)|\,|d\zeta| \tag{2}$$

exists for any arc I, $I\subset\mathbb{T}$:

(ii) The limit

$$\mu(I) = \lim_n \mu^*(I_n) \tag{3}$$

exists for any sequence of closed arcs (I_n) such that $I_1 \subset I_2 \subset \dots$ and $\bigcup_n I_n = I$, I being any open arc;

(iii) $\mu(I)$, defined by (3) for open arcs I, $I \subset \mathbb{T}$, admits a unique extension to a premeasure;

(iv) for any B.-C. set F, whose complementary arcs are I_n, the series $\sum_n \mu(I_n)$ is absolutely convergent;

(v) if we define

$$\sigma_f(F) = 2\pi \log|f(0)| - \sum_n \mu(I_n), \tag{4}$$

for B.-C. sets F, then σ_f admits a unique extension to a finite non-positive Borel measure on every B.-C. set.

DEFINITION 4. The measure σ_f (defined on the set of all Borel sets contained in a B.-C. set) is called *the $\mathcal{X}$-singular measure associated with f*, $f \in \mathcal{H}^2$ (it is assumed that $f(z) \neq 0$ in $\mathbb{D}$).

Proposition 1 follows immediately from the results of [1,2], since $f \in \mathcal{H}^2$ implies

$$\log|f(z)| \leq c_f + \log\frac{1}{1-|z|} . \tag{5}$$

DEFINITION 5. An element f, $f \in \mathcal{H}^2$, is called *weakly invertible (or cyclic)* iff $\operatorname{clos}\{fg : g \in H^\infty\} = \mathcal{H}^2$.

PROPOSITION 2. The following conditions are necessary for an element f, $f \in \mathcal{H}^2$, to be weakly invertible:

(a) $f(z) \neq 0$ $(z \in \mathbb{D})$; (6)

(b) $\sigma_f = 0$. (7)

This proposition follows easily from the main theorem in [2] which gives a description of closed ideals in the topological algebra $A^{-\infty}$ of analytic functions f satisfying

$$|f(z)| \leq c_f(1-|z|)^{-n_f} \qquad (z \in \mathbb{D}).$$

CONJECTURE 1. <u>Conditions (6) and (7) are sufficient for an</u> f, $f \in \mathcal{H}^2$, <u>to be weakly invertible.</u>

CONJECTURE 2. <u>The same conditions also describe weakly invertible elements in any Bergman space</u> $\mathcal{H}^p$ $(1 \leq p < \infty)$ <u>of analytic functions</u> f <u>with the norm</u>

$$\|f\| = \left(\frac{1}{\pi}\int_{\mathbb{D}} |f(x+iy)|^p \, dx\, dy\right)^{\frac{1}{p}}.$$

REFERENCES

1. K o r e n b l u m B. An extension of the Nevanlinna theory. - Acta Math. 1975, 135, 187-219.
2. K o r e n b l u m B. A Beurling-type theorem. - Acta Math. 1977, 138, p.265-293.

BORIS KORENBLUM

Department of Mathematics,
State University of New York
at Albany
1400 Washington Avenue
Albany, New York, 12222, USA

* * *

COMMENTARY

Both Conjectures a r e s u p p o r t e d by the results cited in Commentary to 7.8.

7.11. INVARIANT SUBSPACES OF THE BACKWARD SHIFT OPERATOR IN THE SMIRNOV CLASS

Denote by N_* the Smirnov class i.e. the space of all functions f holomorphic in the unit disc $\mathbb{D}$ and such that $\{\log^+|f_r|\}_{0<r<1}$ is uniformly integrable in $L^1 = L^1(\mathbb{T}, m)$. Here $f_r \overset{def}{=} f(rz)$, m is the normalized Lebesgue measure on the unit circle $\mathbb{T}$. The space N_* can be indentified with the closure of the set of polynomials (in z) in $\log L$, where $\log L$ is the space of all measurable functions f on $\mathbb{T}$ such that $\log(1+|f|)\in L^1$. In $\log L$ the distance ρ is introduced by $\rho(f,g)\overset{def}{=}\int_{\mathbb{T}} \log(1+|f-g|)\,dm$. Let S^* denote the backward shift operator, $S^*f = \frac{f-f(0)}{z}$.

1. INVARIANT SUBSPACES AND RATIONAL APPROXIMATION

PROBLEM. <u>Describe the invariant subspaces of</u> $S^*: N_* \to N_*$.

It should be noted that an analogous problem for the shift operator $S: N_* \to N_*$, $Sf = zf$, can be reduced easily to the famous Beurling theorem describing the invariant subspaces of $S: H^2 \to H^2$ (see [5]).

THE PROBLEM is connected with the description of the closure in N_* of the linear span of the Cauchy kernels $\left\{\frac{1}{1-\bar{\zeta}z}\right\}_{\zeta\in F}$, F being a closed subset of $\mathbb{T}$. In [2] and [3] analogous problems are solved for the case of the Hardy spaces H^p $(0<p<1)$. In this case the real variable characterization of H^p (see [7]) plays an important role.

CONJECTURE 1. <u>If</u> F <u>has no isolated points then the closure of the linear span of</u> $\left\{\frac{1}{1-\bar{\zeta}z}\right\}_{\zeta\in F}$ <u>is the set of all functions</u> $f\in N_*$ <u>such that</u> $z\bar{f}\in N_*$ <u>and</u> f <u>has an analytic continuation to</u> $\hat{\mathbb{C}}\setminus F$.

The case $F=\mathbb{T}$ is considered in [5].

2. EXAMPLES OF S^*-INVARIANT SUBSPACES. Let $X\subset N_*$ and let I be an inner function. Set $I^*(X) \overset{def}{=} \{f\in X : zI\bar{f}\in N_*\}$. Denote by $\sigma(I)$ the spectrum of I (see [1]). Let F be a closed subset of $\mathbb{T}$, $F\supset\sigma(I)\cap\mathbb{T}$. We say that a function $k: F\to \mathbb{N}\cup\{\infty\}$ is I-admissible if $k(\zeta)=\infty$ for all $\zeta\in\sigma(I)\cap\mathbb{T}$ and for all non-isolated points $\zeta\in F$. Denote by

$I^*(N_*, F, k)$ the set of all functions $f \in I^*(N_*)$ having a meromorphic continuation $\tilde{f}$ to $\hat{\mathbb{C}} \setminus F$ such that ζ is a pole of $\tilde{f}$ of order at most $k(\zeta)$ for all $\zeta \in F$ with $k(\zeta) \neq \infty$. It is easy to see that $I^*(N_*, F, k)$ is an invariant subspace of $S^*: N_* \to N_*$.

Let E be an invariant subspace of $S^*: N_* \to N_*$, $E \neq N_*$. Then $E \cap H^\infty$ is an invariant subspace of $S^*: H^\infty \to H^\infty$ and $E \cap H^\infty \neq H^\infty$. Hence $E \cap H^\infty = I^*(H^\infty)$ for some inner function $I = I_E$ (see [6]). Let us construct a closed subset $F = F_E$ of $\mathbb{T}$ and an I-amdissible function $k = k_E$ as follows

$$F \overset{def}{=} \{\zeta \in \mathbb{T} : (1-\bar{\zeta}z)^{-1} \in E\}, \quad k(\zeta) \overset{def}{=} \sup\{n \in \mathbb{N} : (1-\bar{\zeta}z)^{-n} \in E\}.$$

CONJECTURE 2. $E = I^*(N_*, F, k)$.

In case $F = \mathbb{T}$ the conjecture is true, see Corollary 5.2.3 in [5].

Results of the following section imply the inclusion $E \subset I^*(N_*)$.

3. CYCLIC VECTORS OF S^*. Let P_+ denote the Riesz projection. The following proposition can be obtained from Remark 1 in [6].

PROPOSITION. Let (X, Y) be a Smirnov dual pair having properties 1° and 2° (see [6]). Suppose that $X \supset H^\infty$, $Y \subset L^1/H^1_- \overset{def}{=} P_+(L^1)$ and $(f, P_+g) = \int f\bar{g}\,dm$ if $f \in H^\infty$, $g \in L^1$, $P_+g \in Y$.

Let E be an invariant subspace of $S^*: X \to X$. Then if $E \cap H^\infty \subset I^*(H^\infty)$ for some inner function I, then $E \subset I^*(X)$.

The proposition allows to generalize Theorem 5.2.4 in [5]. Let Ω be a function holomorphic in $\mathbb{D}$, $|\Omega| \geqslant 1$ everywhere in $\mathbb{D}$ and $\Omega \in N_*$. Using the PROPOSITION for the Smirnov dual pair $(\Omega H^\infty, P_+(\Omega^{-1}L^1))$, $(f, P_+g) \overset{def}{=} \int f\bar{g}\,dm$, we can prove the following

THEOREM. Let X be a Hausdorff topological vector space, $X \subset N_*$, $S^*X \subset X$. Suppose that X has the following property:

$$f_n \in X,\ g \in X,\ f \in N_*,\ f_n \to f \text{ a.e.},\ |f_n| \leqslant |g| + 1 \text{ a.e.} \Rightarrow$$
$$\Rightarrow f \in X,\ \lim_{n\to\infty} f_n = f \text{ (in } X).$$

Let E be an invariant subspace of $S^*: X \to X$, $E \cap H^\infty = I^*(H^\infty)$ for some inner function I. Then $E \subset I^*(X)$.

COROLLARY 1. If E is an invariant subspace of $S^*: N_* \to N_*$, $E \neq N_*$, then $E \subset I^*(N_*)$, where $I = I_E$.

COROLLARY 2. <u>The closure of the linear span of the family</u>

$\{S^{*n}f\}_{n\geqslant 0}$ $(f\in N_*)$ <u>does not coincide with</u> N_* <u>iff</u> f <u>has a pseudocontinuation</u> (see [1] for the definition).

4. DISTRIBUTION OF VALUES OF FUNCTIONS IN $\mathbb{1}^*(N_*)$.

Denote by $f^*:(0,1)\to\mathbb{R}$ the decreasing rearrangement of $|f|$, where f is a measurable function on $\mathbb{T}$. In [4] (see also [5]) it is proved that if $f\in\mathbb{1}^*(N_*)$ and $\overline{\lim_{t\to 0}} f^*(t)\cdot t = 0$, then $f\equiv 0$.

CONJECTURE 3. <u>If</u> $f\in\mathbb{1}^*(N_*)$ <u>and</u> $\underline{\lim_{t\to 0}} f^*(t)\cdot t = 0$,

<u>then</u> $f\equiv 0$.

Results of [4],[8] imply that if $f\in\mathbb{1}^*(N_*)$, $\overline{\lim_{t\to 0}} f^*(t)\cdot t < +\infty$,

then $\underline{\lim_{t\to 0}} t f^*(t) = \overline{\lim_{t\to 0}} f^*(t)\cdot t$.

REFERENCES

1. Никольский Н.К. Лекции об операторе сдвига. М., Наука, 1980.

2. Александров А.Б. Аппроксимация рациональными функциями и аналог теоремы М.Рисса о сопряженных функциях для пространств L^ρ с $\rho\in(0,1)$. - Матем.сборн.,1978, 107, № I, 3-19.

3. Александров А.Б. Инвариантные подпространства оператора обратного сдвига в пространстве $H^\rho(\rho\in(0,1))$. - Записки научн.семин.ЛОМИ, 1979, 92, 7-29.

4. Александров А.Б. Об $\mathcal{A}$-интегрируемости граничных значений гармонических функций. - Матем.заметки, 1981, 30, № I,59-72.

5. Aleksandrov A.B. Essays on non locally convex Hardy classes. - Lect.Notes Math., 1981, 864, 1-89.

6. Александров А.Б. Инвариантные подпространства операторов сдвига. Аксиоматический подход. - Записки научн.семин.ЛОМИ, 1981, II3, 7-26.

7. Coifman R.R. A real variable characterization of H^ρ . - Studia Math., 1974, 51, N 3, 269-274.

8. Hruščëv S.V., Vinogradov S.A. Free interpolation in the space of uniformly convergent Taylor series. - Lect.Notes Math. 1981, v.864, 171-213.

A.B.ALEKSANDROV   СССР, 198904, Ленинград, Петродворец
(А.Б.АЛЕКСАНДРОВ)   Библиотечная пл.2. Математико-механический
                     факультет ЛГУ

7.12. DIVISIBILITY PROBLEMS IN $A(\mathbb{D})$ AND $H^\infty(\mathbb{D})$.

We offer two problems on divisibility, one in the disc algebra $A(\mathbb{D})$, the other in $H^\infty(\mathbb{D})$. The first is this. For which $X\subset\overline{\mathbb{D}}$, $X\neq\emptyset$, do we have

$$A_X=\bigcap_{\zeta\in X}A_\zeta \tag{1}$$

where A_X is the ring of fractions

$$\{f/g : f,g\in A(\mathbb{D}),\ g \text{ vanishes nowhere in } X\}$$

and A_ζ is the local ring of fractions

$$\{f/g : f,g\in A(\mathbb{D}),\ g(\zeta)\neq 0\}?$$

We might point out that (1) holds if $X\cap\mathbb{T}$ is closed. (PROOF. Put $Y=X\cap\mathbb{T}$, $\Delta=X\cap\mathbb{D}$, and let $\gamma\in$ the right side of (1). Then

$$\gamma\in\bigcap_{\zeta\in Y}A_\zeta .$$

It is proved in [1] that (1) holds if X is closed in $\mathbb{D}$; hence $\gamma=F/G$ where $F,G\in A(\mathbb{D})$ and G vanishes nowhere in Y. Let $\zeta\in\Delta$; then $\gamma=f/g$ where $f,g\in A(\mathbb{D})$, $g(\zeta)\neq 0$. We have $Fg=fg$; thus $o(G,\zeta)\leqslant o(F,\zeta)$ where $o(\varphi,\zeta)$ is the order of vanishing of φ at ζ. This proves that the Blaschke product B made up of the zeros of G in Δ (counting multiplicities) divides F. Put $F_1=F/B$, $G_1=G/B$. Then $F_1, G_1\in A(\mathbb{D})$ and $\gamma=F_1/G_1$. ●).

Let R be an integral domain, F its field of fractions, and Δ the space of all prime ideals p in R, $p\neq R$. Let Q be an ideal in R; then Q is said to be an ideal of denominators if

$$Q=\{g\in R : g\gamma\in R\}$$

where $\gamma\in F$. Suppose R is such that every ideal of denominators is principal; then if $X\subset\Delta$, $X\neq\emptyset$, we have

$$R_X = \bigcap_{p \in X} R_p$$

where R_X is the ring of fractions

$$\{f/g : f, g \in R,\ g \notin \cup X\}$$

and R_p the local ring of fractions

$$\{f/g : f, g \in R,\ g \notin p\}.$$

This is easy to prove. Now in $H^\infty(\mathbb{D})$ every ideal of denominators is principal. (Easy to prove, but not obvious.) Thus if X is any nonempty set in the maximal ideal space of $H^\infty(\mathbb{D})$, then

$$H^\infty_X = \bigcap_{\zeta \in X} H^\infty_\zeta$$

where

$$H^\infty_X = \{f/g : f, g \in H^\infty,\ \hat{g} \text{ vanishes nowhere in } X\}$$
$$H^\infty_\zeta = \{f/g : f, g \in H^\infty,\ \hat{g}(\zeta) \neq 0\}.$$

This suggests that (1) might hold for every nonempty subset of $\bar{\mathbb{D}}$, although unlike $H^\infty(\mathbb{D})$ not every ideal of denominators in $A(\mathbb{D})$ is principal. E.g. if $\zeta \in \mathbb{T}$ and

$$P_\zeta = \{f \in A(\mathbb{D}) : f(\zeta) = 0\}, \tag{2}$$

then P_ζ is an ideal of denominators, but it is not principal.

Here is our second PROBLEM. <u>Let</u> Q <u>be a prime ideal in</u> $H^\infty(\mathbb{D})$, $Q \neq 0$, $H^\infty(\mathbb{D})$. <u>Suppose</u> Q <u>is finitely generated. Do we then have</u>

$$Q = \{f \in H^\infty(\mathbb{D}) : f(\zeta) = f'(\zeta) = \ldots = f^{(n)}(\zeta) = 0\}$$

<u>where</u> $\zeta \in \mathbb{D}$ <u>and</u> $n \geqslant 0$? (Yes if Q is maximal [2].) In $A(\mathbb{D})$ we have the following.

Let P and Q be prime ideals in $A(\mathbb{D})$ with $P \neq Q$ and $P \subset Q \subset P_\zeta$ where $\zeta \in \mathbb{T}$ (P_ζ = the right side of (2)). Then the $A(\mathbb{D})$ -module Q/P is not finitely generated, i.e.

$Q \neq P$ a finitely generated ideal in $A(\mathbb{D})$.

This is a corollary to Nakayama's lemma. It suggests that the answer to Problem 2 is yes, but then the proof would have to be different.

REFERENCES

1. F o r e l l i F. A note on ideals in the disc algebra. - Proc. Amer.Math.Soc. 1982, 84, 389-392.
2. T o m a s s i n i G. A remark on the Banach algebra $LH^{\infty}(\mathbb{D}^N)$ - Boll.Un.Mat.Ital. 1969, 2, 202-204.

FRANK FORELLI

Department of Mathematics
University of Wisconsin-Madison
Madison, Wisconsin 53706
USA

7.13. A REFINEMENT OF THE CORONA THEOREM

The usual methods for proving corona type theorems [1,2] use existence of solutions with bounded radial limits for equations $\frac{\partial u}{\partial \bar z} = f$ when $|f|\,dx\,dy$, or something similar to $|f|\,dx\,dy$, is a Carleson measure. Our problem is a variant of the corona theorem for which apparently no Carleson measure is in sight.

PROBLEM. <u>Suppose</u> $f, f_1, f_2 \in H^\infty$ <u>with</u> $|f(z)| \leqslant |f_1(z)| + |f_2(z)|$ <u>for all</u> $z \in \mathbb{D}$. <u>Must there be</u> $g_1, g_2 \in H^\infty$ <u>with</u> $g_1 f_1 + g_2 f_2 = f^2$?

The answer is known to be yes if the exponent 2 is replaced by 3 (or 2 $+\varepsilon$ if f has no zeroes) and no if 2 is replaced by 1 (or 2 $-\varepsilon$). See [2]. The answer is also yes if g_1, g_2 are only required to be in H^1. This is by a $\bar\partial$ argument using the estimate

$$\iint\limits_{z\in\mathbb{D}:\,4^{-1}\leqslant\left|\frac{f_1(z)}{f_2(z)}\right|\leqslant 4} \left|\left(\frac{f_1}{f_2}\right)'(z)\right|^2 (1-|z|)\,dx\,dy < \infty,$$

which comes from the Ahlfors-Shimizu form of the first fundamental theorem. If $\mu \stackrel{def}{=} \chi(z)\left|\left(\frac{f_1}{f_2}\right)'(z)\right|^2(1-|z|)\,dx\,dy$ (χ = characteristic function of $\left\{z\in\mathbb{D}: 4^{-1}\leqslant\left|\frac{f_1(z)}{f_2(z)}\right|\leqslant 4\right\}$) were always a Carleson measure our problem could be answered affirmatively by arguments in [2]; but in general μ is not Carleson.

REFERENCES

1. C a r l e s o n L. The corona theorem. - Proc. 15th Scandinavian Congress, Springer-Verlag, Lect. Notes in Math, 1970, 118, 121-132
2. G a r n e t t J. Bounded Analytic Functions. New York, Academic Press, 1981.

T. WOLFF

Department of Mathematics 253-37
California Institute of Technology
Pasadena, CA 91125, USA

EDITORS' NOTE. The results similar to the ones mentioned in the Problem were obtained by V.A.Tolokonnikov [3] in the following slightly different setting: for which increasing functions α does the inequality

$$|f| \leq \alpha((\sum_{k\geq 1} |f_k|^2)^{1/2}) ; \quad f, f_k \in H^\infty$$

imply $f = \sum_{k\geq 1} f_k g_k$ with $g_k \in H^\infty$, $\sup_{\mathbb{D}} \sum_{k\geq 1} |g_k|^2 < \infty$?

See also [4].

3. Т о л о к о н н и к о в В.А. Оценки в теореме Карлесона о короне. Идеалы алгебры H^∞, задача Секефальви-Надя. – Записки научн.семин. ЛОМИ, 1981, 113, 178–198.
4. Т о л о к о н н и к о в В.А. Интерполяционные произведения Бляшке и идеалы алгебры H^∞. – Записки научн.семин.ЛОМИ, 1983, 126, 196–201.

7.14. old INVARIANT SUBSPACES OF THE SHIFT OPERATOR IN SOME SPACES OF ANALYTIC FUNCTIONS

1. Let X be a Banach algebra of functions analytic in the unit disk $\mathbb{D}$ (with pointwise addition and multiplication). Assume that $X \subset C_A^{(m)}$ and let $n \overset{def}{=} max\{m : X \subset C_A^{(m)}\} < \infty$. Let $E^{(n)}(f) = \{\zeta : \zeta \in \mathbb{T} ; \; f^{(j)}(\zeta) = 0 \;, \; 0 \leqslant j \leqslant n\}$.

DEFINITION. A closed subset E of the unit circle $\mathbb{T}$ is called D_n - set for the algebra X if for any function f, $f \in X$ with $E^{(n)}(f) \supset E$ there is a sequence $\{f_k\}_{k \geqslant 1}$ of functions in X, satisfying the following conditions:

(i) $|f_k(z)| \leqslant C_k [dist(z, E)]^{n+1}, \; z \in clos\,\mathbb{D}$;

(ii) $\lim\limits_{k} \| f_k f - f \|_X = 0$.

It follows from [1] that every Beurling-Carleson set *) is a D_n--set for a number of standard algebras of analytic functions. In particular, it is true for the algebra $H^p_{n+1} = \{f : f^{(n+1)} \in H^p\}$, $1 < p < \infty$.

QUESTION 1. <u>Is every Beurling-Carleson set</u> a D_n <u>-set for the algebra</u> H^1_{n+1} , $n \geqslant 0$?

REMARK. If $E = \bigcup\limits_{k=1}^{m} E_k$, where $mE_k = 0$, $k = 1, \ldots, m$, and the lengths of complementary intervals of each E_k tend to zero exponentially fast then (as proved in [2]) E is a D_n -set for H^1_{n+1} and the sequence $\{f_k\}_{k \geqslant 1}$ can be chosen not depending on f , $f \in H^1_{n+1}$.

2. Let $A^p = A^p(\mathbb{D})$ be the Bergman space in the unit disc $\mathbb{D}$ (i.e. the space of all functions in $L^p(\mathbb{D})$ analytic in $\mathbb{D}$).

QUESTION 2. <u>Is there a closed subspace</u> G , $G \subset A^p$, $G \neq A^p$ <u>invariant under the shift operator</u> S , $S(f)(z) \overset{def}{=} z f(z)$, <u>and not finitely generated?</u>

QUESTION 3. <u>Let</u> G , $G \subset A^p$ <u>be an invariant subspace of</u> S . <u>Assume that for every</u> ζ , $\zeta \in \mathbb{D}$ <u>there is a function</u> f , $f \in G$,

*) See the definition in 9.3.- Ed

$f(\zeta)\neq 0$. <u>Is it true that</u> G <u>is generated by one function</u> g , <u>i.e.</u> $G=V(z^n g: n\geqslant 0)$?

REMARK. Questions 2 and 3 arise in the following way. Let $\mathbb{D}^2$ be the unit polydisk in $\mathbb{C}^2$ and $H^p(\mathbb{D}^2)$ be the Hardy space in $\mathbb{D}^2$ (see[3]). Denote by D_2 the "diagonal operator" in $H^p(\mathbb{D}^2)$, i.e. $D_2 f(\zeta) = f(\zeta,\zeta)$, $\zeta\in\mathbb{D}$, and denote by S_i the operator

$$S_i(f)(\zeta_1,\zeta_2)=\zeta_i f(\zeta_1,\zeta_2),\ \zeta_i\in\mathbb{D},\ f\in H^p(\mathbb{D}^2),\ i=1,2\ .$$

It is known that $D_2(H^p(\mathbb{D}^2)) = A^p$, $0<p<\infty$ (see [4],[5]) and there exist $S_{1,2}$ -invariant and non-finitely generated subspaces ([3], p.67). Moreover there is an invariant subspace generated by two functions in $H^2(\mathbb{D}^2)$ without common zeros in $\mathbb{D}^2$ and this subspace cannot be generated by any of its elements (see[6]).

REFERENCES

1. Ш а м о я н Ф.А. Структура замкнутых идеалов в некоторых алгебрах функций, аналитических в круге и гладких вплоть до его границы. - Докл.АН Арм.ССР, 1975, 60, № 3, 133-136.
2. Ш а м о я н Ф.А. Построение одной специальной последовательности и структура замкнутых идеалов в некоторых алгебрах аналитических функций. - Изв.АН Арм.ССР, Математика, 1972, УII, № 6, 440-470.
3. R u d i n W. Function Theory in Polydiscs. Benjamin, New York, 1969.
4. H o r o w i t z C., O b e r l i n D. Restrictions of functions to diagonal of $\mathbb{D}^2$. - Indiana Univ.Math.J. 1975, 24, N 7, 767-772.
5. Ш а м о я н Ф.А. Теорема вложения в пространствах n-гармонических функций и некоторые приложения. - Докл.АН Арм.ССР,1976, XII, № 1, 10-14.
6. J a c e w i c z Ch.Al. A nonprincipal invariant subspace of the Hardy space on the torus. - Proc.Amer.Math.Soc. 1972, 31, 127-129.

F.A.SHAMOYAN
(Ф.А.ШАМОЯН)

СССР, 375200, Ереван, 19,
ул.Барекамутян, 24б
Институт математики АН Арм.ССР

.15.
old

BLASCHKE PRODUCTS AND IDEALS IN C_A^∞.

Let A be the space of functions analytic in the open unit disc $\mathbb{D}$ and continuous in $clos\,\mathbb{D}$; and let $C_A^\infty = \{f \in A : f^{(n)} \in A, n=0,1,\ldots\}$. lthough the sets of uniqueness for C_A^∞ have been described [1], 2], [3], [4], and the closed ideal structure of C_A^∞ is known [5], there re still some open questions concerning the relationship of Blaschke roducts with closed ideals in C_A^∞. I pose two problems. Let I, $I \subset C_A^\infty$, denote a closed ideal and let B denote a Blaschke product hich divides some non-zero C_A^∞ function.

(1) For which B is it true that

$$BI \overset{def}{=\!=} \{Bf : f \in I\} \subset C_A^\infty \ ?$$

(2) If B is the g.c.d. (greatest common divisor) of the Blaschke factors of the non-zero functions in I, when is $(\frac{1}{B})I \overset{def}{=\!=} \{f \in C_A^\infty : Bf \in I\}$ a closed ideal in C_A^∞ ?

Note that the corresponding problems for singular inner functions are easier and are solved in section 4 of [5].

To discuss the problems for Blaschke products we need some notation. Let

$$Z^n(I) = \bigcap_{f \in I} \{z \in clos\,\mathbb{D} : f^{(k)}(z) = 0,\ k=0,\ldots,n\}, \quad n=0,1,\ldots,$$

and let

$$Z^\infty(I) = \bigcap_{n=0}^{\infty} Z^n(I)\ , \quad Z(I) = \{Z^n(I)\}_{n=0}^{\infty} \ .$$

If $I(Z(I))$ denotes the closed ideal of all f, $f \in C_A^\infty$, with $f^{(n)}(z) = 0$ for $z \in Z^n(I)$, $n = 0,1,\ldots$, then the closed ideal structure theorem says $I = SI(Z(I))$, where S is the g.c.d. of the singular inner factors of the non-zero functions in I.

DEFINITION. A sequence $\{z_j\} \subset \mathbb{D}$ has *finite degree of contact* at E, $E \subset \partial\mathbb{D}$, if there exist k,

$k>0$, and ε, $\varepsilon>0$, such that $1-|z_j|\geq \varepsilon\rho(z_j/|z_j|, E)^k$ for all j. (Here ρ denotes the Euclidean metric.)

The following unpublished theorem of B.A.Taylor and the author provides solutions to problems (1) and (2) in a special case.

THEOREM. (a) Assume $Z^0(I)=Z^\infty(I)$. In order that $BI\subset C_A^\infty$ it is necessary and sufficient that the zeros of B have finite degree of contact at $Z^\infty(I)$. If $BI\subset C_A^\infty$, then multiplication by B is continuous on I, BI is closed, and the inverse operation is continuous.

(b) Assume $Z^0(I)\cap\partial\mathbb{D}=Z^\infty(I)$. Let B be the g.c.d. of the Blaschke factors of the non-zero functions in I. In order that $(1/B)I$ be closed it is necessary and sufficient that the zeros of B have finite degree of contact at $Z^\infty(I)$.

THE PROOF of sufficiency in (a) is primarily a computation of the growth of the derivatives of B near $Z^\infty(I)$. The computation has also been done by James Wells [6]. The proof of necessity in (a) requires the construction of outer functions. (One can assume without loss of generality that the g.c.d. of the singular inner factors of the non-zero functions in I is $\mathbb{1}$.) In section 3 of [5] it is demonstrated that there is an outer function F, $F\in C_A^\infty$, vanishing to infinite order precisely on $Z^\infty(I)$, and such that $\log|F(e^{i\theta})|=-\omega(-\log\tilde\rho(\theta))$, where $\tilde\rho$ is continuous, const.$\rho(e^{i\theta})\leq\tilde\rho(\theta)\leq$ const.$\rho(e^{i\theta})$, and ω is a positive increasing infinitely differentiable function on $\mathbb{R}$ which can be chosen so that $\lim_{x\to+\infty} x^{-1}\omega(x)=+\infty$ as slowly as desired. An appeal to the closed ideal structure theorem places $F^{(n)}\in I_\infty$ for all n. Now, since BF and BF' are assumed to belong to C_A^∞, $B'F=(BF)'-BF'\in C_A^\infty$. Thus

$$\log|B'(e^{i\theta})|=-\log|F(e^{i\theta})|+O(1)=\omega(-\log\tilde\rho(\theta))+O(1)$$

for all choices of ω. Hence, for some $k>0$,

$$\log|B'(e^{i\theta})|\leq -k\log\rho(e^{i\theta}, Z^\infty(I))+O(1)$$

or $|B'(e^{i\theta})|=O(\rho(e^{i\theta}, Z^\infty(I))^{-k})$. A computation shows that this implies that the zeros of B have finite degree of contact at $Z^\infty(I)$.

The last assertion follows from the closed graph theorem.

To prove sufficiency in(b), let

$$J=\{f\in C_A^{\infty}: f^{(n)}(z)=0,\ z\in Z^{\infty}(I),\ n=0,1,\dots\}.$$

Then $Z^0(J)=Z^{\infty}(J)=Z^{\infty}(I)$ and $(1/B)I\subset J$. Applying (a) to J, one concludes $(1/B)I$ is closed. To prove necessity in (b), let

$$K=\{f\in C_A^{\infty}: f^{(n)}(z)=0,\ z\in Z^0(I)\cap\partial\mathbb{D},\ n=0,1,\dots\}.$$

(Again, one can ignore singular inner factors.) Then $Z^0(K)=$ $=Z^{\infty}(K)=Z^0(I)\cap\partial\mathbb{D}$ and by the closed ideal structure theorem $(1/B)I\supset K$. Thus $BK\subset I\subset C_A^{\infty}$; and so, applying (a) to K, the zeros of B have finite degree of contact at $Z^{\infty}(K)=Z^0(I)\cap\partial\mathbb{D}$. ●

Let us consider problem (1) in the more general case where $Z^0(I)\subset\partial\mathbb{D}$ but $Z^0(I)\neq Z^{\infty}(I)$ in the light of the above results. From the computation referred to in the proof of sufficiency in THEOREM (a), it is clear that if the zeros of B have finite degree of contact at $Z^{\infty}(I)$, then $BI\subset C_A^{\infty}$; however, it is not difficult to construct examples to show that this condition is not necessary. On the other hand, THEOREM (a) along with the closed ideal structure theorem implies that a necessary condition for $BI\subset C_A^{\infty}$ is that the zeros of B have finite degree of contact with $Z^0(I)$; however, this condition is clearly not sufficient. It appears that the sets $Z^n(I)$, $0<n<\infty$, play a role in determining whether or not $BI\subset C_A^{\infty}$.

Similar remarks apply to problem (2). That is, if the zeros of B have finite degree of contact at $Z^{\infty}(I)$, then $(1/B)I$ is closed; and, if $(1/B)I$ is closed, then the zeros of B have finite degree of contact at $Z^0(I)\cap\partial\mathbb{D}$.

In regard to problem (2), it is not always the case that $(1/B)I$ is closed. In fact, it is possible to construct a closed ideal I where the zeros of B, the g.c.d. of the Blaschke factors of the non-zero functions in I, do not have finite degree of contact at $Z^0(I)\cap\partial\mathbb{D}$ and, hence, $(1/B)I$ is not closed.

We note that if B is a Blaschke product which divides a non-zero C_A^{∞} function, then there is a Carleson set E, $E\subset\partial\mathbb{D}$, such that the zeros of B have finite degree of contact at E . In fact one can take $E = clos\{z/|z| : B(z) = 0\}$; see Theorem 1.2 of [3].

REFERENCES

1. Коренблюм Б.И. О функциях голоморфных в круге и гладких вплоть до его границы.-Докл.АН СССР,1971,200,№ I, 24-27.

2. C a u g h r a n J.G. Zeros of analytic function with infinitely differentiable boundary values. - Proc.Amer.Math.Soc. 1970, 24, 700-704.

3. N e l s o n D. A characterization of zero sets for C_A^∞ . - Mich.Math.J. 1971, 18, 141-147.

4. T a y l o r B.A., W i l l i a m s D.L. Zeros of Lipschitz functions analytic in the unit disc. - Mich.Math.J. 1971, 18, 129-139.

5. T a y l o r B.A., W i l l i a m s D.L. Ideals in rings of analytic functions with smooth boundary values. - Can.J.Math. 1970, 22, 1266-1283.

6. W e l l s J. On the zeros of functions with derivatives in H^1 and H^∞ . - Can.J.Math. 1970, 22, 342-347.

DAVID L.WILLIAMS

Department of Mathematics
Syracuse University,
Syracuse, New York, 13210,
USA

7.16. CLOSED IDEALS IN THE ANALYTIC GEVREY CLASS

Let $\mathbb{D}$ denote the open unit disc in $\mathbb{C}$. The (analytic) Gevrey class of order α is the class of holomorphic functions f in $\mathbb{D}$ such that

$$|f^{(n)}(z)| \leqslant C_f Q_f^n n! n^{n/\alpha}.$$

The class G_α is endowed with a natural topology under which it becomes a topological algebra. So it is natural to ask for the structure of the closed ideals of G_α. For $\alpha \geqslant 1$ the class G_α is quasianalytic and so this question is trivial in this case.

For $0<\alpha<1$, this characterization should be along the lines of previous works on this topic ([2] and [3]) concerning the classes A^k, $k=0,1,\dots,\infty$. Namely, for a closed ideal I one considers for $k=0,1,\dots,\infty$

$$Z^k(I)=\{z\in\mathbb{D}: f^{(j)}(z)=0 \text{ for all } f \text{ in } I,\ j=0,\dots,k\}$$

and

S = greatest common divisor of the inner parts of functions in I

The precise QUESTION is then stated as follows: <u>is every closed ideal I characterized by the sequence $Z^k(I)$ and S_I in the sense</u>

$$I=\{f\in G_\alpha : S_I \text{ divides } f \text{ and } Z^k(f)\subset Z^k(I)\}\ ?$$

Hruščëv's paper [1] is basic in this context for it contains the characterization of the sets of uniqueness for G_α, i.e. the sets E which can appear as $Z^\infty(I)$. The imitation of the proof of [2] or [3] for the A^k-case just gives the result in special cases which include the restriction $\alpha < 1/2$.

REFERENCES

1. H r u š č ë v S.V. Sets of uniqueness for the Gevrey classes. - Ark.för Mat., 1977, 15, 235-304.

2. К о р е н б л ю м Б.И. Замкнутые идеалы кольца A^n. – Функц. анал. и его прил., 1972, 6, № 3, 38–52.
3. T a y l o r B.A. and W i l l i a m s D.L. Ideals in rings of analytic functions with smooth boundary values, Canad.J.Math., 1970, 22, 1266–1283.

J.BRUNA

Universitat autònoma
de Barcelona
Secció matemàtiques
Bellaterra (Barcelona)
España

.17. COMPLETENESS OF TRANSLATES OF A GIVEN FUNCTION IN A WEIGHTED SPACE

old

Let $L^1_\varphi(\mathbb{R})$ be the Banach space of measurable functions f with $f\varphi \in L^1(\mathbb{R})$, the norm in L^1_φ being defined by: $\|f\| = \int_{\mathbb{R}} |f|\,\varphi$. Here the weight φ is a measurable function satisfying

$$1 \leqslant \varphi(t+\tau) \leqslant \varphi(t)\,\varphi(\tau), \quad \forall t, \tau \in \mathbb{R}. \tag{1}$$

The multiplication being defined by

$$(f_1 * f_2)(t) \stackrel{def}{=\!=} \int_{\mathbb{R}} f_1(t-\tau)\, f_2(\tau)\, d\tau$$

the space becomes a Banach algebra without unit; that follows from (1).

Condition (1) ensures the existence of finite limits $\lim\limits_{t\to\pm\infty} \dfrac{\log \varphi(t)}{t} = \alpha_\pm$ and so for f in L^1_φ the Fourier transform $\mathcal{F}f = \frac{1}{2\pi}\int f(t)\, e^{itz}\, dt$ turns out to be continuous in the strip $\Pi = \{z : \alpha_- \leqslant \operatorname{Im} z \leqslant \alpha_+\}$ and analytic in $int\,\Pi$. The maximal ideal space for L^1_φ is homeomorphic to Π ([1]), and any maximal ideal $M(z_0)$ $(z_0 \in \Pi)$ has the form:

$$M(z_0) = \{ f \in L^1_\varphi : \mathcal{F}f(z_0) = 0 \}.$$

Let us consider together with L^1_φ its closed subalgebra

$$L^1_\varphi(\mathbb{R}_+) = \{ f \in L^1_\varphi(\mathbb{R}) : f(t) = 0, \quad \forall t \in \mathbb{R}_- \}.$$

The maximal ideal space of $L^1_\varphi(\mathbb{R}_+)$ is homeomorphic to the half-plane $\operatorname{Im} z \leqslant \alpha_+$, and the Fourier transform of the function f, $f \in L^1_\varphi(\mathbb{R}_+)$ turns out to be continuous in this half-plane and analytic in its interior.

THE PROBLEM we treat here is the following: <u>let $\mathfrak{M}$ be a family of functions in $L^1_\varphi(\mathbb{R})$</u> ($\mathfrak{M}$ <u>may consist of a single element</u>) <u>and let $I_{\mathfrak{M}}$ be the space spanned by all translates of functions from $\mathfrak{M}$. What are the conditions on $\mathfrak{M}$ for $I_{\mathfrak{M}}$ to coincide with $L^1_\varphi(\mathbb{R})$, i.e. when every function in $L^1_\varphi(\mathbb{R})$ can be approximated in $L^1_\varphi(\mathbb{R})$</u>

by linear combinations of translates of functions in $\mathfrak{M}$?

The problem for $\varphi = 1$ ($L^1_\varphi(\mathbb{R}) = L^1(\mathbb{R})$) was stated by N.Wiener [2], he proved that $I_{\mathfrak{M}} = L^1(\mathbb{R})$ if and only if the Fourier transforms of $\mathfrak{M}$ have no common zero on $\mathbb{R}$. Since $I_{\mathfrak{M}}$ is the smallest closed ideal of L^1 containing $\mathfrak{M}$, Wiener's theorem means that a closed ideal of $L^1(\mathbb{R})$ is contained in no maximal ideal if and only if it is equal to $L^1(\mathbb{R})$.

A.Beurling [3] discovered the validity of Wiener's theorem for the space $L^1_\varphi(\mathbb{R})$ if the weight φ satisfies condition (1) and

$$\int_{\mathbb{R}} \frac{\log \varphi(t)}{1+t^2}\, dt < \infty \tag{2}$$

Later it turned out ([1]) that simple "Banach algebra" arguments prove both Wiener's and Beurling's theorem and work in a more general case of any regular Banach algebra. The regularity of L^1_φ is ensured by (2).

The Wiener-type theorem for non-regular Banach algebras were obtained by B.Nyman [4], who proved the following theorem for the case $\varphi(t) = e^{\alpha|t|}$, $\alpha > 0$: $I_{\mathfrak{M}} = L^1_\varphi(\mathbb{R})$ iff Fourier transforms of $\mathfrak{M}$ have no common zero in the strip $|\operatorname{Im} z| \leq \alpha$ and

$$0 = \sup_{f \in \mathfrak{M}} \overline{\lim_{x \to \infty}} \frac{\log|\mathcal{F}f(x)|}{\exp \frac{\pi x}{\alpha}} = \sup_{f \in \mathfrak{M}} \overline{\lim_{x \to -\infty}} \frac{\log|\mathcal{F}f(x)|}{\exp \frac{\pi x}{\alpha}} . \tag{3}$$

Scrutinizing conditions (3) we can see that in the algebra $L^1_\varphi(\mathbb{R})$ there are closed proper ideals contained in no maximal ideal. Such ideals will be called *prime ideals corresponding to infinity points of the strip* $|\operatorname{Im} z| \leq \alpha$.

Independently Nyman's result was rediscovered by B.Korenblum, who described completely the prime ideals corresponding to infinity points of the strip $|\operatorname{Im} z| \leq \alpha$. The question concerning Wiener-type theorems in algebras $L^1_\varphi(\mathbb{R})$, where the weight φ satisfies (1) and

$$\int_{\mathbb{R}} \frac{\log \varphi(t)}{1+t^2}\, dt = \infty , \tag{4}$$

$\alpha_+ = \alpha_- = 0$, is still open.

The methods used earlier don't work in this case. The author does not know a necessary and sufficient condition for the validity of approximation theorem even under very restrictive conditions of regularity of the weight φ (for example $\varphi(t) = exp(\frac{|t|}{\log(1+|t|)})$).

For a weight φ , satisfying (1), (4). one can find chains of prime ideals corresponding to infinity points (see, for example, [6]). The reason for the existence of prime ideals of this sort is that by (4) there are the functions in $L^1_\varphi(\mathbb{R})$ with the greatest possible rate of decrease of Fourier transform (see [7], [8]). It remains unknown whether all prime ideals are of this sort.

There is a similar question for the algebra $L^1_\varphi(\mathbb{R}_+)$. We have the following theorem [9]: Let φ satisfy (1), (2), and let $I_{\mathfrak{M}}$ be the closure of the linear span of all right translates of $\mathfrak{M}$.

Then $I_{\mathfrak{M}} = L^1_\varphi(\mathbb{R})$ if and only if the following two conditions are fulfilled:

1) there is no interval I adjacent to the origin and such that all functions in $\mathfrak{M}$ vanish a.e. on I ,

2) Fourier transforms of functions in $\mathfrak{M}$ have no common zero in $\operatorname{Im} z \leq 0$.

It is worthwhile to note, that this case is simpler than the case of $L^1_\varphi(\mathbb{R})$ because there are no prime ideals of the above type. (The case when $\varphi = e^{\alpha t}$ doesn't differ from the case when $\varphi = 1$). We conjecture that the previous assertion remains true also for the case when φ satisfies condition (4). In particular the following conjecture can be stated:

CONJECTURE. <u>Let</u> φ <u>satisfy (1)</u>, $\varphi \in L^1_\varphi(\mathbb{R}_+)$ <u>and</u> $g \in L^\infty_\varphi(\mathbb{R}_+)$. <u>If the Fourier transform</u> $\mathcal{F}f$ <u>doesn't vanish in</u> $\operatorname{Im} z \leq 0$ <u>and if</u> $\varlimsup_{y \to -\infty} \frac{\log|\mathcal{F}f(iy)|}{|y|} = 0$, <u>then the equality</u>

$$\int_{\mathbb{R}_-} f(t)\, g(t+\tau)\, dt = 0 \quad \forall \tau > 0$$

<u>implies</u> $g(t) = 0$ <u>a.e. on</u> $\mathbb{R}_+$.

There are some reasons to believe the conjecture is plausible. We have no possibility to describe them in detail here, but we can note that condition (2) is much stronger than the condition

$$\int_{\mathbb{R}_+} \frac{\log \varphi(t)}{1+t^{3/2}}\, dt < \infty$$

which is a well-known condition of "non-quasianalyticly" of $L^1_\varphi(\mathbb{R}_+)$. A more detailed motivation of the above problem and a list of related problems of harmonic analysis can be found in [10].

In conclusion we should like to call attention to a question on the density of right translates in $L^1(\mathbb{R})$. Let $f \in L^1(\mathbb{R})$, let I_f be the closure in $L^1(\mathbb{R})$ of linear span of all right translates of f. It is easy to prove that $I_f = L^1(\mathbb{R})$ implies that

$$\int_{\mathbb{R}} \frac{\log|\mathcal{F}f(x)|}{1+x^2}\,dx = -\infty$$

and that $\mathcal{F}f$ is nowhere zero. However these conditions are not sufficient. There are some sufficient conditions but unfortunately they are far from being necessary. I think that it deserves attention to find necessary and sufficient conditions.

REFERENCES

1. Гельфанд И.М., Райков Д.А., Шилов Г.Е. Коммутативные нормированные кольца, М., Ф.-М., 1960.
2. Винер Н. Интеграл Фурье и некоторые его приложения, М., Ф.-М., 1963.
3. Beurling A. Sur les integrales de Fourier absolument convergentes et leur application a une transformation fonctionelle. Congres des Math.Scand., Helsingfors, 1938.
4. Nyman B. On the one-dimentional translations group and semigroup in certain function spaces. Thesis, Uppsala, 1950.
5. Коренблюм Б.И. Обобщение тауберовой теоремы Винера и гармонический анализ быстрорастущих функций. - Труды Моск.матем. об-ва, 1958, 7, 121-148.
6. Vretblad A. Spectral analysis in weighted L^1-spaces on $\mathbb{R}$. - Ark.Math., 1973, 11, 109-138.
7. Джрбашян М.М. Теоремы единственности для преобразований Фурье и для бесконечно дифференцируемых функций. - Изв.АН Арм.ССР, сер.ф.-м., 1957, 10, № 6, 7-24.
8. Бабенко К.И. О некоторых классах пространств бесконечно дифференцируемых функций. - Докл.АН СССР, 1960, 132, № 6, 1231--1234.
9. Гурарий В.П., Левин Б.Я. О полноте системы сдвижек в пространстве $L(0,\infty)$ с весом. - Зап.мех.-мат.ф-та ХГУ и ХМО, 1964, 30, сер.4, 178-185.

0. Г у р а р и й В.П. Гармонический анализ в пространствах с весом. - Труды Моск.матем.об-ва, 1976, 35, 21-76.

V.P.GURARII
(В.П.ГУРАРИЙ)

СССР, 142432, Черноголовка,
Московская область, Филиал
ин-та Химической Физики АН СССР

7.18. TWO PROBLEMS OF HARMONIC ANALYSIS IN WEIGHTED SPACES

We consider the space $L^\infty_\varphi(\mathbb{R})$ of measurable functions on $\mathbb{R}$ with the norm $\|g\| = \operatorname{ess\,sup} |g(t)|/\varphi(t)$. The weight φ is supposed to be measurable and to satisfy the conditions

(a) $1 \leqslant \varphi(t+\tau) \leqslant \varphi(t)\varphi(\tau), \quad \forall t,\tau \in \mathbb{R}$;

(b) $\lim\limits_{t\to\pm\infty} \dfrac{\log\varphi(t)}{t} = 0.$

Assign to each function $g \in L^\infty_\varphi(\mathbb{R})$ the smallest w^*-closed subspace of $L^\infty_\varphi(\mathbb{R})$ (denoted by B_g) invariant under all translations and containing g. The set

$$\Lambda_g = \{x \in \mathbb{R} : \exp(-itx) \in B_g\}$$

is called the spectrum of g.

Denote

$$L^\infty_\varphi(\mathbb{R}_+) = \{g \in L^\infty_\varphi(\mathbb{R}) : g(t) = 0 \quad \text{for } t<0\}.$$

For each g_+ in $L^\infty_\varphi(\mathbb{R}_+)$ a spectrum $\Lambda^+_{g_+}$ is defined as follows:

$$\Lambda^+_{g_+} = \{z : \operatorname{Im} z \leqslant 0,\ \exp(-itz) \in B^+_{g_+}\}.$$

Here $B^+_{g_+}$ is the smallest w^*-closed subspace of $L^\infty_\varphi(\mathbb{R}_+)$ invariant under translations to the left and containing g_+.

The spectrum Λ_g (resp. $\Lambda^+_{g_+}$) is a closed subset of the real line (resp. of the lower halfplane).

The spectrum $\Lambda^+_{g_+}$ (or Λ_g) is said to be "simple" if the only functions in $B^+_{g_+}$ (resp. B_g) that have one-point spectrum, are exponentials times constants.

PROBLEM 1. Describe the subsets Λ of $\mathbb{R}$ with the following property: every function $g_+ \in L^\infty_\varphi(\mathbb{R}_+)$ with a "simple" spectrum $\Lambda^+_{g_+} \subset \Lambda$ admits an extension g to the whole of $\mathbb{R}$ so that $g \in L^\infty_\varphi(\mathbb{R})$ and

$$\Lambda_g = \Lambda^+_{g+} . \tag{1}$$

If $\varphi(t) \equiv 1$ the set $\mathbb{Z}$ of all integers is an example of such set. Indeed, if $g_+ \in L^\infty(\mathbb{R}_+)$ and $\Lambda^+_{g_+} \subset \mathbb{Z}$ then the theorem on spectral synthesis in $L^\infty(\mathbb{R}_+)$ proved in [1] implies that g_+ lies in the w^*-closure of the trigonometric polynomials with frequencies in $\mathbb{Z}$. Thus g_+ admits a 2π-periodic extension g to the whole of $\mathbb{R}$, and clearly (1) holds for this g. There also exist more refined examples.

When treating the spectral synthesis in $L^\infty_\varphi(\mathbb{R}_+)$, the following problem might be useful.

PROBLEM 2. Let $f \in L^1_\varphi(\mathbb{R}_+)$, $g \in L^\infty_\varphi(\mathbb{R}_+)$ and suppose $t^k \in L^\infty_\varphi(\mathbb{R}_+)$ for $0 \leqslant k \leqslant \omega$ (ω is a positive integer or the symbol ∞). Suppose also that

$$\int_0^\infty t^k f(t)\,dt = 0, \quad 0 \leqslant k < \omega \tag{2}$$

and

$$\int_0^\infty f(t)\, g(t+\tau) = C, \qquad \forall \tau > 0 \tag{3}$$

where C is a constant. Describe weights φ such that (2) and (3) imply $C = 0$.

For the weight φ, $\varphi(t) \equiv 1$ this implication has been proved in [1]. If $\varphi(t) = 1 + |t|$, a proof has been proposed by E.L.Suris.

Some considerations concerning Problems 1 and 2 are implicit in [2].

REFERENCES

1. Г у р а р и й В.П. Спектральный синтез ограниченных функций на полуоси. – Функц.анал. и его прил., 1969, 3, вып.4, 34-48.
2. Г у р а р и й В.П. Гармонический анализ в пространствах с весом. – Труды Моск.матем.об-ва, 1976, 35, 21-76.

V.P.GURARII
(В.П.ГУРАРИЙ)

СССР, 142432, Черноголовка,
Московская обл., Отделение
ин-та химической физики АН СССР

.19. A CLOSURE PROBLEM FOR FUNCTIONS ON $\mathbb{R}_+$.
old

A w e i g h t f u n c t i o n w is here a positive, ounded, decreasing function on $\mathbb{R}_+$, satisfying $x^{-1}\log w(x) \to -\infty$, s $x \to \infty$. L_w is the Banach space of functions f on $\mathbb{R}_+$ with $w \in L^1(\mathbb{R}_+)$. For every a, $a \in \mathbb{R}_+ \cup \{0\}$ the translation T_a, de-ined by

$$T_a f(x) = \begin{cases} 0 & , \ 0 < x \leqslant a \\ f(x-a), & x > a \end{cases},$$

s a contraction in L_w. A_w is the set of all f, $f \in L_w$, which do not vanish almost everywhere near 0, and B_w is the set of c y c l i c elements in L_w, i.e. elements f such that the translates $T_a f$, $a > 0$, span a dense subspace. Obviously $B_w \subset A_w$.

<u>Is</u> $$B_w = A_w \ ?$$

Some light is thrown on this problem by the corresponding problem on $\mathbb{Z}_+ \cup \{0\}$. A w e i g h t s e q u e n c e is a positive decreasing sequence $w = (w_n)_{n \geqslant 0}$, satisfying $n^{-1}\log w_n \to -\infty$ as $n \to \infty$. ℓ_w is the Banach space of sequences $c = (c_n)_{n \geqslant 0}$ with $cw = (c_n w_n)_{n \geqslant 0} \in \ell^1(\mathbb{Z}_+ \cup \{0\})$ and the translations T_m, are defined as above, giving contractions of ℓ_w. A_w is the set of c, $c \in \ell_w$, with $c_0 \neq 0$, B_w is the set of cyclic elements. $B_w \subset A_w$, and we ask whether $B_w = A_w$. This time results are easier to obtain. Let us say that w is of submultiplicative type if $w_{m+n} \leqslant C w_m w_n$, $m, n \in \mathbb{Z}_+$, for some constant C. In that case, ℓ_w is a unital Banach algebra under convolution, with $\ell_w \setminus A_w$ as its only maximal ideal, and $B_w = A_w$ follows from elementary Banach algebra theory. It should be observed that the submulitplicativity condition is an assumption on the regularity of w, not a restriction of its growth at ∞. If this condition is not fulfilled, there are cases, when $B_w = A_w$ [1], and other cases when $B_w \neq A_w$ [2], [1].

In an analogous way we say that a weight function w is of submultiplicative type, if $w(x-y) \leqslant C\, w(x)\, w(y)$, $x, y \in \mathbb{R}_+$, for some constant C. Using the results of Nikolskii and Styf it is easy to produce weight functions w, of non-submultiplicative type, for which $B_w = A_w$. But if we from now on restrict the attention to weight functions of submulitplicative type, we can in no single case answer the question whether $B_w = A_w$. It is tempting to conjecture, in analogy to the discrete case, that the answer is affirmative for every w. Now again we have a convolution Banach algebra, but the absence of a unit prevents us from carrying over the arguments from the discrete case. A vague indication that the answer perhaps is yes, at least if w tends to zero rapidly at infinity, is given by the circumstance that $B_w = A_w$ if the corresponding problem is formulated in the limiting case when w is non-negative and vanishing for large x. (This follows from Titchmarsh's theorem).

It is a direct consequence of Hahn-Banach's theorem that $f \in B_w$ if and only if the convolution equation

$$\int_0^\infty \varphi(x+y) f(y)\, dy = 0, \quad x \in \mathbb{R}_+$$

has the zero functions as only solution with $\varphi / w \in L^\infty(\mathbb{R}_+)$. Thus $B_w \neq A_w$ if and only if there exists a $f \in A_w$ such that the equation has a non-zero solution. Maybe function theory, in particular the theory of special functions, can provide an example showing that $B_w = A_w$ for at least some w.

Here are some s u f f i c i e n t c o n d i t i o n s for $f \in B_w$

1. $f \in B_w$ if $f \in A_w$ and $\int_0^\infty |f(x)| e^{bx} dx < \infty$ for some $b, b \in \mathbb{R}$.

(This follows directly from the results in [3] or [4], and is valid also for w of non-submultiplicative type.)

2. Suppose $-\log w$ is convex and $x^{-2} \log w \to -\infty$, as $x \to \infty$. Let $f \in L_w$ and suppose that for some f_1, $f_1 \in L^1(\mathbb{R}_+)$, with compact support and coinciding with f near 0,

$$\left| \int_0^\infty f_1(x) e^{-yx} dx \right| \geq \exp\left\{ -Cy \left[w^{-1}\left(\left[\int_0^\infty w(x) e^{yx} dx \right]^{-1} \right) \right]^{-1} \right\}$$

for large y, $y \in \mathbb{R}_+$, where C is a constant and w^{-1} denotes the inverse of w. Then $f \in B_w$. (In particular, $\log w(x) \sim -x^p$, $p > 2$, yields the right hand member $\sim \exp\{-Cy^\alpha\}$, for some C, where

$\alpha = (p-2)/(p-1)$

3. Suppose $-\log w$ is convex and $(x \log x)^{-1} \log w \to -\infty$, as $x \to \infty$. Let $f \in L_w$ and suppose that f is of bounded variation near zero with $f(+0) \neq 0$. Then $f \in B_w$.

REFERENCES

1. S t y f B. Closed translation invariant subspaces in a Banach space of sequences, summable with weights Uppsala University, Dept. of Math., Report 1977:3
2. Н и к о л ь с к и й Н.К. Об инвариантных подпространствах взвешенных операторов сдвига. - Матем.сб., 1967, 74, № 2, 171-190.
3. N y m a n B. On the one-dimensional translation group and semi-group in certain function spaces Uppsala, 1950
4. Г у р а р и й В.П. Спектральный синтез ограниченных функций на полуоси. - Функц.анал. и его прил., 1969, 3, № 4, 34-48.

YNGVE DOMAR

Uppsala Universitet
Matematiska Institutionen
Sysslomansgatan 8
75223 Uppsala, Sweden

* * *

COMMENTARY

The proofs of Propositions 1-3 can be found in [5]. In an important paper [6] it is shown that $A_w = B_w$ if $\log w$ is eventually concave and $(\log x)^{1/2} = o(\log|x^{-1}\log w(x)|)$ for $x \to \infty$. An analogous theorem is proved under the same hypotheses for the spaces L^p_w, $1 \leq p < \infty$. These results are derived from a general theorem on convolution equations in L^p_w-spaces which is a strong form of the famous Titchmarsh theorem.

Concerning all these and many other problems on translation invariant subspaces and ideals in $L^p_w(\mathbb{R}_+)$, $L^p_w(\mathbb{R})$ see also [7], a very informative book.

REFERENCES

5. D o m a r Y. Cyclic elements under translation in weighted L^1 spaces on $\mathbb{R}^+$. - Ark.mat. 1981, 19, N 1, 137-144.

6. D o m a r Y. Extensions of the Titchmarsh convolution theorem with applications in the theory of invariant subspaces - Proc London Math.Soc.(3), 1983, 46, 288-300.

7. Radical Banach Algebras and Automatic Continuity, Proceedings, Long Beach 1981, Ed.by J.M.Bachar, W.G.Bade, P.C.Curtic Jr., H.G.Dales, and M.P.Thomas, - Lect.Notes.in Math., 1983, 975,

7.20. old TRANSLATES OF FUNCTIONS OF TWO VARIABLES

In [1] and [2] the following theorem is proved: if $f \in L^1(\mathbb{R}_+)$ and $f(x)=0$ for $x<0$, then the system of functions $\{f(x-h): h \in \mathbb{R}_+\}$ is dense in $L^1(\mathbb{R}_+)$ if and only if the following conditions are fulfilled:

1. The function

$$\mathcal{F}(z) = \int_0^\infty f(x)\, e^{ixz}\, dx$$

doesn't vanish in $\operatorname{Im} z \geqslant 0$.

2. There is no $\delta > 0$ such that $f(x)=0$ a.e. on $(0, \delta)$.

PROBLEM: Let $f \in L^1(P_1)$, $P_1 = \{(x,y) \in \mathbb{R}^2 : x>0, y>0\}$ and $f \equiv 0$ in $\mathbb{R}^2 \setminus P_1$. Find necessary and sufficient conditions for the system $\{(f(x-h, y-k) : h \geqslant 0, k \geqslant 0\}$ to be dense in $L^1(P_1)$.

REFERENCES

1. N y m a n B. On the one-dimensional translation group and semigroup in certain function spaces. Uppsala, 1950.

2. Г у р а р и й В.П., Л е в и н Б.Я. О полноте системы сдвижек в пространстве $\mathcal{L}(0,\infty)$ с весом. - Зап.Харьк.матем.о-ва, I960, 30, сер.4.

B.Ya.LEVIN
(Б.Я.ЛЕВИН)

СССР, 3I0I64, Харьков
пр.Ленина 47
Физико-технический институт
низких температур АН УССР

7.21. ALGEBRA AND IDEAL GENERATION IN CERTAIN RADICAL BANACH ALGEBRAS

Let $\mathbb{C}[[z]]$ denote the algebra of formal power series over $\mathbb{C}$. We say that a sequence of positive reals $\{w(n)\}$ is a *radical algebra weight* provided the following hold:

(1) $w(0)=1$ and $0<w(n)\leq 1$ for all $n\in\mathbb{Z}_+$.

(2) $w(m+n)\leq w(m)w(n)$, for all $m,n\in\mathbb{Z}_+$.

(3) $\lim\limits_{n\to\infty} w(n)^{1/n}=0$.

If these conditions hold it is routine to check that

$$\ell^1(w(n))=\Big\{y=\sum_{n=0}^{\infty}y(n)z^n:\sum_{n=0}^{\infty}|y(n)|w(n)<\infty\Big\}$$

is both a subalgebra of $\mathbb{C}[[z]]$ and a radical Banach algebra with identity adjoined. The norm is defined in the natural way: $\|y\|=\sum_{n=0}^{\infty}|y(n)|w(n)$. The multiplication is given by the usual convolution of formal power series. We shall generally refer to $\ell^1(w(n))$ as a radical Banach algebra and $\{w(n)\}$ as simply a "weight". Let $A=\ell^1(w(n))$ in all the following. Besides A itself, there are obvious proper closed ideals in A:

$$M(n)=\Big\{\sum_{k=0}^{\infty}y(k)z^k\in A: y(0)=y(1)=\ldots=y(n-1)=0\Big\}$$

for $n=1,2,\ldots$, and, of course, the zero ideal. Such closed ideals are referred to as *standard ideals*. Any other closed ideals are denoted *non-standard ideals*. Note that the unique maximal ideal in A is $M(1)=M$.

We first discuss the problem of polynomial generation. Let $x=\sum_{n=1}^{\infty}x(n)z^n$ be an element of A with $x(1)\neq 0$. One says that x *generates a non-standard closed subalgebra* if the smallest closed subalgebra containing x is properly contained in M. Since this algebra is the losed linear span of polynomials in x , we could equivalently ay

4) $\overline{\text{span}}\{x^{k+1}\}_{k=0}^{\infty}$ is properly contained in $M(1)$.

The requirement that $x(1)\neq 0$ is necessary, otherwise (4) is vacuous. If the weight is very well behaved there are positive results which show that non-standard closed subalgebras are not present [3]. On the other hand, it was shown [5, Theorem 3.11] that, for certain star-shaped weights, non-standard closed subalgebras exist (A weight W is star-shaped if $\{W(n)^{1/n}\}$ is non-increasing). Hence one problem is the following.

PROBLEM 1. <u>Characterize the radical algebra weights</u> W <u>such that</u> $\ell^1(W(n))$ <u>has non-standard closed subalgebras</u>.

We next consider the problem of ideal generation. Whether $\ell^1(W(n))$ has only standard closed ideals or not is the problem whether each non-zero element x generates a standard closed ideal or not. If we let T be the operator of right translation on A , we could equivalently say [1, Lemma 4.5] *)

(5) $\overline{\text{span}}\{T^k x\}_{k=0}^{\infty}$ contains a power of z,

for each non-zero x in A . If $\ln W$ is a concave function it is well known [1, Theorem 4.1]**) that all closed ideals are standard. More generally, it can be shown [4,Corollary 3.6] that if W is star-shaped and $W(n)^{1/n}$ is $O\left(\frac{1}{n^a}\right)$ for some $a>0$ then all closed ideals are standard. This is in contrast to the fact that star-shaped weights can support non-standard closed subalgebras. Apparently Šilov first posed the problem whether or not there exists any radical algebra weight W such that $\ell^1(W(n))$ contains a non-standard ideal. The answer is affirmative [6, Theorem] for certain semi-multipli-

---

*) This is also a part of Lemma 1 in Никольский Н.К.,Известия АН СССР, серия матем., 1968, 32, 1123-1137. - Ed.

**) This is also a part of Theorem 2 in Никольский Н.К., Вестник ЛГУ серия матем.мех. и астрон., 1965, № 7, 68-77. - Ed.

cative weights [5, Definition 2.1]. These are weights where $w(m+n)$ actually equals $w(m)w(n)$ for many values of m, n in $\mathbb{Z}_+$. Hence we propose the following

PROBLEM 2. Characterize the radical algebra weights w such that $\ell^1(w(n))$ has non-standard ideals.

Even substantial necessary conditions on the weight w for the existence of a non-standard ideal would be welcome.

Finally we remark that one can consider related radical algebras $\ell^1(Q_+, w)$ built upon Q_+ rather than $\mathbb{Z}_+$ (we again require (1)-(3) for m, n in Q_+). Define for x non-zero in $\ell^1(Q_+, w)$

$$a(x) = \inf\{r : x(r) \neq 0,\ r \in Q_+\}.$$

Also define

$$M = \{y \in \ell^1(Q_+, w) : y(0) = 0\}.$$

We pose the final problem.

PROBLEM 3. Does there exist some $\ell^1(Q_+, w)$ containing an element x, $a(x) = 0$, such that the closed ideal generated by x is properly contained in M?

Preliminary results on this problem can be found in [2].

REFERENCES

1. Grabiner S. Weighted shifts and Banach algebras of power series. - American J.Math., 1975, 97, 16-42.
2. Gronbaek N. Weighted discrete convolution algebras. "Radical Banach Algebras and Automatic Continuity", Proceedings, Long Beach, 1981, Lect.Notes Math., N 975.
3. Söderberg D. Generators in radical weighted ℓ^1, Uppsala University Department of Mathematics Report 1981:9.
4. Thomas M.P. Approximation in the radical algebra $\ell^1(w_n)$ when $\{w_n\}$ is star-shaped. "Radical Banach Algebras and Automatic Continuity", Proceedings, Long Beach, Lect.Notes in Math., N 975.

- T h o m a s M.P. A non-standard closed subalgebra of a radical Banach algebra of power series. - J.London Math.Soc., to appear.
- T h o m a s M.P. A non-standard closed ideal of a radical Banach algebra of power series, submitted to Bull.Amer.Math.Soc.

MARC THOMAS

Mathematics Department
California State College
at Bakersfield
9001 Stockdale Hwy.
Bakersfield, CA 93309
USA

7.22.
old

HARMONIC SYNTHESIS AND COMPOSITIONS

Let $\mathcal{F}\ell^1$ be the algebra of all absolutely convergent Fourier series on the circle $\mathbb{T}$:

$$f(\xi)=\sum_{n\in\mathbb{Z}}\hat{f}(n)\xi^n,\quad \xi\in\mathbb{T}\ ;\ \|f\|_1\stackrel{def}{=\!=}\sum|\hat{f}(n)|<\infty\ .$$

Let $f\in\mathcal{F}\ell^1$ and suppose $f(\mathbb{T})=[-1,1]$.

We say f admits t h e h a r m o n i c s y n t h e s i s (=h.s.) if there is a sequence $\{\varphi_n\}\subset\mathcal{F}\ell^1$ such that $\|\varphi_n-f\|_1\longrightarrow 0$ and $f^{-1}(0)\subset \operatorname{Int}\varphi_n^{-1}(0)$, $n=1,2,\ldots$. The algebra $\mathcal{F}\ell^1$ contains functions not admitting h.s. though every sufficiently smooth function admits h.s.

QUESTION 1. <u>Let</u> f <u>admit h.s. Is it possible to choose functions</u> φ_n <u>in the above definition so that</u> $\varphi_n=F_n\circ f$, F_n <u>being some functions on</u> $[-1,1]$?

Denote by $[f]$ the set of all functions F on $[-1,1]$ such that $F\circ f\in\mathcal{F}\ell^1$. This set is a Banach algebra with the norm $\|F\|_{[f]}=$ $=\|F\circ f\|_1$. It contains the identity function $F(x)\equiv x$. Now we can reformulate our question.

QUESTION 2. <u>Let</u> f <u>admit h.s. Is it possible to approximate</u> $F(x)\equiv x$ <u>in the algebra</u> $[f]$ <u>by functions vanishing near the point</u> $x=0$?

If $[f]\subset C^1[-1,1]$ (this embedding has to be continuous by the Banach theorem), then the functional $\delta':F\longrightarrow F'(0)$ separates X from functions in $[f]$ vanishing at a vicinity of zero. So the following question is a particular case of our problem.

QUESTION 3. <u>Let</u> f <u>admit h.s. Is it possible that</u> $[f]\subset C^1[-1,1]$?

Obviously $\|e^{2\pi i n x}\|_{C^1[-1,1]}\asymp n$, $n\longrightarrow\pm\infty$, and so we have a further specialization.

QUESTION 4. <u>Is it possible to construct a function</u> $f\in\mathcal{F}\ell^1$ <u>admitting h.s. with</u>

$$\lim_{n} \frac{1}{n} \| e^{inf} \|_1 > 0 \ ?$$

Now (1983) very little is known about the structure of the ring $[f]$. The theorems of Wiener-Levy type ([1] Ch.VI [2]) give some sufficient conditions for the inclusion $F \subset [f]$, but these conditions are much stronger than C^1-smoothness of F. On the other hand, let the function $t \longrightarrow f(e^{it})$ be even on $[-\pi, \pi]$ and strictly monotone on $[0, \pi]$. Thus any even function on $[-\pi, \pi]$ has a form $F \circ f$ and our Question 1 has the affirmative answer. Hence $[f] \not\subset C^1$ and all known theorems of Wiener-Levy type are a priori too rough for this f. Kahane [3] has constructed examples of functions f with $[f] \subset C^\infty[-1,1]$. Thus, the ring $[f]$ is quite mysterious.

A possible way to answer our questions is the following. If $[f] \subset C^1$ then the functional $\delta' : F \longrightarrow F'(0)$ is well-defined on $[f]$ and generates a functional $\delta'(f)$ on the subalgebra $[[f]] = \{ F \circ f : F \in [f] \} \subset \mathcal{F}\ell^1$. If there is an extension of this functional to $\mathcal{F}\ell^1$ with $\langle \delta'(f), g \rangle = 0$ for $f^{-1}(0) \subset \operatorname{Int} g^{-1}(0)$, then f cannot admit h.s. It is in this way that Malliavin's lemma on the absence of h.s. has been proved. Namely,

if (in addition) $\int_0^\infty u \| e^{inf} \|_{(\mathcal{F}\ell^1)^*} du < \infty$ and

$\int_{\mathbb{T}} dm \int_{-\infty}^{\infty} e^{iuf} du = 1$, then Malliavin's functional

$$g \longrightarrow \int_{\mathbb{T}} dm \int_{-\infty}^{\infty} iue^{iuf} g \, du, \quad g \in \mathcal{F}\ell^1$$

gives the desired extension of $\delta'(f)$.

The author thanks professor Y.Domar for a helpful discussion in 1978 in Leningrad.

REFERENCES

1. K a h a n e J.-P. Séries de Fourier absolument convergentes. Springer, Berlin, 1970.

2. Д ы н ь к и н Е.М. Теоремы типа Винера-Леви и оценки для операторов Винера-Хопфа. - Математические исследования, 1973, 8, № 3, 14-25.

3. K a h a n e J.-P. Une nouvelle réciproque du théorème de Wiener - Lévy. - C.R.Acad.Sci.Paris, 1967, 264, 104-106.

E.M.DYN'KIN

(Е.М.ДЫНЬКИН)

СССР, 197022, Ленинград
ул.проф.Попова, 5
Ленинградский Электротехнический институт им.В.И.Ульянова (Ленина)

7.23. DEUX PROBLÈMES CONCERNANT LES SÉRIES TRIGONOMÉTRIQUES

old

1. Soit $\sum_{n\in\mathbb{Z}} a_n e^{int}$ une série trigonométrique dont les coefficients tendent vers 0 et dont les sommes partielles tendent vers 0 sur un ensemble fermé $F\subset\mathbb{T}$:

$$\lim_{N\to\infty} \sum_{n=-N}^{N} a_n e^{int} = 0 \qquad \text{quand } t\in F,$$

$$\lim_{|n|\to\infty} a_n = 0 .$$

Soit $\mu\in M^+(F)$ une mesure positive portée par F,

$$\mu \sim \sum_{n\in\mathbb{Z}} \hat{\mu}(n) e^{int}$$

telle que

$$\sum_{n\in\mathbb{Z}} |\hat{\mu}(n)||a_{-n}| < +\infty .$$

<u>A-t-on nécessairement</u>

$$\sum_{n\in\mathbb{Z}} \hat{\mu}(n) a_{-n} = 0 \ ?$$

Une réponse positive (dont je doute) donnerait une nouvelle preuve de l'existence d'ensembles $U(\varepsilon)$ de Zygmund de mesure pleine.

2. <u>Soit</u> $f\in L^p(\mathbb{T})$, $f\sim\sum_{n\in\mathbb{Z}} \hat{f}(n) e^{int}$. <u>Peut-on approcher</u> f <u>dans</u> $L^p(\mathbb{T})$ <u>par des polynômes trigonométriques</u> $P=\sum_{finie} \hat{P}(n) e^{int}$ <u>tels que</u> $\hat{f}(m)=\hat{f}(n) \Longrightarrow \hat{P}(m)=\hat{P}(n)$?

La question a été posée par W.Rudin [1] pour $p=1$ (la réponse est alors négative [2]). Pour $p=2$, la réponse positive est évidente. Pour $p=\infty$, la question n'a d'intérêt que si on suppose f continue (la réponse est négative). La question est ouverte pour $1<p<2$ et $2<p<\infty$.

BIBLIOGRAPHIE

1. R u d i n W. Fourier analysis on groups. N.Y., Interscience, 1962.

2. K a h a n e J.-P. Idempotents and closed subalgebras of $L^1(\mathbb{T})$ - In: Funct. algebras, 198-207, ed. T. Birtel, Proc. Intern. Symp. Tulane Univ., 1965, Chicago, Scott-Forestmann, 1966.

J.-P. KAHANE

Université de Paris-Sud,
Mathématique, Bâtiment 425, Centre
d'Orsay 91405, Orsay Cedex, France

* * *

COMMENTAIRE

La réponse au second problème est négative (voir [3] pour $1<p<2$ et [4,5] pour $2<p<\infty$).

BIBLIOGRAPHIE

3. R i d e r D. Closed subalgebras of $L^1(\mathbb{T})$. - Duke Math.J., 1969, 36, N 1, 105-115.
4. O b e r l i n D.M. An approximation problem in $L^p[0,2\pi]$, $2<p<\infty$. - Studia Math., 1981, 70, N 3, 221-224.
5. B a c h e l i s G.F., G i l b e r t J.E. Banach algebras with Rider subalgebras. Preprint, 1979.

CHAPTER 8

APPROXIMATION AND CAPACITIES

M o s t problems of our Collection may be viewed as approximation problems. That is why selection principles in this Chapter are even more vague and conventional than in the others. Problems collected under the above title illustrate, nevertheless, some important tendencies of modern Approximation Theory.

Some Problems below are closely related to the ideas of the preceding Chapter. This is, of course, not a mere coincidence, the approximation being really the core of spectral analysis-synthesis. An attentive reader will not be deceived by the seemingly scattered contents of items 8.1, 8.3, 8.4, 8.8, which can be given a unified interpretation from the (broadly conceived) "spectral" point of view. What really matters is, after all, not w h a t or by w h a t m e a n s we approximate (by rational functions or by exponentials with prescribed frequencies, by weighted polynomials or by δ-measures within a spectral subspace), but the intrinsic sense, the aim, and the motive impelling to the approximation, i.e. singling out elementary harmonics (with respect to an action) and subsequent recovering of the object they are generated by.

The variant of spectral synthesis mentioned in Problem 8.1 is aimed at L^p -approximation by solutions of elliptic differential equations (in particular, by analytic and harmonic functions). The same can be said about Problems 8.8–8.10 Problem 8.9 deals also with some estimates of the derivative of a conformal mapping. Such estimates are useful in connection with "the weak invertibility" (see Chapter 7 again) and especially with the "crescent effect" discovered by M.V.Keldysh.

Problems 8.5–8.7 are interesting variants of the classical uniform approximation (in the spirit of Mergelyan - Vitushkin - Arakelyan).

Padé approximations, an intensively growing branch of rational approximations, is presented in Problems 8.11 and 8.12 (this direction seems to be promising in connection with some operator-theoretic aspects.See Problem 4.9).

The best approximation à la Tchebysheff, the eternal theme of Approximation Theory, emerges in Problem 8.13 (as in Problem 5.1 amid Hankel operators and s-numbers).

Problem 8.14 concerns some ideas arising in Complex Analysis under the influence of the Theory of Banach Algebras.

But all this explains only the first half of the title. As to the second, it is a manifestation of close connections of many modern approximation problems with potential theory. Items 8.1, 8.9, 8.10,8.15 –8.18 make an extensive use of various kinds of capacities though all of them have in mind (or are inspired by) some approximation theoretic problems.

"The capacitary ideology" appears here also in connection with other themes, namely with the solvability of boundary value problems for elliptic equations (see the "old" Problem 8.20, its Commentary being a new problem article), with metric estimates of capacities (8.15–8.19, 8.21) and with removable singularities of analytic functions (8.15–8.19).

Sobolev spaces play an essential role in many approximation problems of this Chapter. In Problem 8.22 they are considered in heir own right.

Five problems (8.15-8.19) dealing with removable singularities of b o u n d e d analytic functions (or, what is the same, with analytic capacity) formed a separate chapter in the first edition. Here we reproduce the translation of some fragments from its preface.

"Analysts became interested in sets of removable singularities of bounded analytic functions in the eighties of the last century, attracted by the very possibility to formulate problems in the new set-theoretic language. This interest being still alive today (as it is witnessed by this Chapter whose five sections have a non-void and even a fairly large intersection), modified its spirit many times during the past century. Now connected with the classification of Riemann surfaces then with extremal problems of Function Theory it was born again in early sixties after Vitushkin's works on rational approximations [...].

The problem of relations between analytic capacity and length was the theme of active debates during the Yerevan Conference on Analytic Functions (September 1960) when L.D.Ivanov pointed out in his conference talk the role which irregular plane sets (in the Besicovich sense) are likely to play in the theory of removable sungularities of bounded analytic functions. But an essential new progress (namely, the proof of the Denjoy conjecture*) became possible in the last (1977) year only, after the remarkable achievement by A.Calderón, namely, after his proof of the L^p-continuity of the Cauchy singular integral operator on a smooth curve. The whole Chapter is written under the influence of the Calderón theorem. May be, thanks to it the time is near when the geometric nature of singularities of bounded analytic functions will be completely understood."

The historical information concerning analytic capacity is given

*) See Problem 8.15.

in 8.15 and 8.16. We should like to add the article Uryson P.S. Sur une fonction analytique partout continue. - Fund.Math.1922, 4, 144-150 (the Russian translation in the book Урысон П.С. Труды по топологии и другим областям математики, т.I., М.Л.ГИТТЛ, 1951, 3-100).

Capacitary motives can be heard also in other Chapters. The classical logarithmic capacity appears (rather unexpectedly) in the item 1.10 devoted to the isomorphic classification of spaces of analytic functions . The analytic capacity (the main subject of Problems 8.15-8.19 influenced by the recent progress in singular integrals, see Chapter 6) takes part in the purely operator-theoretic item 4.36. The use of capacities in the Operator Theory is not at all a novelty or a surprise. Spectral capacities describing the sets carrying non-trivial spectral subspaces, the exquisite classification of the uniqueness sets for various classes of trigonometrical series (the particular case corresponding to the shift $f \mapsto zf$), metric characteristics of spectra in the classification of operators (transfinite diameters et al.), all these are the everyday tools of Spectral Theory and the corresponding connections are well illustrated, e.g., by Chapter 4.

.1.
ld

SPECTRAL SYNTHESIS IN SOBOLEV SPACES

Let X be a Banach space of functions (function classes) on $\mathbb{R}^d$. We have in mind the Sobolev spaces W_s^p, $1 \leq p < \infty$, $s \in \mathbb{Z}_+$ r the spaces obtained from Sobolev spaces by interpolation (Bessel otential spaces L_s^p, $s > 0$, and Besov spaces $B_s^{p,q}$, $s > 0$). hen the dual space X' is a space of distributions. We say that a losed set K in $\mathbb{R}^d$ admits X-spectral synthe- is if every T in X' that has support in K can be approximated arbitrarily closely in X' by linear combinations of measures nd derivatives of order $< s$ of measures with support in K.

PROBLEM. Do all closed sets admit X-spectral synthesis for he above spaces?

The PROBLEM can also be given a dual formulation. If ν is a easure with support in K such that a partial derivative $D^k \nu$ belongs to X', then one can define $\int D^k f\, d\nu$ for all f in X. Then K admits X-spectral synthesis if every f such that $\int D^k f\, d\nu = 0$ for all such ν and all such multiindices k can oe approximated arbitrarily closely in X by test functions that vanish on some neighborhood of K.

The PROBLEM is of course analogous to the famous spectral synthesis problem of Beurling, but in the case of W_s^2 this terminology was introduced by Fuglede. He also observed that the so called fine Dirichlet problem in a domain D for an elliptic partial differential equation of order $2s$ always has a unique solution if and only if the complement of D admits W_s^2-spectral synthesis. See [1 ; IX, 5.1].

In the case of W_1^p the PROBLEM appeared and was solved in the work of V.P.Havin [2] and T.Bagby [3] in connection with the problem of approximation in $L^{p'}$ by analytic functions. For W_1^2 the solution appears already in the work of Beurling and Deny [4]. In fact, in these spaces all closed sets have the spectral synthesis property. This result, which can be extended to L_s^p, $0 < s < 1$, depends mainly on the fact that these spaces are closed under truncations.

When $s > 1$ this is no longer true, and the PROBLEM is more complicated. Using potential theoretic methods the author [5] has given sufficient conditions for sets to admit spectral synthesis in $W_s^p(\mathbb{R}^d)$, $s \in \mathbb{Z}_+$. These conditions are so weak that they are satisfied for all closed sets if $p > max(d/2, 2-1/d)$, thus in particular if

$p=2$ and $d=2$ or 3. There are also some still unpublished results for L^p_s and $B^{p,p}_s$ showing for example that sets that satisfy a cone condition have the spectral synthesis property.

Otherwise, for general spaces the author is only aware of the work of H.Triebel [6],where he proved, extending earlier results of Lions and Magenes, that the boundary of a C^∞ domain admits spectral synthesis for L^p_s and $B^{p,p}_s$.

REFERENCES

1. Schulze B.-W., Wildenhain G. Methoden der Potentialtheorie fur elliptische Differentialgleichungen beliebiger Ordnung. Berlin, Akademie-Verlag, 1977.
2. Хавин В.П. Аппроксимация в среднем аналитическими функциями. - Докл.АН СССР, 1968, 178, 1025-1028.
3. Bagby T. Quasi topologies and rational approximation. - J. Funct.Anal.,1972, 10, 259-268.
4. Beurling A., Deny J. Dirichlet spaces. - Proc.Nat. Acad.Sci., 1959, 45, 208-215.
5. Hedberg L.I. Two approximation problems in function spaces. - Ark.mat.,1978, 16, 51-81.
6. Triebel H. Boundary values for Sobolev-spaces with weights. Density of $\mathcal{D}(\Omega)$. - Ann.Sc.Norm.Sup.Pisa,1973, 3, 27, 73-96.

LARS INGE HEDBERG

Department of Mathematics
University of Stockholm
Box 6701
S-11385 Stockholm, Sweden

* * *

COMMENTARY BY THE AUTHOR

For the Sobolev spaces W^p_s , $1<p<\infty$, $s\in\mathbb{Z}_+$, the problem has been solved. In fact, all closed sets admit spectral synthesis for these spaces. See L.I.Hedberg [7], L.I.Hedberg and T.H.Wolff [8], and concerning the Dirichlet problem also T.Kolsrud [9].

REFERENCES

7. Hedberg L.I. Spectral synthesis in Sobolev spaces, and uniqueness of solutions of the Dirichlet problem. - Acta Math., 1981, 147, 237-264.

8. H e d b e r g L.I., W o l f f T.H. Thin sets in nonlinear potential theory. - Ann.Inst.Fourier (Grenoble), 1983, 33, N 4 (to appear).

9. K o l s r u d T. A uniqueness theorem for higher order elliptic partial differential equations. - Math.Scand., 1982, 51, 323-332.

* * *

EDITORS' NOTE. 1) The works [7] and [8] are of importance not only in connection with the Problem but in a much wider context representing an essential breakthrough in the general nonlinear potential theory.

2) When $d=1$ some details concerning the problem of synthesis in $W_s^p(\mathbb{R})$, $W_s^p(\mathbb{T})$ are contained in the following papers: J.-P. Kahane, Séminaire N.Bourbaki, 1966, Nov.; Akutowicz E.G., C.R.Acad. Sci., 1963, 256, N 25, 5268-5270; Ann.Scient.École Norm.Sup., 1965, 82, N 3, 297-325; Ill.J.Math., 1970, 14, N 2, 198-204; Осадчий Н.М. Укр.мат.ж., 1974, 26, № 5, 669-670.

3) If $X\subset C(\mathbb{R}^d)$ then the spectral synthesis holds: every ideal of X is divisorial, i.e. is the intersection of primary ideals. However, the identification of divisors generating closed ideals is a non trivial task. This problem is the theme of articles by L.G.Hanin (Л.Г.Ханин): Геометрическая классификация идеалов в алгебрах дифференцируемых функций двух переменных, in the book "Исследования по теории функций многих вещественных переменных", Ярославль, изд-во ЯГУ, 1982, 122-144; "Геометрическая классификация идеалов в алгебрах дифференцируемых функций", Докл. АН СССР, 1980, 254, № 2, 303-307.

8.2. APPROXIMATION BY SMOOTH FUNCTIONS IN SOBOLEV SPACES

Let $G \subset \mathbb{R}^2$ be a bounded domain whose boundary is a Jordan curve. Put

$$W^{k,p}(G) = \{ f : D^\alpha f \in L^p(G),\ 0 \leq |\alpha| \leq k \} .$$

This is the usual Sobolev space defined on G.

<u>Is</u> $C^\infty(\mathbb{R}^2)\,|\,G$ <u>dense in</u> $W^{k,p}(G)$, $1 \leq k,\ p < \infty$?

(The corresponding question for a disc minus a slit has a negative answer). The only thing I know is that this can be verified when $k = 1$ and $p = 2$ (Use conformal mapping). To the best of my knowledge this question was first raised by C.Amick.

PETER W.JONES

Institut Mittag-Leffler
Auravägen 17
S-182 62 Djursholm
Sweden

Usual Address:

Dept.of Mathematics
University of Chicago
Chicago, Illinois 60637
USA

.3. SPLITTING AND BOUNDARY BEHAVIOR IN CERTAIN H^2 SPACES
ld

Let μ be a finite Borel measure with compact support in $\mathbb{C}$. Even for very special choices of μ the structure of $H^2(\mu)$, the $L^2(\mu)$ -closure of the polynomials, can be mysterious. We consider measures $\mu = \nu + W\,dm$, where ν is carried by $\mathbb{D}$ and W is in $L^1(m)$. If $\log W$ is in $L^1(m)$, $H^2(\mu)$ is well understood and behaves like the classical Hardy space $H^2(m)$ [1]. We assume that ν is circularly symmetric, having the simple form $d\nu = G(r)\,r\,dr\,d\theta$, where $G > 0$ on $[0,1]$. Hastings [2] gave an example of such a measure with $W > 0$ m-a.e. such that $H^2(\mu) = H^2(\nu) \oplus L^2(W\,dm)$; we say then that $H^2(\mu)$ *splits*. A modification of this example will show that given *any* W with $\int \log W\,dm = -\infty$, G can be chosen to be positive and non-increasing on $[0,1]$ such that $H^2(\mu)$ splits. Suppose G is smooth and there exist C, $C>0$, and d, $0<d<2$, so that

$$G(r)^d \leq -G'(r) \leq \frac{1}{G(r)^C} \tag{1}$$

for $0 \leq r < 1$. Suppose further that for some ε, $\varepsilon > 0$,

$$\int_0^1 (1-r)^{-1-\varepsilon}\, G(r)\, dr < +\infty . \tag{2}$$

THEOREM 1. [3] <u>Let G satisfy (1) and (2). Suppose that $\int_\Gamma \frac{1}{W}\,dm < \infty$ for some arc Γ of $\mathbb{T}$, and that $W=0$ on a set of positive measure in $\mathbb{T} \setminus \Gamma$. Then $H^2(\mu)$ splits of and only if</u>

$$\int_{1-\delta}^1 \log\log \frac{1}{G(r)}\, dr = \infty \tag{3}$$

<u>for small δ</u>.

This theorem settles the question of splitting only when W is well-behaved. Conditions similar to (3) were introduced by Keldyš and Džrbašjan, and have been used by several authors in the study of other closure problems (cf. [4] and [5]).

QUESTION 1. Can W be found such that splitting occurs when the integral in (3) is finite, or even when $G \equiv 1$?

For λ in $\mathbb{D}$ the point evaluation $p \to p(\lambda)$ is a bounded linear functional on $H^2(\mu)$ (at least for those μ that we are considering); let $E^{\mu}(\lambda)$ denote its norm. If $E^{\nu}(\lambda)$ is analogously defined, then $E^{\mu}(\lambda) \leq E^{\nu}(\lambda)$ (an upper bound for E^{μ}). It is easy to show that $H^2(\mu)$ splits if and only if $E^{\mu}(\lambda) = E^{\nu}(\lambda)$ for all λ in $\mathbb{D}$. At the orther extreme, there is always an asymptotic lower bound for E^{μ} [6]: $\varliminf_{r \to 1} (1-r^2) E^{\mu}(re^{i\theta})^2 \geq 1/W(\theta)$ m-a.e.

Sometimes equality holds a.e. on an arc Γ of $\mathbb{T}$:

$$\lim_{r \to 1} (1-r^2) E^{\mu}(re^{i\theta})^2 = \frac{1}{W(\theta)} \quad m\text{-a.e. on } \Gamma \tag{4}$$

e.g. if $\log W$ is in $L^1(m)$, then (4) holds with $\Gamma = \mathbb{T}$. One verifies that if W does not vanish a.e. on Γ and if (4) holds, then $H^2(\mu)$ cannot split.

THEOREM 2. [7] Suppose that $\int_{\Gamma} \log W \, dm > -\infty$ and

$$\int_0^1 \log \frac{1}{G(r)} \, dr < +\infty \, . \tag{5}$$

Then (4) holds, every f in $H^2(\mu)$ has boundary values $\tilde{f}(e^{i\theta})$ m-a.e. on Γ, $f = \tilde{f}$ m-a.e. on Γ, and $\int_J \log|f| \, dm > -\infty$ whenever J is a closed arc interior to Γ and $f \not\equiv 0$ in $H^2(\mu)$. Every zero set for $H^2(\mu)$ with no limit points outside of J is a Blaschke sequence.

The hypothesis on W is weaker than that in Theorem 1, (5) is stronger than finiteness of the integral in (3) and the conclusion is stronger than the "only if" conclusion of Theorem 1 by an unknown amount. Eq.(4) can fail if the hypothesis on W is removed. Fix α, $0 < \alpha < 1$, and let

$$G(r) = \exp\left(-\frac{1}{(1-r)^{\alpha}}\right) \, ,$$

G satisfies (5). Define $\Omega(\theta,\delta)=(\pi/\delta)\, m\{x: x\in[\theta-\delta,\theta+\delta],$ $W(x)\leqslant \exp(-\delta^{-\alpha})\}$ and note that $0\leqslant\Omega\leqslant 1$.

THEOREM 3. ([3]). If G is as in (6), there exist constants a , $b>0$ with $E^{\mu}(re^{i\theta})> a\exp\left(b\,\frac{\Omega(\theta,1-r)}{(1-r)^{\alpha}}\right)$ for all $re^{i\theta}$ in $\mathbb{D}$.

If $\int_{\Gamma}\log W\,dm>-\infty$, then $\Omega(\theta,\delta)=O(\delta^{\alpha})$ as $\delta\to 0$ m-a.e. on Γ and Theorem 3 yields no information near Γ . On the other hand, for any d, $d>1$, one can construct W , $W>0$ a.e. with $\Omega(\theta,\delta)>(\mathrm{const})(-\log\delta)^{-d}$ for δ small and all θ [3]. Thus (4) can fail even if (5) holds.

QUESTION 2. Assume that the integral in (3) is finite, or even that G is given by (6), or that $G\equiv 1$. Is there a measurable set E , $E\subset\mathbb{T}$, with

$$H^2(\mu)=H^2(\nu+\chi_{\mathbb{T}\setminus E}W\,dm)\oplus L^2(\chi_E W\,dm),$$

where the first summand consists of "analytic" functions? Might such an E contain any arc on which $\Omega(\theta,\delta)$ (or a suitable analogue) tends to zero sufficiently slowly as $\delta\to 0$? If there is no such E with $mE>0$, exactly how can the various conclusions of Theorem 2 fail, if indeed they can?

QUESTION 3. Let $W(\theta)$ be smooth with a single zero at $\theta=0$. Assuming the integral in (3) is finite, describe the invariant subspaces of the operator "multiplication by z " on $H^2(\mu)$ in terms of the rates of decrease of $W(\theta)$ near 0 and $G(r)$ near 1.

Perhaps more complete results can be obtained than in the similar situation discussed in [8].

Finally we mention that the study of other special classes may be fruitful. Recently A.L.Volberg has communicated interesting related results for measures $\nu+W\,dm$, where ν is supported on a radial line segment. (See [10], [13] in the reference list after Commentary. - Ed.)

REFERENCES

1. C l a r y S. Quasi-similarity and subnormal operators. - Doct. Thesis, Univ.Michigan, 1973.
2. H a s t i n g s W. A construction of Hilbert spaces of analytic functions. - Proc.Amer.Math.Soc., 1979, 74, N 2, 295-298.
3. K r i e t e T. On the structure of certain $H^2(\mu)$ spaces. - Indiana Univ.Math.J., 1979, 28, N 5, 757-773.
4. B r e n n a n J.E. Approximation in the mean by polynomials on non-Caratheodory domains. - Ark.Mat. 1977, 15, 117-168.
5. М е р г е л я н С.Н. О полноте систем аналитических функций. - Успехи матем.наук, I953, 8, № 4, 3-63.
6. K r i e t e T., T r e n t T. Growth near the boundary in $H^2(\mu)$ spaces. - Proc.Amer.Math.Soc. 1977, 62, 83-88.
7. T r e n t T. $H^2(\mu)$ spaces and bounded evaluations. Doct. Thesis, Univ.Virginia, 1977.
8. K r i e t e T., T r u t t D. On the Cesaro operator. - Indiana Univ.Math.J. 1974, 24, 197-214.

THOMAS KRIETE

Department of Math.
University of Virginia
Charlottesville, Virginia
22903, USA

* * *

COMMENTARY

THEOREM (A.L.Vol'berg) <u>There exists</u> W , $W>0$ <u>a.e. on</u> $\mathbb{T}$ <u>such that</u> $H^2(\mu)$ <u>splits even for</u> $G=\mathbb{1}$.

The theorem gives an affirmative answer to QUESTION 1. It may be seen from the proof that $\Omega(\theta,\delta)$ tends to zero rather rapidly for every θ . The proof follows an idea of N.K.Nikolskii [9], p.243.

PROOF. It is sufficient to construct a function W , $W>0$ a.e. on $\mathbb{T}$ and a sequence of polynomials $\{P_n\}_{n\geq 1}$ such that $\lim_n (P_n|\mathbb{T}, P_n|\mathbb{D})=(\mathbb{0},\mathbb{1})$ in the Hilbert space $L^2(Wdm)\oplus L^2(\mathbb{D},dxdy)$.

Let $\{\delta_n\}_{n\geq 1}$ be any sequence of positive numbers satisfying

$\sum_n \delta_n < 1$, $\delta_n \downarrow 0$, and let $\Gamma_n \stackrel{def}{=} \{e^{it} : |t| \leq \pi \cdot \delta_n\}$.

Pick any smooth outer function h_n with the constant modulus $|h_n| = \varepsilon_n$ on $\mathbb{T} \setminus \Gamma_n$ ($\varepsilon_n \downarrow 0$) and such that $h_n(0) = 1$. The last condition implies the existence of an integer N_n such that

$$\iint_{\mathbb{D}} |h_n(z^{N_n}) - 1|^2 dx\,dy < \frac{1}{n} \cdot \qquad (7)$$

Consider now the set $e_n = \{\zeta \in \mathbb{T} : \zeta^{N_n} \in \Gamma_n\}$. It is clear that $m e_n = \delta_n$ and therefore $m(\bigcap_{n \geq 1} \bigcup_{k \geq n} e_k) = 0$. This implies that the increasing family of sets $S_k = \mathbb{T} \setminus \bigcup_{k \geq n} e_k = \bigcap_{k \geq n} \mathbb{T} \setminus e_k$ almost exhausts the unit circle: $\lim_k m S_k = 1$. Now we are in a position to define the weight W :

$$W(\zeta) = \begin{cases} 1, & \zeta \in S_1 \\ \min(1, \frac{1}{kC_k^2}), & \zeta \in S_k \setminus S_{k-1},\ k = 2, \ldots, \end{cases}$$

where C_n stands for $\|h_n\|_\infty$.

Set $H_n(z) \stackrel{def}{=} h_n(z^{N_n})$ and note that $|H_n| = \varepsilon_n$ on S_n because $S_n \subset \mathbb{T} \setminus e_n$. Clearly $\lim_n C_n = +\infty$. These imply

$$\int_{\mathbb{T}} |H_n|^2 W dm = \int_{S_n} |H_n|^2 dm + \frac{1}{n} m(\mathbb{T} \setminus S_n) \leq \varepsilon_n^2 + \frac{1}{n} \cdot$$

The last inequality together with (7) yields obviously the desired conclusion. ●

THEOREM 1 in the text of the problem can be strengthened. Suppose that the function G satisfies some regularity conditions and there exists an arc Γ with $\int_\Gamma \frac{dm}{W} < +\infty$. Then $H^2(\mu)$ splits iff

$$\int^1 \log \log \frac{1}{G(r)} dr = +\infty,$$

$$\int_{\mathbb{T}} \log W dm = -\infty .$$

The new point here is that we do not require for W to be identically zero on a set of the positive length. See [10] for the proof.

QUESTION 2 can be also answered affirmatively. Recall that a closed subset E of $\mathbb{T}$ satisfies (by definition) the Carleson condition if $\sum_{\nu} m(\ell_\nu) \log \frac{1}{m(\ell_\nu)} < +\infty$. Here $\{\ell_\nu\}$ stands for the family of all complementary intervals of E. Let $\mathcal{A}$ be the family of all closed E, $mE>0$ which do not contain subsets of positive length satisfying the Carleson condition.

Suppose again that $G \equiv 1$.

THEOREM (S.V.Hruščëv). Let $E \in \mathcal{A}$. Then there exists a positive weight w such that

$$H^2(\mu) = H^2(\nu + \chi_{\mathbb{T}\setminus E} W dm) \oplus L^2(\chi_E W dm),$$

where the first summand does not split.

It has been shown in [11] that such sets E do exist. For example, any set of Cantor type having positive Lebesgue measure and not satisfying the Carleson condition does the job.

PROOF. Pick a closed set E in $\mathcal{A}$ and consider an auxiliary region $\mathcal{D}$ having the smooth boundary as it is shown on the figure

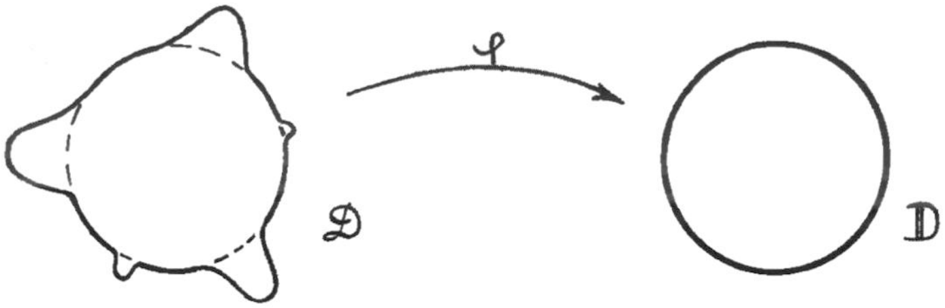

The region $\mathcal{D}$ abuts on $\mathbb{T}$ precisely at the points of E and its boundary Γ has at these points the second order of tangency. Let φ be a conformal mapping of $\mathcal{D}$ onto $\mathbb{D}$. Then φ does not distort the Euclidean distance by the Kellog's theorem (see [12],

.411). It follows that $\varphi(E)\in\mathcal{A}$. By theorem 4.1 in [11] there xists a sequence of polynomials $\{P_n\}_{n\geq 1}$ satisfying

$$\lim_n P_n = 0 \qquad \text{uniformly on } \varphi(E) \;;$$

$$\lim_n P_n = 1 \qquad \text{uniformly on compact subsets of } \mathbb{D} \;;$$

$$|P_n(z)| \leqslant \frac{Const}{(1-|z|)^{1/16}} .$$

Using the Kellog's theorem again, we see that the sequence $\{f_n\}_{n\geq 1}$, $f_n(z) = P_n(\varphi(z))$ satisfies the following

$$\lim_n f_n = 0 \quad \text{uniformly on } E \;; \tag{8}$$

$$\lim_n f_n = 1 \quad \text{uniformly on compact subsets of } \mathcal{D} \tag{9}$$

$$|f_n(z)| \leqslant \frac{Const}{dist(z,E)^{1/8}}, \quad z\in clos\,\mathbb{D} . \tag{10}$$

Define

$$w(t) = \begin{cases} 1, & t\in E \\ (m\ell)^{1/4}, & t\in\ell, \ \ell \text{ being a complementary interval of } E . \end{cases}$$

The function $g(t) = dist(t,E)^{-1/4}\cdot \sum_\ell (m\ell)^{1/4}\cdot \chi_\ell$ is evidently summable on $\mathbb{T}$ and dominates $|f_n|^2\cdot w$. Together with (9) this implies

$$\lim_n \int_{\mathbb{T}\setminus E} |f_n-1|^2\cdot w\, dm = 0$$

by the Lebesgue theorem on dominated convergence. Besides, (9) and (10) yield

$$\lim_n \iint_{\mathbb{D}} |f_n-1|^2 dx\, dy = 0$$

(see [11]). Finally, $\lim_n \int_E |f_n|^2 w\, dm = 0$, see (8).

The space $H^2(\nu + \chi_{\mathbb{T}\setminus E} W\, dm)$ does not split because $\inf_{t\in \ell} w(t) = m(\ell)^{1/4} > 0$ for every complementary interval ℓ . This can be deduced either from theorem 2 cited in the text of the problem or from theorem 3.1. of [11]. ●

Note that an appropriate choice of E provides the additional property of the weight W in the theorem:

$$\int_{\mathbb{T}} \left(\log \frac{1}{W}\right)^{\rho} dm < +\infty$$

for every ρ, $\rho < 1$. Pick a Cantor type set E in $\mathbb{A}$ satisfying $\sum_{\ell} \ell \left(\log \frac{1}{\ell}\right)^{\rho} < +\infty$ for $\rho < 1$.

The construction of the theorem can be extended for other weights G satisfying (5). Such a splitting cannot occur if $\int_E \log W\, dm > -\infty$ and E, $mE > 0$ satisfies the Carleson condition (see theorem 3.1 in [11]).

REFERENCES

9. Никольский Н.К. Избранные задачи весовой аппроксимации и спектрального анализа. - Труды Мат.ин-та им.В.А.Стеклова АН СССР, 1974, 120.
10. Вольберг А.Л. Логарифм почти-аналитической функции суммируем. - Докл.АН СССР, 1982, 265, № 6, с.1297-1302.
11. Хрущёв С.В. Проблема одновременной аппроксимации и стирание особенностей интегралов типа Коши. - Труды Мат.ин-та им.В.А. Стеклова АН СССР, 1978, 130, с.124-195.
12. Голузин Г.М. Геометрическая теория функций комплексного переменного. М., "Наука", 1966.
13. Вольберг А.Л. Одновременная аппроксимация полиномами на окружности и внутри круга. - Зап.научн.семин.ЛОМИ, 1978, 92, 60-84.

8.4. ON THE SPAN OF TRIGONOMETRIC SUMS IN WEIGHTED L^2 SPACES
old

Let $\Delta = \Delta(\gamma)$ be an odd non-decreasing bounded function of γ on the line $\mathbb{R}$, let $Z(\Delta) = L^2(\mathbb{R}, d\Delta)$ and let $Z^T(\Delta)$ denote the closure in $Z(\Delta)$ of finite trigonometric sums $\sum c_j e^{i\gamma t_j}$ with $|t_j| \leq T$. It is readily checked that $Z^{T_1}(\Delta) \subset Z^{T_2}(\Delta)$ for $T_1 \leq T_2$ and that $\bigcup_{T \geq 0} Z^T(\Delta)$ is dense in $Z(\Delta)$. Let

$$T_0(\Delta) = \inf \{ T > 0 : Z^T(\Delta) = Z(\Delta) \}$$

with the understanding that $T_0(\Delta) = \infty$ if the equality $Z^T(\Delta) = Z(\Delta)$ is never attained. The following 3 examples indicate the possibilities:

(1) if $\Delta(\gamma) = \int_0^\gamma (\xi^2+1)^{-1} d\xi$ then $T_0 = \infty$;

(2) if $\Delta(\gamma) = \int_0^\gamma e^{-|\xi|} d\xi$ then $T_0 = 0$;

(3) if Δ is a step function with jumps of height $1/(n^2+1)$ at every integer n, then $T_0 = \pi$.

PROBLEM. <u>Find formulas for</u> T_0, <u>or at least bounds on</u> T_0, <u>in terms of</u> Δ.

DISCUSSION. Let Δ' denote the Radon-Nikodym derivative of Δ with respect to Lebesgue measure. It then follows from a well known theorem of Krein [1] that $T_0 = \infty$ as in example (1) if

$$\int_{-\infty}^{\infty} \frac{\log \Delta'(\gamma)}{\gamma^2+1} d\gamma > -\infty .$$

A partial converse due to Levinson-McKean implies that if Δ is absolutely continuous and if $\Delta'(\gamma)$ is a decreasing function of $|\gamma|$ and $\int_{-\infty}^{\infty} \frac{\log \Delta'(\gamma)}{\gamma^2+1} d\gamma = -\infty$ (as in example (2)), then $T_0 = 0$. A proof of the latter and a discussion of example (3) may be found in Section 4.8 of [2]. However, apart from some analogues for the case in which Δ is a step function with jumps at the integers, these two theorems seem to be the only general results available for computing T_0 directly from Δ. (There is an explicit formula for T_0 in terms of the solution to an inverse spectral problem, but this is of little practical value because the computations involved are typically not manageable.)

The problem of finding T_0 can also be formulated in the language of Fourier transforms since $Z^T(\Delta)$ is a proper subspace of $Z(\Delta)$ if and only if there exists a non-zero function $f \in Z(\Delta)$ such that

$$\tilde{f}(t) = \int_{-\infty}^{\infty} e^{i\gamma t} \cdot f(\gamma)\, d\Delta(\gamma) = 0$$

for $|t| \leq T$. Thus

$$T_0 = \inf \{ T > 0 : \tilde{f}(t) = 0 \quad \text{for } |t| \leq T \Rightarrow f = 0 \quad \text{in } Z(\Delta) \}.$$

Special cases of the problem in this formulation have been studied by Levinson [3] and Mandelbrojt [4] and a host of later authors. For an uptodate survey of related results in the special case that Δ is a step function see [5]. The basic problem can also be formulated in $L^p(\mathbb{R}, d\Delta)$ for $1 \leq p \leq \infty$. A number of results for the case $p = \infty$ have been obtained by Koosis [6], [7] and [8].

REFERENCES

1. К р е й н М.Г. Об одной экстраполяционной проблеме А.Н.Колмогорова. - Докл.АН СССР, 1945, 46, 306-309.
2. D y m H., M c K e a n H.P. Gaussian Processes, Function Theory and the Inverse Spectral Problem, New York, Academic Press, 1976.
3. L e v i n s o n N. Gap and Density Theorems. Colloquium Publ., 26, New York, Amer.Math.Soc., 1940.
4. M a n d e l b r o j t S. Séries de Fourier et Classes Quasi-analytiques. Paris, Gauthier-Villars, 1935.
5. R e d h e f f e r R.M. Completeness of sets of complex exponentials. - Adv.Math. 1977, 24, 1-62.
6. K o o s i s P. Sur l'approximation pondérée par des polynômes et par des sommes d'exponentielles imaginaires. - Ann.Sci.Ec.Norm. Sup., 1964, 81, 387-408.
7. K o o s i s P. Weighted polynomial approximation on arithmetric progressions of intervals or points. - Acta Math., 1966, 116, 223-277.
8. K o o s i s P. Solution du problème de Bernstein sur les entiers. - C.R.Acad.Sci.Paris,Ser.A 1966, 262, 1100-1102.

HARRY DYM Department of Mathematics The Weizmann Institute of Science Rehovot, Israel

.5. DECOMPOSITION OF APPROXIMABLE FUNCTIONS

Let $H(\mathcal{D})$ be the space of all analytic functions in some open subset $\mathcal{D}$ of the extended complex plane $\mathbb{C}$. Let $\mathcal{D}^*$ denote the one point compactification of $\mathcal{D}$.

If F is relatively closed subset of $\mathcal{D}$, $A_{\mathcal{D}}(F)$ is the functions on F being uniform limits on F by sequences from $H(\mathcal{D})$. The problem of characterizing $A_{\mathcal{D}}(F)$ was raised by N.U.Arakelyan some years ago [1]. A closely related question was raised in [2].

Recently we obtained the following characterization of $A_{\mathcal{D}}(F)$ for a large class of sets $\mathcal{D}$:

$$A_{\mathcal{D}}(F) = C_{na}(F \cup \Omega(F)) + H(\mathcal{D}) \tag{1}$$

where $C_{na}(F \cup \Omega(F))$ is the space of analytic functions on $F \cup \Omega(F)$ with a continuous extension to the Riemann sphere, and where $\Omega(F)$ is the smallest open subset of $\mathcal{D} \setminus F$ such that $\mathcal{D}^* \setminus (F \cup \Omega(F))$ is arcwise connected. For details see [4].

PROBLEM 1: <u>Obtain a decomposition like (1) for any proper nonempty open subset of the Riemann sphere.</u>

PROBLEM 2: <u>Obtain decompositions like (1) when $\mathcal{D}$ is the unit disc</u> $\{|z|<1\}$ <u>and</u> $H(\mathcal{D})$ <u>is replaced by other function spaces in</u> $\mathcal{D}$.

REMARK: A positive answer to Problem 1, will immediately give a solution to problem 9.6 in [0] in light of the results about $C_{na}(F)$ in [3].

REFERENCES

0. Anderson J.M., Barth K.F., Brannan D.A. Research Problems in Complex Analysis. - Bull.London Math.Soc., 1977, 9, 152.
1. Arakeljan N.U. Approximation complexe et propriétés des fonctions analytiques. - Actes Congrès intern.Math., 1970, 2, Gauthier-Villars / Paris, 595-600.
2. Brown L., Shields A.L. Approximation by analytic functions uniformly continuous on a set. - Duke Math.Journal, 1975, 42, 71-81.

3. S t r a y A. Uniform and asymptotic approximation. - Math.Ann., 1978, 234, 61- 68.
4. S t r a y A. Decomposition of approximable functions.

ARNE STRAY

Agder Distriktshogskole
Postboks 607,
N-4601 Kristiansand
Norway

.6. A PROBLEM OF UNIFORM APPROXIMATION BY FUNCTIONS ADMITTING QUASICONFORMAL CONTINUATION

The following subalgebras of the Banach space $C(K)$ of all ontinuous functions on a compact set K, $K\subset\mathbb{C}$, are important in he theory of rational approximation. These are the algebra $A(K)$ f all functions in $C(K)$ holomorphic in the interior of K and he algebra $R(K)$ consisting of uniform limits of rational func- ions continuous on K.

For $\varepsilon>0$ let $K_\varepsilon \overset{def}{=} K+D(\varepsilon)$, $D(\varepsilon)\overset{def}{=}\{z\in\mathbb{C}: |z|<\varepsilon\}$. Consider the Beltrami equation in K_ε

$$f_{\bar z}=\mu(z)f_z, \tag{1}$$

μ being a measurable function such that

$$\operatorname*{ess-sup}_{K_\varepsilon}|\mu|\le k<1.$$

A continuous function f is said to be a generalized solution of (1) if its generalized derivatives (in the sense of the distribution theory) belong to L^1 locally and satisfy (1) a.e. on K_ε.

Clearly $f|K\in C(K)$ for such a solution and it is known that $f|K\in A(K)$ provided $\mu\equiv 0$ on K [1].

Fix $k<1$ and consider a set $B_\varepsilon(K)$ of all restrictions $f|K$, where f ranges over the family of generalized solutions of (1) in K_ε with $\mu\equiv 0$ on K_ε. Let $B(K)$ be the closure of $\bigcup_{\varepsilon>0} B_\varepsilon(K)$ in $C(K)$. Then clearly

$$R(K)\subset B(K)\subset A(K).$$

PROBLEM 1. <u>Is there</u> K <u>such that</u> $R(K)\ne B(K)$?

An affirmative answer to the question would entail the following problem.

PROBLEM 2. <u>Find necessary and sufficient conditions on</u> K <u>for</u>

a) $B(K)=C(K)$

<u>and for</u>

b) $B(K)=A(K)$.

Suppose $k=0$. Then a complete solution of problem 2 is given by Vitushkin's theorem [2],[3]. The case $k>0$ corresponds to the problem of approximation by functions admitting a quasi-conformal continuation.

One of possible ways to solve problem 1 consists in the construction of a "Swiss cheese" satisfying $R(K)\neq C(K)$, $B(K)=C(K)$.

These problems were posed for the first time at the International Conference on Approximation Theory (Varna, 1981).

REFERENCES

1. L e h t o O., V i r t a n e n K.I. Quasiconformal Mappings in the Plane. Springer-Verlag, Berlin. Heidelberg. New-York, 1973.
2. В и т у ш к и н А.Г. Аналитическая ёмкость множеств в задачах теории приближений. -Успехи матем.наук,I967,22,№5, I4I-I99.
3. Z a l c m a n L. Analytic Capacity and Rational Approximation. Lect.Notes in Math., 1968, 50.

V.I.BELYI

(В.И.БЕЛЫЙ)

СССР, 340048, Донецк 48,
Университетская 77,
Институт прикладной
математики и механики

.7. TANGENTIAL APPROXIMATION

Let F be a closed subset of the complex plane $\mathbb{C}$ and let $\mathcal{F}$ nd G be two spaces of functions on F. The set F is said to e a s e t o f t a n g e n t i a l a p p r o x i m a t i o n f functions in the class $\mathcal{F}$ by functions in the class G if for ach function $f \in \mathcal{F}$ and each positive continuous ε on F, there s a function $g \in G$ with

$$|f(z)-g(z)|<\varepsilon(z),\ z\in F.$$

arleman's theorem [1] states that the real axis is a set of angential approximation of continuous functions by entire functi- ns Hence, tangential approximation is sometime called Carleman ap- roximation

PROBLEM: <u>For given classes of functions</u> $\mathcal{F}$ <u>and</u> G, <u>characterize the sets of tangential approximation</u>.

Of course, this problem is of interest only for certain classes $\mathcal{F}$ and G. We shall use the following notations:

$H(\mathbb{C})$: entire functions

$M_F(\mathbb{C})$: meromorphic functions on $\mathbb{C}$ having no poles on F.

$H(F)$: functions holomorphic on (some neighbourhood of) F.

$U(F)$: uniform limits on F, of functions in $H(F)$.

$A(F)$: functions continuous on F and holomorphic on F°.

$C(F)$: continuous complex-valued functions on F.

Each of these classes is included in the one below it. We consider each problem of tangential approximation which results by choosing G as one of the first three classes and choosing $\mathcal{F}$ as one of the last three. Thus each square in the following table corresponds to a problem

G \ $\mathcal{F}$	$U(F)$	$A(F)$	$C(F)$
$H(\mathbb{C})$		[4]	[2]
$M_F(\mathbb{C})$			[3]
$H(F)$			[3]

<u>The blank squares correspond to open problems</u>. For partial results on the central square, see [5]. In [3], the conditions stated characterize those sets of tangential approximation for the classes $\mathcal{F}=C(F)$ and $G=M_F(\mathbb{C})$. One easily checks that these conditions are also necessary and sufficient for the case $\mathcal{F}=C(F)$ and $G=H(F)$ The first column was suggested to us by T.W.Gamelin and T.J.Lyons.

One can formulate similar problems for harmonic approximation. The most general harmonic function in a neighbourhood of an isolated singularity at a point $y\in\mathbb{R}^n$, $n\geq 2$ can be written in the form

$$u(x)=p_0\cdot K(x-y)+\sum_{k=1}^{\infty} p_k(x-y)|x-y|^{2-n-2k}+\sum_{k=0}^{\infty} q_k(x-y)$$

where

$$K(x)=\begin{cases}\log|x| & \text{if } n=2\\ |x|^{2-n} & \text{if } n\geq 3\end{cases}$$

and p_k, q_k are homogeneous harmonic polynomials of degree k , $k\geq 0$. The singularity of u is said to be non-essential if $p_k=0$, $k\geq k_0$. An essentially harmonic function on an open set $\Omega\subset\mathbb{R}^n$ is a function which is harmonic in Ω except possibly for non-essential singularities.

Let F be a closed set in $\mathbb{R}^n$, $n\geq 2$. We introduce the following notations:

$h(\mathbb{R}^n)$: functions harmonic on $\mathbb{R}^n$.

$m_F(\mathbb{R}^n)$: essentially harmonic functions on $\mathbb{R}^n$ having no singularities on F.

$h(F)$: functions harmonic on (some neighbourhood of) F.

$u(F)$: uniform limits on F , of functions in $h(F)$.

$a(F)$: functions continuous on F and harmonic on $\overset{\circ}{F}$.

$c(F)$ continuous real-valued functions on F .

As in the complex case, we have <u>a table of problems</u>.

$G \setminus \mathcal{F}$	$\mathcal{U}(F)$	$\mathcal{A}(F)$	$C(F)$
$h(\mathbb{R}^n)$			[6]
$m_F(\mathbb{R}^n)$			[7]
$h(F)$			[8]

REFERENCES

1. C a r l e m a n T. Sur un théorème de Weierstrass. - Ark.Mat. Astronom.Fys. 1927, 20B, 4, 1-5.
2. К е л д ы ш М.В. Л а в р е н т ь е в М.А. Об одной задаче Карлемана. - Докл.АН СССР, 1939, 23, № 8, 746-748.
   М е р г е л я н С.Н. Равномерные приближения функций комплексного переменного. - Успехи матем.наук, 1952, 7, вып.2 (48), 31-123. (English. Translations Amer.Math.Soc.1962, 3, 294-391).

   А р а к е л я н Н.У. Равномерные и касательные приближения аналитическими функциями. - Изв.АН Арм.ССР, сер.матем., 1968, 3, № 4-5, 273-286.
3. Н е р с е с я н А.А. О равномерной и касательной аппроксимации мероморфными функциями. - Изв.АН Арм.ССР, сер.матем., 1972, 7, № 6, 405-412.
   R o t h A. Meromorphe Approximationen. - Comment.Math.Helv. 1973, 48, 151-176.
   R o t h A. Uniform and tangential approximations by meromorphic functions on closed sets. - Canad.J.Math.1976, 28, 104-111.
4. Н е р с е с я н А.А. О множествах Карлемана. - Изв.АН Арм.СССР, сер.матем., 1971, 6, № 6, 465-471.
5. B o i v i n A. On Carleman approximation by meromorphic functions. - Proceedings 8th Conference on Analytic Functions,Blazejewko, August 1982, Ed.J.Lawrynowicz (to appear).
6. Ш а г и н я н А.А. О равномерной и касательной гармонической аппроксимации непрерывных функций на произвольных совокупностях. - Матем.заметки 1971, 9, вып.2, 131-142. (English: Mat.Notes 1971, 9, pp.78-84).
7. G a u t h i e r P.M. Carleman approximation on unbounded sets by harmonic functions with Newtonian singularities. - Proceedings 8th Conference on Analytic Functions, Blazejewko, August 1982, Ed.J.Lawrynowicz (to appear).

8. L a b r è c h e M. De l'approximation harmonique uniforme. Doctoral Dissertation Université de Montréal, 1982.

ANDRÉ BOIVIN

PAUL M. GAUTHIER

Département de Mathématiques et de Statistique
Université de Montréal
C.P. 6128, Succursale "A"
Montréal, Québec
H3C 3J7
CANADA

3.8. THE INTEGRABILITY OF THE DERIVATIVE OF A CONFORMAL MAPPING

ld

Let Ω be a simply connected domain having at least two boundary points in the extended complex plane and let ϕ be a conformal mapping of Ω onto the open unit disk $\mathbb{D}$. In this note we pose the following QUESTION: <u>For which numbers</u> p <u>is</u>

$$\iint_{\Omega} |\phi'|^p \, dx\,dy < +\infty \,?$$

For $p=2$ the integral is equal to the area of the disk and is therefore finite. In general, it is known to converge for $4/3 < p < 3$ and if Ω is the plane slit along the negative real axis then it obviously diverges for $p = 4/3$ and $p = 4$. These facts are consequences of the Koebe distortion theorem and were first discovered by Gehring and Hayman (unpublished) for $p < 2$ and by Metzger [1] for $p > 2$. Recently, the author has succeeded in proving that the upper bound 3 can be increased. The following theorem summarizes the known results.

THEOREM 1. <u>There exists a number</u> τ, $\tau > 0$, <u>not depending on</u> Ω, <u>such that</u>

$$\iint_{\Omega} |\phi'|^p \, dx\,dy < +\infty$$

<u>if</u> $4/3 < p < 3 + \tau$.

For a wide class of regions, including "starlike" and "close-to-convex" domains, $p = 4$ is the correct upper bound (cf. [2], Theorem 2). Quite likely, $\iint |\phi'|^p \, dx\,dy < +\infty$ for $4/3 < p < 4$ in all cases but, unfortunately, the argument in [2] will not give this result.

Here is a SKETCH OF THE PROOF OF THEOREM 1. We shall assume that $x_0 \in \Omega$, $\phi(x_0) = 0$ and we shall denote by $\delta(z)$ the Euclidean distance from the point z to $\partial\Omega$. It is easy to see, using polar coordinates, that

$$\iint_{\Omega} |\phi'|^p \, dx\,dy = \int_0^1 2\pi r \int_{|\phi|=r} |\phi'|^{p-2} \, d\omega_r \, dr .$$

where $d\omega_r$ is harmonic measure on the curve $|\phi| = r$ relative to x_0. Moreover, it follows from the Koebe distortion theorem that $|\phi'(z)| \simeq K \dfrac{1-|\phi(z)|}{\delta(z)}$ near $\partial\Omega$ and, consequently,

$\iint |\phi'|^p \, dx\, dy < \infty$ if and only if

$$\int_0^1 (1-r)^{p-2} \int_{|\phi|=r} \frac{d\omega_r}{\delta(z)^{p-2}} \, dr < +\infty .$$

Thus, Theorem 1 is now an immediate consequence of the following lemma on the growth of the integral $\int \delta(z)^{-\lambda} d\omega_r$ as $r \to 1$.

LEMMA 1. <u>There exists a constant</u> ρ, $\rho > 0$, <u>such that if</u> $\lambda > 1/2$ <u>then</u>

$$\int_{|\phi|=r} \frac{d\omega_r}{\delta(z)^{\lambda}} = 0\left[\frac{1}{(1-r)^{2\lambda-\rho}}\right] .$$

Of course, if we could prove the lemma for all ρ, $\rho < 1$, then we could prove Theorem 1 for $4/3 < p < 4$. So far, however, this has still not been done. The proof of the lemma is based on an idea of Carleson [3], which he expressed in connection with another problem. The QUESTION is the following: <u>On a Jordan curve is harmonic measure absolutely continuous with respect to</u> α-<u>dimensional Hausdorff measure for every</u> $\alpha, \alpha < 1$? On the one hand, according to the Beurling projection theorem (cf. [4], p.72), the question can be answered affirmatively if $\alpha \leqslant 1/2$. On the other hand, Lavrent'ev [5], McMillan and Piranian [6] and Carleson [3] have shown by means of counterexamples that absolute continuity does not always occur if $\alpha = 1$. In addition, Carleson was able to show that the upper bound 1/2 in Beurling's theorem can be increased. It is interesting to speculate on the extent to which it is possible to observe a similarity between the two problems. For example, it is well known (cf. [7], p.44) that harmonic measure is absolutely continuous with respect to 1-dimensional Hausdorff measure if there are no points ξ on $\partial\Omega$ for which

$$\lim_{z \to \xi} \sup_{z \in \Omega} \arg (z - \xi) = + \infty \tag{1}$$

$$\lim_{z \to \xi} \inf_{z \in \Omega} \arg (z - \xi) = - \infty . \tag{2}$$

he QUESTION arises: <u>if this condition is satisfied must</u> $\iint|\phi'|^p dxdy <$ $< \infty$ for $4/3 < p < 4$? At this time the answer is not known. Be-ore proceeding to the solution of the general problem it apparently emains to answer this more modest question.

To the best of my knowledge, the question about the integrabili-y of the derivative of a conformal mapping arose in connection with several problems in approximation theory. We shall mention only one of these and then indicate an application of Theorem 1. Our problem as first posed by Keldyš in 1939 (cf. [8] and [9], p.10) and he ob-tained the first results in this direction. Further progress has been achieved in the works of Džrbašjan [10], Šaginjan [11], Maz'ja and Havin [12], [13] and the author [14], [15], [2]. A complete discussion of the results obtained up to 1975 can be found in the surveys of Mergeljan [9], Mel'nikov and Sinanjan [16].

Let us assume that D , $\mathcal{U}$ are two Jordan domains in the comp-lex plane, $\mathcal{U} \subset D$, and let $\Omega = \operatorname{Int}(D \setminus \mathcal{U})$. We shall denote by $H^p(\Omega), p \geq 1$, the closure of the set of all polynomials in the space $L^p(\Omega, dxdy)$ and we shall denote by $L^p_a(\Omega)$ the subspace consisting of those functions f, $f \in L^p(\Omega)$, which are analytic in Ω . Clear-ly, $H^p \subset L^p_a$. An interesting question concerns the possibility of equality in this inclusion. It is well known that in order for H^p and L^p_a to coincide the determining factor is the "thinness" of the region Ω near multiple boundary points (i.e. near points of $\partial D \cap \cap \partial \mathcal{U}$). Here is a result which gives a quantitative description of that dependence. The proof is based in part on Theorem 1 (cf. [2] and [15], pp.143-148).

THEOREM 2. <u>Let</u> $\delta(z)$ <u>be the distance from</u> z <u>to</u> $\mathbb{C} \setminus D$ <u>and let</u> $d\omega$ <u>be harmonic measure on</u> $\partial\mathcal{U}$ <u>relative to the domain</u> $\mathcal{U}$. <u>There exists an absolute constant</u> τ, $\tau > 0$, <u>not depending on</u> Ω , <u>such that if</u>

$$\int_{\partial \mathcal{U}} \log \delta(z)\, d\omega(z) = -\infty ,$$

<u>then</u> $H^p(\Omega) = L^p_a(\Omega)$ <u>for all</u> p, $p < 3 + \tau$.

The QUESTION remains: <u>is</u> $p = 4$ <u>the upper bound or is the theo-rem true for all</u> p, $p < +\infty$?

REFERENCES

1. M e t z g e r T.A. On polynomial approximation in $A_q(\mathcal{D})$. - Proc.Amer.Math.Soc.,1973, 37, 468-470.
2. B r e n n a n J. The integrability of the derivative in conformal mapping. - J.London Math.Soc.,1978, 18, 261-272.
3. C a r l e s o n L. On the distortion of sets on a Jordan curve under conformal mapping. - Duke Math.J., 1973, 40, 547-559.
4. M c M i l l a n J.E. Boundary behavior under confromal mapping. - Proc. of the N.R.L. Conference on classical function theory, Washington D.C. 1970, 59-76.
5. Л а в р е н т ь е в М.А. О некоторых граничных задачах в теории однолистных функций. - Матем.сб., I963, № I, 8I5-844.
6. McM i l l a n J.E., P i r a n i a n G. Compression and expansion of boundary sets. - Duke Math.J.,1973, 40, 599-605.
7. McM i l l a n J.E. Boundary behavior of a conformal mapping. - Acta Math.,1969, 123, 43-67.
8. К е л д ы ш М.В. Sur l'approximation en moyenne quadratique des fonctions analytiques. - Матем.сб.,I939,47,№ 5, 39I-402.
9. М е р г е л я н С.Н. О полноте систем аналитических функций. - Успехи матем.наук,I953, 8, № 4, 3-63.
I0. Д ж р б а ш я н М.М. Метрические теоремы о полноте и представимости аналитических функций. Докт.диссертация, Ереван, I948.
II. Ш а г и н я н А.Л. Об одном признаке неполноты системы аналитических функций. - Докл.АН Арм.ССР, I946, У, № 4, 97-I00.
I2. М а з ь я В.Г., Х а в и н В.П. Об аппроксимации в среднем аналитическими функциями. - Вестн.Ленингр.ун-та, сер.матем.,мех. и астрон., I968, № I3, 62-74.
I3. М а з ь я В.Г., Х а в и н В.П. Приложения (p,ℓ) - емкости к нескольким задачам теории исключительных множеств. - Матем.сб., I973, 90, № 4, 558-59I.
I4. B r e n n a n J. Invariant subspaces and weighted polynomial approximation. - Ark.Mat.,1973, 11, 167-189.
I5. B r e n n a n J. Approximation in the mean by polynomials on non-Carathéodory domains. - Ark.Mat.,1977, 15, 117-168.
I6. М е л ь н и к о в М.С., С и н а н я н С.О. Вопросы теории приближений функций одного комплексного переменного. - В кн.: Современные проблемы математики, т.4, Итоги науки и техники, Москва, ВИНИТИ, I975, I43-250.

J.BRENNAN

University of Kentucky
Lexington, Kentucky 40506
USA

.9. WEIGHTED POLYNOMIAL APPROXIMATION

Let Ω be a bounded simply connected domain in the complex plane $\mathbb{C}$, let dxdy denote two-dimensional Lebesgue measure and let $w(z) > 0$ be a bounded measurable function defined on Ω. For each p, $1 \leq p < \infty$, we shall consider two spaces of functions:

(i) $H^p(\Omega, w\,dxdy)$, the closure of the polynomials in $L^p(\Omega, w\,dxdy)$

(ii) $L^p_a(\Omega, w\,dxdy)$, the set of functions in $L^p(\Omega, w\,dxdy)$ which are analytic in Ω.

If w is bounded away from zero locally it is easy to see that L^p_a is a closed subspace of L^p and that $H^p \subseteq L^p_a$. It is an OLD PROBLEM to determine: <u>for which regions</u> Ω <u>and weights</u> w <u>is</u> $H^p(\Omega, w\,dxdy) = L^p_a$? Whenever this happens the polynamials are said to be c o m p l e t e in L^p_a.

As the problem suggests, completeness depends both on the region Ω and on the weight w. In this article, however, we shall be primarily concerned with the role of w when no restrictions are placed on Ω, save simple connectivity. The main difficulty then stems from the fact that Ω may have a nonempty i n n e r b o u n d a r y; that is, there may be points in $\partial\Omega$ which belong to the interior of $\overline{\Omega}$ (viz., a Jordan domain with a cut or incision in the form of a simple arc from an interior point to a boundary point). Roughly speaking, $H^p(\Omega, w\,dxdy) = L^p_a$ if $w(z) \longrightarrow 0$ sufficiently rapidly at every point of the inner boundary. But, this is not the only factor that must be considered and in order to avoid certain snags we shall make the additional ssumption that w is constant on the level lines of some conformal mapping φ of Ω onto the open unit disk $\mathbb{D}$ (i.e., $w(z) = W(1 - |\varphi(z)|)$, where $W(t) \longrightarrow 0$ as $t \to 0$). Put another way, w depends only on Green's function. With this requirement the problem becomes conformally invariant and every significant result going back to the early 1940's and the seminal work of Keldyš [1] makes use of this or some equivalent fact. Additional information and background material on the completeness question can be found in the survey article of Mergeljan [2], in the author's papers [3], [4] and in the references cited therein.

In the ensuing discussion Ω_∞ will denote the unbounded component of $\mathbb{C} \setminus \overline{\Omega}$ and ω is harmonic measure on $\partial\Omega$ relative to some convenient point in Ω. For weights which depend only on Green's function the author [4] has obtained the following result:

THEOREM 1. Suppose that $w(t) \downarrow 0$ as $t \downarrow 0$ and that $\omega(\partial\Omega_\infty) > 0$. Then there exists a universal constant $\tau > 0$ such that

$$\int_0 \log \log \frac{1}{w(t)}\, dt = +\infty$$

implies $H^p(\Omega, w\,dx\,dy) = L^p_a$ whenever $1 \le p < 3 + \tau$.

Since there are only two restrictions, one on Ω and one on the L^p-class, TWO QUESTIONS arise:

(1) Can the assumption $\omega(\partial\Omega_\infty) > 0$ be removed?

(2) Is the theorem true for all p, $1 \le p < \infty$?

If $w(t) \downarrow 0$ in a sufficiently regular fashion then the answer to both questions is yes. For example, if $w(t) = e^{-h(t)}$ and $t^2 h(t) \uparrow +\infty$ as $t \downarrow 0$ the divergence of the integral $\int_0 \log h(t)\, dt$ is sufficient to guarantee that $H^p(\Omega, w\,dx\,dy) = L^p_a$ for all p, $1 \le p < \infty$, even when $\omega(\partial\Omega_\infty) = 0$.

In order to give an indication of how the hypotheses are used, here is a brief outline of the proof of Theorem 1. For each $q > 1$ we shall denote by C_q the capacity naturally associated with the Sobolev space $W^{1,q}$ and Λ_α will stand for α-dimensional Hausdorff measure. A comparison of these two set functions together with their formal definitions can be found in the survey article of Maz'ja and Havin [5].

Let g be any function in $L^q(\Omega, w\,dx\,dy)$, $q = p/(p-1)$, with the property that $\int Q g w\,dx\,dy = 0$ for all polynomials Q and form the Cauchy integral

$$f(\zeta) = \int_\Omega \frac{g w(z)}{z - \zeta}\, dx\,dy\,, \quad z = x + iy\,.$$

Evidently, f vanishes identically in Ω_∞ and so, by "continuity", $f = 0$ a.e. $-C_q$ on $\partial\Omega_\infty$. To establish the completeness of the polynomials we have only to prove that $f = 0$ a.e. $-C_q$ on the rest of the boundary as well, this approach having first been suggested by Havin [6] (cf. also [3] and [4]). The argument is then carried out n two stages; one verifies that:

STEP 1. $f=0$ a.e. with respect to harmonic measure on $\partial\Omega$;

STEP 2. $f=0$ a.e. with respect to the capacity C_q on $\partial\Omega$.

Moreover, in the process it will be convenient to transfer the problem from Ω to $\mathbb{D}$ by means of conformal mapping. With $\psi=\varphi^{-1}$ let $F=f(\psi)$. For each $\varepsilon>0$ let $\Omega_\varepsilon=\{z\in\Omega:|\varphi(z)|<1-\varepsilon\}$ and put

$$f_\varepsilon(\zeta)=\int_{\Omega_\varepsilon}\frac{gw(z)}{z-\zeta}\,dx\,dy \qquad \text{and } F_\varepsilon=f_\varepsilon(\psi).$$

Thus, F and F_ε are both defined on $\mathbb{D}$ and F_ε is analytic near $\partial\mathbb{D}$.

STEP 1. By choosing $\tau>0$ and sufficiently small we can find a corresponding $\delta>0$ such that the following series of implications are valid for any Borel set $E\subseteq\partial\Omega$:

$$C_q(E)=0\overset{(i)}{\Longrightarrow}\Lambda_{1/2+\delta}(E)=0\overset{(ii)}{\Longrightarrow}\omega(E)=0. \qquad (1.1)$$

Here $q=p/(p-1)$ and $p<3+\tau$. The first implication (i) is essentially due to Frostman [7]. Although he considered only Newtonian capacity, his argument readily extends to the nonlinear capacities which enter into the completeness problem (cf. Maz'ja-Havin [5]). The second assertion (ii) is a consequence of a very deep theorem of Carleson [8]. Because $f=0$ a.e. $-C_q$ on $\partial\Omega_\infty$ and $\omega(\partial\Omega_\infty)>0$ it follows that $f=0$ on some boundary set of positive harmonic measure. Consequently, taking radial limits, $F=0$ on a set of positive arc length on $\partial\mathbb{D}$.

We may now suppose that $w(t)=e^{-h(t)}$, where $h(t)\uparrow+\infty$ as $t\downarrow 0$. Then, using the fact that $\int|\varphi'|^p dx\,dy<\infty$ for $p<3+\tau$ (cf. [9] and [10], the latter being reprinted in this collection, Problem 8.8) it is an easy matter to check that

(1) $\displaystyle\int_{\partial\mathbb{D}}|F-F_\varepsilon|\,d\theta\leqslant e^{-ch(\varepsilon)}$

(2) $\displaystyle\int_{|z|=r}|F_\varepsilon|\,d\theta\leqslant K,\quad 1-\varepsilon\leqslant r\leqslant 1,$

where c and K are constants independent of δ. Because $F=0$ on a set of positive $d\theta$ measure on ∂D and $\int_0 \log h(t)\,dt = +\infty$, if follows that $F=0$ a.e. $-d\theta$. The argument here is based on a modification of Beurling's ideas [11] and it states, in essence, that those functions on ∂D which can be sufficiently well approximated by analytic functions retain the uniqueness property of the approximating family. As a general priciple, of course, this goes back to Bernstein [12]. The upshot is $f=0$ a.e. $-d\omega$ on $\partial\Omega$ and Step 1 is complete.

STEP 2. At this point we are required to show that from $f=0$ a.e. $-d\omega$ it can be concluded that $f=0$ a.e. $-C_q$ on $\partial\Omega$ which runs counter to the known relationship (1.1) between harmonic measure and capacity. We shall be content here to simply note that the reasoning is based on an argument from the author's article [13] and that essential use is made of the fact that Step1 is valid for e v e r y annihilator q.

In light of what has now been said one IMPORTANT QUESTION remains: <u>Is the divergence of the log log-integral necessary for completeness to occur; that is, if</u> $\int_0 \log\log \frac{1}{W(t)}\,dt < +\infty$ <u>is</u> $H^p(\Omega, w\,dxdy) \neq L^p_a$? In case the inner boundary of Ω contains an isolated "smooth" arc and $W(t) \downarrow 0$ as $t \downarrow 0$ the answer is yes and the proof is a simple adaptation of an argument of Domar [14].

REFERENCES

1. K e l d y c h M. Sur l'approximation en moyenne par polynômes des fonctions d'une variable complexe. - Матем.сборник, I945, 58, № I, I-20.
2. М е р г е л я н С.Н. О полноте систем аналитических функций. - Успехи матем.наук, I953, 8, № 4, 3-63.
3. B r e n n a n J. Approximation in the mean by polynomials on non-Caratheodory domains. - Ark.Mat., 1977, 15, 117-168.
4. B r e n n a n J. Weighted polynomial approximation, quasianalyticity and analytic continuation. - Preprint.
5. М а з ь я В.Г., Х а в и н В.П. Нелинейная теория потенциала. - Успехи матем.наук, I972, 27, № 6, 67-I38.
6. Х а в и н В.П. Аппроксимация аналитическими функциями в среднем. - Докл.АН СССР, I968, I78, № 5, I025-I028.
7. F r o s t m a n O. Potentiel d'équilibre et capacite des en-

sembles. - Meddel.Lunds Univ.Mat.Sem., 1935, N 3, 1-118.

8. Carleson L. On the distortion of sets on a Jordan curve under conformal mapping. - Duke Math.J.,1973, 40, 547-559.

9. Brennan J. The integrability of the derivative in conformal mapping. - J.London Math Soc.,1978, 18, 261-272.

0. Brennan J. Интегрируемость производной конформного отображения. - Зап.научн.семин.ЛОМИ, 1978, 81, 173-176.

1. Beurling A. Quasianalyticity and general distributions. Lecture Notes, Stanford Univ., 1961.

2. Bernstein S.N. "Léçons sur les Propriétés Extrémales et la Meilleure Approximation des Fonctions Analytiques d'une Variable Réelle",Gauthier-Villars, Paris, 1926.

13. Brennan J. Point evaluations, invariant subspaces and approximation in the mean by polynomials. - J.Functional Analysis,1979, 34, 407-420.

14. Domar Y. On the existence of a largest subharmonic minorant of a given function. - Ark.Mat.,1958, 3, 429-440.

J.BRENNAN

University of Kentucky
Lexington, Kentucky, 40506
USA

8.10. APPROXIMATION IN THE MEAN BY HARMONIC FUNCTIONS

We discuss analogues of the Vituškin approximation theorem [10] for mean approximation by harmonic functions. We assume that p is fixed, $1<p<\infty$. We let X be a compact subset of $\mathbb{R}^n$ of positive Lebesgue measure, and we assume $n \geqslant 3$. If $x \in \mathbb{R}^n$, let $B_r(x)=\{y \in \mathbb{R}^n : |y-x|<r\}$. All functions will be real-valued. If $\ell \in \{0,1\}$, let $\mathcal{P}_\ell$ denote the vector space of all polynomials on $\mathbb{R}^n$ which are homogeneous of degree ℓ, with inner product

$$\{P_1,P_2\}=\sum_{|\alpha|=\ell} c_\alpha^{(1)} c_\alpha^{(2)} \qquad \text{if} \quad P_i(\xi) \equiv \sum_{|\alpha|=\ell} c_\alpha^{(i)} \xi^\alpha .$$

If $k \in \{1,2\}$ is fixed, define the (positive) function $G_k \in L^1(\mathbb{R}^n, loc)$ as the inverse Fourier transform of $\hat{C}_k(\xi) = (1+|\xi|^2)^{-k/2}$, and for each $A \subset \mathbb{R}^n$ define the Bessel capacity

$b_{k,p'}(A) = \inf\{\|f\|_{L^{p'}(\mathbb{R}^n)} :\ f$ measurable and $f \geqslant 0$ on $\mathbb{R}^n$, $G_k * f \geqslant 1$ on $A\}$; if $kp'<n$, there exists a constant $C>0$ such that $C^{-1} \leqslant b_{k,p'}(B_r(0))/r^{n/p'-k} \leqslant C$ for $0<r\leqslant 1$. See [7], [8].

We say that X has the L^p h a r m o n i c a p p r o x i m a t i o n p r o p e r t y (L^p h.a.p.) provided that for each $\varepsilon>0$, and each function $f \in L^p(X)$ which is harmonic on the interior $\operatorname{int} X$, there exists a harmonic function u on an open neighborhood of X such that $\|u-f\|_{L^p(X)}<\varepsilon$.

THEOREM 1. <u>If any one of the following conditions holds, then X has the L^p h.a.p.</u>

a) ([8],[2]) $p'>n$.

b) ([8]) $p'<n$ <u>and there exists a constant</u> $\eta>0$ <u>such that</u>

$$b_{1,p'}(B_r(x)\setminus X) \geqslant \eta r^{n/p'-1} \quad \text{if} \quad x \in \partial X \quad \text{and} \quad 0<r\leqslant 1 .$$

c) ([4],[5]) <u>For each</u> $k \in \{1,2\}$ <u>one of the following two conditions is met:</u> (i) $kp'>n$ <u>or</u> (ii) $kp' \leqslant n$ <u>and there exists a set</u> E_k <u>with</u> $b_{k,p'}(E_k)=0$ <u>such that</u>

$$\int_0^1 (r^{k-n/p'} b_{k,p'}(B_r(x)\setminus X))^p \, dr/r = \infty \qquad \underline{\text{if}} \quad x \in \partial X \setminus E_k .$$

See also [5 , Theorem 6]; it follows from [8, Theorem 2.7] that the condition in c) for $k=2$ is necessary for the L^p h.a.p., but Hedberg has pointed out that the condition in c) for $k=1$ is not necessary (see [1 , Section 2] .) To characterize the sets having the L^p h.a.p. we define other capacities. We use the notation $\langle T, \Phi \rangle$ to denote the action of the distribution T of compact support on the function $\Phi \in C^\infty(\mathbb{R}^n)$. Let $E(x) = c_n/|x|^{n-2}$ be a fundamental solution for Δ . Let A be a subset of the open set $\Omega \subset \mathbb{R}^n$. If $H \in \mathcal{P}_0 \setminus \{0\}$, we define $\gamma_{p,H}(A,\Omega) = \sup_T |\langle T,H\rangle|$, where the supremum is taken over all (real) distributions T on $\mathbb{R}^n$ such that the support of T is a compact subset of A , $E*T \in L^1(\mathbb{R}^n, loc)$ and $\|E*T\|_{L^p(\Omega)} \leq 1$. If $H \in \mathcal{P}_1 \setminus \{0\}$, we define $\gamma_{p,H}(A,\Omega) = \sup_T |\langle T,H\rangle|$, where the supremum is taken over all distributions T on $\mathbb{R}^n$ satisfying the following four conditions: (i) the support of T is a compact subset of A ; (ii) $\langle T,1\rangle = 0$; (iii) $\langle T,P\rangle = 0$ for each $P \in \mathcal{P}_1$ satisfying $\{H,P\} = 0$; (iv) $E*T \in L^p(\mathbb{R}^n, loc)$ and $\|E*T\|_{L^p(\Omega)} \leq 1$. For references to related capacities of Harvey-Polking, Hedberg, and Maz'ja, see [1, Section 2, Remark 3].

The capacity $\gamma_{p,1}$ is closely related to the Bessel capacity $b_{2,p'}$. Moreover, if $k \in \{1,2\}$, one can prove that there exists a constant $C>0$ such that

$$\gamma_{p,H}(A, B_2(x)) \geq C\{H,H\}^{1/2} b_{k,p'}(A) \tag{1}$$

for $H \in \mathcal{P}_{2-k} \setminus \{0\}$, $x \in \mathbb{R}^n$, and $A \subset B_1(x)$ a Borel set. The next result follows from the proof of [1, Theorem 2.1], with obvious changes since here our functions are real-valued; the proof is constructive, extending techniques of Lindberg [6] which are based on those of Vituškin [10].

THEOREM 2. *The following conditions are equivalent.*

a) X *has the h.a.p.*

b) *If* $H \in \mathcal{P}_0 \cup \mathcal{P}_1 \setminus \{0\}$, *and if* G *and* Ω *are open subsets of* $\mathbb{R}^n$ *satisfying* $G \Subset \Omega \Subset \mathbb{R}^n$, *then* $\gamma_{p,H}(G \setminus \operatorname{int} X, \Omega) = \gamma_{p,H}(G \setminus X, \Omega)$.

c) *There exist numbers* $\eta > 0$ *and* $\rho > 0$ *such that*

$$\gamma_{p,H}(B_r(x) \setminus \operatorname{int} X, B_{2r}(X)) \leqslant \eta\, \gamma_{p,H}(B_r(H) \setminus X, B_{2r}(x))$$

if $H \in \mathcal{P}_0 \cup \mathcal{P}_1 \setminus \{0\}$, $x \in \mathbb{R}^n$ *and* $0 < r \leqslant 1$.

From Theorem 2 it is possible to deduce part a) of Theorem 1 (see [1, Section 2]) and part b) of Theorem 1; however, we have not been able to deduce part c) of Theorem 1, and this forms a motivation for the first problem below. We also note that a motivation for problems 1 and 3 is provided by corresponding results in the theory of mean approximation by analytic functions; see the references in the first paragraph of [1]. A number of other papers related to the present note are also given in the references in [1].

PROBLEMS. 1. *Can one characterize the compact sets having the* L^p *h.a.p. by means of conditions of Wiener type*? Specifically, let us say that X has property $(^*)$ provided that for each $k \in \{1,2\}$ one of the following two conditions is met: (i) $kp' > n$ or (ii) $kp' \leqslant n$ and there exists a set E_k with $b_{k,p'}(E_k) = 0$ such that

$$\int_0^1 r^{k-n/p'} \gamma_{p,H}(B_r(x) \setminus X, B_2(x))^p \, dr/r = \infty \quad \text{if } H \in \mathcal{P}_{2-k} \setminus \{0\}$$

and $x \in \partial X \setminus E_k$.

If X has the L^p h.a.p., then X has property $(^*)$; this follows from Theorem 2, (1) and the Kellogg property [5, Theorem 2]. Our question is whether the converse holds: *if* X *has property* $(^*)$, *does it have the* L^p *h.a.p.*? We remark that if this question were answered affirmatively, then part c) of Theorem 1 would follow by use of (1).

2. If $p=2$ and $n \geqslant 5$, a different criterion for the L^p h.a.p. follows from work of Saak [9]. What is the relation between Saak's work and Theorem 2?

3. If $H \in \mathcal{P}_0 \cup \mathcal{P}_1 \setminus \{0\}$ and Ω is an open set, then $\gamma_{p,H}(\cdot,\Omega)$ is an increasing set function defined on the subsets of Ω. Can one characterize the compact sets having the L^p h.a.p. by means of increasing set functions which are countably subadditive and have the property that all Borel sets are capacitable? (To say that a set E is capacitable with respect to a set function γ means that $\gamma(E)=\sup\{\gamma(K): K$ compact, $K\subset E\}=\inf\{\gamma(G): G$ open, $G\supset E\}$.) See [3],[7]. For the case $p=2$, see [9, Lemma 2].

REFERENCES

1. B a g b y T. Approximation in the mean by solutions of elliptic equations. - Trans.Amer.Math.Soc.
2. Б у р е н к о в В.И. О приближении функций из пространства $W_p^m(\Omega)$ финитными функциями для произвольного открытого множества Ω. - Труды Матем.ин-та им.В.А.Стеклова АН СССР,1974, 131, 51-63.
3. C h o q u e t G. Forme abstraite du théorème de capacitabilité. - Ann.Inst.Fourier (Grenoble), 1959, 9, 83-89.
4. H e d b e r g L.I. Spectral synthesis in Sobolev spaces, and uniqueness of solutions of the Dirichlet problem. - Acta Math., 1981, 147, 237-264.
5. H e d b e r g L.I., W o l f f T.H. Thin sets in nonlinear potential theory.Stockholm, 1982. (Rep.Dept.of Math.Univ. of Stockholm, Sweden, ISSN 0348-7652, N 24).
6. L i n d b e r g P. A constructive method for L^p-approximation by analytic functions. - Ark.för Mat., 1982, 20, 61-68.
7. M e y e r s N.G. A theory of capacities for functions in Lebesgue classes. - Math.Scand., 1970, 26, 255-292.
8. P o l k i n g J.C. Approximation in L^p by solutions of elliptic partial differential equations. - Amer.J.Math., 1972, 94, 1231-1244.
9. С а а к Э.М. Ёмкостный критерий для области с устойчивой задачей Дирихле для эллиптических уравнений высших порядков. - Ма-

тем.сб., I976, I00(I42), № 2 (6), 20I-209.

I0. В и т у ш к и н А.Г. Аналитическая емкость множеств в задачах теории приближений. - Успехи матем.наук, I967, 22, вып.6, I4I--I99.

THOMAS BAGBY

Indiana University
Department of Mathematics
Bloomington, Indiana 47405, USA

EDITORS' NOTE. Many years before the appearance of [6] the constructive techniques of Vitushkin was applied to the L^p-approximation by analytic functions by S.O.Sinanyan (Синанян С.О. Аппроксимация аналитическими функциями и полиномами в среднем по площади. - Матем.сб., I966, 69, № 4, 546-578.)See also the survey Мельников М.С., Синанян С.О. Вопросы теории приближений функций одного комплексного переменного. - В кн: "Итоги науки и техники". Современные проблемы математики, т.4, Москва, I975, изд-во ВИНИТИ, I43-250.

.11. old

RATIONAL APPROXIMATION OF ANALYTIC FUNCTIONS

1. Local approximations. Let

$$\mathfrak{f}(z)=\sum_{n\geq 0} f_n/z^n,\quad |z|>R_{\mathfrak{f}}\quad \left(R_{\mathfrak{f}}=\overline{\lim_n}\,|f_n|^{1/n}<\infty\right); \tag{1}$$

and let f be a complete analytic function corresponding to the element $\mathfrak{f}$. For any $n\in\mathbb{N}$ define $\nu_n(\mathfrak{f})=\sup\{\nu(\mathfrak{f}-r):r\in\mathcal{R}_n\}$ where $\nu(g)$ is the multiplicity of the zero of g at ∞, $\mathcal{R}_n$ is the set of all rational functions of degree at most n.

For any n there exists a unique function π_n, $\pi_n\in\mathcal{R}_n$, such that $\nu_n(\mathfrak{f})=\nu(\mathfrak{f}-\pi_n)$. It is called the n-th diagonal Padé approximant to the series (1). Let a, $a>1$ be an arbitrary fixed number and let $|\cdot|=a^{-\nu(\cdot)}$; then π_n is the function of the best approximation to $\mathfrak{f}$ in $\mathcal{R}_n$ with respect to the metric: $\lambda_n(\mathfrak{f})=|\mathfrak{f}-\pi_n|=\inf\{|\mathfrak{f}-r|:r\in\mathcal{R}_n\}$. See [1], [2] for a more detailed discussion on the Padé approximants (the definition in [2] slightly differs from the one given above).

For any power series (1) we have

$$\nu_n(\mathfrak{f})\geq n,\ n\in\mathbb{N};\quad \nu_n(\mathfrak{f})\geq 2n,\ n\in\Lambda,\ \Lambda\subset\mathbb{N}, \tag{2}$$

Λ being an infinite subset of $\mathbb{N}$ depending on $\mathfrak{f}$.

A functional analogue of the well-known Thue-Siegel-Roth theorem (see [3], Theorem 2, (i)) can be formulated in our case as follows: is $\mathfrak{f}$ is an element of an algebraic nonrational function then for any $\varkappa$, $\varkappa>2$, the inequality $\nu_n(\mathfrak{f})>\varkappa n$ holds only for a finite number of indices n. From this it follows easily that in our case

$$\lim n^{-1}\nu_n(\mathfrak{f})=2 \tag{3}$$

Apparently, this theorem is true for more general classes of analytic functions.

CONJECTURE 1. <u>If $\mathfrak{f}$ is an element of a multi-valued analytic function f with a finite set of singular points then (3) is valid</u>.

In connection with CONJECTURE 1 we note that if

$$\overline{\lim_n}\ n^{-1}\nu_n(f) = +\infty \tag{4}$$

then f is a single-valued analytic function; but for any $A, A>0$ the inequality $\overline{\lim_n}\ n^{-1}\nu_n(f) > A$ is compatible with the fact that f is multi-valued (the first assertion is contained essentially in [4],[5], the second follows from the results of Polya [6]).

Everything stated above can be reformulated in terms of sequences of normal indices of the diagonal Padé approximations (see[7], [1]). In essence the question is about possible lacunae in the sequence of the Hankel determinants

$$F_n = \begin{vmatrix} f_1 & f_2 & \dots & f_n \\ f_2 & f_3 & \dots & f_{n+1} \\ \cdot & \cdot & \cdot & \cdot \\ f_n & f_{n+1} & & f_{2n} \end{vmatrix}, \quad n \in \mathbb{N}.$$

Thus (3) means that the sequence $\{F_n\}$ has no "Hadamard lacunae" and (4) means that $\{F_n\}$ has "Ostrowski lacunae" (in the terminology of [8]). Apparently many results on lacunary power series (see [8]) have their analogues for diagonal Padé approximations.

2. Uniform approximation. We restrict ourselves by the corresponding approximation problems on discs centered at infinity for the functions satisfying (1). Let $R > R_f$, $E = \{z : |z| \geqslant R\}$ (f is holomorphic on E) and $\mathbb{D}_R = \{z : |z| < R\}$. Denote by $\rho_n(f)$ the best approximation of f on E by the elements of $\mathcal{R}_n$:

$$\rho_n(f) = \inf\{\|f - r\|_E : r \in \mathcal{R}_n\},$$ $\|\cdot\|$ is the supnorm on E.

Let $\mathfrak{F}$ be the set of all compacta, F, $F \subset \mathbb{D}_R$ (with the connected complement) such that f admits a holomorphic (single-valued) continuation on $\mathbb{C} \setminus F$. Denote by $C_{\mathbb{D}}(F)$ the Green capacity of F with respect to $\mathbb{D}_R$ (the capacity of the condenser (E, F)) and define

$$a_f = \sup\left\{\exp\frac{1}{C_{\mathbb{D}}(F)} : F \in \mathfrak{F}\right\}.$$

For every f

$$\overline{\lim_n}\ \rho_n(f)^{1/n} \leqslant a_f^{-1}. \tag{5}$$

This inequality follows from the results of Walsch ([9], ch.VIII).

CONJECTURE 2. <u>For any</u> f

$$\underline{\lim_n}\ \rho_n(f)^{1/n} \leqslant a_f^{-2}. \tag{6}$$

Inequalities (5), (6) are similar to inequalities (2). To clarify (here and further) the analogy with the local case one should pass in section 1 from ν_n to the best approximations λ_n. In particular, equality (3) will be written as

$$\lim_n \lambda_n(f)^{1/n} \leqslant a^{-2}.$$

CONJECTURE 3. <u>If</u> f <u>is an element of an analytic function</u> f <u>which has a finite set of singular points then</u>

$$\lim_n \rho_n(f)^{1/n} = a_f^{-2}. \tag{7}$$

If under the hypothesis of this conjecture f is a single-valued analytic function, both parts of (7) are obviously equal to zero.

CONJECTURE 3 can be proved for the case when all singular points of f lie on $\mathbb{R}$ (for the case of two singular points see [10]).

In contradistinction to the local case the question of validity of (7) remains open for the algebraic functions also.

REFERENCES

1. P e r r o n O. Die Lehre von den Kettenbrüchen, II, Stuttgart, 1957.
2. B a k e r G.A. Essentials of Padé Approximant, New-York, "AP", 1975.

3. U c h i y a m a S. Rational approximations to algebraic functions. - Jornal of the Faculty of Sciences Hokkaido University, Ser.I, 1961, vol.XV, N 3,4, 173-192.

4. Г о н ч а р А.А. Локальное условие однозначности аналитических функций. - Матем.сб., I972, 89, I48-I64.

5. Г о н ч а р А.А. О сходимости аппроксимаций Паде. - Матем.сб., I973, 92, I52-I64.

6. P o l y a G. Untersuchungen über Lücken und Singularitäten von Potenzreihen. - Math.Z., 1929, 29, 549-640.

7. Г о н ч а р А.А. О сходимости аппроксимаций Паде для некоторых классов мероморфных функций. - Матем.сб., I975, 97, 605 - - 627.

8. B i e b e r b a c h L. Analytische Fortsetzung. Berlin - Heidelberg, Springer-Verlag, 1955.

9. W a l s h J.L. Interpolation and approximation by rational functions in the complex domain. AMS Coll.Publ., 20, Sec.ed.1960.

IO. Г о н ч а р А.А. О скорости рациональной аппроксимации некоторых аналитических функций. - Матем.сб., I978, I05, I47-I63.

A.A.GONCHAR
(А.А.ГОНЧАР)

СССР, II7966, Москва
ул.Вавилова 42,
МИАН СССР.

.12. old

A CONVERGENCE PROBLEM ON RATIONAL APPROXIMATION IN SEVERAL VARIABLES

1. The one-variable case, $z \in \mathbb{C}$. Let me first give the background in the one-variable case. Let $f(z) = \sum c_j z^j$, $z \in \mathbb{C}$, be a formal power series and P/Q , $Q \not\equiv 0$, a rational function in one variable z of type (n, ν) , i.e. P is a polynomial of degree $\leq n$ and Q of degree $\leq \nu$. It is in general not possible to determine P/Q so that it interpolates to f of order at least $n+\nu+1$ at the origin (i.e. having the same Taylor polynomial of degree $n+\nu$ as f). However, given n and ν , we can always find a unique rational function P/Q of type (n,ν) such that P interpolates to fQ of order at least $n+\nu+1$ at the origin, i.e. $(fQ-P)(z) = O(z^{n+\nu+1})$. This function P/Q , the $[n,\nu]$-P a d é approximant to f , was first studied systematically by Padé in 1892; see [1]. In 1902 Montessus de Ballore [2] proved the following theorem which generalizes the well-known result on the circle of convergence for Taylor series.

THEOREM. Suppose f is holomorphic at the origin and meromorphic in $|z| < R$ with ν poles (counted with their multiplicities). Then the $[n,\nu]$ -Padé approximant to f , P_n/Q_n , converges uniformly to f , with geometric degree of convergence, in those compact subsets of $|z| < R$ which do not contain any poles of f .

With the assumption in the theorem it can also be proved that P_n/Q_n diverges outside $|z| = R$ if R is chosen as large as possible [3, p.269] and that the poles of P_n/Q_n converge to the poles of f in $|z| < R$. Furthermore, when n is sufficiently large, P_n/Q_n is the unique rational function of type (n,ν) which interpolates to f at the origin of order at least $n+\nu+1$. Montessus de Ballore's original proof used Hadamard's theory of polar singularities (see [4]). Today, several other, easier proofs are known; see for instance [5], [6], [7] and [8].

Padé approximants have been used in a variety of problems in numerical analysis and theoretical physics, for instance in the numerical evaluation of functions and in order to locate singularities of functions (see [1]). One reason for this is, of course, the fact that the Padé approximants of f are easy to calculate from the power series expansion of f . In recent years there has been an increasing interest in using analogous interpolation procedures to approximate functions of several variables (see [9]). I propose the problem to investigate in which sense it is possible to generalize Montessus de Ballore's theorem to several variables.

2. The two-variable case, $z=(z_1,z_2)$; $z_1,z_2\in\mathbb{C}$.
We first generalize the definition of Padé approximants to the two-variable case. Let $f(z)=\sum c_{jk}z_1^j z_2^k$ be a formal power series and let P/Q , $Q\not\equiv 0$, be a rational function in two variables z_1 and z_2 of type (n,ν) , i.e. P is a polynomial in z_1 and z_2 of degree $\leq n$ and Q of degree $\leq \nu$. By counting the number of coefficients in P and Q we see that it is always possible to determine P and Q so that, if $(Qf-P)(z)=\sum d_{jk}z_1^j z_2^k$, then $d_{jk}=0$ for $(j,k)\in S$, where S , the interpolation set, is a chosen subset of $\mathbb{N}\times\mathbb{N}$ with $\frac{1}{2}(n+1)(n+2)+\frac{1}{2}(\nu+1)(\nu+2)-1$ elements. There is no natural unique way to choose S but it seems reasonable to assume that $\{(j,k):j+k\leq n\}\subset S$ and that $(j,k)\in S\Rightarrow$ $\Rightarrow(l,m)\in S$ if $l\leq j$ and $m\leq k$. In this way we get a *rational approximant* P/Q of type (n,ν) to f corresponding to S . With a suitable choice of S , P/Q is unique [7 , Theorem 1.1]. The definition, elementary properties, and some convergence results have been considered for these and similar approximants in [9], [10] and [7]. The possibility to generalize Montessus de Ballore's theorem has been discussed in [6], [7] and [11] but the results are far from being complete.

PROBLEM 1. In what sense can Montessus de Ballore's theorem be generalized to several variables?

It is not clear what class of functions f one should use. We consider the following concrete situation. Let $f=F/G$, where F is holomorphic in the polydisc $\{z=(z_1,z_2):|z_i|<R_i,\ i=1,2\}$ and G is a polynomial of degree ν , $G(0)\neq 0$. By the method described above we obtain for every n a rational approximant P_n/Q_n of type (n,ν) to f corresponding to some chosen interpolation set $S=S_n$. In what region of $\mathbb{C}^2$ does P_n/Q_n converge to f ? Partial answers to this problem are given in [7] and [11] (in the latter with a somewhat different definition of the approximants). If $\nu=1$, explicit calculations are possible and sharp results are easy to obtain [7 , Section 4]. These show that in general we do not have convergence in $\{z:|z_i|<R_i,\ i=1,2\}\setminus\{z:G(z)=0\}$. This proves that the general analogue of the Montessus de Ballore's theorem is not true. It may be

dded, that it is easy to prove - by just using Cauchy's estimates - hat there exist rational functions $r_{n\nu}$ of type (n,ν) interpolaing to f at the origin of order at least $n+1$ and converging niformly, as $n \to \infty$, to f in compact subsets of $\{z:|z_i|<R_i\}\setminus\{z:G(z)=0\}$.

A disadvantage, however, of $r_{n\nu}$ compared to the rational aproximants defined above is that $r_{n\nu}$ is not possible to compute rom the Taylor series expansion of f (see [7 , Theorem 3.3]).

In the one-variable case the proof of Montessus de Ballore's heorem is essentially finished when you have proved that the poles f the Padé approximants converge to the poles of f . In the seveal-variable case, on the other hand, there are examples [7 , Section , Counterexample 2], when the rational approximants P_n / Q_n do ot converge in the whole region $\{z:|z_i|<R_i\}\setminus\{z:G(z)=0\}$ n spite of fact that the singularities of P_n / Q_n converge to the ingularities of F/G . This motivates:

PROBLEM 2. <u>Under what conditions does</u> Q_n <u>converge to</u> G ?

The choice of the interpolation set S_n is important for the onvergence. For instance, if $\nu=1$ and $R_1=R_2=\infty$, we get conergence in $\mathbb{C}^2\setminus\{z:G(z)=0\}$ with a suitable choice of S_n 7, Section 4]. On the other hand if we change just one point in S_n- ithout violating the reasonable choices of S_n indicated in the deinition of the rational approximants - we get examples [7, Section , Counterexample 1], where we do not have convergence in any polyisc around $z=0$.

PROBLEM 3. <u>How is the convergence</u> $P_n/Q_n \to f$ <u>influenced by the choice of the interpolation set</u> S_n ?

Since we do not get a complete generalization of Montessus de Ballore's theorem it is also natural to ask:

PROBLEM 4. <u>If the sequence of rational approximants does not converge, is there a subsequence that converges to</u> f ? (Compare [7, Theorem 3.4])

Finally, I want to propose the following conjecture.

CONJECTURE. $P_n/Q_n \to f$ <u>in</u> $\mathbb{C}^2\setminus\{z:G(z)=0\}$ <u>if</u> $R_1=R_2=\infty$ <u>and the interpolation set</u> S_n <u>is suitably chosen</u>. (Compare [11, Corollary 2] and the case $\nu=1$ referred to just after PROBLEM 2 above.)

REFERENCES

1. B a k e r C.A. Essentials of Padé Approximants. New York, Academic Press, 1975.
2. d e M o n t e s s u s d e B a l l o r e R. Sur les fractions continues algébriques. - Bull.Soc.Math.France,1902, 30, 28-36.
3. P e r r o n O. Die Lehre von den Kettenbrüchen. Band II. Stuttgart, Teubner, 1957.
4. G r a g g W.B. On Hadamard's theory of polar singularities. - In: Padé approximants and their applications (Graves-Morris, P.R., e.d.), London, Academic Press, 1973, 117-123.
5. S a f f E.B. An extension of Montessus de Ballore's theorem on the convergence of interpolation rational functions. - J. Approx.T., 1972, 6, 63-68.
6. C h i s h o l m J.S.R., G r a v e s - M o r r i s P.R. Generalization of the theorem of de Montessus to two-variable approximants. - Proc.Royal Soc.Ser.A., 1975, 342, 341-372.
7. K a r l s s o n J., W a l l i n H. Rational approximation by an interpolation procedure in several variables.- In: Padé and rational approximation (Saff, E.B. and Varga, R.S., eds.), New York, Academic Press, 1977, 83-100.
8. Г о н ч а р А.А. О сходимости обобщенных аппроксимаций Паде мероморфных функций. - Матем.сб., 1975, 98, 4, 563-577.
9. C h i s h o l m J.S.R. N -variable rational approximants. - In: Padé and rational approximation (Saff, E.B. and Varga, R.S., eds.), New York, Academic Press, 1977, 23-42.
10. Г о н ч а р А.А. Локальное условие однозначности аналитических функций нескольких переменных. - Матем.сб., 1974, 93, № 2, 296-313.
11. G r a v e s - M o r r i s P.R. Generalizations of the theorem of de Montessus using Canterbury approximant. - In: Padé and rational approximation (Saff, E.B. and Varga, R.S., eds.), New York, Academic Press, 1977, 73-82.

HANS WALLIN

U m e å U n i v e r s i t y
S-90187 Umeå, Sweden

* * *

COMMENTARY BY THE AUTHOR

In a recent paper A.Cuyt (A Montessus de Ballore theorem for multivariate Padé approximants, Dept. of Math., Univ. of Antwerp, Belgium, 1983) considers a multivariate rational approximant P/Q to f where P and Q are polynomials of degree $n\nu+n$ and $n\nu+\nu$, respectively, such that all the terms of P and Q of degree less than $n\nu$ vanish. It is then possible to determine P and Q so that $fQ-P$ has a power series expansion where the terms of degree $\leq n\nu+n+\nu$ are all zero. For this approximant P/Q she proves the following theorem where $P/Q = P_n/Q_n$ and P_n and Q_n have no common non-constant factor: Let $f=F/G$ where F is holomorphic in the polydisc $\{z: |z_i| < R_i\}$ and G is a polynomial of degree $\leq \nu$, $G(0) \neq 0$, and assume that $Q_n(0) \neq 0$ for infinitely many n. Then there exists a polynomial $Q(z)$ of degree ν such that $\{z: G(z)=0\} \subset \{z: Q(z)=0\}$ and a subsequence of $\{P_n/Q_n\}$ that converges uniformly to f on compact subsets of $\{z: |z_i| < R_i\} \setminus \{z: Q(z)=0\}$.

8.13. old BADLY-APPROXIMABLE FUNCTIONS ON CURVES AND REGIONS

Let X be a compact Hausforff space and A a uniform algebra on X : that is, A is uniformly closed, separates points, and contains the constants. For example, if $X \subset \mathbb{C}^n$ then we might take $A = P(X)$, the uniform limits on X of polynomials. We say that a function φ, $\varphi \in C(X)$, is b a d l y - a p p r o x i m a b l e (with respect to A) to mean

$$\|\varphi - f\|_\infty \geqslant \|\varphi\|_\infty \qquad \text{for all } f \in A,$$

where $\|\cdot\|_\infty$ is the supremum norm over X. The problems discussed here concern finding concrete descriptions of the badly-approximable functions for some classical function algebras. They are the functions that it is useless to try to approximate.

In this section, we let G be a bounded domain in $\mathbb{C}$, with boundary X, and let $A^*(X)$ be the algebra of boundary values of continuous functions on $G \cup X$ that are analytic in G. In case G is the open unit disc, then $A^*(X)$ is the "disc algebra" (regarded as consisting of functions on X and not on G).

POREDA'S THEOREM. [1]. If X consists of a simple closed Jordan curve, then φ, $\varphi \in C(X)$, is badly-approximable with respect to $A^*(X)$ if and only if φ has nonzero constant modulus, and $\operatorname{ind} \varphi < 0$.

Here, $\operatorname{ind} \varphi$ is the index of φ, defined as the winding number on X of φ around 0.

THEOREM A. [2] If φ, $\varphi \in C(X)$ has nonzero constant modulus and if $\operatorname{ind} \varphi < 0$, then φ is badly-approximable with respect to $A^*(X)$.

THEOREM B. [2] Each badly approximable (with respect to $A^*(X)$) function in $C(X)$ has constant modulus on the boundary of the complement of the closure of G.

THEOREM C. [2] Suppose that X consists of $N+1$ disjoint closed Jordan curves. If φ is badly-approximable with respect to $A^*(X)$, then φ has constant modulus, and $\operatorname{ind} \varphi < N$.

An example was given in [2] to show that the range $0 \leqslant \operatorname{ind} \varphi < N$ is indeterminate, so that one cannot tell from the winding number alone, on such domains, whether or not φ is badly approximable.

PROBLEM I. Find necessary and sufficient conditions for a function φ to be badly-approximable with respect to $A^*(X)$ if X is a finite union of disjoint Jordan curves.

Note: In the case of the annulus, $X=\{z:|z|=r$ or $|z|=1\}$, where $0<r<1$, supposing φ is of modulus 1 on X, it is shown in [2] that φ is badly-approximable with respect to $A^*(X)$ if and only if either $ind\,\varphi<0$ or $ind\,\varphi=0$ and

$$\frac{1}{2\pi}\int_{-\pi}^{\pi} arg(\varphi(e^{i\theta}))\,d\theta-\frac{1}{2\pi}\int_{-\pi}^{\pi} arg(\varphi(re^{i\theta}))\,d\theta\equiv\pi \bmod (2\pi).$$

PROBLEM II. The analogue of Problem I for $R^*(X)$, which is the limits on X of rational functions with poles off G, where one permits G to have infinitely many holes.

PROBLEM III. Characterize the badly-approximable functions with respect to $P(X)$, where X is any compact set in $\mathbb{C}$.

PROBLEM III′. The same as problem III but in the special case $X=clos\,\mathbb{D}$.

Despite appearances, Problem III' is just about as general as Problem III. An answer to Problem III could be called a "co-Mergelyan theorem" since Mergelyan's theorem [3] characterizes the "well-approximable" functions on X.

THEOREM. [4]. If φ is badly-approximable with respect to $P(clos\,\mathbb{D})$ then $\left\|\frac{\partial\varphi}{\partial\bar z}\right\|_\infty\geqslant\frac{3}{4}\|\varphi\|_\infty$ where $\|\cdot\|_\infty$ is the supremum norm over $\mathbb{D}$. The converse is false.

PROBLEM IV. Obtain, for sets X, $X\subset\mathbb{C}^n$, $n\geqslant 2$, any significant result about badly-approximable functions with respect to any algebra like $P(X)$, $A(X)$, or $R(X)$.

REFERENCES

1. Poreda S.J. A characterization of badly approximable functions. - Trans.Amer.Math.Soc. 1972, 169, 249-256.
2. Gamelin T.W., Garnett J.B., Rubel L.A., Shields A.L. On badly approximable functions. - J.Approx.

theory, 1976, 17, 280-296.
3. R u d i n W. Real and Complex Analysis, New York, 1966.
4. K r o n s t a d t E r i c. Private communication, September 1977.
5. L u e c k i n g D.H. On badly approximable functions and uniform algebras. - J.Approx.theory, 1978, 22, 161-176.
6. R u b e l L.A., S h i e l d s A.L. Badly approximable functions and interpolation by Blaschke products. - Proc.Edinburgh Math.Soc. 1976, 20, 159-161.

LEE A.RUBEL

University of Illinois at Urbana-
-Champaign Department of Mathematics
Urbana, Illinois 61801 USA

.14.
old

EXOTIC JORDAN ARCS IN $\mathbb{C}^n$

Let γ be a simple (non-closed) Jordan arc in $\mathbb{C}^n (n\geqslant 2)$, $P(\gamma)$ e the closure in $C(\gamma)$ of polynomials in complex variables, $O(\gamma)$ be the uniform closure on γ of algebra of functions holo-orphic in a neighbourhood of γ. Denote by $A(\gamma)$ a uniform al-ebra on γ such that $P(\gamma)\subset A(\gamma)\subset C(\gamma)$ and let $h_A(\gamma)$ be ts spectrum (maximal ideal space). For an arbitrary compact set K n $\mathbb{C}^n$ the spectrum $h_A(K)$ depends essentially on the choice of he subalgebra $A(K)$. Until recently, however, it seemed plau-ible that for Jordan arcs the spectrum $h_A(\gamma)$ depends on γ nly.

Consider also the algebra $R(\gamma)$ of uniform limits on γ of ational functions with poles off γ, and the algebra $H(\gamma)$ which is the closure in $C(\gamma)$ of the set of all functions holomor-phic in a pseudoconvex neighbourhood of γ. Then we obviously have

$$P(\gamma)\subset R(\gamma)\subset H(\gamma)\subset O(\gamma).$$

In 1968 A.Vitushkin (see [1, 2]) discovered the first example of a rationally convex but not polynomially convex arc γ in $\mathbb{C}^2$. In other words in this example

$$h_R(\gamma)=\gamma, \quad \text{but} \quad h_P(\gamma)\neq\gamma.$$

In 1974 the author (see [2], p.116; [3], p.174) found an example of Jordan arc γ in $\mathbb{C}^2$ which, being holomorphically convex, cannot, nevertheless, coincide with an intersection of holomorphically con-vex domains, i.e. $h_O(\gamma)=\gamma$ but $h_H(\gamma)\neq\gamma$.

A curious problem remains, however, unsolved. Namely, whether $h_R(\gamma)=h_H(\gamma)$ for every Jordan arc.

CONJECTURE 1. <u>There exists a Jordan arc</u> γ <u>in</u> $\mathbb{C}^n$ <u>satisfying</u>

$$h_R(\gamma)\neq\gamma, \quad h_H(\gamma)=\gamma.$$

Consider now the algebra $A(K,S)$ of all functions continuous on the Riemann sphere S and holomorphic outside a compact set K, $K\subset S$. To prove conjecture 1 it is sufficient, for example, to prove the following statement which simultaneously strengthens the classical results of J.Wermer (see [4], [6]) and R.Arens (see [5], [6]).

CONJECTURE 2. <u>There exists a Jordan arc γ on S such that $A(\gamma, S)$ contains a finitely generated subalgebra with the spectrum</u> S.

All known exotic Jordan arcs in $\mathbb{C}^n$ are of positive two-dimensional Hausdorff measure. It would be very interesting therefore to prove that there is no exotic arc of zero two-dimensional Hausdorff measure.

PROBLEM. <u>Suppose that a simple (non-closed) Jordan arc γ in $\mathbb{C}^n$ has zero 2-dimensional Hausdorff measure. Is γ polynomially convex</u> (i.e. $h_p(\gamma)=\gamma$)?

Recall that H.Alexander [7] has proved that every rectifiable simple arc in $\mathbb{C}^n$ is polynomially convex.

REFERENCES

1. Витушкин А.Г. Об одной задаче Рудина. - Докл.АН СССР, 1973, 213, № 1, 14-15.
2. Хенкин Г.М., Чирка Е.М. Граничные свойства голоморфных функций нескольких комплексных переменных. - В кн.: Современные проблемы математики, 4, М., ВИНИТИ, 1975, 13-142.
3. Wells R.O. Function theory on differentiable submanifolds. - In: Contributions to analysis. A collection papers dedicated to Lipman Bers, 1974, Academic Press, INC, 407-441.
4. Wermer J. Polynomial approximation on an arc in $\mathbb{C}^3$. - Ann.Math., 1955, 62, N 2, 269-270.
5. Arens R. The maximal ideals of certain function algebras. - Pacific J.Math., 1958, 8, 641-648.
6. Gamelin Th.W. Uniform algebras. Prentice-Hall, INC, N.J., 1969.
7. Alexander H. Polynomial approximation and hulls in sets of finite linear measure in $\mathbb{C}^n$. - Amer.J.Math., 1971, 93, N 1, 65-74.

G.M.HENKIN

(Г.М.ХЕНКИН)

СССР, 117418, Москва
ул.Красикова 32,
Центральный экономико-
математический институт АН СССР

.15. REMOVABLE SETS FOR BOUNDED ANALYTIC FUNCTIONS
old

Suppose E is a compact subset of an open set V, $V \subset \mathbb{C}$. hen E is said to be removable, or a Painlevé null set [1], if very bounded analytic function on $V \setminus E$ extends to be analytic on V. This is easily seen to be a property of the set E and not V. ainlevé [2] asked for a necessary and sufficient condition for a ompact set E to be removable. The corresponding problem for harmoic functions has been answered in terms of logarithmic capacity and ransfinite diameter. Ahlfors [3] has restated the question in terms f the following extremal problem. Let

$$\gamma(E) = \sup\left\{ \lim_{z\to\infty} |z(f(z) - f(\infty))| : f \text{ is analytic on } \mathbb{C}\setminus E \text{ and } |f(z)| \le 1 \right\}$$

e the analytic capacity of E. Then E is removable if and only f $\gamma(E)=0$. A geometric solution to this problem would have appliations in rational approximation and cluster-value theory. See, for xample, [4] and [5]. Also [6] contains an interesting historical ccount.

It is known that Hausdorff measure is not "fine" enough to characterize removable sets. Painlevé (and later Besicovitch [7]) proved that if the 1-dimensional Hausdorff measure, $H^1(E)$, is zero then $\gamma(E)=0$. It is also classical that if $H^{1+\varepsilon}(E)>0$, for some $\varepsilon>0$, then $\gamma(E)>0$. However examples, [8], [9], show that it is possible for $H^1(E)>0$ and $\gamma(E)=0$ *).

If ℓ_θ is the ray from the origin with argument θ, let $|P_\theta(E)|$ denote the Lebesgue measure of the orthogonal projection of E on ℓ_θ. Let

$$CR(E) = \int_0^{2\pi} |P_\theta(E)|\, d\theta / \pi .$$

This quantity first arose in connection with the solution of the Buffon needle problem as given by Crofton [11] in 1868. If the diameter of E is less than 1, it is the probability of E falling on a system of parallel lines one unit apart. See [12] for an interesting geometric interpretation. Vitushkin [4] asked if $CR(E)=0$ is

*) See also pp. 346-348 of the book [10] - Ed.

equivalent to $\gamma(E)=0$. It is not hard to see that if $H^1(E)=0$, then $CR(E)=0$. Marstrand [13] has proved that if $H^{1+\varepsilon}(E)>0$ then $CR(E)>0$. In order to answer Vitushkin's question, one thus needs to consider only sets of Hausdorff dimension 1 .

A special case is the following theorem asserted by Denjoy [14] in 1909.

THEOREM. If E is a compact subset of a rectifiable curve Γ then $\gamma(E)=0$ if and only if $H^1(E)=0$.

Although his proof has a gap, Ahlfors and Beurling [1] noted that it is correct if Γ is a straight line. They extended this result to analytic curves Γ . Ivanov [16] proved it for curves slightly smoother than C^1 . Davie [17] proved that it sufficed to assume Γ is a C^1 curve. Recently, A.P.Calderón [18] proved that the Cauchy integral operator, for C^1 curves, is bounded on L^p , $1<p<\infty$. Denjoy's conjecture is a corollary of this theorem. Here is an OUTLINE OF THE PROOF.

Let $\mathcal{D}$ be a finitely connected planar domain bounded by C , a union of rectifiable arcs $C_1,\dots,C_k$. Let F_n map the unit disk conformally onto C_h^c and let $C_k^\tau = F_n(|z|=\tau)$. We say that f , analytic in $\mathcal{D}$, is in $E^2(\mathcal{D})$ if and only if $\sup\limits_{\tau>\tau_0}\int_{\bigcup_n C_n^\tau} |f(z)|^2\,|dz|<\infty$ and define $\|f\|_{E^2}^2=\int_C |f(z)|^2\frac{|dz|}{2\pi}$, where C is traced twice if it is an arc.

LEMMA 1. ([19]). If C consists of finitely many analytic curves, then

$$\gamma(C)^{1/2} = \sup(|f'(\infty)| : f\in E^2(\mathcal{D}),\ \|f\|_{E^2}\le 1).$$

In this classical paper, Garabedian introduces the dual extremal problem: $\inf(\|g\|_{E^1} : g\in E^1(\mathcal{D}),\ g(\infty)=1)$ to obtain the above relation. It was noticed by Havinson [15] that the result remains true for rectifiable arcs. If $g\in L^2(C)$, let $G(z)=\int_C \frac{g(\zeta)}{\zeta-z}\frac{d\zeta}{2\pi i}$.

LEMMA 2. ([20]). If C is the union of finitely many C^1 - curves and if the Cauchy integral has boundary values G^*, $G^*\in L^2(C)$, then $G\in E^2(\mathcal{D})$.

This follows by writing $G = \sum G_n$ where each G_n is analytic off one of the contours in C. Then use the well-known fact that $\int_{|z|=r} |f(z)|^2 |dz|$ increases with r if $f \in H^2(\mathbb{D})$.

LEMMA 3. ([21]). Let C be a C^1 curve. If for all g, $g \in L^2(C)$, we have $G \in E^2(\mathcal{D})$, then the length and capacity of a subset E of C are simultaneously positive or zero.

This follows by approximating the set E by a subset $\tilde{E}$ of C consisting of finitely many subarcs, then applying Lemma 1 to the characteristic function of $\tilde{E}$.

Thus by Calderón's theorem, Denjoy's conjecture is true for C^1 curves. Davie's result finishes the proof. Incidentally an older theorem of [25], p.267, immediately implies Davie's result. ●

About the same time that Besicovitch rediscovered Painlevé's theorem (see above), he proved one of the fundamental theorems of geometric measure theory. A set E is said to be *regular* if it is contained in a countable union of rectifiable curves. A set E is said to be *irregular* if

$$\limsup_{r \to 0} \frac{H^1(E \cap B(x,r))}{r} \neq \liminf_{r \to 0} \frac{H^1(E \cap B(x,r))}{r} \quad \text{for } H^1\text{-a.a. } x, x \in E,$$

where $B(x,r) = \{y : |y - x| \leq r\}$. Besicovitch [22] proved that if $H^1(E) < \infty$ then $E = E_1 \cup E_2$ where E_1 is regular and E_2 is irregular. Later [23], he showed that if E is irregular, then the orthogonal projection of E in almost all directions has zero length. Thus if $H^1(E) < \infty$ and $CR(E) > 0$ there is a rectifiable curve Γ so that the length of $E \cap \Gamma$ is positive. Since Denjoy's conjecture is true, $\gamma(E)$ is positive whenever $H^1(E) < \infty$ and $CR(E) > 0$.

All examples where the analytic capacity is known concur with Vitushkin's conjecture. For instance, let $\tilde{E}$ be the cross product of the Cantor set, obtained by removing middle halves, with itself. It is shown in [9] that $\gamma(E) = 0$. For each x, $x \in E$, one can find annuli centered on x which are disjoint from $\tilde{E}$ and proportional in size to their distance from x. Thus $\tilde{E}$ is irregular and $CR(E) = 0$. We remark that the projection of $\tilde{E}$ on a line with slope $1/2$ is a full segment. Another relevant example is the cross product of the usual Cantor tertiary set with itself, call it F. The Hausdorff dimension of F is greater than one so that $CR(F) > 0$

and $\gamma(F)>0$. However every subset $\tilde{F}$ of F with $H^1(\tilde{F})<\infty$ is irregular and hence satisfies $CR(\tilde{F})=0$. This shows we cannot easily reduce the problem to compact sets E with $H^1(E)<\infty$.

If $\gamma(E)>0$, one possible approach to prove $CR(E)>0$ is to consider the set $\tilde{E}=\left\{\frac{1+e^{i\theta}}{2}z : z\in E, 0\leq\theta\leq 2\pi\right\}$. A point ζ is not in $\tilde{E}$ if and only if the line passing through z and whose distance to the origin is $|z|$, misses the set E . It is not hard to see $CR(E)>0$ if and only if $\tilde{E}$ has positive area. Uy [24] *) has recently shown that a set F has positive area if and only if there is a Lipschitz continuous function which is analytic on $\mathbb{C}\setminus F$. So one might try to construct such a function for the set $\tilde{E}$. A related question was asbed by A.Beurling. He asked, if $\gamma(E)>0$ and if E has no removable points, then must the part of the boundary of the normal fundamental domain (for the universal covering map) on the unit circle have positive length? This was shown to fail in [26].

Finally, I would like to mention that I see no reason why $CR(E)$ is not comparable to analytic capacity. In other words, does there exist a constant K with $1/K\cdot CR(E)\leq\gamma(E)\leq K\cdot CR(E)$? If this were true, it would have application to other problems. For example, it would prove that analytic capacity is semi-subadditive.

REFERENCES

1. A h l f o r s L.V., B e u r l i n g A. Conformal invariants and function-theoretic null sets. - Acta Math., 1950, 83, 101-129
2. P a i n l e v é P. Sur les lignes singulières des fonctions analytiques. - Ann.Fac.Sci. Toulouse, 1888, 2.
3. A h l f o r s L.V. Bounded analytic functions. - Duke Math.J., 1947, 14, 1-11.
4. В и т у ш к и н А.Г. Аналитическая емкость множеств в задачах теории приближений. - Успехи матем.наук,1967,22,№6, 141-199.
5. Z a l c m a n L. Analytic capacity and Rational Approximation - Lect.Notes Math., N 50, Berlin, Springer, 1968.
6. C o l l i n g w o o d E.P., L o h w a t e r A.J. The Theory of Cluster Sets. Cambridge, Cambridge U.P., 1966.
7. B e s i c o v i t c h A. On sufficient conditions for a function to be analytic and on behavior of analytic functions in the neighborhood of non-isolated singular points. - Proc.London Math. Soc., 1931, 32, N 2, 1-9.

*) See [27] for a short proof. - Ed.

8. Витушкин А.Г. Пример множества положительной длины, но нулевой аналитической емкости.-Докл.АН СССР, 1959, 127, 246-249.
9. Garnett J. Positive length but zero analytic capacity - Proc.Amer.Math.Soc., 1970, 24, 696-699.
10. Иванов Л.Д. Вариации множеств и функций. М., "Наука", 1975.
11. Crofton M.W. On the theory of Local Probability. - Philos. Trans.Roy.Soc., 1968, 177, 181-199.
12. Sylvester J.J. On a funicular solution of Buffon's "Problem of the needle" in its most general form. - Acta Math, 1891, 14, 185-205.
13. Marstrand J.M. Fundamental geometrical properties of plane sets of fractional dimensions. - Proc.London Math.Soc., 1954, 4, 257-302.
14. Denjoy A. Sur les fonctions analytiques uniformes à singularités discontinues. - C.R.Acad.Sci.Paris, 1909, 149, 258-260.
15. Хавинсон С.Я. Об аналитической емкости множеств, совместной нетривиальности различных классов аналитических функций и лемме Шварца в произвольных областях. - Матем.сб., 1961, 54, № 1, 3-50.
16. Иванов Л.Д. Об аналитической емкости линейных множеств. - Успехи матем.наук, 1962, 17, 143-144.
17. Davie A.M. Analytic capacity and approximation problems. - Trans.Amer.Math.Soc., 1972, 171, 409-444.
18. Calderón A.P. Cauchy integrals on Lipschitz curves and related operators. - Proc.Nat.Acad.Sci. USA, 1977, 74, 1324-1327
19. Garabedian P.R. Schwarz's lemma and the Szegö kernel function. - Trans.Amer.Math.Soc., 1949, 67, 1-35.
20. Хавин В.П. Граничные свойства интегралов типа Коши и гармонически сопряженных функций в областях со спрямляемой границей. - Матем.сб., 1965, 68, 499-517.
21. Хавин В.П., Хавинсон С.Я. Некоторые оценки аналитической емкости. - Докл.АН СССР, 1961, 138, 789-792.
22. Besicovitch A. On the fundamental geometrical properties of linearly measurable plane sets of points I. - Math.Ann., 1927, 98, 422-464. II: Math.Ann., 1938, 115, 296-329.
23. Besicovitch A. On the fundamental geometrical properties of linearly measurable plane sets of points III. - Math.Ann., 1939, 116, 349-357.
24. Uy N. Removable sets of analytic functions satisfying a Lipschitz condition. - Ark.Mat., 1979, 17, 19-27.
25. Federer H. Geometric measure theory. Springer-Verlag, Ber-

lin, 1969.

26. M a r s h a l l D.E. Painlevé null sets. Colloq. d'Analyse Harmonique et Complexe. Ed.: G.Detraz, L.Gruman, J.-P.Rosay. Univ. Aix-Marseill I, Marseill, 1977.

27. Х р у щ ё в С.В. Простое доказательство теоремы об устранимых особенностях аналитических функций, удовлетворяющих условию Липшица. - Зап.науч.сем.ЛОМИ, I98I, II3, I99-203.

DONALD E.MARSHALL Department of Mathematics, University of Washington, Seattle, Washington 98195 USA

Research supported in part by National Foundation Grant No MCS 77-01873

.16.
old

ON PAINLEVÉ NULL SETS

Suppose that E is a compact plane set and that N is an open neighbourhood of E. A set is called a P a i n l e v é n u l l s e t (or P.N. set) if every function regular and bounded in $N \setminus E$ can be analytically continued onto E. In this case we also say that E has z e r o a n a l y t i c c a p a c i t y.

The problem of the structure of P.N. sets has a long history. Painlevé proved that if E has linear (i.e. 1 - dimensional Hausdorff) measure zero, then E is a P.N. set, though it seems that this result was first published by Zoretti [1]. Painlevé's theorem has been rediscovered by various people including Besicovitch [2] who proved that if f is continuous on E, as well as regular outside E, and if E has finite linear measure, then f can be analytically continued onto E. Denjoy [3] conjectured that if E lies on a rectifiable curve, then E is a P.N. set if and only if E has linear measure zero. He proved this result for linear sets. Ahlfors and Beurling [4] proved Denjoy's conjecture for sets on analytic curves and Ivanov [5] for sets on sufficiently smooth curves. Davie [6] has shown that it is sufficient to prove Denjoy's conjecture for C^1 curves. On the other hand Havin and Havinson [7] and Havin [8] showed that Denjoy's conjecture follows if the Cauchy integral operator is bounded on L^2 for C^1 curves. This latter result has now been proved by Calderón [9] so that Denjoy's conjecture is true. I am grateful to D.E.Marshall [10] for informing me about the above results.

Besicovitch [11] proved that every compact set E of finite linear measure is the union of two subsets E_1, E_2. The subset E_1 lies on the union of a finite or countable number of rectifiable Jordan arcs. It follows from the above result that E_1 is not a P.N. set unless E_1 has linear measure zero. The set E_2 on the other hand meets every rectifiable curve in a set of measure zero, has projection zero in almost all directions and has a linear density at almost none of its points. The sets E_1 and E_2 were called respectively r e g u l a r and i r r e g u l a r by Besicovitch [11]. Since irregular sets behave in some respects like sets of measure zero I have tentatively conjectured [12, p.231] that they might be P.N. sets. Vitushkin [13] and Garnett [14] have given examples of irregular sets which are indeed P.N. sets, but the complete conjecture is still open.

A more comprehensive conjecture is due to Vitushkin [15 , p.147]

He conjectures that E is a P.N. set if and only if E has zero projection in almost all directions.

It is not difficult to see that a compact set E is a P.N. set if and only if for every bounded complex measure distributed on E, the function

$$F(z)=\int_E \frac{d\mu(\zeta)}{z-\zeta} \qquad (1)$$

is unbounded outside E [*]. Thus E is certainly not a P.N. set if there exists a positive unit measure μ on E such that

$$\int_E \frac{d\mu(\zeta)}{|z-\zeta|}$$

is bounded outside E i.e. if E has positive linear capacity [16, p.73]. This is certainly the case if E has positive measure with respect to some Hausdorff function h, such that

$$\int_0^1 \frac{h(r)\,dr}{r^2} < \infty$$

([17]). Thus in particular E is not a P.N. set if E has Hausdorff dimension greater than one. While a full geometrical characterization of P.N. sets is likely to be difficult there still seems plenty of scope for further work on this intriguing class of sets.

REFERENCES

1. Z o r e t t i L. Sur les fonctions analytiques uniformes qui possèdent un ensemble parfait discontinu de points singuliers. - J.Math.Pures Appl., 1905, 6, N 1, 1-51.
2. B e s i c o v i t c h A. On sufficient conditions for a function to be analytic and on behavior of analytic functions in the neighborhood of non-isolated singular points. - Proc.London Math. Soc., 1931, 32, N 2, 1-9.
3. D e n j o y A. Sur les fonctions analytiques uniformes à singularités discontinues. - C.R. Acad.Sci.Paris, 1909, 149, 258--260.

[*] See Ed. note at the end of the section. - Ed.

4. Х а в и н с о н С.Я. Об аналитической емкости множеств, совместной нетривиальности различных классов аналитических функций и лемме Шварца в произвольных областях. - Матем.сб., 1961, 54, № 1, 3-50.

5. И в а н о в Л.Д. О гипотезе Данжуа. - Успехи матем.наук, 1964, 18, 147-149.

6. D a v i e A.M. Analytic capacity and approximation problems. - Trans.Amer.Math.Soc., 1972, 171, 409-444.

7. Х а в и н В.П., Х а в и н с о н С.Я. Некоторые оценки аналитической емкости. - Докл.АН СССР, 1961, 138, 789-792.

8. Х а в и н В.П. Граничные свойства интегралов типа Коши и гармонически сопряженных функций в областях со спрямляемой границей. - Матем.сб., 1965, 68, 499-517.

9. C a l d e r ó n A.P. Cauchy integrals on Lipschitz curves and related operators. - Proc.Natl.Acad.Sci. USA, 1977, 74, 1324-1327.

10. M a r s h a l l D.E. The Denjoy Conjecture. Preprint, 1977.

11. B e s i c o v i t c h A. On the fundamental geometrical properties of linearly measurable plane sets of points I. - Math. Ann., 1927, 98, 422-464. II: Math .Ann., 1938, 115, 296-329.

12. H a y m a n W.K., K e n n e d y P.B. Subharmonic Functions Vol.1. London - N.Y., Academic Press, 1976.

13. В и т у ш к и н А.Г. Пример множества положительной длины, но нулевой аналитической емкости. - Докл.АН СССР, 1969, 127, 246-249.

14. G a r n e t t J. Positive length but zero analytic capacity. - Proc.Amer.Math.Soc., 1970, 24, 696-699.

15. В и т у ш к и н А.Г. Аналитическая емкость множеств в задачах теории приближений. - Успехи матем.наук, 1967, 22, № 6, 141-199.

16. C a r l e s o n L. Selected problems on exceptional sets. - Van Nostrand Math.stud., N 13, Toronto, Van Nostrand, 1967.

17. F r o s t m a n O. Potentiel d'équilibre et capacité des ensembles avec quelques applications à la théorie des fonctions. - Medded.Lunds,Univ.Mat.Sem., 1935, 3, 1-118.

18. В и т у ш к и н А.Г. Об одной задаче Данжуа. - Изв.АН СССР, сер.матем., 1964, 28, № 4, 745-756.

19. В а л ь с к и й Р.Э. Несколько замечаний об ограниченных аналитических функциях, представимых интегралом типа Коши-Стилтьеса. - Сиб.матем.ж., 1966, 7, № 2, 252-260.

W.K.HAYMAN

Imperial College, Department of Mathematics, South Kensington, London SW7 England

EDITORS' NOTE. As far as we know the representability of all functions bounded and analytic off E and vanishing at infinity by "Cauchy potentials" (1) is guaranteed when E has finite Painlevé's length whereas examples show that this is no longer true for an arbitrary E ([18],[19]). We think THE QUESTION of existence of potentials (1) bounded in $\mathbb{C} \setminus E$ (provided E is not a P.N. set) is one more interesting problem (see also § 5 of [4]).

.17.
old

ANALYTIC CAPACITY AND RATIONAL APPROXIMATION

Let E be a bounded subset of $\mathbb{C}$ and $B(E,1)$ be the set of all functions f in $\hat{\mathbb{C}}$ analytic on $\hat{\mathbb{C}}\setminus E$ and with $f(\infty)=0$, $\sup_{\mathbb{C}} |f| \leqslant 1$. Put $A(E,1)=\{f \in B(E,1):\ f$ is continuous on $\mathbb{C}\}$. The number

$$\gamma(E) \overset{def}{=} \sup_{f\in B(E,1)} \lim_{z\to\infty} |zf(z)|$$

is called the analytic capacity of E. The number

$$\alpha(E) \overset{def}{=} \sup_{f\in A(E,1)} \lim_{z\to\infty} |zf(z)|$$

is called the analytic C-capacity of E.

The analytic capacity has been introduced by Ahlfors [1] in connection with the Painlevé problem to describe sets of removable singularities of bounded analytic functions. Ahlfors [1] has proved that these sets are characterized by $\gamma(E)=0$. However, it would be desirable to describe removable sets in metric terms.

CONJECTURE 1. <u>A compact set E, $E\subset\mathbb{C}$, has zero analytic capacity iff the projection of E onto almost every direction has zero length</u> ("almost every" means "a.e. with respect to the linear measure on the unit circle). Such an E is called irregular provided its linear Hausdorff measure is positive.

If the linear Hausdorff measure of E is finite and $\gamma(E)=0$ then the average of the measures of the projections of E is zero. This follows from the Calderon's result [2] and the well-known theorems about irregular sets (see [3], p.341-348). The connection between the capacity and measures is described in detail in [4].

The capacitary characteristics are most efficient in the approximation theory [5],[6],[7],[8]. A number of approximation problems leads to an unsolved question of the semiadditivity of the analytic capacity:

$$\gamma(E\cup F) \leqslant C\,[\gamma(E)+\gamma(F)]\ ,$$

where C is an absolute constant and E, F are arbitrary disjoint compact sets.

Let $A(K)$ denote the algebra of all functions continuous on a compact set K, $K \subset \mathbb{C}$, and analytic in its interior. Let $R(K)$ denote the uniform closure of rational functions with poles off K and, finally, let $\partial^{\circ} K$ be the inner boundary of K, i.e. the set of boundary points of K not belonging to the boundary of a component of $\mathbb{C} \setminus K$. Sets K satisfying $A(K) = R(K)$ were characterized in terms of the analytic capacity [6]. To obtain geometrical conditions of the approximability a further study of capacities is needed.

CONJECTURE 2. <u>If</u> $\alpha(\partial^{\circ} K) = 0$ <u>then</u> $A(K) = R(K)$.

The affirmative answer to the question of semiaddivity would yield a proof of this conjecture. Since $\alpha(E) = 0$ provided E is of finite linear Hausdorff measure this would also lead to the proof of the following statement.

CONJECTURE 3. <u>If the linear Hausdorff measure of</u> $\partial^{\circ} K$ <u>is zero</u> (K <u>being a compact subset of</u> $\mathbb{C}$) <u>then</u> $A(K) = C(K)$.

The last equality is not proved even for K's with $\partial^{\circ} K$ of zero linear Hausdorff measure.

It is possible however that the semiadditivity problem can be avoided in the proof of CONJECTURE 3.

The semiadditivity of the capacity has been proved only in some special cases ([9],[10-13]), e.g. for sets E and F separated by a straight line. For a detailed discussion of this and some other relevant problems see [14].

REFERENCES

1. A h l f o r s L.V. Bounded analytic functions. - Duke Math.J., 1947, 14, 1-11.
2. C a l d e r ó n A.P. Cauchy integrals on Lipschitz curves and related operators. - Proc.Nat.Acad.Sci., USA, 1977, 74, 1324-1327.
3. И в а н о в Л.Д. Вариации множеств и функций, М., "Наука", 1975.
4. G a r n e t t J. Analytic capacity and measure. - Lect.Notes Math., 297, Berlin, Springer, 1972.
5. В и т у ш к и н А.Г. Аналитическая емкость множеств в задачах

теории приближений. - Успехи Матем.наук, 1967, 22, № 6, 141-199.
6. Мельников М.С., Синанян С.О. Вопросы теории приближения функций одного комплексного переменного. - В кн.: Современные проблемы математики т.4, Москва, ВИНИТИ, 1975, 143--250.
7. Zalcman L. Analytic capacity and Rational Approximation. - Lect.Notes Math., 50, Berlin, Springer, 1968.
8. Gamelin T.W.. Uniform algebras, N.J., Prentice-Hall, Inc. 1969.
9. Davie A.M. Analytic capacity and approximation problems. - Trans.Amer.Math.Soc., 1972, 171, 409-444.
10. Мельников М.С. Оценка интеграла Коши по аналитической кривой. - Матем.сб., 1966, 71, № 4, 503-514.
11. Витушкин А.Г. Оценка интеграла Коши. - Матем.сб., 1966, 71, № 4, 515-534.
12. Широков Н.А. Об одном свойстве аналитической емкости. - Вестник ЛГУ, сер.матем., мех., астрон., 1971, 19, 75-82.
13. Широков Н.А. Некоторые свойства аналитической емкости. - Вестник ЛГУ, сер.матем., мех., астрон., 1972, 1, 77-86.
14. Besicovitch A. On sufficient conditions for a function to be analytic and on behaviour of analytic functions in the neighbourhood of non-isolated singular points. - Proc.London Math.Soc., 1931, 32, N 2, 1-9.

A.G.VITUSHKIN
(А.Г.ВИТУШКИН)

СССР, 117966, Москва,
ул.Вавилова, 42,
МИАН СССР

M.S.MEL'NIKOV
(М.С.МЕЛЬНИКОВ)

СССР, 117234, Москва,
Механико-математический
факультет Московского
университета

8.18. ON SETS OF ANALYTIC CAPACITY ZERO
old

Let K be a compact plane set and $A_\infty(K)$ the space of all functions analytic and bounded outside K endowed with the sup-norm. Define a linear functional L on $A_\infty(K)$ by the formula

$$L(f)=\frac{1}{2\pi i}\int\limits_{|z|=r} f(z)\,dz,$$

with $r > max\{|z| : z\in K\}$. The norm of L is called the analytic capacity of K. We denote it by $\gamma(K)$. The function γ is invariant under isometries of $\mathbb{C}$. Therefore it would be desirable to have a method to compute it in terms of Euclidean distance. E.P.Dolzenko has found a simple solution of a similar question related to the so-called d-capacity, [1]. But for γ the answer is far from being clear. I would like to draw attention to three conjectures.

CONJECTURE 1. There exists a positive number C such that for any compact set K

$$\gamma(K)\geqslant C\int\limits_{\mathbb{T}} \mu(K,\zeta)\,dm(\zeta),$$

where $\mu(K,\zeta)$ denotes the length of the projection of K onto the line through 0 and $\zeta\in\mathbb{T}$.

CONJECTURE 2. There exists a positive number C such that for any compact set K

$$\gamma(K)\leqslant C\int\limits_{\mathbb{T}} \mu(K,\zeta)\,dm(\zeta).$$

These CONJECTURES are in agreement with known facts about analytic capacity. For example, it follows immediately from CONJECTURE 1 that $\gamma(K)>0$ if K lies on a continuum of finite length and has positive Hausdorff length. In turn, CONJECTURE 2 implies that $\gamma(K)=0$ provided the Favard length of K equals to zero. At last, let K be a set of positive Hausdorff h-measure (a survey of literature on the Hausdorff measures can be found in [3]). If $\int_0 h(t)/t^2\,dt<\infty$ then the Favard length of K is positive. This ensures the existence of a compact K_1, $K_1 \subset K$, such that $\mu_h(K_1)>0$ and the function f, $f(z) \overset{def}{=} \int_{K_1} \frac{d\mu_h(z)}{\zeta - z}$, is continuous on $\mathbb{C}$, μ_h being the Hausdorff h-measure. Hence $\gamma(K) \geq \gamma(K_1) > 0$ which also easily follows from CONJECTURE 2.

CONJECTURE 3. <u>For any increasing function</u> $h:(0,+\infty)\to(0,+\infty)$ <u>with</u> $\int_0 h(t)/t^2\, dt = \infty$ <u>there exists a set</u> K <u>satisfying</u> $\mu_h(K)>0$ <u>and</u> $\gamma(K)=0$.

To corroborate this CONJECTURE I shall <u>construct a function</u> h <u>and a set</u> E <u>such that</u> $\lim_{t\to 0} h(t)/t = 0$, $\mu_h(E)>0$ <u>but</u> $\gamma(E)=0$.

Assign to any sequence $\varepsilon = \{\varepsilon_n\}$, $\varepsilon_n \downarrow 0$, a compact set $E(\varepsilon)$. Namely, let $Q_0(\varepsilon) = [0,1]$. If $Q_n(\varepsilon)$ is the union of 2^n disjoint segments $\overset{j}{\Delta}_n$, of length $\delta_n(\varepsilon)$ then $Q_{n+1}(\varepsilon)$ is the union of 2^n sets $\overset{j}{\Delta}_n \setminus \hat{\overset{j}{\Delta}}_n$, $\hat{\overset{j}{\Delta}}_n$ being the interval of length $\frac{1}{2}\delta_n(\varepsilon)(1-\varepsilon_n)$ concentric with the segment $\overset{j}{\Delta}_n$. Put

$$E(\varepsilon) = \bigcap_{n\geq 1} Q_n^2(\varepsilon), \quad Q_n^2(\varepsilon) \overset{def}{=} Q_n(\varepsilon)\times Q_n(\varepsilon)$$

and let ε^c be a constant sequence, $(\varepsilon^c)_n = c$. Finally let $E = E(\varepsilon^0)$.

It is known (see [3]) that $\gamma(E)=0$. This implies the existence of a function φ such that

$$\lim_{t\to 0}\varphi(t)=0 \quad \text{and} \quad \gamma(E(\varepsilon))<\varphi(t)$$

for any sequence ε satisfying $\varepsilon_1 < t$. Then $E(\varepsilon)$ has the desired properties for properly chosen ε, as will be shown later. To choose ε pick numbers $d_1>0$ and $n_1 \in \mathbb{N}$ such that $\varphi(d_1) < 1/2$, $\gamma(Q_{n_1}^2(\varepsilon^{d_1})) < 1/2$ and $(1+d_1)^{n_1} > 2$. Set $\varepsilon_j = d_1$ for $j = 1, 2, \dots, n_1$. Proceeding by induction, pick d_{k+1} to provide the inequality $\varphi(d_{k+1}) < \beta_k = \frac{1}{k+2}\prod_{j=1}^{k}(1+d_j)^{-n_j}$

and next pick n_{k+1} such that

$$(1+d_{k+1})^{n_{k+1}} > 2 \quad \text{and} \quad \gamma\left[Q^2_{n_{k+1}}(\varepsilon^{d_{k+1}})\right] < \beta_k .$$

Set now $\varepsilon_j = d_{k+1}$ for $j = N_k+1, \ldots, N_{k+1}$ $(N_s \overset{def}{=} n_1 + \ldots + n_s)$.

The sequence ε defines a function ψ_ε equal to 4^{-n} at $\delta_n(\varepsilon)$, $n \in \mathbb{N}$ and linear on each segment $[\delta_j(\varepsilon), \delta_{j-1}(\varepsilon)]$. It is easy to verify that $\lim_{t \to 0+} \psi_\varepsilon(t)/t = 0$ and $E(\varepsilon)$ has positive ψ_ε -measure. It remains to check only that $\gamma[E(\varepsilon)] = 0$.

For this purpose let $f \in A_\infty(E(\varepsilon))$ and let

$$\overset{j}{f}_n(z) = \frac{1}{2\pi i} \int_\gamma \frac{f(\zeta)d\zeta}{\zeta - z}, \quad z \notin E(\varepsilon) \cap (\overset{j_1}{\Delta}_n \times \overset{j_2}{\Delta}_n),$$

where contour γ embraces $E(\varepsilon) \cap (\overset{j_1}{\Delta}_n \times \overset{j_2}{\Delta}_n)$ and separates it from z. The set $Q_n^2(\varepsilon)$ being the union of 4^n squares $\overset{j_1}{\Delta}_n \times \overset{j_2}{\Delta}_n$ with the side $\delta_n(\varepsilon)$ and lying at the distance at least $2\delta_n(\varepsilon)\frac{1-\varepsilon_n}{1+\varepsilon_n}$ one from another, it is clear that $\overset{j}{f}_n$ are uniformly bounded and

$$|L(f)| \leqslant 4^n \cdot \sup_{m,j} \|\overset{j}{f}_m\| \cdot \gamma[E(\varepsilon) \cap (\overset{j_1}{\Delta}_n \times \overset{j_2}{\Delta}_n)] .$$

This implies

$$|L(f)| \leqslant 4^{N_k} \cdot \sup_{m,j} \|\overset{j}{f}_m\| \frac{\prod_{j=1}^{k} (1+d_j)^{n_j}}{4^{N_k}} \cdot \gamma[Q^2_{n_{k+1}}(\varepsilon^{d_{k+1}})] \leqslant$$

$$\leqslant \frac{1}{k+1} \sup_{m,j} \|\overset{j}{f}_m\|$$

and finally $L(f) = 0$. ●

REFERENCES

1. Д о л ж е н к о Е.П. О "стирании" особенностей аналитических функций. – Успехи матем.наук, 1963, 18, № 4, 135–142.

. C a l d e r ó n A.P. Cauchy integrals on Lipschitz curves and related operators. - Proc.Nat.Acad.Sci.USA, 1977, 74, 1324-1327.

. R o g e r s C.A. Hausdorff measures. Cambridge, Cambridge University Press, 1970.

. G a r n e t t J. Positive length but zero analytic capacity. - Proc.Amer.Math.Soc., 1970, 24, 696-699.

L.D.IVANOV

(Л.Д.ИВАНОВ)

СССР, I700I3, Калинин

Калининский государственный университет

8.19. old ESTIMATES OF ANALYTIC CAPACITY

Let μ denote the Lebesgue measure on $\mathbb{R}^2 \equiv \mathbb{C}$. In what follows we let z range over $\mathbb{C}$ and r over the interval $(0,1) \subset \mathbb{R}$. We always suppose that $p \geqslant 1$ and $\lambda > 0$. E will be used as a generic notation for compact subsets of $\mathbb{C}$. For any locally integrable (complex-valued) function f on $\mathbb{C}$ we denote by

$$f_{r,z} = \frac{1}{\pi r^2} \int\limits_{\mathbb{D}(r,z)} f \, d\mu$$

its mean value over the disc $\mathbb{D}(r,z) = \{\zeta : \zeta \in \mathbb{C}, |\zeta - z| < r\}$. $\mathcal{L}^{p,\lambda}$ will stand for the class of all functions f on $\mathbb{C}$ that are locally integrable to the power p and satisfy

$$\|f\|_{p,\lambda} \equiv \sup_{r,z} \Big\{ r^{-\lambda} \int\limits_{\mathbb{D}(r,z)} |f - f_{r,z}|^p \Big\}^{1/p} < +\infty$$

(cf. [1] for references on related function spaces). Investigation of removable singularities for holomorphic functions in these classes gives rise naturally to the corresponding capacities $\gamma_{p,\lambda}$ defined as follows (compare [2]). If $A_{p,\lambda}(E,1)$ denotes the class of all $f \in \mathcal{L}^{p,\lambda}$ that are holomorphic in the complement of E (including ∞) and satisfy the conditions

$$f(\infty) = 0, \quad \|f\|_{p,\lambda} \leqslant 1,$$

then

$$\gamma_{p,\lambda}(E) = \sup \{ |f'(\infty)| : f \in A_{p,\lambda}(E,1) \},$$

where $f'(\infty) = \lim\limits_{z \to \infty} z f(z)$.

It is an important feature of these capacities that they admit simple metrical estimates which reduce to those of Melnikov (cf. Chap. V in [3]) for a special choice of the parameters when they yield Dolzenko's result on removable singularities in Hölder classes. Writing $\operatorname{diam} M$ for the diameter of M, $M \subset \mathbb{C}$, and defining for α, $\alpha \geqslant 0$, and ε, $\varepsilon > 0$,

$$I_\alpha^\varepsilon(E) = \inf \sum_n (\operatorname{diam} M_n)^\alpha,$$

here the infimum is taken over all sequences of sets M_n, $M_n \subset \mathbb{C}$ ith $\operatorname{diam} M_n < \varepsilon$ such that $E \subset \bigcup_n M_n$, we may state the following nequalities (cf. [4]).

THEOREM 1. Let $2 \leqslant p+\lambda < \tilde{\lambda} p + 2$, $\tilde{\lambda} = \min(2,\lambda)$ and define $\alpha = 1 + \frac{\lambda - 2}{p}$. Then there are constants c and k such that

$$c I_\alpha^1(E) \leqslant \gamma_{p,\lambda}(E) \leqslant k I_\alpha^1(E) \qquad (1)$$

for all E.

PROBLEM 1. What are the best values of the constants c, k occurring in (1)?

Theorem 1 is of special interest in the case $\lambda = 2$, because it characterizes removable singularities of holomorphic functions of bounded mean oscillation (cf. [5]) as those sets E whose linear measure

$$I(E) = \lim_{\varepsilon \to +0} I_1^\varepsilon(\varepsilon)$$

vanishes. This is in agreement with the example of Vituskin (cf. [6]). The capacity of E corresponding to the broader class of functions of bounded mean oscillation may be positive even though E has zero analytic capacity corresponding to bounded functions which is defined by

$$\gamma(E) = \sup\{|f'(\infty)| : f \in A(E,1)\},$$

where now $A(E,1)$ is the class of all functions holomorphic off E and vanishing at ∞ whose absolute value never exceeds 1. Nevertheless, by the so-called D e n j o y c o n j e c t u r e (which follows from combination of results in [7] , [8] , [9] , [10]) the equivalence

$$\gamma(E) = 0 \Leftrightarrow I(E) = 0$$

is true for E situated on a rectifiable curve. The upper estimate of $\gamma(\cdot)$ by means of $I(\cdot)$ is generally valid (cf. [11]) while the lower estimate of $\gamma(E)$ by means of a multiple of $I(E)$ is possible only for E situated on sets Q of a special shape.

PROBLEM 2. Let $Q \subset \mathbb{C}$ be a compact set. Find geometric conditions on Q guaranteeing the existence of a constant c such that

$$\gamma(E) \geqslant c I(E), \quad E \subset Q. \tag{2}$$

The following theorems 2, 3 may serve as sample results.

THEOREM 2 (cf. [12]). Let $Q = \varphi(\langle 0,1 \rangle)$, where $\varphi : \langle 0,1 \rangle \to \mathbb{C}$ is simple and continuously differentiable, $|\varphi'| = 1$ and

$$\sup_{\tau} \int_0^1 \frac{|\varphi'(t) - \varphi'(\tau)|^2}{|t-\tau|}\, dt = s < +\infty .$$

Then (2) holds and c can be computed by means of s (see also §7 in [11]).

THEOREM 3 (cf. [13]). If Q has only a finite number of components and

$$V(Q) \overset{\mathrm{def}}{=} \sup_{z \in Q} \int_{\mathbb{T}} N_z^Q(\zeta)\, dm(\zeta) < +\infty ,$$

where $N_z^Q(\zeta)$ is the number of points in $Q \cap \{z + t\zeta : t > 0\}$ then (2) holds with $c = (2\pi)^{-1}(2V(Q)+1)^{-1}$.

If Q is a straight-line segment then $V(Q) = 0$ and Pommerenke's equality $\gamma(E) = \frac{1}{4}\ell(E)$ holds (cf. [13]). This leads to the following

PROBLEM 3. Is it possible to improve c in Theorem 3 to $\frac{1}{4}(2V(Q)+1)^{-1}$?

REMARK. It was asserted in [4] that Theorem 3 holds with this alue of the constant. Dr. J.Matyska kindly pointed out that there as a numerical error in the original draft of the corresponding roof in [14].

REFERENCES

1. P e e t r e J. On the theory of $\mathcal{L}^{p,\lambda}$ -spaces. - J.Funct.Anal., 1969, 4, 71-87.
2. H a r v e y R., P o l k i n g J. A notion of capacity which characterizes removable singularities. - Trans.Amer.Math.Soc., 1972, 169, 183-195.
3. М е л ь н и к о в М.С., С и н а н я н С.О. Вопросы теории приближения функций одного комплексного переменного. - В кн.: Современные проблемы математики т.4, Москва, ВИНИТИ, I975, I43--250.
4. K r á l J. Analytic capacity. - In: Proc.Conf."Elliptische Differentialgleichungen" Rostock 1977.
5. J o h n F., N i r e n b e r g L. On functions of bounded mean oscillations. - Comm.Pure Appl.Math., 1961, 14, 415-426.
6. В и т у ш к и н А.Г. Пример множества положительной длины, но нулевой аналитической ёмкости. - Докл.АН СССР, I959, I27, 246--249.
7. C a l d e r ó n A.P. Cauchy integrals on Lipschitz curves and related operators. - Proc.Natl.Acad.Sci. USA, 1977, 74, 1324-1327.
8. D a v i e A.M. Analytic capacity and approximation problems. - Trans.Amer.Math.Soc., 1972, 171, 409-444.
9. Х а в и н В.П., Х а в и н с о н С.Я. Некоторые оценки аналитической емкости. - Докл.АН СССР, I96I, I38, 789-792.
10. Х а в и н В.П. Граничные свойства интегралов типа Коши и гармонически сопряженных функций в областях со спрямляемой границей. - Матем.сб., I965, 68, 499-5I7.
11. G a r n e t t J. Analytic capacity and measure. - Lect.Notes Math., 297, Berlin, Springer, 1972.
12. И в а н о в Л.Д. О гипотезе Данжуа. - Успехи матем.наук, I964, I8, I47-I49.
13. P o m m e r e n k e Ch. Über die analytische Kapazität. - Arch.Math., 1960, 11, 270-277.

14. F u k a J., K r á l J. Analytic capacity and linear measure. - Czechoslovak Math.J., 1978, 28 (103), N 3, 445-461.

JOSEF KRÁL

Matematický ústav ČSAV,
Žitná 25, 11567, Praha 1,
ČSSR

3.20. old ÜBER DIE REGULARITÄT EINES RANDPUNKTES FÜR ELLIPTISCHE DIFFERENTIALGLEICHUNGEN

In den letzten Jahren wurde dem Kreis von Fragen, die um das klassische Kriterium von Wiener über die Regularität eines Randpunktes in Bezug auf harmonische Funktionen gruppiert sind, viel Aufmerksamkeit geschenkt [1,2]. Nach dem Satz von Wiener ist die Stetigkeit im Punkt O, $O \in \partial\Omega$, der Lösung des Dirichletproblems für die Laplace-Gleichung im n-dimensionalen Gebiet Ω ($n>2$) unter der Bedingung, daß auf $\partial\Omega$ eine in O stetige Funktion gegeben ist, äquivalent zur Divergenz der Reihe

$$\sum_{k\geq 1} 2^{(n-2)k} \operatorname{cap}(C_{2^{-k}} \setminus \Omega).$$

Hierbei ist $C_\rho = \{x: \ x \in R^n \ , \ \rho/2 \leq |x| \leq \rho\}$ und $\operatorname{cap} K$ die harmonische Kapazität der kompakten Menge K.

Diese Behauptung wurde (manchmal nur der Teil der Hinlänglichkeit) auf verschiedene Klassen von linearen und quasilinearen Gleichungen zweiter Ordnung ausgedehnt (eine Charakterisierung dieser Untersuchungen und Bibliographie kann man im Buch [3] finden). Was die Gleichungen höherer als zweiter Ordnung betrifft, so gab es für sie bis zur letzten Zeit keine Resultate, die analog zum Satz von Wiener sind. In der Arbeit [4] des Autors wird das Verhalten der Lösung des Dirichletproblems für die Gleichung $\Delta^2 u = f$ mit homogenen Randbedingungen, wobei $f \in C_0^\infty(\Omega)$ ist, in der Umgebung einer Randpunktes untersucht. In [4] wird gezeigt, daß für $n = 5,6,7$ die Bedingung

$$\sum_{k\geq 1} 2^{k(n-4)} \operatorname{cap}_2 (C_{2^{-k}} \setminus \Omega) = \infty \ , \qquad (1)$$

wobei cap_2 die sogenannte biharmonische Kapazität ist, die Stetigkeit der Lösung im Punkt O garantiert. Für $n=2,3$ folgt die Stetigkeit der Lösung aus dem Einbettungssatz von S.L.Sobolev, aber im Fall $n=4$, der ebenfalls in [4] analysiert wird, hat die Bedingung für die Stetigkeit eine andere Gestalt.

HYPOTHESE 1. <u>Die Bedingung $n<8$ ist nicht wesentlich</u>.

Dem Autor ist nur ein Argument für diese Annahme bekannt. Für alle n ist die Lösung der betrachteten Aufgabe für einen beliebigen

Kugelsektor im Eckpunkt stetig. Die Einschränkung $n<\infty$ tritt nur bei einem der Lemmata auf, auf denen die Beweise in [4] beruhen Sie ist aber notwendig für dieses Lemma. Es geht hierbei um die Eigenschaft des Operators Δ^2, positiv mit dem Gewicht $|x|^{4-n}$ zu sein. Diese Eigenschaft erlaubt es, für $n=5,6,7$ folgende Abschätzung der Greenschen Funktion des biharmonischen Operators in einem beliebigen Gebiet anzugeben:

$$|G(x,y)| \leqslant c(n)\,|x-y|^{4-n} \tag{2}$$

wobei $x,y \in \Omega$ und $c(n)$ eine nur von n abhängige Konstante ist.

HYPOTHESE 2. Die Abschätzung (2) gilt auch für $n \geqslant 8$.

Es versteht sich, daß man analoge Fragen auch für allgemeinere Gleichungen stellen kann. Ich möchte die Aufmerksamkeit des Lesers aber auf eine Aufgabe lenken, die auch für den Laplace-Operator nicht gelöst ist. Nach [5], [6] genügt eine harmonische Funktion, deren verallgemeinerte Randwerte einer Hölder-Bedingung im Punkt O genügen, derselben Bedingung in diesem Punkt, falls

$$\lim_{N\to\infty} N^{-1} \sum_{N\geqslant k\geqslant 1} 2^{k(n-2)} \operatorname{cap}(C_{2^{-k}} \setminus \Omega) > 0 . \tag{3}$$

Es wäre interessant, folgende Annahme zu rechtfertigen oder zu widerlegen.

HYPOTHESE 3. Die Bedingung (3) ist notwendig.

Wir wenden uns zum Schluß nichtlinearen elliptischen Gleichungen zweiter Ordnung zu. Wie in [7] gezeigt wurde, ist der Punkt O regulär für die Gleichung $\operatorname{div}(|\operatorname{grad} u|^{p-2}\operatorname{grad} u)=0,\ 1<p<n$, falls

$$\sum_{k\geqslant 1} \left[2^{k(n-p)}\, p\text{-}\operatorname{cap}(C \setminus \Omega)\right]^{1/(p-1)} = \infty \tag{4}$$

ist, wobei $p\text{-}\operatorname{cap}(K)=\inf\{\|\operatorname{grad} u\|^p_{L^p(\mathbb{R}^n)} : u\in C_0^\infty(\mathbb{R}^n),\ u\geqslant 1 \text{ in } K\}$ ist. Dieses Resultat wurde unlängst in der Arbeit [8] auf die sehr allgemeine Klasse von Gleichungen $\operatorname{div}\vec{A}(x,u,\operatorname{grad} u)=B(x,u,\operatorname{grad} u)$ übertragen. Da die Bedingung (4) für $p=2$ mit dem Kriterium von Wiener zusammenfällt, ist es natürlich, folgende Hypothese aufzustellen

HYPOTHESE 4. Die Bedingung (4) ist notwendig.

In [9] wurden Beispiele behandelt, die zeigen, daß die Bedin-
ung (4) in einem gewissen Sinne genau ist. Für die HYPOTHESE 4 spre-
hen auch neuere Ergebnisse über die Stetigkeit nichtlinearer Poten-
iale [10], [11].

* * *

KOMMENTAR DES VERFASSERS.
(FÜNF JAHRE SPÄTER)

"Anscheinend ist noch keine der formulierten Aufgaben gelöst" – mit diesen Worten hatte der Autor vor, die Erörterung des obenangeführten Textes zu beginnen. Aber als der Kommentar fast fertig war, hörte dieser Satz auf, wahr zu sein. Es tauchte folgendes GEGENBEISPIEL ZUR HYPOTHESE 3 auf.

Es sei $\mathcal{B}_\nu = \{x \in \mathbb{R}^n : |x| < \rho_\nu\}$, $n>2$, $\log_2 \log_2 \rho_\nu^{-1} = \nu$ und Ω die Vereinigung der Kugelschichten $\mathcal{B}_\nu \setminus \mathcal{B}_{\nu+1}$, $\nu = 0,1,\ldots$, die durch Öffnungen ω_ν in den Sphären $\partial\mathcal{B}_\nu$, $\nu \geq 1$, verbunden sind. Die Öffnung ω_ν stellt eine geodätische Kugel mit beliebigem Mittelpunkt und dem Radius $2^{-1/\rho_\nu}$ dar.

Es ist klar, daß die Kapazität der Menge $C_{2^{-k}} \setminus \Omega$ nur für $k = 2^\nu, 2^{\nu-1}$ von Null verschieden ist und daß für diese k die Ungleichung $\operatorname{cap}(C_{2^{-k}} \setminus \Omega) \geq c\, 2^{-k(n-2)}$ gilt. Deshalb ist für $N>1$

$$c_1 \log_2 N \leq \sum_{N \geq k \geq 1} 2^{k(n-2)} \operatorname{cap}\left(C_{2^{-k}} \setminus \Omega\right) \leq c_2 \log_2 N .$$

Folglich divergiert die Reihe von Wiener für das betrachtete Gebiet, aber die Bedingung (3) ist nicht erfüllt.

Wir zeigen, daß trotzdem eine beliebige in Ω harmonische Funktion, deren Randwerte der Hölder-Bedingung im Punkt O genügen, ebenfalls der Hölder-Bedingung im Punkt O genügt.

Es sei u Lösung des Dirichletproblems $\Delta u = 0$ in Ω, $u - \varphi = 0$ auf $\partial\Omega$, wobei φ eine stetige Funktion ist, die der Bedingung $\varphi(x) = O(|x|^\beta)$, $\beta > 0$, genügt. Man kann annehmen, daß $1 \geq \varphi(x) \geq 0$ ist. Wir bezeichnen mit x einen beliebigen Punkt des Gebiets Ω und mit ν eine Zahl, für die $\rho_{\nu-1} \geq |x| \geq \rho_\nu$ ist. Es sei $\varphi_\nu = 0$ außerhalb von $\operatorname{clos}(\mathcal{B}_{\nu-1})$ und $\varphi_\nu = \varphi$ in $\mathcal{B}_{\nu-1}$. Ferner sei u_ν eine harmonische Funktion in Ω, die auf $\partial\Omega$ mit φ_ν übereinstimmt. Wegen $0 \leq \varphi_\nu \leq c\rho_{\nu-1}^\beta$ auf $\partial\Omega$ ist

$0 \leq u_\nu(x) \leq C\rho_{\nu-1}^{\beta} \leq C|x|^{\beta/2}$. Wir führen eine in der Kugel $\mathcal{B}_{\nu-2}$ harmonische Funktion h_ν ein, die auf $\partial\mathcal{B}_{\nu-2}$ gleich $u-u_\nu$ ist. Offenbar ist $0 \leq h_\nu \leq 1$ auf $\omega_{\nu-2}$ und $0 \leq h_\nu \leq C\rho_{\nu-2}^{\beta}$ auf $\partial\mathcal{B}_{\nu-2} \setminus \omega_{\nu-2}$. Stellt man h_ν in der Gestalt eines Poisson-Integrals dar so erhält man hieraus die Ungleichung $0 \leq h_\nu \leq$ $\leq C(\rho_{\nu-2}^{\beta} + 2^{-(n-1)/\rho_{\nu-2}}\rho_{\nu-2}^{1-n})$ in $\mathcal{B}_{\nu-1}$. Da nach dem Maximumprinzip $0 \leq u-u_\nu \leq h_\nu$ ist, erhält man $0 \leq u(x) - u_\nu(x) \leq C|x|^{\beta/4}$. Folglich genügt die Funktion u im Punkt 0 der Hölder-Bedingung mit dem Exponenten $\beta/4$.

Modifiziert man das konstruierte Beispiel, so kann man leicht zeigen, daß man auf Grund der Geschwindigkeit des Wachstums der Partialsummen der Reihe von Wiener keine unteren Abschätzungen einer harmonischen Funktion mit Null-Randbedingungen in der Nähe des Punktes 0 machen kann. Indem man nämlich die Art und Weise der Konvergenz der Radien ρ_ν und der Durchmesser der Öffnungen ω_ν gegen Null vorschreibt, kann man eine beliebig schnelle Konvergenz der Funktion u gegen Null im Punkt 0 bei beliebig langsamer Divergenz der Reihe von Wiener erreichen. Somit ist die HYPOTHESE 3 widerlegt, aber desto interessanter bleibt die Frage nach den den Rand charakterisierenden notwendigen und hinreichenden Bedingungen für die Hölder-Stetigkeit einer beliebigen harmonischen Funktion mit Hölder-stetigen Randwerten.

Was die anderen Fragen betrifft, die vor fünf Jahren gestellt wurden, so gibt es auf sie bisher noch keine Antwort *). In der letzten Zeit wurden mit ihnen verbundene neue Informationen gewonnen, die, wenn man es richtig betrachtet, nicht so sehr in die Tiefe wie in die Breite gehen.

Im Zusammenhang mit den HYPOTHESEN 1 und 2 erwähnen wir die Arbeit [12], in der die grundlegenden Ergebnisse des Artikels [4] (die ausführlich in [13] dargestellt sind) auf die erste Randwertaufgabe für die polyharmonische Gleichung $(-\Delta)^m u = f$ übertragen wurden. Leider verlangte auch hier die Methode, die auf der Eigenschaft des Operators $(-\Delta)^m$, $m>2$, beruht, positiv mit dem Gewicht $\Gamma(x-y)$ zu sein, wobei Γ die Fundamentallösung ist, alle Dimensionen mit Ausnahme der folgenden drei zu opfern: $n = 2m, 2m+1, 2m+2$. Als fragwürdigen Ausgleich gestattet uns dies, die folgenden beiden Hypothesen zu formulieren, die sich an die HYPOTHESEN 1 und 2 anschließen.

*) Anmerhung bei der Korrektur: I.W.Skrypnik teilte soeben auf der Tagung "Nichtlineare Probleme der Mathematischen Physik" (13. April, LOMI, Leningrad) mit, daß er die Notwendigkeit der Bedingung (4) für $p \geq 2$ bewiesen hat. Damit ist die HYPOTHESE 3 teilweise gestützt.

HYPOTHESE 1'. Für $m > 2, n \geq 2m+3$ ist die Gleichung

$$\sum_{k\geq 1} 2^{k(n-2m)} \operatorname{cap}_m (C_{2^{-k}} \setminus \Omega) = \infty ,$$

wobei cap_m die m-harmonische Kapazität ist, hinreichend für die Stetigkeit der Lösung des Dirichletproblems mit Nullrandbedingungen der Gleichung $(-\Delta)^m u = f, f \in C_0^\infty(\Omega)$ im Punkt 0.

HYPOTHESE 2'. Für $m > 2$, $n \geq 2m+3$ gilt für die Greensche Funktion G_m des Operators $(-\Delta)^m$ die Abschätzung

$$| G_m (x,y)| \leq c | x-y |^{2m-n},$$

wobei die Konstante c von n und m aber nicht vom Gebiet abhängt.

In der letzten Zeit wurden neue Erkenntnisse über das Verhalten der Lösung der ersten Randwertaufgabe für stark elliptische Gleichungen der Ordnung $2m$ in der Nähe konischer Punkte erhalten. Im allgemeinen (s. [14]) haben die Hauptglieder der Asymptotik solcher Lösungen in der Umgebung des Eckpunktes des Konus die Gestalt

$$c|x|^\lambda \sum_{k=0}^{N} (\log |x|)^k \varphi_k (x/|x|) \tag{5}$$

Dabei ist λ Eigenwert des Dirichletproblems für einen gewissen polynomial vom Spektralparameter abhängigen elliptischen Operator in dem Gebiet, das durch den Konus auf der Einheitssphäre ausgeschnitten wird. Die Funktion (5) hat genau dann ein endliches Dirichlet-Integral, wenn $\operatorname{Re}\lambda > m - n/2$. Sie ist des weiteren stetig und genügt sogar einer Hölder-Bedingung, falls $\operatorname{Re}\lambda > 0$ ist. Wenn im Band $0 > \operatorname{Re}\lambda > m - n/2$ Eigenwerte des genannten Operators existieren, dann besitzt die Ausgangsrandwertaufgabe verallgemeinerte Lösungen, die in einer beliebigen Umgebung des Eckpunktes des Konus unbeschränkt sind, und von einer Regularität nach Wiener kann man selbst bei einem konischen Punkt nicht reden. Es zeigt sich ([15], [16]), daß wir auf solche unerwarteten Erscheinungen schon bei stark elliptischen Gleichungen zweiter Ordnung mit konstanten Koeffizienten

$$P_2(D_x) u \equiv \sum_{i,j=1}^{n} a_{ij} \frac{\partial^2 u}{\partial x_i \partial x_j} = 0 , \quad n > 3 , \tag{6}$$

stoßen, falls nicht alle Koeffizienten reell sind. In [16] (eine ausführliche Darstellung erscheint evtl. in "Математический сборник") wird das homogene Dirichletproblem außerhalb eines dünnen Konus $K_\varepsilon=\{x=(y,x_n)\in\mathbb{R}^n: x_n>0,\ yx_n^{-1}\in\omega_\varepsilon\}$ untersucht, wobei ε ein kleiner Parameter, $\omega_\varepsilon=\{y\in\mathbb{R}^{n-1}: y\varepsilon^{-1}\in\omega\}$ und ω ein Gebiet im $\mathbb{R}^{n-1}$ ist.

Es wird die Lösung $u(x)=|x|^{\lambda(\varepsilon)}\phi(\varepsilon, x/|x|)$ des stark elliptischen Systems $P_{2m}(D_x)u(\varepsilon,x)=0$ betrachtet, wobei $P_{2m}(\xi)$ eine Matrix mit homogenen Polynomen der Ordnung $2m$ als Elementen und $\lambda(\varepsilon)=o(1)$ für $\varepsilon\to+0$ ist. Hauptergebnis ist eine asymptotische Formel für den Eigenwert $\lambda(\varepsilon)$, welche für den einfachsten Fall der Gleichung (6) die Gestalt

$$\lambda(\varepsilon)=\varepsilon^{n-3}\left\{\frac{n-3}{n-2}\frac{|S^{n-2}|}{|S^{n-1}|}\operatorname{cap}_{P_2(D_y,0)}(\omega)\frac{(\det\|\bar{a}_{jk}\|_{j,k=1}^{n})^{(n-3)/2}}{(\det\|\bar{a}_{jk}\|_{j,k=1}^{n-1})^{(n-2)/2}}+o(1)\right\}$$

hat. Hierbei ist $|S^k|$ die Oberfläche der $(k+1)$-dimensionalen Einheitskugel und $\operatorname{cap}_{P_2(D_y,0)}$ eine komplexwertige Funktion des Gebiets ω, welche eine Verallgemeinerung der harmonischen Kapazität darstellt:

$$\operatorname{cap}_{P_2(D_y,0)}(\omega)=\frac{1}{(n-3)|S^{n-2}|}\int_{\mathbb{R}^{n-1}\setminus\omega}\sum_{j,k=1}^{n-1}a_{jk}\frac{\partial w}{\partial y_k}\frac{\partial\bar{w}}{\partial y_j}\,dy,$$

wobei w eine im Unendlichen verschwindende Lösung der Gleichung $P_2(D_y,0)\,w(y)=0$ in $\mathbb{R}^{n-1}\setminus\omega$ ist, die auf $\partial\omega$ gleich 0 ist. Nach [16] kann man die Koeffizienten a_{jk} so wählen, daß die Ungleichung $0>\operatorname{Re}\lambda>(2-n)/2$ erfüllt ist. Im Fall $n=3$ gilt

$$\lambda(\varepsilon)=(2|\log\varepsilon|)^{-1}(1+o(1))\qquad\text{für}\quad\varepsilon\to+0.$$

Folglich erfüllt jede verallgemeinerte Lösung die Hölder-Bedingung, falls der Öffnungswinkel des Konus K_ε genügend klein ist. Es ist nicht ausgeschlossen, daß die Forderung nach einem kleinen Öffnungswinkel unwesentlich ist. Dies ist gleichbedeutend mit folgendem Satz

HYPOTHESE 5. Für $n=3$ ist ein konischer Punkt für einen beliebigen elliptischen Operator $P_2(D_x)$ mit komplexen Koeffizienten regulär nach Wiener.

Für den biharmonischen Operator im $\mathbb{R}^n$ und für die Systeme von Lamé und Stokes im $\mathbb{R}^3$ wurden derartige Ergebnisse in [17], [18] erhalten.

REFERENCES

1. Wiener N. The Dirichlet problem. – J.Math. and Phys. 1924, 3, 127-146.
2. Wiener N. Certain notions in potential theory – J.Math and Phys. 1924, 3, 24-51.
3. Ландис Е.М. Уравнения второго порядка эллиптического и параболического типа, М., Наука, 1971.
4. Мазья В.Г. О поведении вблизи границы решений задачи Дирихле для бигармонического оператора. – Докл.АН СССР, 1977, 18, № 4, 15-19.
5. Мазья В.Г. О регулярности на границе решений эллиптических уравнений и конформного отображения. – Докл.АН СССР, 1963, 152, № 6, 1297-1300.
6. Мазья В.Г. О поведении вблизи границы решения задачи Дирихле для эллиптического уравнения второго порядка в дивергентной форме. – Матем.заметки, 1967, № 2, 209-220.
7. Мазья В.Г. О непрерывности в граничной точке решений квазилинейных эллиптических уравнений. – Вестн.ЛГУ, 1970, 25, 42-55 (поправка: Вестн.ЛГУ 1972, 1, 158).
8. Gariepy R., Ziemer W.P. A regularity condition at the boundary for solutions of quasilinear elliptic equations. – Arch.Rat.Mech.Anal., 1977, 67, N 1, 25-39.
9. Кроль И.Н., Мазья В.Г. Об отсутствии непрерывности и непрерывности по Гельдеру решений квазилинейных эллиптических уравнений вблизи нерегулярной точки. – Труды Моск.Матем.о-ва, 1972, 26, 73-93.
10. Hedberg L. Non-linear potentials and approximation in the mean by analytic functions. – Math.Z., 1972, 129, 299-319.
11. Adams D.R., Meyers N. Thinness and Wiener criteria for non-linear potentials. – Indiana Univ.Math.J., 1972, 22, 169-197.
12. Мазья В.Г., Дончев Т. О регулярности по Винеру граничной точки для полигармонического оператора. – Докл.Болг.АН,

I983, 36, № 2.
I3. M a z ' y a V.G. Behaviour of solutions to the Dirichlet problem for the biharmonic operator at the boundary point, Equadiff IV, Lect.Notes Math., 1979, 703, p 250-262
I4. К о н д р а т ь е в В.А. Краевые задачи для эллиптических уравнений в областях с коническими или угловыми точками. Труды Моск. матем.о-ва, I967, I6, 209-292.
I5. М а з ь я В.Г., Н а з а р о в С.А., П л а м е н е в с к и й Б.А. Отсутствие теоремы типа Де Джорджи для сильно эллиптических уравнений с комплексными коэффициентами. - Зап.науч.семин.ЛОМИ, I982, II5, I56-I68.
I6. М а з ь я В.Г., Н а з а р о в С.А., П л а м е н е в с к и й Б.А. Об однородных решениях задачи Дирихле во внешности тонкого конуса. - Докл.АН СССР, I982, 266, № 2, 28I-284.
I7. М а з ь я В.Г., П л а м е н е в с к и й Б.А. О принципе максимума для бигармонического уравнения в области с коническими точками. - Изв.ВУЗов, I98I, № 2, 52-59.
I8. М а з ь я В.Г., П л а м е н е в с к и й Б.А. О свойствах решений трехмерных задач теории упругости и гидродинамики в областях с изолированными особенностями. - В сб.: Динамика сплошной среды, Новосибирск, I98I, вып.50, 99-I2I.

V.G.MAZ'YA
(В.Г.МАЗЬЯ)

СССР, I98904, Ленинград,
Петродворец, Ленинградский
государственный университет,
Математико-механический
факультет

.21. THE EXCEPTIONAL SETS ASSOCIATED WITH THE BESOV SPACES

For α real and $0<p,q<\infty$, we will use Stein's notation Λ_α^{pq} for the familiar Besov spaces of distributions on $\mathbb{R}^n$; see [P] nd [S] for details. The purpose of this note is to generally survey nd point out open questions concerning the general problem of determining all the inclusion relations between the classes $\mathcal{E}_{\alpha,p,q}$, $\alpha>0$, of exceptional sets naturally associated with the spaces Λ_α^{pq} for various choices of the parameters α,p,q; c.f.[AMS]. These exceptional sets can be described as sets of Besov capacity zero. Let $A_{\alpha,p,q}(K)=\inf\{|u|_{\alpha,p,q}^p : u\in\mathcal{D}$ and $u\geq 1$ on $K\}$, $\mathcal{D}$ some fixed smooth dense class in the spaces Λ_α^{pq}, $|u|_{\alpha,p,q}$ the norm (quasi-norm) of u in Λ_α^{pq}, K compact. $A_{\alpha,p,q}$ is extended to all subsets of $\mathbb{R}^n$ as an outer capacity. Then $E\in\mathcal{E}_{\alpha,p,q}$ iff $A_{\alpha,p,q}(E)=0$. Thus OUR PROBLEM is: <u>given an arbitrary compact set</u> K <u>such that</u> $A_{\alpha,p,q}(K)=0$, <u>for which</u> β, r,s <u>does it follow that</u> $A_{\beta,r,s}(K)=0$? When this holds, we will write $A_{\beta,r,s}\ll A_{\alpha,p,q}$. The symbol $\approx$ will mean that both directions, $\ll$ and $\gg$, hold.

Now when $1\leq p,q<\infty$, there is quite a bit that can be said about this problem. First of all, one can restrict attention to $1\leq p\leq n/\alpha$. Functions in Λ_α^{pq} for $p>n/\alpha$ are all equivalent to continuous functions and hence $A_{\alpha,p,q}(E)>0$ iff $E\neq\emptyset$. Continuity also occurs, for example, when $p=n/\alpha$ and $q=1$. Secondly, in the range $1<p\leq n/\alpha$, $1\leq q<\infty$, there appear to be presently four methods for obtaining inclusion relations. They are:

1. If $\Lambda_\alpha^{pq}\subset\Lambda_\beta^{rs}$ (continuous embedding), then clearly $A_{\beta,r,s}\ll A_{\alpha,p,q}$. Such embeddings occur, but not very often. However, since $\mathcal{L}_{\alpha+\varepsilon}^p\subset\Lambda_\alpha^{pq}\subset\mathcal{L}_{\alpha-\varepsilon}^p$, $\varepsilon>0$, with $\mathcal{L}_\alpha^p$ denoting the usual class of Bessel potentials of L^p functions on $\mathbb{R}^n$ (see [S]), and since the inclusion relations for the exceptional sets associated with the Bessel potentials are all known [AM], it is easy to see that $A_{\beta,r,s}\ll A_{\alpha,p,q}$ when $\beta r<\alpha p$ (no additional restrictions on s and q) and that the reverse implication is false.

2. Using the min-max theorem, it is possible to give a dual for-

mulation of the Besov capacities: $A_{\alpha,p,q}(K)^{1/p} = \sup\{<\mu,1>: \mu\in(\Lambda^{p'q'}_{-\alpha})^+,$ $\operatorname{supp}\mu\subset K$ and $|\mu|_{-\alpha,p',q'}\leqslant 1\}$, $\frac{1}{p'}=1-\frac{1}{p}$. Thus it suffices to prove inclusion relations between the positive cones in the dual spaces. This method is facilitated by the characterization: $\mu\in(\Lambda^{p'q'}_{-\alpha})^+$ iff

$$\int_0^1\Big\{\int[r^{\alpha p-n}\mu(B(x,r))]^{1/p-1}\,d\mu(x)\Big\}^{q'/p'}\frac{dr}{r}<\infty$$

for $p,q>1$, and

$$\sup_{0<r\leq 1}\int[r^{\alpha p-n}\mu(B(x,r))]^{1/p-1}\,d\mu(x)<\infty$$

for $p>1$, $q=1$.

3. The reason that one cannot expect all inclusion relations to follow from the first two methods is the simple fact that the capacitary extremals (in the primal and dual problems) generally have additional regularity. One can take advantage of this by comparing the Besov capacities $A_{\alpha,p,p}$ to the Bessel capacities $B_{\alpha,p}$, i.e. the capacities associated with $\mathcal{L}^p_\alpha$. Recently, P.Nilsson observed that the positive cones in $\Lambda^{p'p'}_{-\alpha}$ and $\mathcal{L}^{p'}_{-\alpha}$ coincide. Hence $A_{\alpha,p,p}\approx B_{\alpha,p}$. And again since all the inclusion relations for the Bessel exceptional sets are known (a result that relies on the regularity of the Bessel extremals), it follows that the corresponding relations carry over to $A_{\alpha,p,p}$. (The equivalence of $A_{\alpha,p,p}$ and $B_{\alpha,p}$ for all $p>1$ has been known since T. Wolff's recent proof of the Kellogg property in non-linear potential theory; see [HW], also [A_1]).

4. It is possible to apply the method of smooth truncation to the class of Bessel potentials of non-negative functions that belong to the mixed norm space $L^{pq}(\mathbb{R}^n\times\mathbb{R}^n)$ to obtain still further inclusion relations. This is due to the fact that the Besov capacities can be viewed as restrictions of such mixed norm Bessel capacities to subsets of $\mathbb{R}^n$. We refer the reader to [A_2].

In addition to the above relations, it is also possible to show that $A_{\alpha,1,1}\approx H^{n-\alpha}$ = Hausdorff $n-\alpha$ dimensional measure $(0<\alpha<n)$.

We summarize the results of 1. through 4. in the following diagrams; the cross indicates $A_{\alpha,p,q}(K)=0$ and the shaded region, the pairs $(\frac{1}{r}, \frac{1}{s})$ for which $A_{\beta,r,s}(K)=0$, $\beta r = \alpha p$, is a consequence.

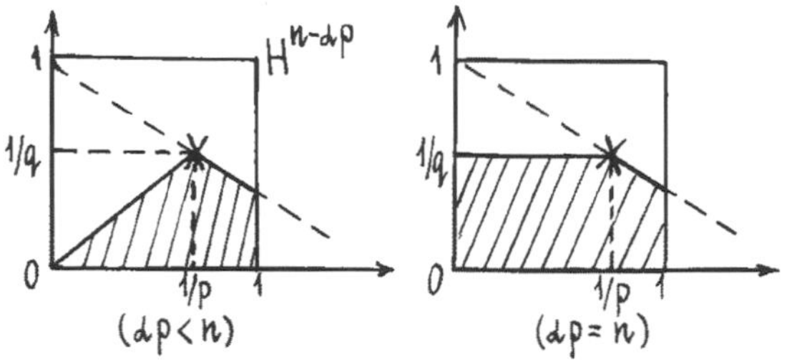

QUESTIONS: That $H^{n-\alpha p}(K)=0$ implies $A_{\alpha,p,q}(K)=0$ for $p \leq q$ is quite easy; <u>is this still true for</u> $q<p$? <u>Do these diagrams represent all inclusion relations? If so, how does one account for the difference in the cases</u> $\alpha p<n$ <u>and</u> $\alpha p = n$?

When $0<p<1$, $0<q<\infty$, very little seems to be known. One obvious thing to try is to compare $A_{\alpha,p,p}$ with $B_{\alpha,p}$ - the latter is now defined using Bessel potentials of the real Hardy spaces H^p on $\mathbb{R}^n$, $0<p<1$. This seems to be a good idea in view of all the recent developments on the structure of Hardy spaces, especially the atomic decomposition. Indeed, it is just such an approach that leads to $H^{n-\alpha p} \ll B_{\alpha,p}$, $\alpha p<n$, $0<p\leq 1$, and then via trace theorems [J_1] to $H^{n-\alpha p} \ll A_{\alpha,p,p}$. However, it is not presently known if $A_{\alpha,p,p} \approx B_{\alpha,p}$ holds for $0<p<1$ though it is probably true. One of the main difficulties now is that the obvious dual capacity is no longer equivalent to the primal one (and the min-max theorem does not apply since the spaces in question are no longer locally convex). This all does, however, suggest comparing the Besov capacities to yet another class of capacities, namely those naturally associated with the Lizorkin-Triebel spaces F_α^{pq} ; see [P] and [J_2] . And since it is known that F_α^{p2} coincides with Bessel potentials of $H^p (0<p<\infty)$ the F -capacities are a natural extension of the Bessel capacities. Thus we might expect some rather interesting results here in view of the things descussed above. However, it should be noted that for fixed α and p , the F

capacities agree with $B_{\alpha,p}$ whenever q satisfies $1 \leq p \leq q < 2$ or $2 \leq q \leq p$. Hence the F-diagram summarizing the inclusion relations for the F-exceptional sets will be considerably different than that for the Λ-exceptional sets $\mathcal{E}_{\alpha,p,q}$. QUESTION: What does it look like?

REFERENCES

[A_1] A d a m s D.R. On the exceptional sets for spaces of potentials. - Pac.J.Math., 1974, 52, 1-5.

[A_2] A d a m s D.R. Lectures on L^p-potential theory. Umeå Univ. Reports, 1981.

[AM] A d a m s D.R., M e y e r s N.G. Bessel potentials. Inclusion relations among classes of exceptional sets. - Ind. U.Math.J., 1973, 221, 873-905.

[AMS] A r o n s z a j n N., M u l l a F., S z e p t y c k i P. On spaces of potentials connected with L^p classes. - Ann. Inst.Fourier, 1962, 13, 211-306.

[HW] H e d b e r g L.I., W o l f f T. Thin sets in nonlinear potential theory. - Ann.Inst.Fourier, 1983, 33.

[J_1] J a w e r t h B. The trace of Sobolev and Besov spaces, $0 < p < 1$. - Studia Math., 1978, 62, 65-71.

[J_2] J a w e r t h B. Some observations on Besov and Lizorkin-Triebel spaces. - Math.Scand., 1977, 40, 94-104.

[P] P e e t r e J. New thoughts on Besov spaces. Duke Univ. Press, 1976.

[S] S t e i n E. Singular integrals and differentiability properties of functions. Princeton U. Press, 1970.

DAVID R. ADAMS

Department of Mathematics
University of Kentucky
Lexington, KY 40506
USA

3.22. COMPLEX INTERPOLATION BETWEEN SOBOLEV SPACES

Put $W^{k,p}(\mathbb{R}^n)=\{f: D^\alpha f\in L^p(\mathbb{R}^n),\ 0\leq\alpha\leq k\}$, the usual Sobolev space. The space $W^{k,\infty}$ seems to be poorly understood. Problem 1.8 gives one example of this. Another example is furnished by considering the complex method of interpolation, $(\cdot,\cdot)_\theta$. Let $1\leq p_0<\infty$.

<u>Is</u> $(W^{k,p_0}(\mathbb{R}^n), W^{k,\infty}(\mathbb{R}^n))_\theta = W^{k,p}(\mathbb{R}^n)$, $\frac{1}{p}=\frac{1-\theta}{p_0}$?

This is easy and true when $n=1$. Using Wolff's theorem [1] it is easy to show that a positive answer for one value of p_0 is equivalent to a positive answer for all values of p_0. The question is also easy to answer if one replaces the L^∞ endpoint by a BMO endpoint. The corresponding problem for the real method of interpolation is solved in [2].

REFERENCES

1. W o l f f T. A note on interpolation spaces. - Lect.Notes Math., 1982, N 908, 199-204. Springer Verlag.
2. D e V o r e R., S c h e r e r K. Interpolation of linear operators on Sobolev spaces. - Ann.Math.,1979, 109, 583-599.

PETER W.JONES

Institut Mittag-Leffler
Auravägen 17
S-182 62 Djursholm
Sweden

Usual Address:
Dept.of Mathematics
University of Chicago
Chicago, Illinois 60637
USA

CHAPTER 9

UNIQUENESS, MOMENTS, NORMALITY

Problems collected in this chapter are variations on the following theme: a "sufficiently analytic" function vanishing "intensively enough" is identically zero. The words in quotation marks get an exact meaning in accordance with every concrete situation. For instance, dealing with the uniqueness of the solution of a moment problem we often exploit traces of the analyticity of the function $\alpha \longrightarrow \int x^{\alpha} d\mu(x)$.

The theme is wide. It encompasses such phenomena as the quasianalyticity and the uniqueness of the moment problem, and borders on normal families (see e.g. Problem 9.5), various refinements of the maximum principle and approximation. Its importance hardly needs any explanation. The Uniqueness marks (more or less explicitly) all contents of this book. After all, every linear approximation problem (and the book abounds in such problems) is a dual reformulation of a uniqueness problem.

Every problem of this Chapter (except for 9.3 and 9.7) deals not only with "the pure uniqueness" but with other topics as well. Problem 9.1 is connected not only with zeros of some function classes but with a moment problem (as is Problem 9.2) and with Fourier - Laplace transforms of measures; in Problem 9.6 the uniqueness in analytic Gevrey classes is considered in connection with peak sets for

Hölder analytic functions. "Old" Problems 9.8 and 9.9 deal (from different points of views) with differential and differential-like operators (both have evoked a great interest, see respective commentary). Problem 9.4 has certain relation to spectral operators and to the "anti-locality" of some convolution operators (in contrast with "the locality" of convolutions discussed in 9.9). Problem 9.5 is a quantitative variation on the title theme and 9.13 gravitates towards spectral analysis-synthesis of Chapter 7. Problem 9.10 is aimed at approximation properties of exponentials and concerns also some aspects of quasianalyticity, as does Problem 9.12. Problem 9.11 deals with an interesting "perturbation" of the "$\int \log|f| > -\infty$"-theorem.

The theme of this chapter emerges in some Problems of other Chapters (3.7, 4.3, 4.4, 4.9, 5.12, 7.7, 7.17, 7.18, 8.4, 10.1, 10.5, 10.6, S.4, S.6).

9.1. old SOME OPEN PROBLEMS IN THE THEORY OF REPRESENTATIONS OF ANALYTIC FUNCTIONS

1. Denote by Ω the set of functions ω, satisfying the following conditions:

1) $\omega(x) > 0, \quad \omega \in C([0,1])$;

2) $\omega(0) = 1, \quad \int_0^1 \omega(x)\,dx < +\infty$;

3) $\int_0^1 |1-\omega(x)|\, x^{-1}\,dx < +\infty$.

In the factorization theory of meromorphic functions in the unit disc $\mathbb{D}$, developed in [1], the following theorem on solvability of the Hausdorff moment problem, proved in [2], played an important rôle: the Hausdorff moment problem

$$\mu_n = \int_0^1 x^n\,d\alpha(x) \qquad (n=0,1,2,\dots)$$

where

$$\mu_0 = 1, \qquad \mu_n = \left(n\int_0^1 \omega(x)\,x^{n-1}\,dx\right)^{-1} \qquad (n=1,2,\dots)$$

and $\omega \in \Omega$, $\omega\uparrow$, has a solution in the class of nondecreasing and bounded functions α on $[0,1]$.

Assuming $\omega_j \in \Omega$ $(j=1,2)$, consider the Hausdorff moment problem of the form

$$\lambda_n = \int_0^1 x^n\,d\beta(x) \qquad (n=0,1,2,\dots) \tag{1}$$

where

$$\lambda_0 = 1, \quad \lambda_n = \left(\int_0^1 \omega_1(x)\,x^{n-1}dx\right)\left(\int_0^1 \omega_2(x)\,x^{n-1}\,dx\right)^{-1} \quad (n=1,2,\dots) \tag{2}$$

CONJECTURE. The moment problem (1)-(2) has a solution in the class of nondecreasing and bounded functions on $[0,1]$, or at least n the class of functions β with bounded variation $\bigvee_{[0,1]}(\beta) < +\infty$ provided the functions ω_j are monotone on $[0,1]$ and ω_1/ω_2 is non-increasing on $[0,1]$.

The proof of this conjecture, which is true in the special case $\omega_1(x) \equiv 1$, would lead to important results on embeddings of classes $N\{\omega_j\}$ (j=1,2) of meromorphic functions in $\mathbb{D}$, considered in [1].

II. Denote by Ω_∞ the set of functions ω, satisfying the following conditions:

1) the function ω is continuous and non-increasing on $[0,\infty)$ $\omega(x)>0$;

2) the integrals

$$\Delta_k = k\int_0^\infty \omega(x)x^{k-1}\,dx, \qquad (k=1,2,\dots)$$

are finite.

Putting $\Delta_0 = 1$, consider entire functions of z :

$$W_\omega^{(\infty)}(z;\xi) = \int_{|\xi|}^\infty \frac{\omega(x)}{x}\,dx - \sum_{k=1}^\infty \Big(\xi^{-k}\int_0^{|\xi|}\omega(x)x^{k-1}\,dx -$$

$$-\bar{\xi}^k\int_{|\xi|}^\infty \omega(x)x^{-k-1}\,dx\Big)\frac{z^k}{\Delta_k} \qquad (0<|\xi|<\infty)$$

and

$$A_\omega^{(\infty)}(z;\xi) = \Big(1-\frac{z}{\xi}\Big)e^{-W_\omega^{(\infty)}(z;\xi)} .$$

Let, finally, $\{z_k\}_1^\infty$ $(0<|z_k|\leqslant|z_{k+1}|<\infty)$ be an arbitrary sequence of complex numbers such that

$$\sum_{k=1}^\infty \int_{|z_k|}^\infty \frac{\omega(x)}{x}\,dx < \infty . \tag{3}$$

CONJECTURE. Under condition (3) the infinite product

$$\pi_\omega(z;z_k) = \prod_{k=1}^{\infty} A_\omega^{(\infty)}(z;z_k)$$

converges on any compact set, not containing points of $\{z_k\}_1^\infty$, provided ω satisfies the additional condition

$$\frac{d\log\omega(x)}{d\log x} \downarrow -\infty, \qquad x\uparrow+\infty.$$

The validity of this conjecture for some special cases and in particular for

$$\omega(x)=\frac{\rho\sigma^{\mu}}{\Gamma(\mu)}\int_x^{\infty} e^{-\sigma t^{\rho}}\, t^{\mu\rho-1}\,dt,$$

where $\rho\,(0<\rho<\infty)$, $\mu\,(0<\mu<\infty)$ and $\sigma\,(0<\sigma<\infty)$ are arbitrary parameters, was proved in [3].

III. Let μ be a complex function on $[0,\infty)$ such that

$$V_\mu(r) \overset{\text{def}}{=\!=} \int_0^{\infty} r^t\,|d\mu(t)| < \infty, \qquad r\geqslant 0. \tag{4}$$

Then it is obvious that the function

$$f_\mu(z) = \int_0^{\infty} z^t\, d\mu(t)$$

is regular on the Riemann surface

$$G_\infty = \{z: |\operatorname{Arg} z| < \infty,\ 0<|z|<\infty\}$$

and that

$$\sup_{|\varphi|<\infty} |f_\mu(re^{i\varphi})| \leqslant V_\mu(r), \qquad r\in[0,\infty).$$

In view of this the following conjecture naturally arises.

CONJECTURE. Let f be an analytic function on G_∞ satisfying

$$\tilde{M}_f(r) \overset{def}{=} \sup_{|\varphi|<\infty} |f(re^{i\varphi})| < \infty, \qquad r \in [0,\infty).$$

Then there exists a function μ_f on $[0,\infty)$ satisfying (4) such that

$$f(z) = \int_0^\infty z^t \, d\mu_f(t), \qquad z \in G_\infty . \tag{5}$$

Note that in the special case, when

$$f(re^{i(\varphi+2\pi)}) = f(re^{i\varphi}), \quad r \in [0,\infty)$$

the function f admits the expansion

$$f(z) = \sum_{k\geq 0} a_k z^k, \quad |z| < \infty.$$

Thus in this case (5) holds with a measure concentrated at the points 0,1,2,... of the axis $[0,\infty)$ only.

IV. Denote by $H^p(\alpha)$ $(0<p<\infty)$, $-1<\alpha<\infty)$ the class of all analytic functions f on $\mathbb{D}$ such that

$$\iint_{\mathbb{D}} (1-r^2)^\alpha |f(re^{i\theta})|^p \, r\,dr\,d\theta < \infty.$$

Let $f \in H^p(\alpha)$ and $\{\alpha_j\}_1^\infty$ $(0<|\alpha_j|\leq|\alpha_{j+1}|<1)$ be the sequence of zeros of f enumerated in accordance with their multiplicity. Denote by $n(t)$ the number of α_j's in the disc $|z|\leq t$ $(0<t<1)$ and let

$$N_\alpha(r) = \int_0^r \frac{n(t)}{t}\,dt \qquad (0<r<1).$$

It is well-known (cf.[4]) that if $f \in H^p(\alpha)$ and $f \not\equiv 0$ then

$$\int_0^1 (1-r)^\alpha e^{pN_\alpha(r)} dr < \infty. \qquad (6)$$

In this connection it is natural to state the following

CONJECTURE. <u>Let $\{\alpha_j\}_1^\infty$ be an arbitrary sequence in $\mathbb{D}$ satisfying (6). Then there exist a sequence of numbers $\{\theta_j\}_1^\infty$ $(0 \leqslant \theta_j < 2\pi)$ and a function f_*, $f_* \not\equiv 0$, $f_* \in H^p(\alpha)$, such that</u>

$$f_*(\alpha_j e^{i\theta_j}) = 0. \qquad (j = 1, 2, \ldots)$$

Note that a statement equivalent to (6) and some other results about zeros of functions of the class $H^p(\alpha)$ (if $\alpha \geqslant 0$) were obtained in [5] much later than [4].

REFERENCES

1. Д ж р б а ш я н М.М. Теория факторизации функций, мероморфных в круге. - Матем.сб., 1969, 79, № 4, 517-615. (Math.USSR, Sbornik, 1969, 8, N 4, 493-591).
2. Д ж р б а ш я н М.М. Обобщенный оператор Римана-Лиувилля и некоторые его применения. - Изв.АН СССР, 1968, 32, 1075-1111. (Math.USSR, Izvestia, 1968, 2, 1027-1064).
3. Д ж р б а ш я н М.М. Об одном бесконечном произведении. - Изв. АН Армянской ССР, Математика, 1978, 13, № 3, 177 - 208. (see also: Soviet Math.Dokl., 1978, 19, N 3, 621-625).
4. Д ж р б а ш я н М.М. К проблеме представимости аналитических функций. - Сообщ.Инст.Мат. и Мех.АН Армянской ССР,1948,2,3-39.
5. H o r o w i t z C. Zeros of functions in the Bergman spaces. - Duke Math.J., 1974, 41, 693-710.

M.M.DJRBASHYAN
(М.М.ДЖРБАШЯН)

СССР, 375019, Ереван
ул.Барекамутян 24б,
Институт математики
АН Арм.ССР

* * *

COMMENTARY

The question posed in section III has a negative answer. We begin with the following observation. The logarithm Log being a biholomorphic map of the Riemann surface G_∞ onto $\mathbb{C}$, it is clear that every function f on G_∞ defines a (unique) entire function F satisfying

$$f(z) = F(\mathrm{Log}\, z), \quad z \in G_\infty .$$

Clearly, f satisfies the assumption $\widetilde{M}_f(r) < +\infty$ if and only if

$$\sup\{|F(z)| : \mathrm{Re}\, z = \log r\} < +\infty . \tag{7}$$

On the other hand $f(z) = \int_0^{+\infty} z^t \, d\mu(t)$, $z \in G_\infty$ if and only if

$$F(z) = \int_0^{\infty} e^{tz} \, d\mu(t), \quad z \in \mathbb{C}. \tag{8}$$

Here, as in section III, μ stands for a complex Borel measure on $[0, +\infty)$ such that $\int_0^\infty e^{ta} |d\mu(t)| < +\infty$ for every $a \in \mathbb{R}$. It follows that μ has a finite full mass (put $a = 0$).

It is shown in [6] (ch III, §4) that there exists a true pseudomeasure S (i.e. a distribution, with the uniformly bounded Fourier transform and not a measure) supported by $E = \{0\} \cup \bigcup_{n=1}^{\infty} \{\frac{1}{n}\}$.

The support of S being compact, the Fourier transform of S coincides with the restriction onto $\mathbb{R}$ of an entire function $\hat{S}$ of exponential type, $|\hat{S}(z)| \leqslant A e^{|z|}$. Moreover, it follows from the L.Schwartz' version of the Paley-Wiener theorem (see [7], Ch.6, §4 for example) that $\sup_{z \in \mathbb{C}_+} |\hat{S}(z)| = \|\hat{S}\|_{L^\infty(\mathbb{R})}$ and that

$$|\hat{S}(z)| \leqslant \|\hat{S}\|_{L^\infty} \cdot \exp\{|\mathrm{Im}\, z|\},$$

if $\mathrm{Im}\, z < 0$. Therefore $F(z) \overset{\text{def}}{=} \hat{S}(iz)$ satisfies (7) but F cannot coincide with the Laplace transform (8) of a finite Borel measure because this contradicts the fact that S is a TRUE pseudomeasure and to the uniqueness theorem for Fourier transforms of distributions. (The solution was found by S.V.Hruščëv).

REFERENCES

6. K a h a n e J.P. Séries de Fourier absolument convergentes. Berlin, Springer-Verlag, 1970.

7. Y o s i d a K. Functional analysis. Berlin, Springer-Verlag, 1965.

.2. MOMENT PROBLEM QUESTIONS

Let $\mathbb{N}_0$ be the non-negative integers and $\mathbb{N}_0^n$ the set of all ulti-indices $\alpha = (\alpha_1, \dots, \alpha_n)$ with each $\alpha_j \in \mathbb{N}_0$. For any $x = (x_1, \dots, x_n) \in \mathbb{R}^n$ write $x^\alpha = x_1^{\alpha_1} \dots x_n^{\alpha_n}$, where $x_j^0 = 1$. enote by $\mathcal{P}_0$ the complex vector space of all polynomials, $x) = \sum a_\alpha x^\alpha$, considered as functions from $\mathbb{R}^n \to \mathbb{C}$. A multi-sequence $\mu_\alpha : \mathbb{N}_0^n \to \mathbb{R}$ is called a moment sequence if there xists a bounded non-negative Borel measure $d\mu(x)$ so that

$$\mu_\alpha = \int_{\mathbb{R}^n} x^\alpha \, d\mu(x).$$

he moment sequence is said to be determined if there xists a unique representing measure. We refer the reader to the reent expository article by B.Fuglede [2] for a discussion of this roblem together with an up to date set of references.

If $n=1$, it is well known that if the moment sequence is deermined, then $\mathcal{P}_0$ is dense in $L^2(d\mu)$. Indeed, more is known. f $\mu_0 = 1$, and $d\mu$ is an extreme point of the convex set of repreenting measures for μ_α, then $\mathcal{P}_0$ is dense in $L(d\mu)$.

In 1978, the theoretical physicist, Professor John Challifour of Indiana University proposed (in private conversation with the uthor) the following question:

QUESTION 1. <u>For</u> $n > 1$, <u>is it still true that if a multi-parameter moment sequence</u> μ_α <u>has a unique representing measure</u> $d\mu$, <u>hen</u> $\mathcal{P}_0$ <u>is dense in</u> $L^2(d\mu)$?

To turn to a second question suppose μ_α and ν_α are moment sequences from $\mathbb{N}_0 \to \mathbb{R}$, $(\mu * \nu)_\alpha$ is the moment sequence formed from the convolution measure $d(\mu * \nu)$, then it was shown in [1] that if $(\mu * \nu)_\alpha$ is a determined moment sequence, then so are the individual moment sequences μ_α and ν_α. Very recently, the statistician, Persi Diaconis of Stanford University proposed (again in private conversation with the author) the following question:

QUESTION II. <u>If</u> μ_α <u>and</u> ν_α <u>are determined moment sequences, is it true that</u> $(\mu * \nu)_\alpha$ <u>is a determined moment sequence</u>?

REFERENCES

1. Devinatz A. On a theorem of Levy-Raikov. - Ann. of Math. Statistics, 1959, 30, 538-586.
2. Fuglede B. The multidimensional moment problem. - Expo. Math., 1983, 1, 47-65.

ALLEN DEVINATZ

Northwestern University
Department of Mathematics
Evanston, IL 60201
USA

EDITORS' NOTE. Christian Berg (Københavns universitets matematiske Institut, 2100 København, Danmark) informed us that he has answered QUESTION 2 in the negative. Moreover, he has constructed a measure μ such that the sequence μ_α is determined, but the sequence $(\mu * \mu)_\alpha$ is not.

.3.
ld

SETS OF UNIQUENESS FOR ANALYTIC FUNCTIONS WITH THE FINITE DIRICHLET INTEGRAL

Let $\mathfrak{X}$ be a class of functions analytic in $\mathbb{D}$. A closed subset E of the closed disc $\operatorname{clos}\mathbb{D}$ is said to be a u n i q u e n e s s s e t for $\mathfrak{X}$ (briefly $E\in\mathcal{E}(\mathfrak{X})$) if

$$f\in\mathfrak{X},\quad f|E=0 \;\Longrightarrow\; f=0.$$

(it is assumed that $f(\zeta)\overset{def}{=}\lim\limits_{r\to 1-0} f(r\zeta)$ at $\zeta\in\mathbb{T}\cap E$).

The structure of $\mathcal{E}(\mathfrak{X})$ -sets is well understood for many important classes $\mathfrak{X}$ (see[1],[2]; [3] contains a short survey). The same cannot be said about the family $\mathcal{E}(\mathcal{D}_A)$, $\mathcal{D}_A$ being the space of all functions analytic in $\mathbb{D}$ with finite Dirichlet integral

$$\frac{1}{\pi}\iint\limits_{\mathbb{D}}|f'(z)|^2\,dx\,dy<+\infty.$$

The description of $\mathcal{E}(\mathcal{D}_A)$ seems to have to do not only with the Beurling - Carleson condition (see (1) below) but with capacity characteristics of sets.

We propose two conjectures concerning subsets of $\mathbb{T}$ in $\mathcal{E}(\mathcal{D}_A)$.

Associate with every closed set $F\subset\mathbb{T}$ a (unique) closed set F^*, $F^*\subset F$ so that $\operatorname{cap}(F\setminus F^*)=0$ (cap stands for the capacity corresponding to the logarithmic kernel) and every non-empty relatively open (in $\mathbb{T}$) part of F^* has positive logarithmic capacity.

CONJECTURE 1. <u>A closed subset</u> F <u>of</u> $\mathbb{T}$ <u>does not belong to</u> $\mathcal{E}(\mathcal{D}_A\cap C_A)$ (C_A <u>stands for the disc-algebra</u>) <u>iff</u>

a) $mF=0$;

b) $\int\limits_{\mathbb{T}}\log\operatorname{dist}(\zeta,F^*)\,dm(\zeta)>-\infty.$

The difficulty of this problem is caused by the fact that functions in $\mathcal{D}_A$ posess no local smoothness on $\mathbb{T}$.

The conjecture agrees with all boundary uniqueness theorems for $\mathcal{D}_A$ we are aware of. These are two.

THEOREM (Carleson [4]). <u>Suppose that</u> $F\subset\mathbb{T}$, $\operatorname{clos}F=F$, $mF=0$ <u>and for some</u> $\alpha>0$ <u>the inequality</u> $C_\alpha(F\cap I(\zeta,\delta))>\mathrm{const}\,\delta$ <u>holds for an arbitrary</u> $\zeta\in F$ <u>and every</u> $\delta>0$, $I(\zeta,\delta)$ <u>being</u>

<u>the arc of length</u> 2δ <u>centered at</u> ζ . <u>Then</u> $F \in \mathcal{E}(\mathcal{D}_A)$ <u>iff</u>

$$\int_{\mathbb{T}} \log \operatorname{dist}(\zeta, F)\, dm(\zeta) = -\infty. \tag{1}$$

Here C_α denotes the capacity corresponding to the kernel $|x-y|^{-\alpha}$. A set F satisfying the conditions of the Carleson theorem coincides with F^* because $C_\alpha(E)>0$ implies $cap(E)>0$.

THEOREM (Maz'ya - Havin [6]). <u>Suppose that</u> $F \subset \mathbb{T}$, $\operatorname{clos} F = F$ <u>and that there exists a family</u> Ω <u>of mutually disjoint open arcs</u> I <u>satisfying</u> $F \subset \bigcup_{I\in\Omega} I$ <u>and</u>

$$\sum_{I\in\Omega} (mI)\cdot \log \frac{cap(F\cap I)}{m(I)} = +\infty. \tag{2}$$

<u>Then</u> $F \in \mathcal{E}(\mathcal{D}_A)$.

Evidently (2) implies $\sum_{I\in\Omega} mI\cdot \log\frac{1}{m(I)} = +\infty$. Any family $\{l_\gamma\}$ of open mutually disjoint arcs , which almost covers an arc I (i.e. $m(I\setminus \bigcup_\gamma l_\gamma)=0$, satisfies $mI\cdot \log\frac{1}{mI} \leqslant \leqslant \sum_\gamma l_\gamma \log \frac{1}{l_\gamma}$. This remark is an easy consequence of the subadditivity of $x\log\frac{1}{x}$. Therefore (2) implies the divergence of the series $\sum_\gamma l_\gamma \log\frac{1}{l_\gamma}$, $\{l_\gamma\}$ being the family of complementary intervals of F^*, provided $mF=0$.

To state the second conjecture consider a class $\mathcal{Y}$ of non-negative functions defined on $\mathbb{T}$. A closed subset E is said to belong to $\mathcal{L}(\mathcal{Y})$ if

$$h\in\mathcal{Y},\quad h|E=0 \;\Longrightarrow \int_{\mathbb{T}} \log h\, dm = -\infty .$$

Let $\mathcal{D}_+$ be the set of all traces on $\mathbb{T}$ of non-negative functions from $W_2^1(\mathbb{C})$ (i.e. the functions in $L^2(\mathbb{C})$ with square summable generalized gradient).

CONJECTURE 2.

$$\mathcal{E}(\mathcal{D}_A) = \mathcal{L}(\mathcal{D}_+)\,. \tag{3}$$

Equality (3) (if true) permits to separate the difficulties connected with the analyticity of functions of $\mathcal{D}_A$ from those of purely real character (such as the investigation of $\int \log h\, dm$ for non-negative h's in W_2^1).

The inclusion $\mathcal{L}(\mathcal{D}_+) \subset E(\mathcal{D}_A)$ is obvious because $|f| \,|\, \mathbb{T} \in \mathcal{D}_+$ for $f \in \mathcal{D}_A$. The proofs of the theorems cited above are based precisely upon this inclusion (and upon Jensen's inequality). Here is another remark suggesting that (3) is a "right" analogue for the Beurling - Carleson theorem. This theorem asserts that $\mathcal{E}(\mathrm{Lip}_A(\alpha)) = \mathcal{L}(\mathrm{Lip}_+\alpha)$. Here $\mathrm{Lip}_A(\alpha) \stackrel{\mathrm{def}}{=\!=} C_A \cap \mathrm{Lip}(\alpha)$ stands for the space of functions in the disc-algebra satisfying the Lipschitz condition of order α .

The well-known Carleson formula for the Dirichlet integral of an analytic function [7] permits to reformulate conjecture 2.

<u>Suppose that for a given</u> $E \subset \mathbb{T}$ <u>there exists a non-zero</u> h <u>in</u> $L^2_+(\mathbb{T})$ <u>satisfying</u>

$$\iint\limits_{\mathbb{T}\times\mathbb{T}} \frac{|h(\xi)-h(\zeta)|^2}{|\xi-\zeta|^2}\, dm(\xi)\, dm(\zeta) < +\infty$$

$$\int\limits_{\mathbb{T}} \log h\, dm > -\infty \tag{4}$$

$$h \,|\, E \equiv 0, \tag{5}$$

<u>then there exists a function</u> $h \in L^2_+(\mathbb{T})$ <u>satisfying</u> (4), (5) <u>and</u>

$$\iint\limits_{\mathbb{T}\times\mathbb{T}} \frac{(h^2(\xi)-h^2(\zeta))(\log h(\xi) - \log h(\zeta))}{|\xi-\zeta|^2}\, dm(\xi)\, dm(\zeta) < +\infty. \tag{6}$$

Some estimates of the Carleson integral in (6) are given in [8]. The sets of $\mathcal{E}(\mathcal{D}_A)$ located in $\mathbb{D}$ have been considered in [4] and [5].

REFERENCES

1. B e u r l i n g A. Ensembles exceptionnels. Acta Math., 1940, 72.
2. C a r l e s o n L. Sets of uniqueness for functions regular in the unit circle. Acta Math., 1952, 87, p.325-345.
3. H r u š č ë v S.V. Sets of uniqueness for the Gevrey classes. Arkiv för Mat., 1977, 6, p.253-304.
4. C a r l e s o n L. On the zeros of functions with bounded Dirichlet integrals. Math.Zeitschrift. 1952, 56, N 3, p.289-295.
5. S h a p i r o H.S. and S h i e l d s A.L. On the zeros of functions with finite Dirichlet integral and some related function spaces. Math.Zeit. 1962, v.80, 217-229.
6. М а з ь я В.Г., Х а в и н В.П. "Приложения (p,l)-ёмкости к нескольким задачам теории исключительных множеств". Матем.сборник I973, 90 (I32), вып.4, 558-59I.
7. C a r l e s o n L. A representation formula for the Dirichlet integral. - Math.Zeit. 1960, 73, N 2, 190-196.
8. А л е к с а н д р о в А.Б., Д ж р б а ш я н А.Э., Х а в и н В.П. "О формуле Карлесона для интеграла Дирихле". Вестник ЛГУ, сер.мат., мех., астр., I979, I9, 8-I4.

V.P.HAVIN
(В.П.ХАВИН)

СССР, I98904 Петродворец,
Математико-механический
факультет ЛГУ

S.V.HRUŠČËV
(С.В.ХРУЩЕВ)

СССР, I9I0II Ленинград Д-II
Фонтанка 27,
ЛОМИ АН СССР

* * *

COMMENTARY

A description of $\mathcal{E}(\mathcal{D}_A)$ can be found in [9] which, unfortunately, is difficult to apply.

Conjecture 1 has been disproved by an ingenious counter-example of L.Carleson [10]. Conjecture 2 remains open.

It is interesting to note that the closely related problem of escription of the interpolating sets for $\mathcal{D}_A$ has been solved in [11]. Namely, a closed set $E\subset\mathbb{T}$ is said to be an interpolation set f $(\mathcal{D}_A \cap C_A)|E = C(E)$. Then E is an interpolation set ff $\operatorname{cap} E = 0$.

REFERENCES

. M a l l i a v i n P. Sur l'analyse harmonique des certaines classes de séries de Taylor.- Symp.Math.Ist.Naz.Alto Mat. London - N.Y., 1977, v.22, p.71-91.

0.C a r l e s o n L. An example concerning analytic functions with finite Dirichlet integrals. - Зап.научн.сем.ЛОМИ, 1979, 92, 283-287.

I1.П е л л е р В.В., Х р у щ ё в С.В. Операторы Ганкеля, наилучшие приближения и стационарные гауссовские процессы. - Успехи Матем.наук, 1982, 37, вып.I, стр.53-124.

9.4. ANALYTIC FUNCTIONS STATIONARY ON A SET, THE UNCERTAINTY PRINCIPLE FOR CONVOLUTIONS, AND ALGEBRAS OF JORDAN OPERATORS

1. The statement of the problem. We say a Lebesgue measurable function φ defined on the circle $\mathbb{T}$ is *stationary* on the set E, $E\subset\mathbb{T}$, if there exists a function ψ absolutely continuous on $\mathbb{T}$ and such that

$$\left.\begin{aligned} &\psi(\zeta)=\varphi(\zeta)\\ &\frac{d}{d\theta}\psi(e^{i\theta})=0\end{aligned}\right\}\qquad \text{a.e. on } E$$

A measurable set E, $E\subset\mathbb{T}$, is said to have *the property* S (in which case we *write* $E\in(S)$) if there is no non-constant function in $H^1(\mathbb{T})$ stationary on E.

PROBLEM. Give a description of sets of the class (S).

We mean a description yielding an answer to the following

QUESTION 1. Does every E with mes $E>0$ belong to (S) ?

There are natural modifications of the PROBLEM. E.g. we may ask the following

QUESTION 2. Suppose $E\subset\mathbb{T}$, $\operatorname{mes} E>0$. Is there a non-constant f in the disc-algebra A (i.e. $f\in C(\mathbb{T})$, $\hat{f}(n)=0$ for all $n\in\mathbb{Z}_-$) stationary on E ? What about $f\in A$ satisfying a Lipschitz condition of order less than one?

Note that every $f\in A$ satisfying the first order Lipschitz condition and stationary on a set of positive length is constant.

Using a theorem of S.V.Hruščev [1], it is not hard to prove [2,3] that a closed set E, $E\subset\mathbb{T}$, has the property S if $\operatorname{mes} E>0$ and if moreover

$$\sum |\ell|\log\frac{1}{|\ell|}<+\infty, \tag{C}$$

the sum being taken over the set of all complementary arcs ℓ of E, $|\ell|\overset{\text{def}}{=}\operatorname{mes}\ell$.

QUESTION 3. Suppose $E\in(S)$. Does E contain a closed

subset E_1 of positive length satisfying (C)?

This question may be, of course, modified in the spirit of QUESTION 2. A deep theorem by S.V.Hruščev (deserving to be known better than it is, [1], Th.4.1 on p 133) suggests the positive answer.

We like our PROBLEM in its own right and feel it is worth solving because it is nice in itself. But there are two "exterior" reasons to look for its solution.

2. The uncertainty principle for convolutions. Let K be a distribution in $\mathbb{R}^n$, X a class of distributions (in $\mathbb{R}^n$). Suppose the convolution $K * f$ has a sense for every $f \in X$. The set $E \in \mathbb{R}^n$ is called a (K, X) - set if

$$f \in X, \quad (K * f) | E = 0, \quad f | E = 0 \Rightarrow f = 0 .$$

(The exact meaning of the convolution $K * f$ and of the restrictions $f | E$, $(K * f) | E$ becomes clear in concrete situations, see [2], [4]).

If the class of (K, X) -sets is sufficiently large (e.g. contains all non-void open sets) then we may say that the operator $f \to K * f$ obeys "the uncertainty principle", namely, the knowledge of both restrictions $f | E$ and $(K * f) | E$ determines f uniquely. For example every set $E \subset \mathbb{R}$ of positive length is an $(\mathcal{H}, L^2)$-set, $\mathcal{H}$ being the Hilbert transform (i.e $(\mathcal{H} * f)(t) \overset{def}{=} \text{v.p.} \frac{i}{\pi} \int_{-\infty}^{+\infty} \frac{f(u)\,du}{t-u}$). Other examples see in [2], [4], [5]. There are interesting situations (e.g. $K(x) = |x|^{-\alpha} (x \in \mathbb{R}^n, 0 < \alpha < n$, X a suitable class of distributions) when we only know that there are many (K, X) -sets but have no satisfactory characterization of such sets. The most interesting is, may be, the case of $K(x) = |x|^{-n+1} (x \in \mathbb{R}^n, n \geq 2)$ closely connected with the Cauchy problem for the Laplace equation.

In an attempt to obtain a large class of relatively simple K's obeying the above "uncertainty principle" and to understand this principle better the kernels with the so-called semirational symbols have been introduced in [2].

Consider a Lebesgue measurable function $k : \mathbb{R} \to \mathbb{C}$ and put $\mathcal{D}_k = \{ f \in L^2(\mathbb{R}) : k \hat{f} \in L^2(\mathbb{R}) \}$, $\hat{f}$ being the Fourier transform of f, and define $K * f$ for $f \in \mathcal{D}_k$ by the identity $(K * f)^\wedge = \hat{f} k$. We call k the symbol of the operator $f \to K * f$. The function k is called semirational if there exists a rational function r such that $k\,|\,(-\infty,0) = r\,|\,(-\infty,0)$, $k(\xi) \neq r(\xi)$ a.e. on a neighbourhood of $+\infty$.

In [2] it was proved that every closed set $E \subset \mathbb{R}$ of positive length satisfying (C) (where ℓ runs through the set of bounded complementary intervals of E) is a $(K, \mathcal{D}_K)$ -set provided k is semirational (a simpler proof see in [3]). It is not known whether condition (C) can be removed. An interesting (and typical) example of a convolution with a semirational symbol is the operator K,

$$(Kf)(t) \overset{def}{=} (\mathcal{H} * f)(t) + ce^{-\varepsilon t} \int_{-\infty}^{t} e^{\varepsilon \tau} f(\tau)\, d\tau \quad (t \in \mathbb{R}, f \in L^2, \varepsilon > 0),$$

a perturbation of the Hilbert transform. We do not know whether every set E with $mes\, E > 0$ is a (K, L^2) -set (though we know it is when $c = 0$ or when E satisfies (C)).

All this is closely connected with our PROBLEM (or better to say with its slight modification).

DEFINITION. 1) A Lebesgue measurable function φ on the line $\mathbb{R}$ is said to be n-s t a t i o n a r y on the set $E, E \subset \mathbb{R}$, if there are functions $\psi_1, \dots, \psi_n \in W_2^1(\mathbb{R})$ (i.e $\in L^2(\mathbb{R})$, absolutely continuous and with the $L^2(\mathbb{R})$ -derivative) such that $\varphi | E = \psi_1 | E$, $\psi_1' | E = \psi_2 | E, \dots, \psi_{n-1}' | E = \psi_n | E$, $\psi_n' | E = 0$.

2) A set $E, E \subset \mathbb{R}$ is said to have t h e p r o p e r t y S_n (or $E \in (S_n)$) if

$$\varphi \in H^2(\mathbb{R}),\ \varphi \ n\text{-stationary on } E \Rightarrow \varphi \equiv 0.$$

It is not hard to see that

$$E = \operatorname{Clos} E,\ E \in (C),\ mes\, E > 0 \Rightarrow E \in (S_n),\ n = 1, 2, \dots$$

and that if $E \in \bigcap_{n=1}^{\infty} S_n$ then E is a $(K, \mathcal{D}_K)$ -set for every semirational k [3].

Moreover if there exists a $\varphi \in H^2(\mathbb{R})$, $\varphi \neq 0$, stationary on the set E, then E is not a $(K, \mathcal{D}_K)$ -set for a semirational k (which may be even chosen so that k agrees with a linear function on $(-\infty, 0)$).

Another circle of problems where n-stationary analytic functions emerge in a compulsory way is connected with

3. Jordan operators. We are going to discuss Jordan operators (J.o.) T of the form

$$T = \mathcal{U} + Q \qquad (*)$$

here $\mathcal{U}$ is unitary, $Q^{n+1}=0, \mathcal{U}Q = Q\mathcal{U}$ (in this case we say T s of order n). It is well known that the spectrum of any such T ies on $\mathbb{T}$ so that T is invertible. Denote by $\mathcal{R}(T)$ the weakly losed operator algebra spanned by T and the identity I We are nterested in conditions ensuring the inclusion

$$T^{-1} \in \mathcal{R}(T). \qquad (**)$$

EXAMPLE. Let E be a Lebesgue measurable subset of $\mathbb{T}$ and H e the direct sum of $(n+1)$ copies of $L^2(\mathbb{T}\setminus E)$. The operator $J = J(E,n)$ defined by the $(n+1)\times(n+1)$ -matrix

$$\begin{pmatrix} z & & & O \\ I & z & & \\ & I & \ddots & \\ O & & I & z \end{pmatrix}$$

(z being the operator of multiplication by the complex variable z) is a J.o. of order n. It is proved in [3] that

$$J^{-1} \in \mathcal{R}(J) \Leftrightarrow E \in (S'_n).$$

Here (S'_n) denotes the class of subsets of $\mathbb{T}$ defined exactly as (S_n) in section 2 but with W_2^1 replaced by the class of all functions absolutely continuous on $\mathbb{T}$.

The special operator $J = J(E,n)$ is of importance for the investigation of J.o. in general, Namely [3], if T is our J.o. (*) of order n and $\mathcal{E}_{\mathcal{U}}$ stands for the spectral measure of $\mathcal{U}$ then

$$\mathcal{E}_{\mathcal{U}}(E) = 0, \quad J^{-1} \in \mathcal{R}(J) \Rightarrow (**)$$

Therefore if $E \in \bigcap_{n=1}^{\infty}(S'_n)$ (in particular if $\operatorname{mes} E > 0$ and $E \in (C)$) then (**) holds whenever $\mathcal{E}_{\mathcal{U}}(E) = 0$

Recall that for a unitary operator T (i.e. when $T = \mathcal{U}, Q = 0$

in (*)) the inclusion (**) is equivalent to the vanishing of $\mathcal{E}_u$ on a set of positive length. A deep approximation theorem by Sarason [6] yields spectral criteria of (**) for a normal T . Our questions concerning sets with the property S and analogous questions on classes (S_n), (S'_n) are related to the following difficult PROBLEM: which spectral conditions ensure (**) for $T = N + Q$ where N is normal and Q is a nilpotent commuting with N ?

REFERENCES

1. Х р у щ ё в С.В. Проблема одновременной аппроксимации и стирание особенностей интегралов типа Коши. - Труды Матем.ин-та АН СССР, 1978, 130, 124-195.
2. Ё р и к к е Б., Х а в и н В.П. Принцип неопределённости для операторов, перестановочных со сдвигом I. - Записки научн.семин. ЛОМИ, 1979, 92, 134-170; II. - ibid., 1981, 113, 97-134.
3. М а к а р о в Н.Г. О стационарных функциях. - Вестник ЛГУ (to be published).
4. H a v i n V.P., J ö r i c k e B. On a class of uniqueness theorems for convolutions. Lect.Notes in Math , 1981, 864, 143-170.
5. Х а в и н В.П. Принцип неопределённости для одномерных потенциалов М.Рисса. - Докл.АН СССР, 1982, 264, № 3, 559-563.
6. S a r a s o n D. Weak-star density of polynomials. - J.reine und angew.Math., 1972, 252, 1-15.

V.P.HAVIN
(В.П.ХАВИН)

СССР, 198904, Ленинград, Петродворец, Ленинградский государственный университет Математико-механический факультет

B.JÖRICKE

Akademie der Wissenschaften der DDR
Zentralinstitut für Mathematik und Mechanik
DDR, 108, Berlin
Mohrenstraße 39

N.G.MAKAROV
(Н.Г.МАКАРОВ)

СССР, 198904, Ленинград, Петродворец, Ленинградский государственный университет
Математико-механический факультет

■.5. PROBLEM IN THE THEORY OF FUNCTIONS

In 1966 I published the following theorem:

<u>There exists a constant</u> $a > 0$ <u>such that any collection of polynomials</u> Q <u>of the form</u>

$$Q(z) = \prod_k \left(1 - \frac{z^2}{z_k^2}\right)$$

<u>with</u>

$$\sum_1^\infty \frac{1}{n^2} \log^+ |Q(n)| \leq a$$

<u>is a normal family in the complex plane</u>.

See Acta Math., 116 (1966), pp.224-277; the theorem is on page 273.

This result can easily be made to apply to collections of polynomials of more general form provided that the sum from 1 to ∞ in its statement is replaced by one over all the non-zero integers. One peculiarity is that the constant $a > 0$ really must be taken quite small for the asserted normality to hold. If a is large enough, the theorem is false.

The results's proof is close to 40 pages long, and I think very few people have been through it. <u>Can one find a shorter and clearer proof?</u> This is my question.

Let me explain what I am thinking of. Take any fixed ρ, $0<\rho<\frac{1}{2}$ and let $\mathcal{D}_\rho$ be the slit domain

$$\mathbb{C} \setminus \bigcup_{-\infty}^{\infty} [n-\rho, n+\rho] .$$

If Q is any polynomial, write

$$Q(n,\rho) = \sup\{|Q(x)|;\ n-\rho \leq x \leq n+\rho\} .$$

By direct harmonic estimation in $\mathcal{D}_\rho$ one can find without too much trouble that

$$\log|Q(z)| \le K_\rho(z) \sum_{-\infty}^{\infty} \frac{\log^+ Q(n,\rho)}{1+n^2} ,$$

where $K_\rho(z)$ depends only on z and ρ . (This is proved in the first part of the paper cited above.) A natural idea is to try to obtain the theorem by making $\rho \longrightarrow 0$ in the above formula. This, however, cannot work because $K_\rho(z)$ tends to ∞ as $\rho \to 0$ whenever z is not an integer. The latter must happen since the set of integers has logarithmic capacity zero.

For polynomials, the estimate provided by the formula is too crude. The formula is valid if, in it, we replace $\log|Q(z)|$ by <u>any</u> function subharmonic in $\mathcal{D}_\rho$ having sufficiently slow growth at ∞ and some mild regularity near the slits $[n-\rho, n+\rho]$. Polynomials, however, are single-valued in $\mathcal{D}_\rho$. This single-valuedness imposes constraints on the subharmonic function $\log|Q(z)|$ which somehow work to diminish $K_\rho(z)$ to something bounded (for each fixed z) as $\rho \to 0$, provided that the sum figuring in the formula is sufficiently small. The PROBLEM here is to <u>see quantitatively how the constraints cause this diminishing to take place</u>.

The phenomenon just described can be easily observed in one simple situation. Suppose that $U(z)$ is subharmonic in $\mathcal{D}_\rho$,that $U(z) \le A|z|$ there, and that $[U(z)]^+$ is (say) continuous up to the slits $[n-\rho, n+\rho]$. If $U(x) \le M$ on each of the intervals $[n-\rho, n+\rho]$, then

$$U(z) \le M + \frac{A}{\pi} \log\left| \frac{\sin \pi z}{\sin \pi\rho} + \sqrt{\frac{\sin^2 \pi z}{\sin^2 \pi\rho} - 1} \right|$$

This estimate is best possible, and the quantity on the right blows up as $\rho \longrightarrow 0$. However, if $U(z) = \log|F(z)|$ where $F(z)$ is a single valued entire function of exponential type $A < \pi$, we have the better estimate

$$U(z) \le M + C_A + A|y|$$

with a constant C_A independent of ρ . The improved

result follows from the theorem of Duffin and Schaeffer. It is no longer true when $A \geqslant \pi$ — consider the functions $F(z) = L \sin \pi z$ with $L \to \infty$.

The whole idea here is to see how harmonic estimation for functions analytic in multiply connected domains can be improved by taking into account those functions' single-valuedness.

PAUL KOOSIS

Institut Mittag-Leffler, Sweden
McGill University, Montreal, Canada
UCLA, Los Angeles, USA

9.6. old PEAK SETS FOR LIPSCHITZ CLASSES

The Lipschitz class A_α, $0<\alpha\leqslant 1$, consists of all functions f analytic in $\mathbb{D}$, continuous on $\operatorname{clos}\mathbb{D}$ and such that

$$|f(\zeta_1)-f(\zeta_2)|\leqslant const\,|\zeta_1-\zeta_2|^\alpha, \qquad \zeta_1,\zeta_2\in\mathbb{T}. \tag{1}$$

A closed set E, $E\subset\mathbb{T}$, is called *peak set for* A_α (in symbols $E\in\mathcal{P}_\alpha$), if there exists a function $f, f\in A_\alpha$ (the so called *peak function*) such that

$$|f|<1 \text{ on } \mathbb{T}\setminus E; \quad f|E\equiv 1.$$

THE PROBLEM is to describe the structure of $\mathcal{P}_\alpha$-sets. B.S.Pavlov [1] discovered a necessary condition; this condition was rediscovered by H.Hutt [2] in a more complicated way. Write $E_\delta = \{\zeta\in\mathbb{T}: dist(\zeta,E)\leqslant\delta\}$.

THEOREM 1 (Pavlov [1], Hutt [2]). <u>If</u> $E\in\mathcal{P}_\alpha$ <u>then</u>

$$m(E_\delta)=O(\delta^\alpha), \quad \delta\to+0. \tag{2}$$

COROLLARY. $\mathcal{P}_1$ <u>-sets are finite</u>.

SKETCH OF THE PROOF. Let f be a peak function for E and $g=\mathbb{1}-f$. Then $g\in A_\alpha$, $\operatorname{Re} g\geqslant 0$ in $\mathbb{D}$ and $g|E\equiv 0$. Hence $\operatorname{Re} 1/g\geqslant 0$ in $\mathbb{D}$ and Herglotz theorem [3] says that $1/g$ is Cauchy integral of a finite measure. Now condition (2) follows from the weak type estimate [4]

$$m\{\zeta:\zeta\in\mathbb{T},\ |1/g|>t\}\leqslant const\cdot t^{-1},\ t>0,$$

and from the evident inequality $|g(\zeta)|\leqslant const\cdot dist(\zeta,E)^\alpha$. ●

Until quite recently only some simple examples of $\mathcal{P}_\alpha$-sets were known [2]; condition (2) holds for these examples "with a reserve". But recent S.V.Hruščev's results [5] on zero sets for Gevrey classes permit us to obtain a very exact sufficient condition. Define the *Gevrey class* G_α as a class of all analytic functions in $\mathbb{D}$ such that $\sup_{\zeta\in\mathbb{D}}|f^{(n)}(\zeta)|\leqslant(const)^{n+1}\cdot(n!)^{1+\frac{1}{\alpha}}$, $n=0,1,\dots$. A set E, $E\subset\mathbb{T}$ is called *zero set for* G_α (or $E\in Z(G_\alpha)$), if there is a non-zero function f, $f\in G_\alpha$, with $f^{(n)}|E\equiv 0$, $n=0,1,\dots$. Hruščev [5] has completely investigated $Z(G_\alpha)$- sets and gave a lot of examples.

THEOREM 2. $Z(G_\alpha) \subset \mathcal{P}_\alpha$

SKETCH OF THE PROOF. It has been shown in [5] that every non-empty E in $Z(G_\alpha)$ defines a positive function U on $\mathbb{T}$ with the following list of properties:

a) $|(\log f)'| \leqslant const\, U^{1+\frac{1}{\alpha}}$;

b) $dist(\zeta, E)^{-\alpha} \leqslant U(\zeta),\ \zeta \in \mathbb{T}$;

c) $0 < c_1 < U(\zeta_1)/U(\zeta_2) < c_2 < +\infty$

provided $\zeta_1, \zeta_2 \in \mathbb{T} \setminus E$ and $|\zeta_1 - \zeta_2| < \frac{1}{2}\, dist(\zeta_1, E)$. Here $f = exp\{-U - i\tilde{U}\}$ stands for the outer function with the modulus $exp(-U)$ on $\mathbb{T}$. It is easy to see that $f \in G_\alpha$, $f^{(n)}|E \equiv 0,\ n = 0, 1, \ldots$ (see [5] for details).

Set $g \stackrel{def}{=\!=} (\log f)^{-1}$ in $clos\, \mathbb{D}$. Then

$$Re\, g(z) \leqslant 0,\quad z \in clos\, \mathbb{D};\quad |g(\zeta)| \leqslant U(\zeta)^{-1} \leqslant dist(\zeta, E)^{\alpha}.$$

Let us prove now that $g \in A_\alpha$. Obviously (1) holds if either ζ_1 and ζ_2 lie in different complementary intervals of E or

$$|\zeta_1 - \zeta_2| > \frac{1}{5} \max\{U(\zeta_1)^{-1/\alpha}, U(\zeta_2)^{-1/\alpha}\}.$$

If $|\zeta_1 - \zeta_2| < \frac{1}{5} U(\zeta_1)^{-1/\alpha}$ then by b) $|\zeta_1 - \zeta_2| < \frac{1}{5}\, dist(\zeta_1, E)$ and by a) and c)

$$|g(\zeta_1) - g(\zeta_2)| \leqslant const\, |\zeta_1 - \zeta_2|\, U(\zeta_1)^{1+\frac{1}{\alpha}}\, U(\zeta_2)^{-2} \leqslant const\, |\zeta_1 - \zeta_2|^{\alpha}.$$

It is clear that $h = (1+g)(1-g)^{-1} \in A_\alpha$ and h is a peak function for E.

Theorem 2 gives a number of examples of $\mathcal{P}_\alpha$ - sets and the following conjecture seems now plausible.

CONJECTURE. $\mathcal{P}_\alpha = Z(G_\alpha)$.

There exist some (not very clear) connections between the free interpolation sets for A_α and G_α [6]; these connections corroborate our conjecture. The A_α-function from theorem 2 is a "logarithm" of some G_α-function. Probably, it is a general rule and it is possible to link two interpolation problems in a direct way.

A possible way to prove our conjecture is a conversion of the proof of Theorem 2 The "strongly vanishing function" $f = \exp\left(-\frac{1}{1-h}\right)$ is of course not in G_α for an arbitrary peak function $h \in A_\alpha$. But in theorem 2 such function f is not arbitrary; it is extremal in some sense [5]. Perhaps, it is possible to obtain $f \in G_\alpha$ for some extremal peak function h. The extremal functions are often analytic on $\mathbb{T} \setminus E$, and such an f may be a smooth function.

The following necessary condition may be a first step to the conjecture: if $E \in \mathcal{P}_\alpha$ then $\operatorname{dist}(\cdot, E)^{-\alpha} \in L^1(\mathbb{T})$.

The description of $\mathcal{P}_\alpha$-sets is interesting for the investigation of the singular spectrum in the Friedrichs model [1,7].

REFERENCES

1. П а в л о в Б.С. Теоремы единственности для функций с положительной мнимой частью. - Проблемы матем.физики, Изд-во ЛГУ, 1970, № 4, 118-125.
2. H u t t H. Some results on peak and interpolation sets of analytic functions with higher regularity. - Uppsala Univ.Dep.Math., Thesis, 1976.
3. П р и в а л о в И.И. Граничные свойства аналитических функций. М.-Л., ГИТТЛ, 1950.
4. Z y g m u n d A., Trigonometric series, Cambridge Univ.Press, London, New York, 1969.
5. H r u š č ё v S.V. Sets of uniqueness for the Gevrey classes. - Ark.för Mat., 1977, 15, N 2, 256-304.
6. Д ы н ь к и н Е.М. Свободная интерполяция в классах Гёльдера. - Матем. сборник, 1979, 109, N 1 (Math.USSR Sbornik, 1980, 37, N 1, 97-117).
7. P a v l o v B.S., F a d d e e v L.D. This Collection, 4.4.

E.M.DYN'KIN
(Е.М.ДЫНЬКИН)

СССР, 197022, Ленинград
ул.проф.Попова, 5
Ленинградский
Электротехнический институт
им.В.И.Ульянова (Ленина)

.7.
ld

A PROBLEM BY R.KAUFMAN

Let G be a bounded Lipschitz domain and $C^{r}(G)$ $(r \geq 1)$ the lass of functions analytic in G, with r-th derivative uniformly ontinuous on G. Do the classes $C^{r}(G)$ have the same zero-sets n $clos\, G$, or on ∂G?

R.KAUFMAN

University of Illinois
at Urbana-Champaign
Department of Mathematics,
Urbana, Illinois, 61801
USA

9.8. old QUASI-ANALYTICITY OF FUNCTIONS WITH RESPECT TO A DIFFERENTIAL OPERATOR

Suppose Ω is a domain in $\mathbb{R}^n$, E is a closed subset of Ω, $\{m_n\}$ is a sequence of positive numbers such that $\sum_n m_n^{*-1/n}=\infty$ (m_n^* is the best monotone majorant for m_n), L is a differential operator of order ℓ with $C^\infty(\Omega)$ coefficients. A function $f\in C^\infty(\Omega)$ is said to belong to the class $\mathcal{L}(m_n)$ if the following inequalities are satisfied:

$$\|L^k f\|_{L^2(\Omega)}\leqslant C m_k^\ell,\quad k=0,1,2,\dots\quad (C=C_f).$$

Denote by Ω_0 the maximal set among the subsets of Ω enjoying the following property:

if a function f_1 , $f_1\in C^\infty(\Omega)$, has a zero of infinite multiplicity on E and satisfies the equation $Lf_1=0$ then $f_1|\Omega_0=0$.

I CONJECTURE that under an appropriate definition of the order ℓ of the operator L the following is true.

CONJECTURE. __If f belongs to the class $\mathcal{L}(m_n)$ and has a zero of infinite multiplicity on E , then $f|\Omega_0=0$__ .

In other words functions quasi-analytic with respect to the operator L behave with respect to the uniqueness theorem as solutions of the homogeneous equation $Lf=0$.

We will SKETCH THE PROOF of the Conjecture IN TWO CASES.

1. Suppose L is an elliptic operator such that the operator $D_t^\ell-L_x$ is elliptic, $E=\{x_0\}$ is a point of Ω , and $m_n=n!$. Consider the solution $g(t,x)$ of the problem $(D_t^\ell-L_x)g=0$, $g(0,x)=f(x)$, $D_t^j g(0,x)=0$ $(j=1,\dots,\ell-1)$. It exists for small t :

$$g(t,x)=\sum_{k=0}^{\infty}\frac{t^{k\ell}}{(k\ell)!}L_x^k f.$$

The function $g(t,x)$ has zero of infinite order in $t=x=0$, hence $g(t,x)\equiv 0$ and $f(x)\equiv 0$.

2. $L=D_{x_1}^2+D_{x_2}^2-D_{x_3}^2$, $\Omega=\mathbb{R}^3$, E is a two-dimensio-

al smooth surface, $f \in C^\infty(\mathbb{R}^3)$ and satisfies $\|L^k f\|_{L^2(\Omega)} \leqslant \leqslant C m_k$, $k = 0, 1, 2, \ldots$ (so that here $\ell = 1$). Denote by L_0 the closure of the operator L defined on the set of functions vanishing of infinite order on E . It is clear that L_0 is a symmetric operator. Suppose the vector φ , $\varphi \neq 0$ is such that $L_0^* \varphi = i\varphi$. Consider for $\Im m\, \lambda > 0$ the vector-valued function $\lambda \to \varphi_\lambda = \varphi + (\lambda - i)\tilde{R}_\lambda \varphi$ where $\tilde{R}_\lambda$ is the resolvent of a self-adjoint extension of the operator L_0 . Then $L_0^* \varphi_\lambda = \lambda \varphi_\lambda$ and

$$|\lambda^k(\varphi_\lambda, f)| = |(L_0^{*k} \varphi_\lambda, f)| = |(\varphi_\lambda, L_0^k \varphi)| \leqslant C m_k , \quad k = 0, 1, 2, \ldots$$

Hence, $(\varphi_\lambda, f) = (\varphi, f) = 0$. Similarly $(\psi, f) = 0$ where $L_0^* \psi = -i\psi$. Hence f belongs to the invariant subspace of the operator L_0 on which L_0 is self-adjoint. By the theorem of Gelfand and Kostjučenko [1] f belongs to the linear span of the generalized eigenfunctions of the operator L_0 , i.e. of solutions of the equation $L_0 g = \lambda g$ vanishing on Ω_0 . Hence $f | \Omega_0 = 0$. ●

REFERENCES

1. Гельфанд И.М., Костюченко А.Г. Разложение по собственным функциям дифференциальных и других операторов. - Докл.АН СССР, 1955,103, 349-352.
2. Березанский Ю.М. Разложение по собственным функциям самосопряженных операторов. Киев, "Наукова думка", 1965.

V.I.MATSAEV
(В.И.МАЦАЕВ)

СССР, 142432 Черноголовка
Московская обл.,
Институт химической
физики АН СССР

* * *

COMMENTARY

The CONJECTURE is confirmed now in various particular cases. Here are some quite recent (unpublished) results.

Let Ω be a domain in $\mathbb{R}^n$, A be a differential operator of order ℓ with $C^\infty(\Omega)$-coefficients , $p \in (1, +\infty]$, $M = \{M_k\}_{k=0}^{+\infty}$ a sequence of positive numbers. Put $m_k = M_k / k!$,

$$\mathcal{A}_p(M,\Omega) \overset{\text{def}}{=} \{ f \in C^\infty(\Omega) : \| A^j f \|_{L^p(\Omega)} \leq C_f Q_f^{jl} M_j^l ,\ j=0,1,\dots \}.$$

Suppose $m_k m_p \leq C m_{k+p}$ ($k, p = 1,2,\dots$) .
Consider a family $d = (d_1,\dots,d_n)$ of positive numbers, $d_1 \leq d_2 \leq \dots \leq d_n$ and a compact set $K, K \subset \Omega$. Put

$$C(M,d,K) \overset{\text{def}}{=} \{ f \in C^\infty(\Omega) : \| \mathcal{D}^\alpha f \|_{C(K)} \leq$$

$$\leq C_{f,K} Q_{f,K}^{|\alpha|} m_{[d_1\alpha_1 + \dots + d_n\alpha_n]} \Gamma(1 + \sum_1^n \alpha_j d_j) \},$$

$[t]$ being the integer part of t . When $d \equiv 1$ we write $C(M,K)$ instead of $C(M,d,K)$.

THEOREM 1 (A.G.Chernyavskii) <u>Suppose</u> that 1) $A = \sum_{|\alpha| + \beta \leq l} a_{\alpha\beta}(t,x) \mathcal{D}_t^\beta \mathcal{D}_x^\alpha$, $a_{\alpha\beta} \in C(M,G)$, <u>where</u> G <u>is the cylinder</u> $\{(t,x) : t \in [-1,1], x \in \mathbb{R}^{n-1}, |x| \leq 1\}$ and $a_{0l}(t,x) \neq 0$ in G ;
2) <u>the class</u> $C(M,G)$ <u>is quasianalytic</u>, i.e.

$$\int_1^\infty \frac{\log T(\tau)}{\tau^2} d\tau = \infty , \quad T(\tau) \overset{\text{def}}{=} \sup_k \frac{\tau^k}{M_k} . \qquad (*)$$

<u>If</u> $u \in \mathcal{A}_\infty(M,G)$ <u>and</u> $\mathcal{D}_t^p A^j(0,x) = 0$ $(0 \leq p \leq l-1, j = 0,1,\dots, |x| \leq$ <u>then</u> $u \equiv 0$ <u>in a neighbourhood of the origin</u>.

Fix now a multiindex λ with decreasing natural coordinates. For every multiindex α put $|\alpha : \lambda| = \sum_1^n \alpha_k / \lambda_k$. Suppose that

$$A = \sum_{|\alpha : \lambda| \leq 1} a_\alpha(x) \mathcal{D}^\alpha ,$$

where $a_\alpha \in C(M,K,d)$ on every compact $K, K \subset \Omega$, and $d_j \overset{\text{def}}{=} \lambda_1 / \lambda_j$ ($j = 1,\dots,n$) . Assume moreover that

$$\sum_{|\alpha : \ell| \leq 1} a_\alpha(x) \xi^\alpha \neq 0 \quad ((x,\xi) \in \Omega \times (\mathbb{R}^n \setminus \{0\})).$$

THEOREM 2. (M.M.Malamud). Under the above conditions

1) $\mathcal{A}_p(M) \subset C(M, d, L_p(\Omega)) \quad (1 < p < \infty)$

(the class $C(M, d, L_p(\Omega))$ is defined like $C(M, d, K)$ above with $\|\cdot\|_{L^p(\Omega)}$ instead of $\|\cdot\|_{C(K)}$).

2) Suppose $\lambda_1 = \ldots = \lambda_m > \lambda_{m+1} \geqslant \ldots > \lambda_n$ and $E = \operatorname{Clos} E \subset \Omega$. If $P_m(E) = P_m(\Omega)$ (P_m being the projection $(x_1, \ldots, x_n) \to \to (0, \ldots, 0, x_{m+1}, \ldots, x_n)$) then

$$(*) \Longleftrightarrow \{u \in \mathcal{A}_p(M, \Omega) : \mathcal{D}_{x_1}^{\alpha_1} \ldots \mathcal{D}_{x_m}^{\alpha_m} u | E = 0, \ 0 \leqslant |\alpha| < \infty\} = \{0\} .$$

In 1) M is not supposed to satisfy (*). The condition concerning $P_m(E)$ cannot be dropped from 2). If A is elliptic (in that case $\ell_1 = \ldots = \ell_n$) then every singleton $\{x_0\}$ can serve as the set E in 2), and $\operatorname{Ker} A$ and $\mathcal{A}_p(M, \Omega)$ are both quasianalytic if $C(M, L_p(\Omega))$ is (this result is due to M.M.Malamud and A.E.Shishkov, see also [3]).

V.P.Palamodov has pointed out (a private communication to the editors) that one can confirm the CONJECTURE in the case of an ultrahyperbolic operator using methods of §17, Ch.VI of [4] .

We conclude by a list of works connected with our theme.

REFERENCES

3. L i o n s J.-L., M a g e n e s P. Problèmes aux limites non-homogènes et applications. Vol.3, Paris, Dunod, 1970.
4. C o u r a n t R. Partial differential equations, N.Y., London, 1962.
5. Л ю б и ч Ю.И., Т к а ч е н к о В.А. Абстрактная проблема квазианалитичности. – Теория функций, функц.анал. и их прил., 1972, 16, 18–29.
6. Ч е р н я в с к и й А.Г. Квазианалитические классы, порожденные гиперболическими операторами с постоянными коэффициентами в $\mathbb{R}^2$. – ibid., 1982, 37, 122–127.
7. Ч е р н я в с к и й А.Г. Об одном обобщении теоремы единственности Хольмгрена. – Сиб.матем.журнал, 1981, 22, № 5, 212–215.
8. K o t a k é T., N a r a s i m h a n M.S. Regularity theorems for fractional powers of a linear elliptic operator – Bull.Soc. Math.France, 1962, 90, 449–471.

9.9. old LOCAL OPERATORS ON FOURIER TRANSFORMS

If f is a square integrable function of a real variable, let $\hat{f}$ denote its Fourier transform. If k is a measurable function of a real variable, the notation $\hat{k}$ is also used to denote a partially defined operator taking $\hat{f}$ into $\hat{g}=\hat{k}*\hat{f}$ whenever f and $g=kf$ are square integrable. The operator $\hat{k}$ is said to be l o c a l if, whenever a function $\hat{f}$ in its domain vanishes almost everywhere in a set of positive measure, the function $\hat{g}=\hat{k}*\hat{f}$ also vanishes almost everywhere in the same set.

THE OPERATOR $\hat{k}$ IS CONJECTURED to be local if k is the restriction to the real axis of an entire function of minimal exponential type.

If the operator $\hat{k}$ is local and if it has in its domain a nonzero function which vanishes almost everywhere in a set of positive measure, then IT IS CONJECTURED that k agrees almost everywhere with the restriction to the real axis of an entire function of minimal exponential type which satisfies the convergence condition

$$\int (1+t^2)^{-1} \log^+ |k(t)|\, dt < \infty \quad . \tag{*}$$

If k is a function which satisfies the convergence condition, if $k \geqslant 1$, and if $\log k$ is uniformly continuous, then IT IS CONJECTURED that for every positive number ε a nonzero function f in the domain of $\hat{k}$ exists which vanishes almost everywhere outside of the interval $(-\varepsilon, \varepsilon)$.

If k is a function which does not satisfy the convergence condition, if $k \geqslant 1$, and if $\log k$ is uniformly continuous, then IT IS CONJECTURED that no nonzero function exists in the domain of k which vanishes in a set of positive measure.

REFERENCE

1. d e B r a n g e s L. Espaces Hilbertiens de Fonctions Entières., Paris, Masson, 1972.

L. DE BRANGES

Purdue University
Department of Math.
Lafayette, Indiana 47907 USA

FROM THE AUTHOR'S SUPPLEMENT, 1983

The problem originates in a theorem on quasi-analyticity due to Levinson [2] . This theorem states that $\hat{f}$ cannot vanish in an interval without vanishing identically if it is in the domain of $\hat{k}$ where k is sufficiently large and smooth. L a r g e means that the integral in (*) is infinite, s m a l l that it is finite. The smoothness condition assumed by Levinson was that $\log^+|k|$ is non-decreasing, but it is more natural [3] to assume that $\log^+|k|$ is uniformly continuous (or satisfies the Lipschitz condition).

A stronger conclusion was obtained by Beurling [4] under the Levinson hypothesis. A function $\hat{f}$ in the domain of $\hat{k}$ cannot vanish in a set of positive measure unless it vanishes identically. The Beurling argument pursues a construction of Levinson and Carleman which is distinct from the methods based on the operational calculus concerned with the concept of a local operator. The Beurling theorem can be read as the assertion that certain operators are local with a trivial domain. It would be interesting to obtain the Beurling theorem as a corollary of properties of local operators with nontrivial domain [...] .

The author thanks Professor Sergei Khrushchev for informing him that a counter-example to our locality conjecture has been obtained by Kargaev.

REFERENCES

2. L e v i n s o n N. Gap and Density theorems. - Amer.Math.Soc., Providence, 1940.
3. d e B r a n g e s L. Local operators on Fourier transforms - Duke Math.J., 1958, 25, 143-153.
4. B e u r l i n g A. Quasianalyticity and generalized distributions, unpublished manuscript, 1961.

* * *

COMMENTARY

P.P.Kargaev has DISPROVED the FIRST and the LAST CONJECTURES.

As to the first, he has constructed an entire function k not only of minimal exponential type, but of z e r o o r d e r such that $\hat{k}$ is not local. Moreover, the following is true.

THEOREM (Kargaev). Let φ be a positive function decreasing to zero on $[0,+\infty)$. Then there exist a function $d: \mathbb{C} \longrightarrow \mathbb{Z}_+$ ("a divisor"), a set $e \subset \mathbb{R}$, and a function $f \in L^2(\mathbb{R})$ with the following properties:

$$\sum_{\lambda \in \mathbb{C}} d(\lambda)\varphi(|\lambda|) < +\infty, \qquad \sum_{\lambda \in \mathbb{C}} \frac{d(\lambda)}{|\lambda|} < +\infty,$$

$$\operatorname{mes} e > 0, \quad \hat{f} \in L^1(\mathbb{R}) \cap L^\infty(\mathbb{R}), \quad k_d \hat{f} \in L^2(\mathbb{R}), \quad f|e = 0,$$

$\hat{k}_d * f$ is bounded away from zero on e, where

$$k_d \stackrel{\mathrm{def}}{=} \prod_{\lambda \in \mathbb{C}} \left(1 - \frac{z}{\lambda}\right)^{d(\lambda)}.$$

Choosing $\varphi(x) = [\log(2+x)]^{-1}$ we see k_d is of zero order, but $\hat{k}_d$ is not local.

The LAST CONJECTURE is disproved by the fact (also found by Kargaev) that there exist real finite Borel measures μ on $\mathbb{R}$ with very large lacunae in $\operatorname{supp}\mu$ (i.e. there is a sequence $\{(a_n, b_n)\}_{n=1}^{\infty}$ of intervals free of $|\mu|$, $b_n < a_{n+1}$, $n=1,2,\dots$ $\frac{b_n - a_n}{a_n}$ tending to infinity as rapidly as we please) and with $\hat{\mu}$ vanishing on a set of positive length. Take $h = \mu * \varphi$, where φ is a suitable mollifier and $k = \exp(\operatorname{dist}(x, \operatorname{supp} h))$. Then $\log k$ is a Lipschitz function and $\int \frac{\log k(x)}{1+x^2}\,dx = +\infty$,

if $\frac{b_n - a_n}{a_n}$ grows rapidly enough. Then the inverse Fourier transform vanishes on a set of positive length and belongs to the domain of $\hat{k}$.

Kargaev's results will soon be published.

The THIRD CONJECTURE is true and follows from the Beurling-Malliavin multiplier theorem (this fact was overlooked both by the author and by the editors). Here is THE PROOF: there exists an entire function f of exponential type ε, $f \neq 0$, satisfying $|fk| \leqslant 1$ on $\mathbb{R}$. Then $\mathcal{F}^{-1}\left(f \frac{\sin \varepsilon x}{x}\right)$ is in the domain of $\hat{k}$.

.10. NON-SPANNING SEQUENCES OF EXPONENTIALS ON RECTIFIABLE PLANE ARCS

Let $\Lambda = (\lambda_n)$ be an increasing sequence of positive numbers with a finite upper density and let γ be a rectifiable arc in $\mathbb{C}$. Let $C(\gamma)$ denote the Banach space of continuous functions on γ with the usual sup-norm. If the relation of order on γ is denoted by $<$ and if z_0 and z_1 are two points on γ such that $z_0 < z_1$ we set

$$\gamma_{z_0 z_1} = \{ z \in \gamma \mid z_0 < z < z_1 \}.$$

The following theorem due to P.Malliavin and J.A.Siddiqi [7] gives a necessary condition in order that the sequence $(e^{\lambda z})_{\lambda \in \Lambda}$ be non-spanning in $C(\gamma)$.

THEOREM. <u>If the class</u>

$$C_{00}(M^{\Lambda}_{n-2}, \gamma_{z_0,z_1}) =$$

$$= \{ f \in C^{\infty}(\gamma_{z_0,z_1}) \mid \| f^{(n)} \|_{\infty} \leq A M^{\Lambda}_{n-2},\ f^{(n)}(z) \to 0 \text{ as } z \to z_i, i = 0,1, \forall n \}$$

<u>is non-empty for some</u> $z_0, z_1 \in \gamma$, <u>where</u>

$$M^{\Lambda}_{n} = \sup_{r \geq 0} \frac{r^n}{G_{\Lambda}(ir)} \text{ and } G_{\Lambda}(z) = \prod_{\lambda \in \Lambda} \left(1 - \frac{z^2}{\lambda^2}\right) \qquad (z = re^{i\theta}),$$

<u>then</u> $(e^{\lambda z})_{\lambda \in \Lambda}$ <u>is non-spanning in</u> $C(\gamma)$.

It had been proved earlier by P.Malliavin and J.A.Siddiqi [6] that if γ is a piecewise analytic arc then the hypothesis of the above theorem is equivalent to the Müntz condition $\sum \lambda_n^{-1} < \infty$. In connection with the above theorem the following problem remains open.

PROBLEM 1. <u>Given any non-quasi-analytic class of functions on</u> γ <u>in the sense of Denjoy-Carleman, to find a non-zero function belonging to that class and having zeros of infinite order at two points of</u> γ.

With certain restrictions on the growth of the sequence $\{\lambda_n\}$, partial solutions of the above problem were obtained by T.Erkamma [3] and subsequently by R.Couture [2], J.Korevaar and M.Dixon [4] and

M.Lundin [5].

Under the hypothesis of the above theorem, A.Baillette and J.A. Siddiqi [1] proved that $(e^{\lambda z})_{\lambda\in\Lambda}$ is not only non-spanning but also topologically linearly independent by effectively constructing the associated biorthogonal sequence. In this connection the following problem similar to one solved by L.Schwartz [8] in the case of linear segments remains open.

PROBLEM 2. <u>To characterize the closed linear span of</u> $(e^{\lambda z})_{\lambda\in\Lambda}$ <u>in</u> $C(\gamma)$ <u>when it is non-spanning</u>.

REFERENCES

1. B a i l l e t t e A., S i d d i q i J.A. Approximation de fonctions par des sommes d'exponentielles sur un arc rectifiable - J.d'Analyse Math., 1981, 40, 263-268.
2. C o u t u r e R. Un théorème de Denjoy-Carleman sur une courbe du plan complexe. - Proc.Amer.Math.Soc., 1982, 85, 401-406.
3. E r k a m m a T. Classes non-quasi-analytiques et le théorème d'approximation de Müntz. - C.R.Acad.Sc.Paris, 1976, 283, 595-597.
4. K o r e v a a r J., D i x o n M. Non-spanning sets of exponentials on curves. Acta Math.Acad.Sci.Hungar, 1979, 33, 89-100.
5. L u n d i n M. A new proof of a Müntz-type Theorem of Korevaar and Dixon. Preprint NO 1979-7, Chalmers University of Technology and The University of Göteborg.
6. M a l l i a v i n P., S i d d i q i J.A. Approximation polynômiale sur un arc analytique dans le plan complexe. C.R.Acad.Sc. Paris, 1971, 273, 105-108.
7. M a l l i a v i n P., S i d d i q i J.A. Classes de fonctions monogènes et approximation par des sommes d'exponentielles sur un arc rectifiable de $\mathbb{C}$, ibid., 1976, 282, 1091-1094.
8. S c h w a r t z L. Études des sommes d'exponentielles. Hermann, Paris, 1958.

J.A.SIDDIQI

Department de Mathématiques
Université Laval
Quebec, Canada, GIK 7P4

.11. WHEN IS $\int_{\mathbb{T}} \log|f|\,dm > -\infty$?

It is a well-known fact of Nevanlinna theory that the inequality n the title holds for boundary values of non-zero holomorphic funcions which belong to the Nevanlinna class in the unit disc. But what an be said about summable functions f with non-zero Riesz projecion $\mathbb{P}_- f \neq 0$? Here $\mathbb{P}_- f \overset{def}{=} \sum_{n>0} \hat{f}(-n)\bar{z}^n$, $|z| \leq 1$.

Given a positive sequence $\{M_n\}_{n\geq 0}$ define

$$C\{M_n\} = \{f \in C^\infty(\mathbb{T}) : \|f^{(n)}\|_\infty \leq C_f Q_f^n M_n\}.$$

t is assumed that
) $M_n^2 \leq M_{n-1} M_{n+1}, n=1,2,\ldots$; b) $\lim_{n\to\infty} \frac{x^n}{M_n} = 0$, $x \geq 0$.
This does not restrict the generality because every Carleman class oincides with one defined by a sequence satisfying a) and b). Let $T(x) = \sup_{n\geq 0} \frac{x^n}{M_n}$, $x > 0$. Then $C\{M_n\}$ is a quasianalytic class iff $\int^\infty \frac{\log T(x)}{x^2}\,dx = +\infty$. In case $C\{M_n\}$ is non-quasianalytic there are, of course, functions f in $C\{M_n\}$ with $\int_{\mathbb{T}} \log|f|\,dm = -\infty$, in fact there exists an f in $C\{M_n\}$ equal to zero on an open subset of $\mathbb{T}$.

QUESTION. <u>Suppose</u> $C\{M_n\}$ <u>is a quasianalytic class and let</u> $f \in L^1(\mathbb{T})$ <u>with</u> $\mathbb{P}_- f \in C\{M_n\}$. <u>Is it true that</u>

$$\int_{\mathbb{T}} \log|f|\,dm > -\infty\ ?$$

Under some additional assumptions on regularity of $\{M_n\}$ the answer is yes [1].

REFERENCE

I. В о л ь б е р г А.Л. Логарифм почти аналитической функции суммируем. - Докл.АН СССР, 1983, 265, 1317-1323.

A.L.VOL'BERG
(А.Л.ВОЛЬБЕРГ)

СССР, 197022, Ленинград
ул.Профессора Попова 5,
Ленинградский Электротехнический
институт

9.12. AN ALTERNATIVE FOR ANALYTIC CARLEMAN CLASSES

Given a sequence of positive numbers $\{M_n\}_{n\geq 0}$ let $C\{M_n\}$ be the Carleman class of infinitely differentiable functions on the unit circle $\mathbb{T}$ satisfying

$$\sup_{z\in\mathbb{T}} |f^{(n)}(z)| \leq C_f Q_f^n M_n$$

for $n=0,1,2,\ldots$ and some positive constants C_f, Q_f.

A class of functions defined on $\mathbb{T}$ is called quasi-analytic if it does not contain any function with $f^{(n)}(\zeta)=0$ for some ζ in $\mathbb{T}$ and every $n=0,1,\ldots$, besides $f\equiv 0$. Otherwise, the class is called nonquasi-analytic. Clearly, each nonquasi-analytic Carleman class contains a nonzero function vanishing on any given proper sub-arc of $\mathbb{T}$. The well-known test of Carleman [1] provides a convenient criterion in terms of $\{M_n\}$ to determine whether $C\{M_n\}$ is quasi-analytic or not.

The analytic Carleman classes $C_A\{M_n\} \stackrel{def}{=} \{f\in C\{M_n\} : \int_{\mathbb{T}} f\bar{z}^k dm = 0,$ $k=-1,-2,\ldots\}$ can also be split into quasi-analytic and nonquasi-analytic ones. There exists an analogue of Carleman's test for such classes [2], but in contradistinction to the classical Carleman classes a nonzero function in $C_A\{M_n\}$, being the boundary values of a bounded holomorphic function in the unit disc, cannot vanish on any subset of $\mathbb{T}$ having positive Lebesgue measure. Nevertheless, for some nonquasi-analytic classes $C_A\{M_n\}$ zero-sets of functions can be rather thick. This is the case, for example, if $M_n=(n!)^{1+1/\alpha}$, $n=0,1,\ldots$, where $0<\alpha<1$ (see [3]). Therefore it looks reasonable to formulate as a conjecture the following alternative.

CONJECTURE 1. <u>For every positive sequence $\{M_n\}_{n\geq 0}$ either the analytic Carleman class $C_A\{M_n\}$ is quasi-analytic or there exist a non-empty perfect subset E of $\mathbb{T}$ and a nonzero function f in $C_A\{M_n\}$ such that $f|E\equiv 0$.</u>

The alternative, if true, would have a nice application to dissipative Schrödinger operators. Consider the class B of all bounded measurable real functions V on $[0,+\infty)$ satisfying

$$\int_0^\infty r|V(r)|\,dr < +\infty .$$

Given $h\in\mathbb{C}$ let ℓ_h be the Schrödinger operator in $L^2(0,+\infty)$ defined by

$$\ell_h y = -y'' + Vy \;, \qquad y'(0) - hy(0) = 0 .$$

The operator ℓ_h is selfadjoint for real V and real h and it can have only a finite number of bound states, i.e. eigenvalues, if $V \in B$. For complex h the situation changes considerably. Now the number of bound states is finite if

$$|V(r)| \leq \exp\{-Cr^{1/2}\}$$

and on the other hand for each α in $(0,1)$ there exist a real-valued potential V satisfying

$$|V(r)| \leq \exp\{-Cr^{\frac{\alpha}{1+\alpha}}\} \qquad (1)$$

and $h \in \mathbb{C}, \operatorname{Im} h > 0$ such that ℓ_h has infinitely many bound states (see [4]). It can be even shown that the family of all closed subsets of $\mathbb{R}$,which may serve as derived sets of the point spectrum of ℓ_h with the potential V satisfying (1), coincides with the family of compact non-uniqueness sets in $(0,+\infty)$ for the Gevrey class $G_\alpha \overset{def}{=\!=} C_A\{M_n\}$, $M_n = (n!)^{1+1/\alpha}$ (see [5]). The above considerations make plausible the following conjecture.

CONJECTURE 2. Let T be a positive function on $[0,+\infty)$ such that $t \longrightarrow \log T(e^t)$ is convex. Then either every Shrödinger operator ℓ_h with the potential V satisfying

$$|V(r)| \leq \frac{const}{T(r)} \;, \qquad r \geq 0 \qquad (2)$$

has only finite number of bound states or there exist V satisfying (2) and $h \in \mathbb{C}, \operatorname{Im} h > 0$, such that the derived set of the point spectrum of ℓ_h is non-empty and perfect.

REFERENCES

1. M a n d e l b r o j t S. Séries adhérentes. Régularisation des suites. Applications. Paris, 1952.
2. R . - S a l i n a s B. Functions with null moments. - Rev.Acad. Ci.Madrid, 1955, 49, 331-368.
3. H r u š č e v S.V. Sets of uniqueness for the Gevrey classes. - Arkiv för Mat., 1977, 15, 253-304.
4. П а в л о в Б.С. О несамосопряжённом операторе Шрёдингера I, II, III. - в кн.: "Пробл.матем.физ.", 1966; 1967; 1968; Вып. 1, 2, 3,

Ленинград, ЛГУ, 102-132; 133-157; 59-80. (English translation: Pavlov B.S. The non-selfadjoint Schrödinger operator I, II, III. - in: Topics in Math.Physics, 1967; 1968; 1969; Consultants Bureau, N.Y., 87-114; 111-134; 53-71.)

5. H r u š č e v S.V. Spectral singularities of dissipative Schrödinger operators with rapidly decreasing potential. - Indiana Univ.Math.J. (to appear).

S.V.HRUŠČËV
(С.В.ХРУЩЁВ)

СССР, 191011 Ленинград, Д-11
Фонтанка 27
ЛОМИ АН СССР

.13. ON A UNIQUENESS THEOREM

old

The symbol $H(\mathcal{D})$, $\mathcal{D}$ being an open set in $\mathbb{C}^n$, denotes the set of all functions analytic in $\mathcal{D}$. Let Ω, $\Omega \subset \mathbb{R}^n$ $(n>1)$ be an arbitrary domain, $\Sigma^{(n)} \overset{def}{=} \{\sigma = (\sigma_1, \sigma_2, \ldots, \sigma_n) : \sigma_j = \pm 1\}$ and let $C = (C_1, C_2, \ldots, C_n) \in \mathbb{R}^n_+$. Define the following sets

$$\mathcal{D}_\sigma(\Omega,c) = \{z = x+iy \in \mathbb{C}^n : x \in \Omega,\ y_j\sigma_j > c_j,\ j=1,\ldots,n\},$$

$$\widetilde{\mathcal{D}}_\sigma(\Omega,c) = clos\, \mathcal{D}_\sigma(\Omega,c),$$

$$\mathcal{D}^K_\sigma(\Omega,c) = \{z \in \mathbb{C}^n : x \in \Omega,\ y_j\sigma_j > c_j,\ j \neq k\},\ k=1,\ldots,n,$$

$$\mathcal{D}(\Omega,c) = \bigcup_{\sigma\in\Sigma^{(n)}} \mathcal{D}_\sigma(\Omega,c), \quad \mathcal{D}^k(\Omega,c) = \bigcup_{\sigma\in\Sigma^{(n)}} \mathcal{D}^k_\sigma(\Omega,c).$$

Suppose that for a function f, $f \in H(\mathcal{D}(\Omega,0))$ the restriction $f|\mathcal{D}_\sigma(\Omega,0) \overset{def}{=} f_\sigma$ is continuous on the set $\widetilde{\mathcal{D}}_\sigma(\Omega,0)$. Then the function g,

$$g(x) = \sum_{\sigma\in\Sigma^{(n)}} sign\,\sigma \cdot f_\sigma(x), \quad x \in \Omega, \quad sign\,\sigma = \sigma_1\cdot\sigma_2\cdot\ldots\cdot\sigma_n$$

is well-defined in Ω. The following uniqueness theorem has been proved in [1].

<u>If there exist</u> $C = (C_1, C_2, \ldots, C_n) \in \mathbb{R}^n_+$ <u>and functions</u> h^k, $h^k \in H(\mathcal{D}^k(\Omega,c))$, $k = 1, \ldots, n$ <u>such that</u>

$$f(z) = \sum_{k=1}^{n} h^k(z), \quad z \in \mathcal{D}(\Omega,c) \quad \underline{\text{then}}\ g \equiv 0 \quad \underline{\text{on}}$$

$$\Omega^C = \{x \in \Omega : dist(x, \partial\Omega) > \|C\| = \sqrt{C_1^2 + C_2^2 + \cdots + C_n^2}\}.$$

Note that the theorem is important for studying homogeneous convolution equations in domains of real $(\mathbb{R}^n)$ or complex $(\mathbb{C}^n)$ spaces (see [1], [2], [3]). One might think that $g \equiv 0$ on Ω, as it occurs in the one-dimensional case. However there exists an example (see [1]), where all conditions of the uniqueness theorem are satisfied, but

$g \neq 0$ in Ω (for sufficiently large $\|C\|$). Hence the appearance of the set Ω^c is therefore inevitable although Ω^c does not seem to be the largest set where $g \equiv 0$.

PROBLEM. Find the maximal open subset of the domain Ω where $g \equiv 0$.

REFERENCES

I. Н а п а л к о в В.В. Об одной теореме единственности в теории функций многих комплексных переменных и однородные уравнения типа свертки в трубчатых областях $\mathbb{C}^n$. - Изв.АН СССР, сер.матем., I976, 40, № I, II5-I32.

2. Н а п а л к о в В.В. Однородные системы уравнений типа свертки на выпуклых областях $\mathbb{R}^n$. - Докл.АН СССР, I974, 2I9, № 4, 804-807.

3. Н а п а л к о в В.В. О решениях уравнений бесконечного порядка в действительной области. - Матем.сб., I977, I02, № 4, 499-5I0.

V.V.NAPALKOV
(В.В.НАПАЛКОВ)

СССР, 450057, Уфа
ул.Тукаева, 50
Башкирский филиал АН СССР
Сектор математики

CHAPTER 10

INTERPOLATION, BASES, MULTIPLIERS

We discuss in this introduction only one of various aspects of interpolation, namely the f r e e (or Carleson) interpolation by analytic functions.

Let X be a class of functions analytic in the open unit disc $\mathbb{D}$. We say that the interpolation by elements of X on a set $E \subset \mathbb{D}$ is free if the set $X|E$ (of all restrictions $f|E$, $f \in X$) can be described in terms not involving the complex structure inherited from $\mathbb{D}$. So, for example, if E satisfies the well-known Carleson condition (see formula (C) in Problem 10.3 below), the interpolation by elements of H^∞ on E is free in the following sense: a n y function, bounded on E, belongs to $H^\infty | E$. The freedom of interpolation for many other classes X means (as in the above example) that the space $X|E$ is ideal (i.e. $\varphi \in X|E, |\psi| \leqslant |\varphi|$ on $E \Longrightarrow \Rightarrow \psi \in X|E$). Sometimes the freedom means something else, as is the case with classes X of analytic functions enjoying certain smoothness at the boundary (see Problem 10.4), or with the Hermite interpolation with unbounded multiplicities of knots (this theme is treated in the book Н.К.Никольский, Лекции об операторе сдвига, Москва, Наука, 1980, English translation, Springer-Verlag, 1984; see also the article Виноградов С.А., Рукшин С.Е., Записки научных семинаров ЛОМИ, 1982, 107, 36–45).

Problems 10.1-10.5 below deal with free interpolation which is also the theme (main or peripheral) of Problems 4.10, 6.9, 6.19, 9.2, 11.6. But the information, contained in the volume, does not exhaust the subject, and we recommend the survey Виноградов С.А., Хавин В.П., Записки научн.семинаров ЛОМИ, 1974, 47, 15-54;1976, 56, 12-58, the book Garnett J., "Bounded analytic functions" and the recent doctoral thesis of S.A.Vinogradov "Free interpolation in spaces of analytic functions", Leningrad, 1982.

There exists a simple but important connection of interpolation (or, in other words, of the moment problem) with the study and classification of biorthogonal expansions (bases). This fact was (at last) widely realized during the past 15-20 years, though it was explicitly used already by S.Banach and T.Carleman. Namely, every pair of biorthogonal families $\mathcal{F}=\{f_\lambda\}_{\lambda\in\sigma}$, $\mathcal{F}'=\{f'_\lambda\}_{\lambda\in\sigma}$ (f_λ are vectors in the space V , f'_λ belong to the dual space, $\langle f_\lambda, f'_\mu\rangle=\delta_{\lambda\mu}$) generates the following interpolation problem: to describe the coefficient space $\mathcal{J}V$ ($\mathcal{J}f\overset{def}{=\!=}\{\langle f,f'_\lambda\rangle\}_{\lambda\in\sigma}$) of formal Fourier expansions $f\sim\sum_\lambda \langle f,f'_\lambda\rangle f_\lambda$. There are also continual analogues of this connection which are of importance for the spectral theory. "Freedom" of this kind of interpolation (or, to be more precise, the ideal character of the space $\mathcal{J}V$) means that $\mathcal{F}$ is an unconditional basis in its closed linear hull. This observation plays now a significant role in the interaction of interpolation methods with the spectral theory, the latter being the principal supplier of concrete biorthogonal families. These families usually consist of eigen- or root-vectors of an operator T (in Function Theory T is often differentiation or the backward shift, the two being isomorphic): $Tf_\lambda=\lambda f_\lambda$, $\lambda\in\sigma$. Thus the properties of the equation $\mathcal{J}f=q$ (q is the given function defined on σ) depend on the amount of m u l t i p l i e r s of $\mathcal{F}$, i.e. of operators $V\mapsto V$ sending f_λ to $\mu(\lambda)f_\lambda$, where μ denotes a function $\sigma\to\mathbb{C}$ or the ultiplier itself). These multipliers μ may turn out to be functions f T ($\mu=\varphi(T)$) and then we come to another interpolation problem given μ, find φ). The solution of this "multiplier" interpolaion problem often leads to the solution of the initial problem $f=g$. Interpolation and multipliers are related approximately in this ay in Problem 10.3, whereas Problem 10.8 deals with Fourier multipiers in their own right. These occur, as is well-known, in numerous roblems of Analysis, but in the present context the amount of mulipliers determines the convergence (summability) properties of stanlard Fourier expansions in the given function space. (By the way, the word "interpolation" in the title of Problem 10.8 has almost nothing to do with the same term in the Chapter title, and means the interpolation o f o p e r a t o r s . We say "almost" because the latter is often and successfully used in free interpolation). We cannot enter here into more details or enlist the literature and refer the reader to the mentioned book by Nikol'skii and to the article Hruščёv S.V., Nikol'skii N.K., Pavlov B.S. in Lecture Notes in Math. 864, 1981.

Problem 10.6 concerns biorthogonal expansions of analytic functions. The theme of bases is discussed also in 10.2 and in 1.7, 1.10, 1.12.

Problem 10.7 represents an interesting and vast aspect of interpolation, namely, its "real" aspect. We mean here extension theorems à la Whitney tending to the constructive description of traces of function classes determined by global conditions.

Free interpolation by analytic functions in $\mathbb{C}^n$ (and by harmonic functions in $\mathbb{R}^n$) is a fascinating area (see, e.g., Preface to Garnett's book). It is almost unexplored, not counting classical results on extensions from complex submanifolds and their refinements. Free interpolation in $\mathbb{C}^n$ is discussed in Problem 10.5.

10.1. old

NECESSARY CONDITIONS FOR INTERPOLATION BY ENTIRE FUNCTIONS

Let ρ be a subharmonic function on $\mathbb{C}$ such that $\log(1+|z|)= = O(\rho(z))$ and let A_ρ denote the algebra of entire functions f such that $|f(z)|\leq A\exp(B\rho(z))$ for some $A,B>0$. Let V denote a discrete sequence of points $\{a_n\}$ of $\mathbb{C}$ together with a sequence of positive integers $\{p_n\}$ (the multiplicities of $\{a_n\}$). If $f\in A_\rho$, $f\not\equiv 0$, then $V(f)$ denotes the sequence $\{a_n\}$ of zeros of f and p_n is the order of zero of f at a_n.

In this situation, there are THREE NATURAL PROBLEMS to study.

I. Zero set problem. Given ρ, describe the sets $V(f)$, $f\in A_\rho$.

II. Interpolation problem. If $\{a_n,p_n\}=V\subset V(f)$ for some f, $f\in A_\rho$, describe all sequences $\{\lambda_{n,k}\}$ which are of the form

$$\lambda_{n,k}=\frac{g^{(k)}(a_n)}{k!},\quad 0\leq k<p_n,\quad n=1,2,\dots \text{ for some } g,\ g\in A_\rho . \tag{I}$$

III. Universal Interpolation problem. If $V\subset V(f)$ for some f, $f\in A_\rho$, under what contitions on V is it true that for e v e r y sequence $\{\lambda_{n,k}\}$ such that $|\lambda_{n,k}|\leq A\exp(B\rho(a_n))$ there exists g, $g\in A_\rho$, satisfying (1).

In case $\rho(z)=\rho(|z|)$ (and satisfies some mild, technical conditions), quite good solutions to problems I-III are known. This work has been carried out by A.F.Leont'ev and others (see e.g. [1] for a survey). However, when ρ is not a function of $|z|$, the general solutions are not known.

T h e p u r p o s e o f t h i s n o t e is to call attention to an interesting special case of III. Consider the case $\rho(z)=|\operatorname{Im} z|+\log(1+|z|^2)$. Then $A_\rho=\hat{\mathcal{E}}'$, the space of all entire functions of exponential type with polynomial growth on the real axis. The space $\hat{\mathcal{E}}'$ is of special interest because, by the Paley-Wiener-Schwartz Theorem, it is the space of Fourier transforms of distributions on $\mathbb{R}$ with compact support. The problems I-III are then dual to some problems about convolution operators on the space $\mathcal{E}=C^\infty(\mathbb{R})$ (see e.g. [1], [2], or [3]).

Specifically, suppose

for some $\varepsilon>0$, $c>0$, $f\in\hat{\mathcal{E}}'$, we have

$$V = \{a_n, p_n\} \subset V(f) \text{ , where } \frac{|f^{(p_n)}(a_n)|}{p_n!} \geq \varepsilon \exp\left(-\frac{\rho(a_n)}{p_n}\right) \qquad (2)$$

$(\rho(z) = |\operatorname{Im} z| + \log(1+|z|^2))$.

Then it is not hard to show that (2) is a sufficient condition that V has the universal interpolation property III. We wish to pose the converse problem.

PROBLEM. Suppose that $V \subset V(F)$ for some F , $F \in \hat{\mathcal{E}}'$, and that V is a universal interpolating sequence: i.e. III holds. Is it true that (2) must hold for some f , $f \in \hat{\mathcal{E}}'(\mathbb{R})$?

In all the cases known to the author where the PROBLEM has answer yes, it is also true that the range of the multiplication operator $M_F : A_\rho \to A_\rho$ given by $M_F(f) = Ff$ is closed. Is the fact that M_F has closed range necessary for a "yes" answer? (In the case $A_\rho = \hat{\mathcal{E}}'$, if M_F has closed range, then the PROBLEM has answer yes, as can be shown by the techniques of [4]). However, the main interest in the PROBLEM is to find if (2) must hold with no additional assumptions on F .

REFERENCES

1. Л е о н т ь е в А.Ф. О свойствах последовательностей линейных агрегатов, сходящихся в области, где порождающая линейные агрегаты система функций не является полной. - Успехи матем.наук, 1956, II, № 5, 26-37.
2. E h r e n p r e i s L. Fourier Analysis in Several Complex Variables. New York, Wiley-Interscience, 1970.
3. П а л а м о д о в В.П. Линейные дифференциальные операторы с постоянными коэффициентами, М., Наука, 1967.
4. E h r e n p r e i s L., M a l l i a v i n P. Invertible operators and interpolation in AU spaces. - J.Math.Pure Appl., 1974, 13, 165-182.
5. Б о р и с е в и ч А.И., Л а п и н Г.П. Об интерполировании целых функций. - Сиб.матем.ж., 1968, 9, № 3, 522-529.

B.A.TAYLOR

Mathematics Department
The University of Michigan
Ann Arbor, Michigan 48109
USA

* * *

COMMENTARY

Papers [6] , [7] contain useful information concerning the Problem.

REFERENCES

6. Berenstein C.A., Taylor B.A. A new look at interpolation theory for entire functions of one variable. - Adv. Math., 1979, 33, N 2, 109-143.

7. Squires W.A. Necessary conditions for universal interpolation in $\hat{\mathcal{E}}'$. - Canad. J. Math., 1981, 33, N 6, 1356-1364 (MR 83g: 30040).

0.2. BASES OF REPRODUCING KERNELS AND EXPONENTIALS

1. Bases of exponential polynomials. For a non-negative integer-valued function k (a divisor) in the complex plane $\mathbb{C}$ let us denote by $\mathcal{E}(k)$ the family $\{\mathcal{E}_\lambda : \lambda \in \mathbb{C}\}$ of exponential polynomial subspaces $\mathcal{E}_\lambda = \{pe^{i\lambda x} : p$ is a polynomial, $\deg p < k(\lambda)\}$.

QUESTION 1. For what divisors k does the family $\mathcal{E}(k)$ form an unconditional basis in the space $L^2(0,a)$, $a > 0$?

"Unconditional basis" is used in the usual sense and means the existence, uniqueness and unconditional convergence of the expansion

$$f = \sum_{\lambda \in \mathbb{C}} p_\lambda e^{i\lambda x}\ ; \quad \deg p_\lambda < p(\lambda), \quad \lambda \in \mathbb{C} \qquad (*)$$

for any function $f \in L^2(0,a)$. It is clear that in this case $k \equiv 0$ off a countable discrete set $\sigma = \operatorname{supp} k$ and the starred expansion turns out to be a generalized Fourier series with respect to the minimal family of subspaces $\mathcal{E}(k)$.

The most interesting problem arises for $k \leqslant 1$ (i.e. $k = \chi_\sigma$, the characteristic function of σ) in which case the reader deals, as a matter of fact, with the well-known problem on exponential bases on intervals of the real axis.

Here is a bit of known information:

1) for $K = \chi_\sigma$, $\sigma \subset \mathbb{C}_+ \overset{def}{=} \{\lambda : \operatorname{Im} \lambda > 0\}$ the Question has been answered in [1]. Namely, σ must be a (Carleson) interpolation subset of $\mathbb{C}_+$ and the function $x \longmapsto \arg \Theta \overline{B}_\sigma(x)$ must satisfy the Devinatz - Widom condition, where B_σ is the Blaschke product with zero set σ and $\Theta = \exp(iaz)$. In case $\sup \operatorname{Im} \sigma < +\infty$ the answer can be reformulated in terms of density of σ. Paper [1] contains also exhausting historical remarks.

2) In the limit case $a = \infty$ (which implies $\operatorname{supp} k \subset \mathbb{C}_+$) no $\mathcal{E}(k)$ forms a basis in $L^2(0,+\infty)$. The right analogue of the problem in such a situatuon is to describe all divisors k for which $\mathcal{E}(k)$ is an unconditional basis in the closed linear span of $\mathcal{E}(k)$ ($\overset{def}{=} \operatorname{span} \mathcal{E}(k)$). This problem has been solved in its complete generality in [2] in terms of the generalized (multiple) Carleson condition.

3) It is not hard to see that for $\operatorname{supp} k \subset \mathbb{C}_+$ the Question is equivalent to a kind of multiple free interpolation problem for entire functions of exponential type $a/2$ (see [1], [3] for details).

2. Bases extending a given basis. Exponential (or exponential-polynomial) bases problem is a special case ($\theta = exp(iaz)$) of the problem on reproducing bases in the model space

$$K_\theta = H^2_+ \ominus \theta H^2_+ \quad ,$$

where H^2_+ stands for the usual Hardy space in $\mathbb{C}_+$ and θ is an inner function. Denote by

$$k_\theta(z,\lambda) = \frac{1-\overline{\theta(\lambda)}\,\theta}{z-\bar{\lambda}} \quad , \quad \mathcal{I}m\lambda > 0$$

the reproducing kernel for K_θ and put $\mathcal{K}_\theta(\sigma) = \{k_\theta(\cdot,\lambda) : \lambda \in \sigma\}$ for $\sigma \subset \mathbb{C}_+$.

QUESTION 2. Let $\mathcal{K}_\theta(\sigma)$ be an unconditional basis in $span\,\mathcal{K}_\theta(\sigma)$. Is it true that there exist unconditional bases $\mathcal{K}_\theta(\sigma')$ in the whole space K_θ containing $\mathcal{K}_\theta(\sigma)$ (i.e. such that $\sigma \subset \sigma' \subset \mathbb{C}_+$)?

QUESTION 2'. Let $a > 0$ and let $\mathcal{E}(k)$ be an unconditional basis in $span\,\mathcal{E}(k)$. Is it true that there exist unconditional bases $\mathcal{E}(k')$ in the whole space $L^2(0,a)$ containing $\mathcal{E}(k)$ (i.e. such that $k \leqslant k'$) ? Is it possible to choose such a k' multiplicity-free (i.e. $k' = \chi_{\sigma'}$) provided $k = \chi_\sigma$?

The second part of Question 2' is a special case of Question 2 ($\theta = exp(iaz)$). The answer to this part of Question is known to be positive (V.I.Vasyunin, S.A.Vinogradov) under some additional assumptions (i.e.a quantitative relation between $\inf \mathcal{I}m\,\sigma$, $\sigma = supp\,k$ and the interpolating constant of σ, see [1]).

3. Existence of a basis.

QUESTION 3. In which model space K_θ does there exist an unconditional basis of the form $\mathcal{K}_\theta(\sigma)$?

Each of the following two questions 3' and 3'' is equivalent to Question 3 (see [1], [3] for the proofs). For which inner functions θ does there exist an interpolating Blaschke product B such that

3') $dist(\theta, BH^\infty) < 1$, $dist(B, \theta H^\infty) < 1$?

or

3'') the Toeplitz operator $T_{\theta\bar{B}} f \overset{def}{=} P_+ \theta \bar{B} f$, $f \in H^2$ is inverible in H^2 ?

It is proved in [4] that, θ being an inner function, there exist interpolating Blaschke products B, B' such that $\|\theta B - B'\|_\infty < 1$. It follows that the space K_θ can be "complemented" by the space K_B with an unconditional basis of reproducting kernels ($\equiv$ of rational fractions in this case) to the space $K_{\theta B} = clos(K_\theta + K_B)$, $K_\theta \cap K_B = \{0\}$ in such a way that $K_{\theta B}$ has also an unconditional basis of the form $\mathcal{K}_{\theta B}(\sigma)$.

A limit case of the problem (the existence of orthogonal bases of the form $k_\theta(\cdot, \lambda)$, $|\lambda| = 1$) is considered in [5].

REFERENCES

1. Hruščëv S.V., Nikol'skii N.K., Pavlov B.S. Unconditional bases of exponentials and of reproducing kernels, Lect.Notes in Math., 1981, v.864, p.214-335.
2. Васюнин В.И. Безусловно сходящиеся спектральные разложения и задачи интерполяции. - Труды матем.ин-та им.В.А.Стеклова АН СССР, 1977, 130, с.5-49.
3. Никольский Н.К. Лекции об операторе сдвига. Москва, Наука, 1980.
4. Jones P.W. Ratios of interpolating Blaschke products. - Pacific J. Math., 1981, v.95, N 2, p.311-321.
5. Clark D.N. On interpolating sequences and the theory of Hankel and Toeplitz matrices. - J.Funct.Anal., 1970, v.5, N 2, p.247-258.

N.K.Nikol'skii
(Н.К.Никольский)

СССР, 191011, Ленинград
Фонтанка 27, ЛОМИ

10.3. MULTIPLICATIVE PROPERTIES OF ℓ_A^p.
old

Let ℓ_A^p be the Banach space of all functions $f=\sum_{k\geq 0}\hat{f}(k)z^k$ holomorphic in the unit disc $\mathbb{D}$ and satisfying

$$\|f\|_{\ell^p}=\Big(\sum_{n\geq 0}|\hat{f}(n)|^p\Big)^{1/p}<+\infty\,,\quad 1\leq p\leq\infty\,,$$

$\big(\|f\|_{\ell^\infty}\overset{def}{=\!=}\sup_{n\geq 0}|\hat{f}(n)|\big)$. It is well-known that ℓ_A^p is not an algebra with respect to the pointwise multiplication of functions if $p\neq 1$. Therefore, when studying the multiplicative structure of ℓ_A^p, the space $M_A^p\overset{def}{=\!=}\{g\in\ell_A^p : gf\in\ell_A^p,\ \forall f\in\ell_A^p\}$ becomes very important. Recall that $M_A^p=M_A^q$, $\frac{1}{p}+\frac{1}{q}=1$, $1\leq p\leq\infty$; ℓ_A^2 coincides with the Hardy class H^2; $M_A^1=\ell_A^1$, $\ell_A^1\subset M_A^p\subset H^\infty$, $1\leq p\leq 2$.

The first conjectures of the paper are closely connected with the theorem of L.Carleson [1] on the interpolation by bounded analytic functions. Given a subset E of $\mathbb{D}$ let R^E denote the restriction operator onto E.

THEOREM [1]. $R^E(H^\infty)=\ell^\infty(E)$ <u>iff</u>

$$\delta(E)\overset{def}{=\!=}\inf\Big\{\prod_{\eta\in E\setminus\{\xi\}}\Big|\frac{\eta-\xi}{1-\bar{\eta}\xi}\Big| : \xi\in E\Big\}>0\,.$$

Note that (C), being necessary for $R^E(M_A^p)=\ell^\infty(E)$, $1<p<2$, is not sufficient *). On the other hand it turns out to be sufficient for $1<p\leq 2$ if E satisfies the Stolz condition (i.e. E is contained in a finite union of domains $S_\lambda(\zeta)\overset{def}{=\!=}\{\eta\in\mathbb{D}:|\zeta-\eta|\leq\lambda(1-|\eta|)\}$, where $1<\lambda<\infty$, $\zeta\in\mathbb{T}$), cf. [2].

Suppose that E satisfies the Stolz condition. Then it is easy to check (see [4]) that

$$\delta(E)>0 \Longleftrightarrow \sigma(E)\overset{def}{=\!=}\inf\Big\{\Big|\frac{\xi-\eta}{1-\bar{\eta}\xi}\Big| : \xi,\eta\in E,\ \xi\neq\eta\Big\}>0 \Longleftrightarrow$$

$$\sigma(E)>0\ \&\ \gamma(E)\overset{def}{=\!=}\sup\Big\{\sum_{\substack{\eta\in E\\ |\eta|\geq|\xi|}}\frac{1-|\eta|}{1-|\xi|} : \xi\in E\Big\}<+\infty\,.$$

The conditions $\sigma(E)>0$ and $\gamma(E)<+\infty$ are important for the problems of interpolation theory in ℓ_A^p as well as in other spaces. [3]. Everything said above makes plausible the following conjecture.

---

*) That (C) does not imply $R^E(M_A^p)=\ell^\infty(E)$, $1<p<2$ can be proved with help of [3].

CONJECTURE 1. $\sigma(E)>0, \gamma(E)<+\infty \Longrightarrow \mathcal{R}^E(M_A^p) = \ell^\infty(E)$.

onjecture 1 is related to

CONJECTURE 2. $\gamma(E)<+\infty \Longrightarrow B^E \in \bigcap_{1<p\le 2} M_A^p$, where $B^E \stackrel{def}{=} \prod_{\eta\in E} \frac{|\eta|}{\eta}\frac{\eta-z}{1-\bar\eta z}$ stands for the Blaschke product generated y E.

CONJECTURE 1 follows from CONJECTURE 2. To see this it is sufficient to apply the Earl theorem [5] about the interpolation by Blaschke products. It is not hard to show that the zero set E of the corresponding Blaschke product can be chosen in this case satisfying $\sigma(E)>0$, $\gamma(E)<+\infty$ (see [6], §4 for details).

It follows from $M_A^p = M_A^q$ $(\frac{1}{p}+\frac{1}{q}=1,\ 1\le p\le\infty)$ that every inner function $I\in M_A^p$ satisfies

$$(I\cdot F\in \ell_A^p,\quad F\in H^1) \Rightarrow F\in \ell_A^p \tag{1}$$

for $1<p\le 2$.

Therefore the proof of CONJECTURE 2 would give new non-trivial examples of Blaschke products I with property (1). In this direction at present, apparently, only the following is known.

1. $I=\exp\frac{z+1}{z-1}$ does not satisfy (1) for $1\le p\le 4/3$ (see [7], [8]).

2. For $1<p<2$ B^E satisfies (1) provided $\gamma(E)<\infty$ and E satisfies the Stolz condition (see [2]).

CONJECTURE 3. (a). Suppose $\gamma(E)<\infty$. Then B^E satisfies (1), $1<p<2$.

(b) If $\gamma(E)<+\infty$ and E satisfies the Stolz condition then B^E satisfies (1) with $p=1$.

Analogous conjectures can be formulated for multipliers of

$$\left\{\int_0^\infty f(t)e^{itz}\,dt : f\in L^p(0,+\infty)\right\}.$$

REFERENCES

1. Carleson L. An interpolation problem for bounded analytic functions. - Amer.J.Math., 1958, 80, N 4, 921-930.
2. Виноградов С.А. Мультипликаторы степенных рядов с последовательностью коэффициентов из ℓ^p. - Зап.научн.семин.ЛОМИ,

1974, 39, 30-40.

3. Виноградов С.А. Базисы из показательных функций и свободная интерполяция в банаховых пространствах с L^p – нормой. - Зап.научн.семин.ЛОМИ, 1976, 65, 17-68.
4. Виноградов С.А., Хавин В.П. Свободная интерполяция в H^∞ и некоторых других классах функций. I. - Зап.научн.семин. ЛОМИ, 1974, 47, 15-54.
5. Earl J.P. On the interpolation of bounded sequences by bounded analytic functions. - J.London Math.Soc., 1970, 2, N 2, 544-548.
6. Виноградов С.А., Хавин В.П. Свободная интерполяция в H^∞ и в некоторых других классах функций. II. - Зап.научн.семин. ЛОМИ, 1976, 56, 12-58.
7. Гурарий В.П. О факторизации абсолютно сходящихся рядов Тейлора и интегралов Фурье. - Зап.научн.семин.ЛОМИ, 1972, 30, 15-32.
8. Широков Н.А. Некоторые свойства примарных идеалов абсолютно сходящихся рядов Тейлора и интегралов Фурье. - Зап.научн.семин. ЛОМИ, 1974, 39, 149-161.

S.A.VINOGRADOV
(С.А.ВИНОГРАДОВ)

СССР, 198904, Ленинград, Петродворец, Ленинградский государственный университет, Математико-механический факультет

* * *

COMMENTARY BY THE AUTHOR

Conjectures 2 and 3 are disproved in [9] (see Corollary 1 in [9] disproving Conjecture 2 and Corollary 2 and Theorem 5 in [9] disproving Conjecture 3). The results of [8], [9] and [10] lead to the following question.

QUESTION. Is there a singular inner function in $\bigcup_{1\le p<2} M_A^p$?

REFERENCES

9. Виноградов С.А. Мультипликативные свойства степенных рядов с последовательностью коэффициентов из ℓ^p -. Докл.АН СССР, 1980, 254, № 6, 1301-1306. (Sov.Math.Dokl., 1980, 22, N 2, 560-565)
10. Вербицкий И.Э. О мультипликаторах пространств ℓ_A^p. - Функц.анализ и его прил., 1980, 14, вып.3, 67-68.

0.4. FREE INTERPOLATION IN REGULAR CLASSES

Let $\mathbb{D}$ denote the open unit disc in $\mathbb{C}$ and let X be a closed subset of $\bar{\mathbb{D}}$. For $0<\alpha<1$, let Λ_α denote the algebra of holomorphic functions in $\mathbb{D}$ satisfying a Lipschitz condition of order α. The set X is called an interpolation set for Λ_α if the restriction map

$$\Lambda_\alpha \longrightarrow \mathrm{Lip}\,(\alpha, X)$$

$$f \longmapsto f\,|\,X$$

is onto. The interpolation sets for Λ_α, $0<\alpha<1$ (and also of other classes of functions) were characterized by Dyn'kin in [3] as those for which the following conditions hold:

The condition (K): if $d(z)=\inf\{|z-w|: w\in X\}$, then for all arcs $I\subset\mathbb{T}$,

$$\sup_{z\in I} d(z)\geqslant c\,|I|$$

where $|I|$ denotes the length of I.

The Carleson condition (C) : $X\cap\mathbb{D}$ must be a sequence (z_n) such that

$$\inf_n \prod_{m\neq n} \frac{|z_m-z_n|}{|1-\bar{z}_n z_m|}>0.$$

In the limit case $\alpha=1$ there are (at least) three different ways of posing the problem:

1. We can simply ask when the restriction map $\Lambda_1\to$ Lip $(1,X)$ is onto, Λ_1 being the class of holomorphic functions in $\mathbb{D}$ satisfying a Lipschitz condition of order 1 .

2. We can also consider the class

$$A_1 = H(\mathbb{D})\cap C^1(\bar{\mathbb{D}})$$

and call X an interpolation set for A^1 if for all $\varphi\in C^1(X)$ (the space of Whitney jets) such that $\bar\partial\varphi=0$ there exists f in A^1 with $f=\varphi$, $f'=\partial f$ on X.

3. Finally one can consider the Zygmund class version of the problem. Let Λ_* denote the class of holomorphic functions in $\mathbb{D}$ having continuous boundary values belonging to the Zygmund class of $\mathbb{T}$. We say that X is an interpolation set for Λ_* if for any φ in the Zygmund class of $\mathbb{C}$ there exists f in Λ_* such that $f=\varphi$ on X.

In [1] and [2], it has been shown that Dyn'kin's theorem also holds for A^1-interpolation sets. For Λ_1 interpolation sets the Carleson condition must be replaced by

(2C) $X \cap \mathbb{D}$ is a union of two Carleson sequences.

Our PROBLEM is the following: <u>which are the interpolation sets for the Zygmund class</u>?

Considering the special nature of the Zygmund class, I am not sure whether the condition describing the interpolation sets for Λ_* (one can simply think about the boundary interpolation, i.e. $X\subset\mathbb{T}$) should be different or not from condition (K). Recently I became aware of the paper [4], where a description of the trace of Zygmund class (of $\mathbb{R}^n$) on any compact set and a theorem of Whitney type are given. These are two important technical steps in the proofs of the results quoted above and so it seems possible to apply the same techniques.

REFERENCES

1. Bruna J. Boundary interpolation sets for holomorphic functions smooth to the boundary and BMO. - Trans.Amer.Math.Soc.,1981, 264, N 2, 393-409.
2. Bruna J., Tugores F. Free interpolation for holomorphic functions regular up to the boundary.-to appear in Pacific J.Math.
3. Дынькин Е.М. Множества свободной интерполяции для классов Гёльдера. - Матем.сборн., 1979, 109 (151), № 1, 107-128 (Math.USSR Sbornik, 1980, 37, 97-117).
4. Jonsson A., Wallin H. The trace to closed sets of functions in $\mathbb{R}^n$ with second difference of order $O(h)$. - J. Approx.theory,1979, 26, 159-184.

J.BRUNA

Universitat autònoma de Barcelona
Secció matemàtiques, Bellaterra (Barcelona)
España

10.5. TRACES OF $H^\infty(\mathbb{B}^N)$-FUNCTIONS ON HYPERPLANES
old

Let $\mathbb{B}^N$ be the unit ball of $\mathbb{C}^N$ ($N>1$) and denote by $H^\infty(\mathbb{B}^N)$ the algebra of all bounded holomorphic functions in $\mathbb{B}^N$. An analytic subset E of $\mathbb{B}^N$ is said to be a zero-set for $H^\infty(\mathbb{B}^N)$ (in symbols: $E\in ZH^\infty(\mathbb{B}^N)$) if there exists a non-zero function f in $H^\infty(\mathbb{B}^N)$ with $E=f^{-1}(0)$; E is said to be an interpolation set for $H^\infty(\mathbb{B}^N)$ (in symbols: $E\in IH^\infty(\mathbb{B}^N)$) if for any bounded holomorphic function φ on E there exists a function f in $H^\infty(\mathbb{B}^N)$ with $f|E=\varphi$. The problem to describe the sets of classes $ZH^\infty(\mathbb{B}^N)$ and $IH^\infty(\mathbb{B}^N)$ proves now to be very difficult. I would like to propose some partial questions concerning this problem; the answers could probably suggest conjectures in the general case.

Let A be a countable subset of $\mathbb{B}^N$. Set $T_A=\{z\in\mathbb{B}^N:(z,a)=|a|^2\}$, $T_A=\bigcup_{a\in A}T_a$.

PROBLEM 1. <u>What are the sets</u> A <u>such that</u> $T_A\in ZH^\infty(\mathbb{B}^N)$?

PROBLEM 2. <u>What are the sets</u> A <u>such that</u> $T_A\in IH^\infty(\mathbb{B}^N)$?

It follows easily from results of G.M.Henkin [1] and classical results concerning the unit disc that the following two conditions are necessary for $T_A\in ZH^\infty(\mathbb{B}^N)$:

$$\sum_{a\in A}\max\left(1-\frac{|a|^4}{|(a,\theta)|^2},0\right)<\infty,\quad \theta\in\mathbb{C}^n,\ |\theta|=1, \tag{1}$$

$$\sum_{a\in A}(1-|a|^2)^N<\infty. \tag{2}$$

These conditions do not seem however to be sufficient.

A necessary (insufficient) condition for $T_A\in IH^\infty(\mathbb{B}^N)$ can also be indicated. Namely, $T_A\in IH^\infty(\mathbb{B}^N)$ implies that there exists $\delta_A>0$ such that for every $a\in A$ the set T_a intersects no ellipsoid $Q_{\delta_A}(a')$ with $a'\neq a$, where

$$Q_\delta(a)\overset{def}{=\!=}\left\{z\in\mathbb{C}^N:\left|\frac{1-|a|^2\delta^2}{\delta(1-|a|^2)}\cdot\frac{(z,a)}{|a|}-\frac{a}{\delta}\,\frac{1-\delta^2}{1-|a|^2}\right|^2+\right. \tag{3}$$

$$+\frac{1-|a|^2\delta^2}{1-|a|^2}\left(|z|^2-\frac{|(z,a)|^2}{|a|^2}\right)<1\Big\}.$$

If A lies in a "sufficiently compact" subset of $\mathbb{B}^N$, it is possible to give complete solutions to Problems 1 and 2. Let $\rho, q\in(0,1)$, $c>0$; the (ρ, c, q)-wedge with the top at a point $e_0\in\partial\mathbb{B}^N$ is, by definition, the union of the ball $\{z\in\mathbb{B}^N: |z|<\rho\}$ and the set $E(e_0)\overset{def}{=}\{z\in\mathbb{B}^N: |\operatorname{Im}(1-(z,e_0))|\leqslant c\operatorname{Re}(1-(z,e_0));$

$$|z|^2-|(z,e_0)|^2\leqslant q(1-|(z,e_0)|^2);\ \operatorname{Re}(1-(z,e_0))\leqslant\frac{1}{c^2+1}\Big\}.$$

The scale of all (ρ, c, q) -wedges in equivalent (in a sense) to that of Fatou-Korányi-Stein wedges [1]. The following theorem holds.

THEOREM. <u>Let</u> A <u>be a subset of a finite union of</u> (ρ, c, q) <u>-wedges with</u> $q<\frac{1}{2}$. <u>Then</u> $T_A\in ZH^\infty(\mathbb{B}^N)$ <u>if and only if</u> $\sum_{a\in A}(1-|a|^2)<\infty$; $T_A\in IH^\infty(\mathbb{B}^N)$ <u>if and only if there exists</u> $\delta>0$ <u>such that</u> T_A <u>intersects no one of the sets</u> $Q_\delta(a')$ <u>with</u> $a'\neq a$, $Q_\delta(a)$ <u>being defined by (3)</u>.

In view of this theorem the following specializations of Problems 1 and 2 are of interest. Let A be a subset of a (ρ,c,q)-wedge, but in contrast to the theorem q can be an arbitrary number from $(0,1)$.

PROBLEM 1'. <u>Is it true that</u> $T_A\in ZH^\infty(\mathbb{B}^N)$ <u>implies</u>

$$\sum_{a\in A}(1-|a|^2)<\infty ?$$

PROBLEM 2'. <u>Is it true that</u> $T_a\cap Q_\delta(a')=\emptyset$ <u>for all</u> $a'\neq a$ <u>implies</u> $T_A\in IH^\infty(\mathbb{B}^N)$?

REFERENCE

1. Х е н к и н Г.М. Уравнение Г.Леви и анализ на псевдовыпуклом многообразии. – Матем.сб., 1977, 102, № 1, 71–108.

N.A.SHIROKOV
(Н.А.ШИРОКОВ)

СССР, 191011, Ленинград
Фонтанка, 27
ЛОМИ

0.6. old REPRESENTATIONS OF FUNCTIONS BY EXPONENTIAL SERIES

1. Let L be an entire function of exponential type with zero divisor $k=k_L$ ($k(\lambda)$ is the zero multiplicity of L at the point λ , $\lambda\in\mathbb{C}$), and let γ the Borel transform of L , namely

$$L(\lambda)=\frac{1}{2\pi i}\int_C \gamma(t)e^{\lambda t}dt \qquad (\lambda\in\mathbb{C})$$

where the closed contour C embraces a closed set $\mathfrak{D}$ containing all singularities of γ . There exists a family $\{\Psi_{k,\lambda}: 0\leqslant k<k(\lambda)\}$ of functions analytic in $\mathbb{C}\setminus\mathfrak{D}$ and biorthogonal to the family $\{z^s e^{\lambda z}: 0\leqslant s<k(\lambda)\}$, so that

$$\frac{1}{2\pi i}\int_C z^s e^{\lambda z}\Psi_{k,\mu}(z)dz=\delta_{\lambda\mu}\cdot\delta_{sk},$$

where $\delta_{\alpha\beta}$ is the Kronecker delta (see the construction of $\Psi_{k,\lambda}$ in [1], p.228). Any analytic function on $\mathfrak{D}$ can be expanded in Fourier series

$$f\sim\sum_{\lambda,\,k(\lambda)>0}\ \sum_{k=0}^{k(\lambda)-1} a_{k,\lambda}z^k e^{\lambda z};\quad a_{k,\lambda}=\frac{1}{2\pi i}\int_C f(z)\Psi_{k,\lambda}(z)\,dz. \tag{1}$$

The following uniqueness theorem is known ([1], p.255): if L has infinitely many zeros and $\mathfrak{D}$ is a convex set then $a_{k,\lambda}\equiv 0\Rightarrow f\equiv 0$. The proof uses in an essential way the convexity of $\mathfrak{D}$.

PROBLEM 1. Does the uniqueness theorem hold without the convexity assumption?

2. Let $\mathfrak{D}$ be the closed convex envelope of the set of singular points of γ and suppose that L has simple zeros only (i.e. $k(\lambda)\leqslant 1$, $\lambda\in\mathbb{C}$). The necessary and sufficient conditions for series (1) to converge to f in the interior of $\mathfrak{D}$ for any f analytic in $\mathfrak{D}$ are the following:

a) $|L'(\lambda)|>e^{[h(\varphi)-\varepsilon]|\lambda|}$, $\varphi\equiv\arg\lambda$, $k(\lambda)>0$, $|\lambda|>r(\varepsilon)$

for any $\varepsilon > 0$;

b) there exist numbers $\rho > 0$ and τ_k , $0 < \tau_k \nearrow \infty$ such that $|L(\lambda)| > e^{\rho|\lambda|}$, $|\lambda| = \tau_k$, $k \geq 1$.

Condition a) ensures the convergence of (1) in int $\mathcal{D}$ and b) implies that the obtained sum equals f .

PROBLEMS 2. Is b) implied by a)?

The negative answer would mean that series (1) generated by f may converge to a function different from f .

3. Suppose that int $\mathcal{D}$ is an unbounded convex domain containing $(-\infty, 0)$. Suppose further that φ ranges over the interval $(-\varphi_0, \varphi_0)$, $0 < \varphi_0 \leq \pi/2$ when $\ell_\varphi \overset{def}{=} \{x+iy : x\cos\varphi + y\sin\varphi - k(\varphi) = 0\}$ ranges through the set of all supporting lines of int $\mathcal{D}$. The possibility for ℓ_{φ_0} to be supporting lines is not excluded but in such a case, evidently, the boundary $\partial\mathcal{D}$ eventually coincides with $\ell_{\pm\varphi_0}$.

Let $h(\varphi) = K(-\varphi)$ and let

$$|L(re^{i\varphi})| < \frac{C}{1+r^{\alpha}} e^{rh(\varphi)}, \alpha > 1, \quad |\varphi| < \varphi_0 \tag{2}$$

(may be with $\varphi = \pm\varphi_0$ in the above mentioned case). All zeros of L are assumed to be simple. Let $\{\Psi_\lambda : k(\lambda) > 0\}$ be the biorthogonal family to $\{e^{\lambda z} : k(\lambda) > 0\}$,

$$\Psi_\lambda(z) \overset{def}{=} \frac{1}{L'(\lambda)} \int_0^{\infty e^{i\varphi}} \frac{L(t)}{t-\lambda} e^{-tz} dt, \quad |\varphi| < \varphi_0, \ k(\lambda) > 0.$$

Condition (2) implies that Ψ_λ are analytic outside $\mathcal{D}$, continuous up to the boundary and bounded (by the constants, which may depend on λ).

Let $B(\mathcal{D})$ be the class of all functions f analytic in int $\mathcal{D}$, continuous in $\mathcal{D}$ and such that

$$f(z) = O\left(\frac{1}{|z|^{\mu}}\right), \mu > 1, z \in \mathcal{D}, z \to \infty.$$

Putting $C = \partial\mathcal{D}$, $\Psi_{k,\lambda} = \Psi_\lambda$, $k(\lambda) > 0$ in (1), associate with very function $f \in B(\mathcal{D})$ its Fourier series. In this section it is onvenient to enumerate the zeros of L, counted with multiplici- ies: $\{\lambda_\nu\}_{\nu \geq 1}$.

We shall be concerned with the convergence of (1) to f in nt $\mathcal{D}$. Suppose that L satisfies the following additional requi- ement. There is a family of closed contours Γ_k $(k \geq 1)$ and a fa- ily of curvilinear annuli containing these contours

$$\mathcal{P}_k = \bigcup_{t \in \Gamma_k} \{z : |z-t| \leq e^{-\varepsilon_k \tau_k}\}, \; \varepsilon_k > 0, \quad \lim_k \varepsilon_k = 0, \; \lim_k \tau_k = \infty \left(\tau_k \overset{def}{=} \min_{t \in \Gamma_k} |t|\right)$$

satisfying

a) for all γ, $\gamma > 0$, and ε, $\varepsilon > 0$,

$$\lim_{k \to \infty} \min_{\lambda \in \Gamma_k'} H(\lambda) = \infty, \text{ where } H(\lambda) = \frac{\log |L(\lambda)|}{|\lambda|};$$

for $k > k(\gamma, \varepsilon)$ the function H is greater than $h(\varphi) - \varepsilon$ on $\mathcal{P}_k'$, greater than $h(\varphi_0) - \varepsilon$ on $\mathcal{P}_k''$, greater than $h(-\varphi_0 + \gamma) - \varepsilon$ on $\mathcal{P}_k'''$ where Γ_k' is the part of Γ_k lying in the complement of the angle $|\varphi| < \varphi_0 + \gamma$, $\mathcal{P}_k'$ is the part of $\mathcal{P}_k$ lying in the angle $|\varphi| < \varphi_0 - \gamma$, $\mathcal{P}_k''$ and $\mathcal{P}_k'''$ are the parts of $\mathcal{P}_k$ lying in the small angles $|\varphi - \varphi_0| \leq \gamma$, $|\varphi + \varphi_0| \leq \gamma$ correspondingly;

b) if the boundary of the curvilinear half-annulus $\mathcal{P}_k' \cup \mathcal{P}_k'' \cup \mathcal{P}_k'''$ is divided into the parts C_k' and C_k'' by Γ_k (moreover let C_k'' be inside Γ), then the lengths of the curves Γ_k', C_k', C_k'' are $\exp[O(1)\tau_k]$ when $k \to \infty$;

c) L has only one zero in the annulus between Γ_k and Γ_{k+1}, namely λ_k.

Under these contitions it has been proved in [2] that if $f \in B(\mathcal{D})$ and E is a compact subset of int $\mathcal{D}$, then

$$\left|f(z) - \sum_{\nu=1}^{n} A_\nu e^{\lambda_\nu z}\right| < e^{-\delta_0 \tau_n}, \; \delta_0 = \delta_0(E) > 0, \; n > n_0 \quad (z \in E),$$

so

$$f = \sum_{\nu \geq 1} a_{\lambda_\nu} e^{\lambda_\nu z}, \quad z \in \mathcal{D}.$$

It was shown in [2] and [3] how the general case (i.e. the case of an arbitrary f analytic in int $\mathcal{D}$) can be be reduced to the case $f \in B(\mathcal{D})$.

PROBLEM 3. <u>Show that for any domain</u> int $\mathcal{D}$ <u>there exists a function</u> L <u>with the properties</u> (1), a), b), c).

REFERENCES

I. Л е о н т ь е в А.Ф. Ряды экспонент. М., Наука, I976.

2. Л е о н т ь е в А.Ф. К вопросу о представлении аналитических функций в бесконечной выпуклой области рядами Дирихле. - Докл. АН СССР, I975, 225, № 5, I0I3-I0I5.

3. Л е о н т ь е в А.Ф. Об одном представлении аналитической функции в бесконечной выпуклой области. - Anal.Math., 1976, 2, I25-I48.

A.F.LEONTIEV
(А.Ф.ЛЕОНТЬЕВ)

СССР, 450057, Уфа
ул.Тукаева, 50
Башкирский филиал АН СССР

0.7. RESTRICTIONS OF THE LIPSCHITZ SPACES TO CLOSED SETS

The Lipschitz space $\Lambda^k_\omega(\mathbb{R}^n)$ is defined by the finiteness of the semi-norm

$$|f|^k_\omega = \sup_{x,h\in\mathbb{R}^n} \frac{|\Delta^k_h f(x)|}{\omega(|h|)} .$$

Here as usual $\Delta^k_h = (\tau_h - 1)^k$ and $\tau_h f(x) = f(x+h)$. The majorant $\omega : \mathbb{R}_+ \to \mathbb{R}_+$ is non-decreasing and $\omega(+0)=0$. Without loss of generality one can suppose that $\omega(t)/t^k$ is non-increasing. Let $\lambda^k_\omega(\mathbb{R}^n)$ be the closure of the set C^∞_0 in $\Lambda^k_\omega(\mathbb{R}^n)$.

This note deals with some problems connected with the space of traces $\Lambda^k_\omega(F) = \Lambda^k_\omega(\mathbb{R}^n)|_F$ and with its separable subspace $\lambda^k_\omega(F)$, where $F\in\mathbb{R}^n$ is an arbitrary closed set. Among spaces under consideration there are well-known classes C^ℓ, $C^{\ell,\alpha}$ and $\Lambda^{\ell+1}$ whose importance is indubitable. Recall that $C^{\ell,\alpha}$ consists of all functions $f\in C^\ell$ with ℓ-th derivatives satisfying Hölder condition of order α. Replacing here Hölder condition by Zygmund condition *) we obtain the definition of the class $\Lambda^{\ell+1}$.

CONJECTURE 1. <u>There exists a linear continuous extension operator</u> $\mathcal{E} : \Lambda^k_\omega(F) \to \Lambda^k_\omega(\mathbb{R}^n)$.

A nonlinear operator of this type exists by Michael's theorem of continuous selection [1]. The lineartiy requirement complicates the matter considerably.

Let us review results confirming our Conjecture 1. Existence of a linear extension operator for the space of jets $W^{\ell,\alpha}(F)$ connected with $C^{\ell,\alpha}(\mathbb{R}^n)$ is proved in the classical Whitney theorem [2]. But the method of Whitney does not work for $\Lambda^{\ell+1}(F)$. Recently the author and P.A.Shwartzman have found a new extension process proving Conjecture 1 for $k=2$ (the case $k=1$ is well-known, see for example [3]). The method is closely connected with the ideology of the local approximation theory [4]).

The following version of Conjecture 1 is intresting in connection with the problem of interpolation of operators in Lipschitz spaces.

---

*) Function g satisfies Zygmund condition if $|\Delta^2_h g| = O(|h|)$.

CONJECTURE 2. Let ω_i, $i=1,2$, be majorants. There exists a linear exstension operator $\mathcal{E}: C(F) \to C(\mathbb{R}^n)$ mapping $\Lambda^k_{\omega_i}(F)$ into $\Lambda^k_{\omega_i}(\mathbb{R}^n)$, $i=1,2$.

The above mentioned extension operators do not possess the required property. If Conjecture 2 turns out to be right we would be able to reduce the problem of calculation of interpolation spaces for the pair $\Lambda^k_{\omega_i}(F)$, $i=1,2$, to a similar problem for $\mathbb{R}^n$.

PROBLEM. Find condition necessary and sufficient for a given function $f \in C(K)$ to be extendable to a $g \in \Lambda^k_\omega(\mathbb{R}^n)$ (or $\lambda^k_\omega(\mathbb{R}^n)$). In other words we ask for a description of the restriction $\Lambda^k_\omega(\mathbb{R}^n)|F$ (or $\lambda^k_\omega(\mathbb{R}^n)|F$).

The problem was solved by Whitney for the space $C^k(\mathbb{R})$ in 1934 (see [5]). In 1980 P.A.Shwartzman solved the problem for the space $\Lambda^2_\omega(\mathbb{R})$ and in the same year A.Jonsson got (independently) a solution for the space $\Lambda^{\ell+1}(\mathbb{R})$ (see [6,7]). The situation is much more complicated in higher dimensions; there is nothing but a non-effective description of functions from the space $\Lambda^k_\omega(F)$ involving a continual family of polynomials, connected by an infinite chain of inequalities (see [8] for power majorant; general case is considered in [9] in another way). Analysis of the articles [5-7] makes possible the following

CONJECTURE 3. Let $N=N(k,\omega,F)$ be the least integer with the following property *): if the restriction of a function $f \in C(F)$ on any subset $H \subset F$ with card $H < N$ is extendable to a function $f_H \in \Lambda^k_\omega(\mathbb{R}^n)$ and $\sup_H |f_H|^k_\omega < \infty$, then f belongs to $\Lambda^k_\omega(F)$.

Define $N(k,n)$ by the formula

$$N(k,n) = \sup_{\omega, F} N(k,\omega,F).$$

Then the number $N(k,n)$ is finite. It is obvious that $N(1,n)=2$; the calculation of $N(k,n)$ for $k>1$ is a very complicated problem. P.A.Shwartzman has proved recently that $N(2,n) = 3\cdot 2^{n-1}$

---

*) One can prove that $N(k,\omega,F) < \infty$.

and using this result has obtained a characteristic of functions from $\Lambda^2_\omega(F)$, $F\subset\mathbb{R}^2$ by means of interpolation polynomials (see [6]). When $n>2$ the number $N(2,n)$ is too large and the possibility of such a description is dubious.

In conclusion we note the connection of the considered problems with a number of other interesting problems in analysis (spectral synthesis of ideals in algebras of differentiable function, H^p space theory etc.)

REFERENCES

1. M i c h a e l E. Continuous selections. - Ann.Math., 1956, 63, 361-382.
2. W h i t n e y H. Analytic extensions of differentiable functions defined in closed sets. - Trans.Amer.Math.Soc., 1934, 36, 63-89.
3. D a n z e r L., G r ü n b a u m B., K l e e V. Helly's theorem and its relatives. - Proc.Symp.pure math., VIII, 1963.
4. Б р у д н ы й Ю.А. Пространства, определяемые с помощью локальных приближений. - Труды ММО, I97I, 24, 69-I32.
5. W h i t n e y H. Differentiable functions defined in closed sets, I. - Trans.Amer.Math.Soc., 1934, 36, 369-387.
6. Ш в а р ц м а н П.А. О следах функций двух переменных, удовлетворяющих условию Зигмунда. - В сб."Исследования по теории функций многих вещественных переменных". - Ярославль, I982, I45 - - I68.
7. J o n s s o n A. The trace of the Zygmund class $\Lambda_k(\mathbb{R})$ to closed sets and interpolating polynomials. - Dept.Math.Umeå,1980, 7.
8. J o n s s o n A., W a l l i n H. Local polynomial approximation and Lipschitz type condition on general closed sets. - Dept. Math.Umeå, 1980, 1.
9. Б р у д н ы й Ю.А., Ш в а р ц м а н П.А. Описание следа функции из обобщенного пространства Липшица на произвольный компакт. - В сб."Исследования по теории функций многих вещественных переменных". Ярославль I982, I6-24.

Yu.A.BRUDNYI

(Ю.А.БРУДНЫЙ)

СССР, I50000, Ярославль,
Ярославский государственный
университет

10.8. MULTIPLIERS, INTERPOLATION, AND $\Lambda(p)$ SETS

Let G be a locally compact Abelian group, with dual group Γ. An operator $T: L_p(G) \to L_p(G)$ will be called a multiplier provided there exists a function $\hat{T} \in L_\infty(\Gamma)$ so that $T(f)^\wedge = \hat{T}\hat{f}$, for all integrable simple functions f. The space of multipliers on $L_p(G)$ is denoted by $M_p(G)$. Let $CM_p(G) = \{T \in M_p(G) : \hat{T} \in C(\Gamma)\}$. In response to a question of J.Peetre, the author has recently shown that for the classical groups, $CM_p(G)$ is not an interpolation space between $M_1(G) = M(G)$ and $CM_2(G) = L_\infty(\Gamma) \cap C(\Gamma)$. More specifically, we obtained the following theorem (see [2]).

THEOREM 1. <u>Let</u> G <u>denote one of the groups</u> $\mathbb{T}^n$, $\mathbb{R}^n$, <u>or</u> $\mathbb{Z}^n$. <u>Then there exists an operator</u> T <u>so that</u>

(a) T <u>is a bounded operator on</u> $L_\infty(\Gamma) \cap C(\Gamma)$.

(b) $T | M(G)$ <u>is a bounded operator on</u> $M(G)$.

(c) $T | CM_p(G)$ <u>is n o t a bounded operator on</u> $CM_p(G), p \in (1,2)$.

Observe that T is i n d e p e n d e n t of p, $1 < p < 2$. Our method of construction makes essential use of certain results concerning $\Lambda(q)$ sets. Recall that a set $E \subseteq \mathbb{Z}$ is said to be of type $\Lambda(q)$ $(1 < q < \infty)$ if whenever $f \in L_1(\mathbb{T})$ and $\hat{f}(n) = 0$ for all $n \notin E$, we have $f \in L_q(\mathbb{T})$. We used the following elegant result of W.Rudin [1].

THEOREM 2. <u>Let</u> $s \geq 2$ <u>be an integer, let</u> N <u>be a prime with</u> $N > s$, <u>and let</u> $M = s^{s-1}N^s - 1$. <u>Then there exists a set</u> $F \subseteq \{0,1,2,\dots,M\}$ <u>so that</u>

(a) F <u>contains exactly</u> N <u>points, and</u>

(b) $\|f\|_{2s} \leq C\|f\|_2$, <u>for every trigonometric polynomial</u> f, <u>with</u> $\hat{f}(n) = 0$ <u>for</u> $n \notin F$ (<u>Such</u> f <u>are called</u> F-<u>polynomials</u>).

(Here C is i n d e p e n d e n t of N).

As a consequence of Theorem 2, Rudin showed that there exist sets of type $\Lambda(2s)$ which are not of type $\Lambda(2s+\varepsilon)$, for all $\varepsilon > 0$ (see [1]).

An obvious conjecture arising from Theorem 1 is the following:

CONJECTURE 1. <u>Let</u> $1 \leq p_1 < p_2 \leq 2$. <u>Then there exists an ope-</u>

ator T so that

(a) T is a bounded operator on CM_{p_2} .

(b) $T|CM_{p_1}$ is a bounded operator on CM_{p_1} .

(c) $T|CM_p$ is n o t bounded operator on CM_p , for all $p \in (p_1, p_2)$.

It is natural to attempt to analyze this conjecture by means of the techniques used to obtain Theorem 1. But it soon becomes evident that such an analysis requires a deep extension of Rudin's theorem. Specifically, we require a result of the following form:

CONJECTURE 2. (The $\Lambda(p)$ Problem). Let $2 < p < \infty$. Then there exists a set of type $\Lambda(p)$ which is not of type $\Lambda(p+\varepsilon)$, for all $\varepsilon > 0$.

This conjecture (which was essentially posed by Rudin) has remained unresolved for nearly a quarter of a century, and is one of the fundamental open questions in harmonic analysis. Its solution will undoubtedly require very subtle new ideas involving estimation in L_p . Conjecture 1 may be just one of the manifold consequences of the $\Lambda(p)$ problem.

Let us attempt to briefly outline one possible approach to the study of Conjecture 2. Let $p = 2s/k$ where $s > 1$ and $k \geq 1$ are integers, and $s > k$. Let $F = F_N$ be the set of Theorem 2, and let $F_{k,N}$ denote the k-fold sum $F + \ldots + F$. In essence, the "piecing together" of the $F_{k,N}$ (for an infinity of N's) provides an example of a set which is not of type $\Lambda(p+\varepsilon)$, for all $\varepsilon > 0$. The difficulty is in proving that $F_{k,N}$ is of type $\Lambda(p)$ (with all constants uniform in N). One may seek to accomplish this by writing an $F_{k,N}$ -polynomial f in a judicious way as a sum of products of F -polynomials, and carefully examining the resultant representation of f . However, new estimation techniques for L_p norms would still be very much a necessity in order to carry out this program.

REFERENCES

1. R u d i n W. Trigonometric series with gaps. - J.Math.Mech., 1960, 9, 203-227.

2. Z a f r a n M. Interpolation of Multiplier Spaces, Amer. J. Math., to appear.

MISHA ZAFRAN

Department of Mathematics
University of Washington
Seattle, WA 98195
USA

CHAPTER 11

ENTIRE, MEROMORPHIC AND SUBHARMONIC FUNCTIONS

This old and ramified theory, the oldest one among those presented in this collection, hardly needs any preface. By the same reason ten papers constituting the Chapter cannot reflect all tendencies existing in the field. But even a brief acquaintance with the contents of the problems shows that the main tendency remains invariable as though more than a quarter of the century, which passed since the appearance of the book by B. Ya.Levin "Zeros of Entire Functions", has shrunk up to an instant. We reproduce here the first paragraph of the preface to this book:

"One of the most important problems in the theory of entire functions is the problem of connection between the growth of an entire function and the distribution of its zeros Many other problems in fields close to complex function theory lead to this problem".

The only discrepancy between then and now, apparently, consists in more deep and indirect study of this problem

A good illustration to the above observation is provided by Problem 11.6. It deals with description of zero-sets of sine-type functions and is important for the purposes of Operator Theory.

Problem 11.2 is, probably, "the most classical" one in the Chapter. The questions posed there look very attractively because their formulations are so simple.

The theory of subharmonic functions is presented by Problems 11.7, 11.8.

Problems 11.3 and 11.4 deal with exceptional values in the spirit of R.Nevanlinna Theory.

Problem 11.10 concerns the limit behaviour of entire functions.

An important class of entire functions of completely regular growth is the subject of Problem 11.5.

"Old" Problem 11.9 by B.Ya.Levin includes three questions on functions in the Laguerre-Pólya class.

Problem 11.1 is rather a problem of approximation theory

The problems 11.1, 11.5, 11.6, 11.8, 11.9 are "old" and the rest are new.

1.1. THE INVERSE PROBLEM OF BEST APPROXIMATION OF BOUNDED UNIFORMLY CONTINUOUS FUNCTIONS BY ENTIRE FUNCTIONS OF EXPONENTIAL TYPE, AND RELATED QUESTIONS

old

Let E be a separable infinite-dimensional Banach space, let $E_0 \subset E_1 \subset \ldots$ be a chain of its finite dimensional subspaces such that $\dim E_n = n$ and $\bigcup_{n \geq 1} E_n$ is dense in E. For $x \in E$ we define the sequence of "deviations" from E_n by

$$d(x,n) = \inf\{\|x-y\| : y \in E_n\}, \quad n = 0,1,2,\ldots .$$

S.N.Bernstein [1] (see also [2]) has proved, that for every sequence $\{d_n\}_{n \geq 0}$ of non-negative numbers such that $d_n \downarrow 0$ there exists $x \in E$, with

$$d(x,n) = d_n, \quad (n = 0,1,2,\ldots).$$

This is a (positive) solution of the inverse problem of best approximation in a separable space in the case of finite dimensional subspaces.

Strictly speaking S.N.Bernstein has treated only the case of $E = C[a,b]$, E_n being the subspace of all polynomials of degree $\leq n-1$, but his solution may be reproduced in general case without any change.

Now let $B(\mathbb{R})$ be the Banach space of all bounded uniformly continuous functions on $\mathbb{R}$ with the sup-norm; let B_σ be its closed subspace consisting of entire functions of exponential type $\leq \sigma$ (or, to be more precise, of their restrictions to $\mathbb{R}$). S.N.Bernstein has shown [3] that many results concerning the best approximation of continuous functions by polynomials have natural analogues in the theory of best approximation in $B(\mathbb{R})$ by elements of B_σ $(0 \leq \sigma < \infty)$.

We define the deviation of f from B_σ by

$$A(f,\sigma) = \inf\{\|f-g\| : g \in B_\sigma\}.$$

A function f being fixed, the function $A(f,\sigma)$ has the following properties: 1. $A(f,\rho) \geq A(f,\sigma)$ for $\rho < \sigma$. 2. $A(f,\sigma) = A(f,\sigma+0)$. 3. $\lim_{\sigma \to \infty} A(f,\sigma) = 0$.

PROBLEM 1. Let a bounded function $\sigma \longrightarrow F(\sigma)$ $(0 \leqslant \sigma < \infty)$ satisfy conditions 1-3. Is there a function f, $f \in B(\mathbb{R})$ such that $A(f,\sigma) \equiv F(\sigma)$?

PROBLEM 2. Let $B_{\sigma-0}$ be the closure of $\bigcup_{\rho<\sigma} B_\rho$. So $B_{\sigma-0}$ is a proper subspace of B_σ. What is its codimension in B_σ ?

PROBLEM 3. Let f be a Bohr almost-periodic function. Is $A(f,\sigma)$ necessarily a jump function?

PROBLEM 4. Let $A(f,\sigma)$ be a jump function. Is f almost-periodic?

REFERENCES

1. Б е р н ш т е й н С.Н. Об обратной задаче теории наилучшего приближения непрерывных функций. - В кн.: Собр.соч., т.2, М., Изд-во АН СССР, 1954, 292-294.
2. Н а т а н с о н И.П. Конструктивная теория функций, М.-Л., ГИТТЛ, 1949.
3. Б е р н ш т е й н С.Н. О наилучшем приближении непрерывных функций на всей вещественной оси при помощи целых функций данной степени. - В кн.: Собр.соч., т.2, М., Изд-во АН СССР, 1954, 371-395.

M.I.KADEC
(М.И.КАДЕЦ)

СССР, 310002, Харьков
Харьковский институт
инженеров коммунального
строительства

* * *

COMMENTARY

Problems 1, 2 and 4 have been solved by A.Gordon (А.Я.Гордон) **who kindly supplied us with the following information.**

THEOREM 1. (A.Gordon). Let F **satisfy 1-3 above. Then there exists** $f \in B(\mathbb{R})$ **satisfying**

$$A(f,\sigma) \equiv F(\sigma), \quad \sigma \geqslant 0.$$

PROOF. Pick a dense sequence $\{\lambda_K\}_{K\geq 0}$ in $(0,+\infty)$ and consider a monotone sequence of positive numbers $\{\varepsilon_K\}_{K\geq 0}$ such that

$$\varepsilon_K < \min(1,\lambda_K)\ , \qquad \lim_K \varepsilon_K = 0 .$$

Let $\{t_K\}_{K\geq 0}$ tend to $+\infty$ so fast that the intervals $I_K \overset{def}{=} \{t\in\mathbb{R}: |t-t_K|\leq 2^{K+1}\varepsilon_K^{-2}\}$ do not overlap. Standard arguments show

$$\sum_{K\geq 0}(1-\varepsilon_K)\cdot\left|\frac{\sin\varepsilon_K(t-t_K)}{\varepsilon_K(t-t_K)}\right| < 1 \tag{1}$$

for each $t\in\mathbb{R}$ and

$$\lim_{t\to+\infty}\sum_{K\geq 0}\left|\frac{\sin\varepsilon_K(t-t_K)}{\varepsilon_K(t-t_K)}\right|\cdot\chi_{\mathbb{R}\setminus I_K}(t) = 0 . \tag{2}$$

Here χ_E stands as usual for the characteristic function of a set E.

Given $\lambda\geq\varepsilon>0$ let

$$\varphi(\lambda,\varepsilon,t) = (1-\varepsilon)\frac{\sin\varepsilon t}{\varepsilon t}\cdot\sin(\lambda-\varepsilon)t .$$

Clearly $\varphi\in B_\lambda$. The desired f is defined as follows:

$$f(t) = \sum_{K\geq 0} F(\lambda_K)\cdot\varphi(\lambda_K,\varepsilon_K,t-t_K) . \tag{3}$$

The series converges absolutely and uniformly on compact subsets of $\mathbb{R}$ (see (1)), which implies $f\in B(\mathbb{R})$.

Fix $\sigma>0$ and consider

$$f_\sigma(t) \overset{def}{=} \sum_{\lambda_K\leq\sigma} F(\lambda_K)\cdot\varphi(\lambda_K,\varepsilon_K,t-t_K) .$$

Since B_σ is closed under bounded and pointwise convergence on $\mathbb{R}$, it is clear that $f_\sigma\in B_\sigma$. On the other hand (1) implies the inequality $\|f-f_\sigma\|_{L^\infty}\leq F(\sigma)$ which in turn yields $A(f,\sigma)\leq F(\sigma)$.

Suppose now $A(f,\sigma)<F(\sigma)$. Then there exist $\lambda>\sigma$, $A(f,\sigma)<F(\lambda)$ and $g\in B_\sigma$, $\|f-g\|_{L^\infty}<F(\lambda)$. Pick a sequence $\{K_n\}_{n\geq 1}$ of integers such that $\lambda_{K_n}\downarrow\lambda$. Property 2 of F and (2) imply that

$$\lim_n f(t+t_{K_n}) = F(\lambda)\cdot\sin\lambda t, \quad t\in\mathbb{R} .$$

We may assume without loss of generality that the sequence $\{g(t+t_{K_n})\}_{n\geq 1}$ converges to $g^*\in B_\sigma$. Obviously

$$\|F(\lambda)\cdot\sin\lambda t - g^*(t)\|_{L^\infty} \leq \|f-g\|_{L^\infty} < F(\lambda) .$$

It follows dist $(\sin\lambda t, B_\sigma) < 1$ in contradiction to [3] . Therefore $A(f,\sigma) \equiv F(\sigma)$. ●

THEOREM 2. (A.Gordon). $\dim(B_\sigma/B_{\sigma-0}) = +\infty$.

PROOF. Let $\{\varepsilon_k\}$ and $\{t_k\}$ be the sequences constructed as above for the constant sequence $\lambda_k \equiv \sigma$. For every bounded sequence $u = \{u_k\}_{k\geqslant 0}$ define

$$f_u(t) \overset{def}{=} \sum_{k\geqslant 0} u_k \varphi(\sigma, \varepsilon_k, t-t_k).$$

Clearly $f_u \in B_\sigma$ and it is easy to check that

$$\mathrm{dist}(f_u, B_{\sigma-0}) = \overline{\lim_k}\, |u_k| \ . \tag{4}$$

Indeed, $\lim_n \varphi(\sigma-1/n, \varepsilon, t) = \varphi(\sigma, \varepsilon, t)$ in $B(\mathbb{R})$. Therefore $\varphi(\sigma,\varepsilon,t)\in B_{\sigma-0}$ which yields $\mathrm{dist}(f_u, B_{\sigma-0}) \leqslant \overline{\lim_k}\, |u_k|$. Using the same arguments as in the proof of Theorem 1, we obtain $\mathrm{dist}(f_k, B_\tau) \geqslant \overline{\lim_k}\, |u_k|$ for $\tau < \sigma$. It follows from (4) that the factor-space $B_\sigma/B_{\sigma-0}$ contains a subspace isometric to ℓ^∞/c_0 and therefore is not separable. ●

Note that the function f_u with u properly choosen gives the negative answer to Problem 4. Indeed, if $\overline{\lim_k}\, |u_k| = 1$, then

$$A(f_u, s) = \begin{cases} 1, & 0 \leqslant s < \sigma \\ 0, & \sigma \leqslant s \end{cases} .$$

1.2. SOME PROBLEMS ABOUT UNBOUNDED ANALYTIC FUNCTIONS

A famous theorem of Iversen [1, p.284a] says that if f is a transcendental entire function then there is a path γ along which f tends to ∞. Thinking about this theorem has led us to formulate the following four problems, which we would like to solve, but cannot yet solve.

1. If f is a transcendental entire function, must there exist a path γ along which every derivative of f tends to ∞? Short of that, how about just having both f and f' tend to infinity along γ? (It follows, say, by Wiman-Valiron theory, that if f is a transcendental entire function, then there exists a sequence (z_n) such that $f^{(k)}(z_n) \longrightarrow \infty$ as $n \longrightarrow \infty$ for each $k = 0, 1, 2, \ldots$, but to obtain a path on which this happens seems much more difficult.)

2. If f is an unbounded analytic function in the open disc $\mathbb{D}$, must there exist a sequence (z_n) of points of $\mathbb{D}$ such that for every $k = 0, 1, 2, \ldots, f^{(k)}(z_n) \to \infty$ as $n \to \infty$? Short of that, how about just having $f(z_n) \to \infty$ and $f''(z_n) \to \infty$? (The authors have shown that one can always find (z_n) so that both $f(z_n) \to \infty$ and $f'(z_n) \to \infty$.) If f grows fast enough, i.e., if either

$$\limsup_{r \to 1-} \frac{\log\log M(r,f)}{-\log(1-r)} = \rho > 0 \tag{1}$$

or

$$\limsup_{r \to 1-} \frac{\log T(r,f)}{\log\log\left(\frac{1}{1-r}\right)} = \infty, \tag{2}$$

then we can show that there exists a (z_n) with $f^{(k)}(z_n) \to \infty$ as $n \to \infty$ for $k = 0, 1, 2, \ldots$, but we do not know how necessary these growth assumptions are. Note that conditions (1) and (2) are not strictly comparable. The proof involving condition (1) quotes a theorem of Valiron, while that involving (2) uses some Nevanlinna theory.)

3. Is there an easy elementary proof, using neither Wiman-Valiron theory nor advanced Nevanlinna theory, that if f is a transcen-

dental entire function then there exists a sequence (z_n) such that

$f^{(k)}(z_n) \longrightarrow \infty$ as $n \to \infty$ for $k = 0, 1, 2, \ldots$?

4. The following possibility is suggested by many examples: the differential equation

$$P_0(y)P_1(y')\ldots P_n(y^{(n)}) = 1,$$

where the P_j are polynomials in one variable, not all constants, has no solutions $y(z)$ that are analytic and unbounded in the unit disc $\mathbb{D}$. For example, the equation $yy'' = 1$ can be solved explicitly by means of the standard substitution $v = y'$, $v' = v\frac{dv}{dy}$, and it is easily seen that it has no unbounded analytic solution in any disc, lending support to the above hypothetical statement. The QUESTION remains whether that statement is true or not.

REFERENCES

1. Titchmarsh E.C. The Theory of Functions 2nd Edition, Oxford 1932.

JAMES LANGLEY
LEE A.RUBEL

Dept.Math.
University of Illinois
Urbana, IL 61801
USA

1.3. COMPARISON OF SETS OF EXCEPTIONAL VALUES IN THE SENSE OF R.NEVANLINNA AND IN THE SENSE OF V.P.PETRENKO

Let f be a meromorphic function in $\mathbb{C}$ and put

$$\beta(a,f)=\varliminf_{r\to\infty}\max_{|z|=r}\log^{+}[f(z),a]^{-1}/T(r,f),\ a\in\hat{\mathbb{C}},$$

where $[a,b]$ is the spherical distance between a and b. Denote by $E_{\beta}(f)\overset{def}{=\!=}\{a\in\hat{\mathbb{C}}:\beta(a,f)>0\}$ the set of exceptional values in the sense of V.P.Petrenko and by $E_N(f)=\{a\in\hat{\mathbb{C}}:\delta(a,f)>0\}$ the set of deficient values of f. It is clear that $E_N(f)\subset E_{\beta}(f)$. The set $E_{\beta}(f)$ is at most countable if f is of finite order [1]. There are examples of f's of finite order with $E_N(f)\neq E_{\beta}(f)$ 1 - 4].

PROBLEM 1. <u>Let $E_1\subset E_2\subset\hat{\mathbb{C}}$ be arbitrary at most countable sets. Does there exist a meromorphic function f of finite order with $E_N(f)=E_1$, $E_{\beta}(f)=E_2$?</u>

It is known that $T(2r,f)=O(T(r,f))$, $r\to\infty$, implies $E_N(f)=E_{\beta}(f)$ [6].

PROBLEM 2. <u>Let f be an entire function of finite order,</u> $\beta(0,f)>0$. <u>Is it true that $\delta(0,f)>0$</u> ?

REFERENCES

1. П е т р е н к о В.П. Рост мероморфных функций, Харьков, "Вища школа", 1978.
2. Г р и ш и н А.Ф. О сравнении дефектов $\delta_p(a)$. - Теория функций, функц.анал. и их прил., Харьков, 1976, № 25, 56-66.
3. Г о л ь д б е р г А.А. К вопросу о связи между дефектом и отклонением мероморфной функции. - Теория функций, функц.анал. и их прил., Харьков, 1978, № 29, 31-35.
4. С о д и н М.Л. О соотношении между множествами дефектных значений и отклонений для мероморфной функции конечного порядка. - Сиб.матем.журнал, 1981, 22, № 2, 198-206.

5. Е р е м е н к о А.Э. О дефектах и отклонениях мероморфных функций конечного порядка (в печати).

A.A.GOL'DBERG
(А.А.ГОЛЬДБЕРГ)

СССР, 290602, Львов
Львовский государственный
университет

A.È.ERËMENKO
(А.Э.ЕРЁМЕНКО)

СССР, 310164, Харьков
пр.Ленина, 47,
Физико-технический институт
низких температур АН УССР

1.4. VALIRON EXCEPTIONAL VALUES OF ENTIRE FUNCTIONS OF COMPLETELY REGULAR GROWTH

Let E_ρ be the class of all entire functions of order ρ, $\frac{1}{2}<\rho<\infty$, and let $\overset{\circ}{E}_\rho$ be its subclass of entire functions of completely regular growth in the sense of B.Ja.Levin and A.Pfluger [1]. Let $E_N(f)$ and $E_V(f)$ be the sets of exceptional values of a function f in the sense of R.Nevanlinna and of G.Valiron respectively. It is known that $\{E_N(f): f\in \overset{\circ}{E}_\rho\}\neq\{E_N(f): f\in E_\rho\}$. Indeed, for every function $f\in \overset{\circ}{E}_\rho$ we have $\operatorname{card} E_N(f)\leqslant[2\rho]^*$, where $[x]^*=\max\{k\in\mathbb{Z}: k<x\}$, [2]. On the other hand there exists $f\in E_\rho$ with the property $\operatorname{card} E_N(f)=\infty$, [3].

PROBLEM. Is it true that $\{E_V(f): f\in E_\rho\}=\{E_V(f): f\in \overset{\circ}{E}_\rho\}$?

There are examples of $f\in \overset{\circ}{E}_\rho$ for which the set $E_V(f)$ has the cardinality of the continuum [4].

REFERENCES

1. Л е в и н Б. Я. Распределение корней целых функций. Москва, ГИТТЛ, I956 (English translation: Levin B.Ja. Distribution of zeros of entire functions. AMS, New York, 1980.)
2. O u m K i - C h o u l. Bounds for the number of deficient values of entire functions whose zeros have angular densities. - Pacif.J.Math. 1969, 29, No.1, 187-202.
3. А р а к е л я н Н.У. Целые функции конечного порядка с бесконечным множеством дефектных значений. - Докл.АН СССР, I966, I70, № 2, 999-I002.
4. Г о л ь д б е р г А.А., Е р ё м е н к о А.Э., О с т р о в с к и й И.В. О сумме целых функций вполне регулярного роста. - Докл.АН УССР, сер."А", I982, № 2, 8-II.

A.A.GOL'DBERG (А.А.ГОЛЬДБЕРГ)
СССР, 290602, Львов, Львовский государственный университет

A.È.ERËMENKO (А.Э.ЕРЁМЕНКО)
СССР, 3I0I64, Харьков, Физико-технический институт низких температур АН УССР

I.V.OSTROVSKII (И.В.ОСТРОВСКИЙ)
СССР, 3I0077, Харьков, Харьковский государственный университет

11.5. old OPERATORS PRESERVING THE COMPLETELY REGULAR GROWTH

We suppose the reader is familiar with notions and theorems of the theory of entire functions of completely regular growth (c.r.g.) (e.g. as presented in chapters II and III of [1]).

In [2] derivatives and integrals of the function f of c.r.g. were considered and the following result was obtained: 1) f' has c.r.g. on all rays $\arg z = \theta$ except maybe for θ with $h_f(\theta) = 0$ *); 2) the integral $\mathcal{F}(z) = \int_0^z f(t)\,dt$ has c.r.g. in $\mathbb{C}$.

Now consider instead of the operator $\mathcal{D} = \frac{d}{dz}$ a more general operator $\varphi(\mathcal{D})$, where φ is an entire function of exponential type. Is there a result similar to that of [2] in this case? For entire functions f of c.r.g. and of order ρ, $\rho < 1$, the answer is given by the following theorem.

THEOREM. Let f be an entire function of c.r.g. and of order ρ, $\rho < 1$, and φ be an entire function of exponential type. Then: 1) the function $\varphi(\mathcal{D})f$ has c.r.g. on all rays, maybe except for rays with $h_f(\theta) = 0$; 2) every solution $\mathcal{F}$ of the equation $\varphi(\mathcal{D})\mathcal{F} = f$ in the class of entire functions of order**) ρ has c.r.g. in $\mathbb{C}$.

The theorem follows from the result of [2] stated above and from the following lemma.

LEMMA. Let f be an entire function of order ρ, $\rho < 1$, φ be an entire function. Then the asymptotic equality holds

$$\varphi(\mathcal{D})\varphi(z) = \frac{f^{m}(z)}{m!}\left[\varphi^{(m)}(0) + o(1)\right], \quad z \overset{\circ}{\to} \infty,$$

where m is the multiplicity of zero of φ at $z=0$, and $z \overset{\circ}{\longrightarrow} \infty$

*) In [2] there is a more explicit and complete chatacterization of the exceptional set of rays; h_f is the indicator function of f.

**) The set of solutions of this sort is nonempty, see A.O.Gelfond [3], p.359.

eans that $z \to \infty$ outside some C°-set.

For the proof of the lemma one needs the following assertion hich is an easy corollary of theorem 2 of [4]. Let g be a mero- orphic function of order $\rho, \rho<1$, R be a fixed number, $R>0$; hen $\frac{g'(z+\eta)}{g(z+\eta)} \to 0 \quad (z \xrightarrow{\circ} \infty)$ uniformly with respect to η, $|\eta|<R$, and moreover, the exceptional C°-set does not de- end on η.

So this assertion and the equality

$$\frac{g(z+\eta)}{g(z)} = \exp\left[\int_0^\eta \left\{\frac{g'(z+t)}{g(z+t)}\right\} dt\right.$$

imply that uniformly with respect to η, $|\eta|<R$

$$\frac{g(z+\eta)}{g(z)} \to 1 \quad (z \xrightarrow{\circ} \infty) \tag{1}$$

(essentially the same was obtained also in [4], p.414).

Now let f, φ be as in the lemma. Denote by Ψ the Borel transform of φ and denote by Γ a circle surrounding all singularities of Ψ. From the equality

$$\varphi^{(k)}(0) = \frac{1}{2\pi i}\int_\Gamma \Psi(\zeta)\zeta^k \, d\zeta \quad (k=0,1,2,\dots) \tag{2}$$

and from the fact that $\varphi^{(k)}(0)=0 \quad (k=0,1,\dots)$ it is clear that

$$\varphi(\mathcal{D})f(z) = \frac{1}{2\pi i}\int_\Gamma \Psi(\zeta) f(z+\zeta)\, d\zeta =$$

$$= \frac{1}{2\pi i}\int_\Gamma \Psi(\zeta)\left\{f(z+\zeta) - f(z) - \dots - \frac{f^{(m-1)}(z)\zeta^{m-1}}{(m-1)!}\right\} d\zeta =$$

$$= f^{(m)}(z)\frac{1}{2\pi i}\int_\Gamma \Psi(\zeta) G_m(z,\zeta)\, d\zeta,$$

where

$$G_m(z,\zeta)=\left\{f(z+\zeta)-f(z)-\ldots-\frac{f^{(m-1)}(z)\zeta^{m-1}}{(m-1)!}\right\}/f^{(m)}(z).$$

But we have

$$G_m(z,\zeta)=\frac{\zeta^m}{(m-1)!}\int_0^1 \frac{\tau^{m-1}f^{(m)}(z+\zeta(1-\tau))}{f^m(z)}\,d\tau.$$

So applying (1) (with $g(z)=f^{(m)}(z)$) we obtain, that

$G_m(z,\zeta)\longrightarrow \zeta^m/m!$ $(z \xrightarrow{\ o\ } \infty)$ uniformly with respect to $\zeta\in\Gamma$. Taking (2) into account we see that the lemma is proved.

For entire functions f of order ρ, $\rho\geq 1$, one cannot expect an analog of [2] as simple as in the case $\rho<1$. It is seen from the following example. Let $\varphi(t)=t^n-1$, $n\geq 3$ and let g be an entire function of exponential type σ, $\sigma<\frac{1}{2}$, which is not a c.r.g. function. Denote by G the solution of the equation $\varphi(D)G=g$ in the class of entire functions of exponential type σ (the set of the solutions is nonempty by the theorem IV of [3], p.361). Set $f(z)=G(z)+\sum_{K=0}^{n-1}\exp\{e^{\frac{2K\pi i}{n}}z\}$. Then f is an entire function of exponential type and of c.r.g. with the positive indicator whereas $g=\varphi(D)f$ is not of c.r.g.

For functions of exponential type the following conjecture seems plausible.

CONJECTURE. Let f be an entire function of exponential type and of c.r.g., and φ has no zeros at the points of the boundary of conjugate diagram of f which are common endpoints of two segments on the boundary of the diagram. Then $\varphi(D)f$ has c.r.g. In particular, if φ has no zeros on the boundary of the conjugate diagram of f, then $\varphi(D)f$ has c.r.g.

For functions f which grow faster than functions of exponential type the answer must be still more complicated, as solutions of the equation $\varphi(D)f=0$ in this class may be not of c.r.g.

The following QUESTION remains unsolved too. Let f be an entire function of c.r.g. and of order ρ, $\rho \geq 1$, φ be an entire function of exponential type. Are there solutions $\mathcal{F}$ of $\varphi(\mathcal{D})\mathcal{F} = f$ which are the entire functions of c.r.g. with respect to the same proximate order as the proximate order of f ? If φ is a polynomial the affirmative answer is an easy consequence of results in [2] and of the integral representation of the solution, however the general case does not follow by passing to the limit

REFERENCES

1. Л е в и н Б.Я. Распределение корней целых функций, М., 1956. (Distribution of zeros of entire functions. Providence, AMS, 1964.)
2. Г о л ь д б е р г А.А., О с т р о в с к и й И.В. О производных и первообразных целых функций вполне регулярного роста. - Теория функций, функц.анал. и их прил., Харьков, 1973, вып.18, 70-81.
3. Г е л ь ф о н д А.О. Исчисление конечных разностей. М., Наука, 1967.
4. M a c i n t y r e A.J., W i l s o n R. The Logarithmic derivatives and flat regions of analytic functions. - Proc.London Math.Soc., 1942, 47, 404-435.

I.V.OSTROVSKII
(И.В.ОСТРОВСКИЙ)

СССР, 310077, Харьков,
ул.Дзержинского 4,
Харьковский государственный
университет

* * *

COMMENTARY BY THE AUTHOR

A partial progress in the last question of the Problem has been made in [5].

For any trigonometrically convex 2π-periodic function h consider the class $[1, h]$ of all entire functions f of exponential type with indicator h_f satisfying $h_f \leq h$.

THEOREM. Let f be a function in $[1, h]$ of completely regular growth on $\ell_\theta = \{z \in \mathbb{C} : \arg z = \theta\}$ and let φ be any entire function of exponential type. Then each solution $\mathcal{F}$ of $\varphi(\mathcal{D})\mathcal{F} = f$ is of completely regular growth on ℓ_θ.

REFERENCE

5. Е п и ф а н о в О.В. О сохранении оператором свертки не вполне регулярного роста функций. Сиб.Матем.журнал, 1979, 20, № 2, 420-422.

11.6. old

ZERO-SETS OF SINE-TYPE FUNCTIONS

An entire function S of exponential type π is called a *sine-type function* (s.t.f.) if there exist positive constants m, M, H such that

$$m < |S(z)| e^{-\pi |\operatorname{Im} z|} < M \quad \text{for} \quad |\operatorname{Im} z| \geqslant H .$$

The class of s.t.f. was introduced in [1]. It found applications in the theory of interpolation by entire functions and for bases of exponentials in $L^2(-\pi,\pi)$.

THEOREM 1 ([2]). Let S be a s.t.f. with simple zeros $\{\lambda_n\}_{n\in\mathbb{Z}}$ satisfying

$$\inf_{k\neq j} |\lambda_k - \lambda_j| > 0 \tag{1}$$

and let $\{c_n\}_{n\in\mathbb{Z}}$ be any sequence in ℓ^p, $1<p<+\infty$. Then the series

$$f(z) = S(z) \sum_n \frac{c_n}{S'(\lambda_n)(z-\lambda_n)}$$

converges in L^p and defines an isomorphism of ℓ^p onto the space $L^p_\pi(\mathbb{R})$ of all entire functions f of exponential type $\leqslant \pi$ such that $\int_{\mathbb{R}} |f|^p\, dx < \infty$.

THEOREM 2 ([1],[3]). Let S satisfy the conditions of theorem 1. Then the system $\{e^{i\lambda_n x}\}_{n\in\mathbb{Z}}$ forms a Riesz basis in $L^2(-\pi,\pi)$.

It had been shown in [1], that $\{e^{i\lambda_n x}\}_{n\in\mathbb{Z}}$ forms a basis in $L^2(-\pi,\pi)$, and later it was proved in [3] that actually it is a Riesz basis. V.È.Katsnel'son [4] has essentially strengthened Theorem 2 and his result can be also formulated in terms of zero-sets of s.t.f. A series of other results in this field has been obtained in [5].

The conditions of simple zeros and (1) can be omitted but this results in a more complicated statements of Theorems 1 and 2.

In connection with the above results the following PROBLEM seems to be interesting: <u>describe the zero-sets</u> $\{\lambda_n\}_{n\in\mathbb{Z}}$ <u>of s.t.f.</u> This problem is, of course, equivalent to the problem of identification of s.t.f. because

$$S(z) = \lim_{R\to\infty} \prod_{|\lambda_n|<R} \left(1-\frac{z}{\lambda_n}\right)$$

for every s.t.f. S .

We cannot take risk of predicting concrete terms in which the problem could be solved but the desired solution should be given with help of "independent parameters". To clarify the last requirement consider an analogous problem for the M.G.Krein class. By the way, this class may be connected with our problem.

An entire function f belongs to the M.G.K r e i n c l a s s if

$$\frac{1}{f(z)} = C + \frac{A}{z} + \sum_n \left(\frac{A_n}{z-\lambda_n} - \frac{A_n}{\lambda_n}\right)$$

with real coefficients $C, A, \{A_n\}, \{\lambda_n\}$ satisfying $\sum_n |A_n| \lambda_n^{-2} < \infty$. The Krein class, which has been introduced in [7], is important for Operator Theory and for the moment problem. It turns out that $\{\lambda_n\}$ is a sequence of zeros of a function in Krein's class iff it can be obtained by the following procedure [6].

Pick an arbitrary domain Ω of the form

$$\Omega = \{\theta \in \mathbb{C}_+ : p\cdot\pi < \mathrm{Re}\,\theta < q\pi\} \setminus \bigcup_{p<k<q} \{\theta : \mathrm{Re}\,\theta = k\pi,\ \ 0 \leqslant \mathrm{Im}\,\theta \leqslant h_k\}$$

where p, q, k are integers (it is allowed that $p=-\infty, q=+\infty$) and $0 \leqslant h_k < \infty$, and map Ω conformally onto $\mathbb{C}_+$ by the mapping φ satisfying $\varphi(\infty) = \infty$. The images of slits I_k of Ω will be disjoint segments $[a_k, b_k]$, $p<k<q$ of the real line. Every choice $\lambda_k \in [a_k, b_k]$ defines a function f in M.G.Krein's class. Here the numbers p , q , h_k , $\lambda_k \in [a_k, b_k]$ play a role of independent parameters.

Single out some results connected with our problem. There exists a necessary and sufficient condition for $\{\lambda_n\}_{n\in\mathbb{Z}}$,

$|\operatorname{Im}\lambda_n| < H$ to be the zero-set of a s.t.function ([5], p.659):

$$\sup_{\xi\in\mathbb{R}} \int_{\mathbb{R}} \frac{(t-\xi)\{n(t)-t\}}{(t-\xi)^2+h^2}\,dt<\infty \quad \forall h>0$$

Here $n(t)$ stands for $\operatorname{sign}(t)\operatorname{Card}\{k: 0<(\operatorname{sign} t)\operatorname{Re}\lambda_k<|t|\}$.

Suppose that $\lambda_k = k + c_k$, $c_k = O(1)$, $|k|\to\infty$. Then $\{\lambda_k\}$ is the zero-set of a s.t.function iff there exists an entire function g of exponential type $\leq \pi$ satisfying $\sup_{t\in\mathbb{R}} |g(t)| < \infty$, $g(k) = (-1)^k c_k$, $k\in\mathbb{Z}$ (see [8], Appendix VI). This condition can be reformulated in terms of special functionals applied to $\{c_k\}$ (see[8]p.591). Perhaps, this observation indicates the right way to the solution ? Anyway, in general it is not true that the zero-set $\{\lambda_n\}$ of s.t.f. satisfies $\lambda_n = n + O(1)$. On the other hand $\lambda_n = n + O(\log|n|)$ for every s.t.f.

In conclusion note that without loss of generality the sequence $\{\lambda_n\}_{n\in\mathbb{Z}}$ can be assumed to be real. A sequence $\{\lambda_n\}_{n\in\mathbb{Z}}$, $|\operatorname{Im}\lambda_n| < H$ is the sequence of zeros of a s.t.f. iff $\{\operatorname{Re}\lambda_n\}$ is (see [9]).

REFERENCES

1. Л е в и н Б.Я. О базисах показательных функций в $L^2(-\pi,\pi)$. - Зап.физ.-матем.ф-та ХГУ и Харьк.матем.об-ва, 1961, 27, 39-48.
2. Л е в и н Б.Я. Интерполяция целыми функциями экспоненциального типа. - Труды ФТИНТ АН УССР, сер."Матем.физика и функц.анализ", 1969, вып.I, 136-146.
3. Г о л о в и н В.Д. О биортогональных разложениях в L^2 по линейным комбинациям показательных функций. - Зап.мех.-матем. ф-та ХГУ и Харьк.матем.об-ва, 1964, 30, 18-29.
4. К а ц н е л ь с о н В.Э. О базисах показательных функций в L^2. - Функц.анал. и его прил., 1971, 5, № 1, 37-47.
5. Л е в и н Б.Я., Л ю б а р с к и й Ю.И. Интерполяция целыми функциями специальных классов и связанные с нею разложения в ряды экспонент. - Изв.АН СССР, сер.матем., 1975, 39, № 3, 657-702.
6. О с т р о в с к и й И.В. Об одном классе целых функций. - Докл.АН СССР, 1976, 229, № 1, 39-41.
7. К р е й н М.Г. К теории целых функций экспоненциального типа. - Изв.АН СССР, сер.матем., 1947, II, № 4, 309-326.

8. Л е в и н Б.Я. Распределение корней целых функций, М., Гостехиздат, 1956.
L e v i n B.Ja. Destributions of zeros of entire functions, Providence, Rhode Island, AMS, Translations of Math.Monographs, v.5, 1964.
9. Л е в и н Б.Я., О с т р о в с к и й И.В. О малых возмущениях множества корней функции типа синуса. - Изв.АН СССР, сер. матем., 1979, 43, №1, 87-110.

B.Ya.LEVIN (Б.Я.ЛЕВИН)
СССР, 310164, Харьков, Физико-технический институт низких температур АН УССР

I.V.OSTROVSKII (И.В.ОСТРОВСКИЙ)
СССР, 310077, Харьков, Харьковский государственный университет

* * *

COMMENTARY

The problem seems to be still unsolved. New related information may be found in

Hruščev S.V., Nikol'skii N.K., Pavlov B.S. Unconditional bases of exponentials and of reproducing kernels. - Lect.Notes in Math., 1981, 864, 214-335.

1.7. AN EXTREMAL PROBLEM FROM THE THEORY OF SUBHARMONIC FUNCTIONS

A closed subset E of $\mathbb{R}$ is said to be relatively dense (in measure) if there exist positive numbers N and δ such that every interval of length N contains a part of E of measure at least δ. In this case we write $E\in E(N,\delta)$. Suppose in addition that all points of E are regular boundary points of the domain $\mathbb{C}\setminus E$. It was proved in [1] that there exists a harmonic function v positive on $\mathbb{C}\setminus E$ and (continuously) vanishing on E. Such a function was constructed in [2] using some special conformal mappings. It can be shown that if we require in addition that $v(\bar{z})=v(z)$ then E determines v uniquely up to a positive constant factor (cf.[3]; see also [4] for a more general result).

If E is relatively dense then a positive limit

$$A=\lim_{y\to+\infty}\frac{v(iy)}{y}<\infty$$

exists. Multiplying by a positive constant we may assume $A=1$. This normalized v will be from now on denoted by v_E.

It was proved in [1] and [2] that v_E is bounded on $\mathbb{R}$ by a constant depending only on N and δ provided $E\in E(N,\delta)$ i.e.

$$E\in E(N,\delta)\Rightarrow \sup_{x\in\mathbb{R}} v_E(x)\leqslant C(N,\delta).$$

PROBLEM. <u>Find the best possible value of</u> $C(N,\delta)$ <u>for given</u> N <u>and</u> δ.

This problem is connected with the following

THEOREM. <u>Suppose that</u>

a) u <u>is subharmonic in</u> $\mathbb{C}$;

b) $u(x)\leqslant 0$ <u>for</u> $x\in E$, $E\in E(N,\delta)$,

c) $\sigma=\overline{\lim_{|z|\to\infty}}\ \frac{u(z)}{|z|}<\infty$.

Then $u(z) \leqslant \sigma v_E(z)$, $z \in \mathbb{C}$. Moreover, the equality at one point $z \in \mathbb{C} \setminus E$ implies $u \equiv \sigma v_E$.

Hence for the class of subharmonic functions u satisfying b) and c) we obtain the estimate

$$\sup_{x \in \mathbb{R}} u(x) \leqslant C(N, \sigma)\sigma .$$

It is easily seen that $C(N,\delta) = NC(1, \delta/N)$, and we can assume $N=1$ and $0 < \delta < 1$ without loss of generality.

It was proved in [4] that

$$\max_{x \in \mathbb{R}} v_{\tilde{E}}(x) = \frac{1}{\pi} \ln \operatorname{ctg} \frac{\pi\delta}{4}$$

where $\tilde{E} = \bigcup_{n \in \mathbb{Z}} \left[n - \delta/2 ,\ n + \delta/2\right]$. I CONJECTURE that

$$C(1,\delta) = \frac{1}{\pi} \operatorname{ctg} \frac{\pi\delta}{4}$$

and that $\max v_E(x)$ attains this value for $E = \tilde{E}$ only.

REFERENCES

1. S c h a e f f e r A.C. Entire functions and trigonometrical polynomials. - Duke Math.J. 1953, 20, 77-88.
2. А х и е з е р Н.И., Л е в и н Б.Я. Обобщение неравенства С.Н.Бернштейна для производных от целых функций. - В кн.: Исследования по современным проблемам теории функций компл.перем., ГИФМЛ, Москва, 1960, 111-165.
3. B e n e d i c k s M. Positive harmonic function vanishing on the boundary of certain domains in $\mathbb{R}^n$. - Arkiv för Math. 1980, 18, N 1, 53-72.
4. Л е в и н Б.Я. Субгармонические мажоранты и их приложения. Всесоюзная конфер. по ТФКП. Харьков, 1971, 117-120.

B.Ya.LEVIN
(Б.Я.ЛЕВИН)

СССР, 310164, Харьков
пр.Ленина 47
ФТИНТ АН УССР

1.8.
old

A PROBLEM ON EXACT MAJORANTS

Let G be a domain on the complex plane ($G=\mathbb{D}$ for example), and let h be a positive function on G. Consider the class B_h of all single-valued functions f analytic in G such that $|f(z)|\leqslant h(z)$ $(z\in G)$ and define a function H by

$$H(z)=\sup_{f\in B_h}|f(z)|,\quad z\in G.$$

OUR PROBLEM <u>is to find conditions on h necessary and sufficient for the equality $h=H$</u>. If $h=H$ the function h will be called a n e x a c t m a j o r a n t (e.m.)

It is clear that for any e.m. h the function $\log h$ is subharmonic. But easy examples show that it is not a sufficient condition. On the other hand the equality $h=|F|$, F being an analytic function, implies that h is an e.m. But this condition is not necessary. When trying to solve the PROBLEM one may impose some additional requirements on h, e.g. suppose that h is continuous in G or even (as the first step) in $\operatorname{clos} G$. Theoretically one may treat this problem using the concept of duality in the theory of extremal problems (cf. e.g. [1]). But I didn't succeed to get a useful information concerning the description of e.m. by this approach. The fact that each e.m. h is also an e.m. in every subdomain of G is likely to be useful in this approach.

Let Q be the class of e.m.'s for G that are continuous in G (or even in $\operatorname{clos} G$). HERE IS ONE OF CONJECTURES concerning the description of Q:

$h\in Q \Longleftrightarrow$ (h <u>is in the closure of functions of the form $|f_1|+\ldots+|f_n|$, f_j being analytic in G</u>), here the closure is either in $C(\operatorname{clos} G)$ (if Q consists of functions continuous in $\operatorname{clos} G$) or in the projective limit of the spaces $C(\operatorname{clos} G_n)$, where the domains G_n exhaust G (if Q consists of functions continuous only in G).

Using the approach of the convex analysis we can formulate the DUAL VERSION OF OUR CONJECTURE: <u>let μ be a real Borel measure on G</u>; <u>does the condition $\int_G |f|\,d\mu\geqslant 0$ for all analytic in G functions f imply $\int_G h\,d\mu\geqslant 0$ for $h\in Q$</u>? We may try to treat the question investigating the measures in the Riesz representation of the subharmonic function $|f|$ (not $\log|f|$!). The answers to the above questions may happen to yield an interesting contribution to the theory of extrema in spaces of analytic functions.

REFERENCE

1. Хавинсон С.Я. Теория экстремальных задач для ограниченных аналитических функций, удовлетворяющих дополнительным условиям внутри области. - Успехи матем.наук, 1963, 18, № 2, 25-98.

S.Ya.HAVINSON
(С.Я.ХАВИНСОН)

СССР, Москва, 121352, Московский инженерно-строительный институт им.В.А.Куйбышева

COMMENTARY

The CONJECTURE has been DISPROVED by A.Gordon (А.Я.Гордон; private communication).

Denote by $C(O)$ the space of all functions continuous in the domain O and by $S(O)$ the closure (with respect to the usual compact convergence topology of $C(O)$) of the set of all sums $|f_1|+\dots+|f_N|$, $f_j\in\mathcal{H}(O)$, where $\mathcal{H}(O)$ stands for the space of all functions holomorphic in O. A.Gordon has shown that the function h, $h(z)\overset{def}{=}\max\{1,2|z|\}$, $z\in\mathbb{D}$, being obviously an exact majorant, does not belong to $S_{\mathbb{D}}\overset{def}{=}S$. HERE IS THE PROOF.

Set $I_r(\psi)\overset{def}{=}\frac{1}{2\pi}\int_0^{2\pi}\psi(re^{i\theta})\,d\theta$. We shall see that if $H\in S$ and $I_r(H)=1$ for all small values of r (say, for $r\in[0,\rho]$) then $I_r(H)=1$ for all $r\in[0,1)$ (whereas $I_r(h)=1$ for $r\le 1/2$, $I_r(h)>1$ for $r>1/2$).

The inclusion $H\in S$ implies the existence of functions $\{f_m^k\}$, $m=1,2,\dots$, $k=1,2,\dots,N_m$ such that $f_m^k\in\mathcal{H}(\mathbb{D})$ and $g_m\overset{def}{=}\sum_{k=1}^{N_m}|f_m^k|$ tend to H in $C(\mathbb{D})$ when $m\to\infty$.
Take $R\in(\rho,1)$. Then $\lim_{m\to\infty} I_R(g_m)=I_R(H)\overset{def}{=}B$ and we may assume (renormalization!) $I_R(g_m)=B$ $(m=1,2,\dots)$. Thus

$$g_m=\sum_{k=1}^{N_m}\lambda_m^k u_m^k\ , \qquad (*)$$

where $\lambda_m^k \overset{def}{=} \dfrac{I_R(f_m^k)}{B}$, $u_m^k \overset{def}{=} \dfrac{|f_m^k|}{\lambda_m^k}$, and so $u_m^k \in \mathcal{M}_R \overset{def}{=} \{u : u \in C(R\cdot\mathbb{D}),\ \sup_{0\le r<R} I_r(u) \le B,\ u = |f|,\ f \in \mathcal{H}(R\cdot\mathbb{D})\}$. Denote by hull $(\mathcal{M}_R)$ the closed convex hull of $\mathcal{M}_R$ in $C(R\cdot\mathbb{D})$ The sets $\mathcal{M}_R$ and hull $(\mathcal{M}_R)$ are compact in $C(R\cdot\mathbb{D})$, elements of $\mathcal{M}_R$ being uniformly bounded and uniformly Lipschitzian on every disc $r\mathbb{D}$, $0<r<R$. We see from (*) that $H \in hull(\mathcal{M}_R)$ It is convenient to rewrite this as follows:

$$F(H) = \int_{\mathcal{M}_R} F(u)\, d\mu(u) \qquad (**)$$

for every $F \in (C(R\cdot\mathbb{D}))^*$, μ being a probability measure on $\mathcal{M}_R$ (see, e.g., Proposition 1.2 in the first chapter of [2]). Equality (**) holds, in particular, for $F = I_r$, $r \in (0, R)$. But $I_r(H) \equiv 1$ $(0 \le r \le \rho)$ and $I_r(u)$ increases with r if $u \in \mathcal{M}_R$. Therefore (see (**)) $1 = I_0(H) = I_\rho(H) =$

$= \int_{\mathcal{M}_R} I_\rho(u)\, d\mu(u) \ge \int_{\mathcal{M}_R} I_0(u)\, d\mu(u) = I_0(H) = 1$, and $I_\rho(u) =$ $= I_0(u)$ μ - a.e. on $\mathcal{M}_R$. But if $u \in \mathcal{M}_R$ then u is the modulus of a function in $\mathcal{H}(R\cdot\mathbb{D})$ and the last equality implies u is constant in $R\cdot\mathbb{D}$. Using (**) with $F = I_r$, $r \in (\rho, R)$ we see $I(H_r) \equiv 1$ on $[0, R)$. ●

A.Gordon remarked that this proof can be slightly modified to yield the following assertion: if $f_1, \dots, f_n \in \mathcal{H}(\mathbb{D})$ and $\max_j |f_j| \not\equiv |f_k|$ $(k = 1, 2, \dots, n)$, then $\max_{1\le j\le n} |f_j| \notin S$.

REFERENCE

2. Phelps R.R. Lectures on Choquet's Theorem. van Nostrand, Princeton, 1966.

11.9. old ENTIRE FUNCTIONS OF LAGUERRE-POLYA CLASS

Laguerre-Polya class $\mathcal{LP}$ plays an important role in the theory of entire functions. This class consists of functions of the form

$$f(z) = e^{-\gamma z^2 + \beta z + \alpha} \cdot \prod_1^{\omega} \left(1 - \frac{z}{a_k}\right) e^{\frac{z}{a_k}} \qquad (\omega \leqslant \infty),$$

where $\gamma \geqslant 0$, $\mathcal{Im}\,\beta = 0$, $\mathcal{Im}\,a_k = 0$, $\sum_1^{\omega} |a_k|^{-2} < \infty$. It is known (see [1]), that this class is the closure (in the sense of the uniform convergence on compact sets) of polynomials with real roots. It follows that $f^{(k)} \subset \mathcal{LP}$, $k = 1, 2, \ldots$ for $f \in \mathcal{LP}$. In 1914 Polya proposed the following conjecture: a real entire function (i.e. an entire f with $f(\mathbb{R}) \subset \mathbb{R}$) such that f and all its derivatives have no zeros off $\mathbb{R}$ is in $\mathcal{LP}$.

There is a plenty of works devoted to this conjecture. The bibliography can be found in [3], [4]. Not long ago S.Hellerstein and J.Williamson solved this problem (in a preprint, see also their works [4], [5]). They have shown that a real entire function f with all the zeros of f, f', f'' real, is in $\mathcal{LP}$.

PROBLEM 1. <u>Prove that a real entire function with all the zeros of f and f' real is in $\mathcal{LP}$.</u>

In [3] it is shown only that loglog $M(r,f) = O(r \log r)$ for a real entire function f such that f and f'' have only real zeros.

Consider a well-known class $\overline{HB}$ of entire functions ω defined by:

a) the zeros of ω lie in the upper half-plane $\mathcal{Im}\, z \geqslant 0$ only;

b) if $\mathcal{Im}\, z < 0$ then $|\omega(z)| \geqslant |\omega(\bar{z})|$.

An arbitrary entire function ω can be represented as

$$\omega = \mathcal{P} + iQ,$$

where $\mathcal{P}$ and Q are real. It is known ([1]) that $\omega \in \overline{HB}$ if and only if for an arbitrary pair of real numbers λ, μ the function

$$\lambda \mathcal{P} + \mu Q$$

has only real zeros.

Applying Hellerstein-Williamson's result we now deduce that if $\omega^{(k)} \in \overline{HB}$, $k = 0, 1, \ldots$, then $\omega \in \mathcal{P}^*$, the class $\mathcal{P}^*$ being defined by:

c) $\omega \in \overline{HB}$

d) $\omega(z) = e^{-\gamma z^2 + \beta z + \alpha} \cdot \prod_1^{\omega} (1 - \frac{z}{a_k}) \cdot e^{\frac{z}{a_k}}, \quad (\omega \leq \infty),$

with $\gamma \geq 0$ and $\sum_1^{\omega} |a_k|^{-2} < \infty$.

If d) holds, c) is equivalent to the following condition:

$$\operatorname{Im}\beta + \sum_1^{\omega} \operatorname{Im}(a_k^{-1}) \geq 0, \quad \operatorname{Im} a_k \geq 0 .$$

It is known that $\mathcal{P}^*$ is the closure of the set of polynomials having all their zeros in $\operatorname{Im} z \geq 0$ (see [1], for example). So we have $\omega^{(k)} \in \mathcal{P}^*$ for $\omega \in \mathcal{P}^*$.

PROBLEM 2. <u>Prove that</u> $\omega \in \mathcal{P}^*$ <u>if all zeros of</u> $\omega^{(k)}$, $k = 0, 1, \ldots$ <u>are in the upper half-plane</u> $\operatorname{Im} z \geq 0$.

A similar problem can be formulated for entire functions of several complex variables. For simplicity we assume $n = 2$. A polynomial is called an $\overline{HB}$ -polynomial if it has no zeros in $\Omega = \{(z, w) : \operatorname{Im} z < 0, \operatorname{Im} w < 0\}$. The closure of the set of $\overline{HB}$ -polynomials will be denoted by $\mathcal{P}_2^*$ (the information about $\overline{HB}$ -polynomials and about the class $\mathcal{P}_2^*$ can be found in [1], Ch.9).

PROBLEM 3. <u>Prove that an entire function</u> ω <u>belongs to</u> $\mathcal{P}_2^*$ <u>if this function and all its derivatives have no zeros in</u> Ω .

REFERENCES

1. Л е в и н Б.Я. Распределение корней целых функций. М., ГИТТЛ, 1956.
2. P o l y a G. Sur une question concernant les fonctions entières. - C.R. Acad.Sci.Paris 1914, 158.
3. Л е в и н Б.Я., О с т р о в с к и й И.В. О зависимости роста целой функции от расположения корней ее производных. - Сиб. матем.журн.1960, I, № 3, 427–455.
4. H e l l e r s t e i n S., W i l l i a m s o n J. Derivatives of entire functions and a question of Polya. - Trans.Amer. Math.Soc., 1977, 227, 227-249.
5. H e l l e r s t e i n S., W i l l i a m s o n J. Derivatives of entire functions and a question of Polya.-Bull.Amer.Math.

Soc., 1975, 81, 453-455.

B.Ya.LEVIN

(Б.Я.ЛЕВИН)

СССР, 3I0I64, Харьков
пр.Ленина 47,
Физико-технический институт
низких температур АН УССР

1.10. CLUSTER SETS AND A PROBLEM OF A.F.LEONT'EV

We use notations from [1].

In connection with some interpolation problems A.F.Leont'ev proposed the following PROBLEM [2].

Let $f \in A(\rho)$, $h_f(\varphi) > 0$, $\varphi \in [0, 2\pi]$. Suppose that the derivative f' is of completely regular growth on the set of zeros of f, i.e. for the sequence $z_k = r_k e^{i\varphi_k}$ of all zeros of f we have

$$\frac{\ln |f'(z_k)|}{r_k^\rho} - h_{f'}(\varphi_k) \to 0 \tag{1}$$

(the zeros are supposed to be simple). Is it true that f is of completely regular growth ($f \in A_{reg}$)?

The following proposition is a corollary of the results in [3]. Suppose f satisfies (1). Denote $h(z) = \sup\{v(z) : v \in Fr[f]\}$ ($h_f(\varphi) = h(e^{i\varphi})$ is the indicator of f). Denote by μ_v the mass distribution associated with v. Then for every $v \in Fr[f]$ we have

$$v(z) = h(z), \quad z \in \operatorname{supp} \mu_v . \tag{2}$$

CONJECTURE 1. <u>Let</u> $Fr[f]$ <u>be the cluster set of</u> f. <u>Suppose that every</u> $v \in Fr[f]$ <u>satisfies</u> (2), <u>and</u>

$$h(z) > 0, \quad z \in \mathbb{C}. \tag{3}$$

<u>Then</u> $Fr[f]$ <u>consists of the single element</u> h, i.e. $f \in A_{reg}$.

If the Conjecture is true then the Leont'ev's problem has a positive solution.

Condition (3) is essential. To see this denote by $h_0(\varphi)$ the indicator of Mittag - Leffler function

$$h_0(\varphi) = \begin{cases} \cos \rho\varphi, & |\varphi| \le \frac{\pi}{2\rho} \\ 0, & |\varphi| > \frac{\pi}{2\rho} \end{cases}$$

and consider the set $\Lambda_0 = \{ a h_0 r^\rho : d_1 \leq a \leq d_2 \}$, $d_1 < d_2$. There exists a function $f_0 \in \mathring{A}(\rho)$ with $Fr[f_0] = \Lambda_0$ [4]. Every function $v \in Fr[f_0]$ satisfies condition (2) but $f_0 \notin A_{reg}$ because Λ_0 contains not only h. Condition (3) is not satisfied.

The condition „$\forall v \in Fr[f]$" is also essential as the following example (pointed out by M.L.Sodin) shows. Let $v(z) = |z|^\rho$ for $|z| \leq 1$ and $v(z) = \rho \ln|z| + 1$ for $|z| > 1$. It is easily verified that v is subharmonic in $\mathbb{C}$. Denote $\Lambda = \{ v_t : 0 < t < \infty \}$. It is clear that every function $v_t \in \Lambda$ satisfies (2) with $h(z) = \sup\{ v_t : t \in (0,\infty)\} = |z|^\rho$. The set Λ satisfies all conditions of theorem 1 from [1] except $Fr_0 \cap Fr_\infty \neq \emptyset$. Thus the set Λ is not the cluster set for any entire function.

REFERENCES

1. A z a r i n V.S. Two problems on asymptotic behaviour of entire functions. - This book, S.9.
2. Л е о н т ь е в А.Ф. Об условиях разложимости аналитических функций в ряды Дирихле. - Изв.АН СССР, сер.матем., 1972, 36, № 6, 1282-1295.
3. Г р и ш и н А.Ф. О множествах регулярности роста целой функции. - Теор. функций, функциональн. анал. и их прил., 1983, Харьков, вып. 40, 41.
4. А з а р и н В.С. Об асимптотическом поведении субгармонических функций конечного порядка. - Матем. сборн., 1979, 108 (150), № 2, 147-167 (Engl.Transl. - Math. USSR Sborn., 1980, 36, N 2, 135-154).

V.S.AZARIN
(В.С.АЗАРИН)

СССР, 310050, Харьков, Харьковский институт инженеров железнодорожного транспорта

A.È.ERËMENKO
(А.Э.ЕРЁМЕНКО)

СССР, 310164, Харьков, Физико-технический институт низких температур АН УССР

A.F.GRISHIN
(А.Ф.ГРИШИН)

СССР, 310077, Харьков, Харьковский государственный университет

CHAPTER 12

$\mathbb{C}^n$

Our stock of $\mathbb{C}^n$-problems being very poor, we just arrange them in author's alphabetical order (see also 1.4, 1.7, 1.13, 1.14, 5.10, 6.5, 7.1-7.3, 7.14, 8.12, 8.14, 9.13, 10.5, S.10).

12.1. POLYNOMIALLY CONVEX HULLS

We shall denote Hausdorff one-dimensional measure ("linear measure") by $\mathcal{H}_1$. For X a compact subset of $\mathbb{C}^n$, $\hat{X}$ will be its polynomially convex hull: $\{z\in\mathbb{C}^n : |p(z)|\leqslant \sup\{|p(\zeta)|,\ \zeta\in X\}$ for all polynomials p in $\mathbb{C}^n$. The unit ball in $\mathbb{C}^n$ will be denoted by B. As usual, $P(X)$ will be the uniform closure in $C(X)$ of the polynomials.

In [1] it was shown that if X is (or lies in) a connected set of finite linear measure, then $\hat{X}\setminus X$ is a one-dimensional analytic variety. Recently V.M.Golovin [4] claimed that the connectedness assumption could be dropped. We find his argument unconvincing and shall list a special case as a first question.

PROBLEM 1. <u>Does</u> $\mathcal{H}_1(X)=0$ <u>imply that</u> X <u>is polynomially convex</u> (i.e., $\hat{X}=X$)?

Known methods to solve this kind of problem involve the classical F. and M.Riesz theorem for Jordan domains with rectifiable boundaries. One way to treat Problem 1 would be to generalize this. Namely, let Ω be a bounded domain in $\mathbb{C}$ with $\mathcal{H}_1(\partial\Omega)<\infty$ (do n o t assume that Ω is even finitely connected). Suppose that the o u t e r boundary Γ of Ω is a Jordan curve. Let $z_0\in\Omega$ and let μ be a Jensen measure for the algebra $P(\bar{\Omega})$ supported on $\partial\Omega$ with respect to z_0.

PROBLEM 2. <u>Is</u> $\mu|\Gamma$ <u>absolutely continuous with respect to</u> $\mathcal{H}_1|\Gamma$?

The F. and M.Riesz theorem is the case $\partial\Omega=\Gamma$. A variant is:

PROBLEM 2′. <u>Let</u> u <u>be subharmonic on</u> Ω <u>and u.s.c. on</u> $\bar{\Omega}$. <u>Let</u> $E\subset\Gamma$ <u>with</u> $\mathcal{H}_1(E)>0$. <u>If</u> $u(z)\to-\infty$ <u>as</u> $z\in\Omega\to E$, <u>does it follow that</u> $u\equiv-\infty$?

Examples of non-polynomially convex sets X which are totally disconnected have long been known; a recent example was given by Vitushkin [6]. It is known that such a set cannot lie in a torus $\mathbb{T}^n$.

PROBLEM 3. <u>Find a set</u> $X\subset\partial B$ <u>which is totally disconnected such that</u> $0\in\hat{X}$.

One possible approach is to approximate such a set $X\subset\mathbb{C}^2$ by sets $V\cap\partial B$ where V is an analytic (or algebraic) curve in $\mathbb{C}^2$ passing through the origin. Then $V\cap\partial B$ would be required to have arbitrarily small components. On the other hand, this will not be possible if there is a lower bound on the size of these components - it is known that the s u m of their lengths, $\mathcal{H}_1(\partial V \cap B)$, is at least 2π .

PROBLEM 4. <u>Is there a lower bound for</u> $\{\mathcal{H}_1(\gamma) : \gamma$ a connected component of $V \cap \partial B\}$<u>?</u>

In [3], V.K.Belošapka conjectured, for „$\mathcal{H}_1$" replaced by "diameter", that one is a lower bound. He showed that if B is replaced by the bidisc $\mathbb{D}^2$ there is some component of diameter at least one.

There exist sets $X \subset \mathbb{C}^2$ such that $\hat{X} \setminus X$ is a non-empty but contains no analytic structure. This phenomenon was discovered by Stolzenberg [5]. A recent example of such a set X has been given by Wermer [7] with the additional property that $X \subset \mathbb{T} \times \mathbb{D}$.

PROBLEM 5. <u>Find a set</u> $X \subset \mathbb{T}^2$ <u>such that</u> $0 \in \hat{X}$ <u>and</u> $\hat{X} \setminus X$ <u>contains no analytic structure</u>.

One interesting property of such a set would be that it could be "reflected" in $\mathbb{T}^2$ which would then become a "removable singularity".

The Stolzenberg and Wermer sets both arise from limits of analytic varieties. A well-known question asks if this must necessarily hold. Our last problem is a special case of this.

PROBLEM 6. <u>Let</u> $X \subset \mathbb{T} \times \mathbb{D}$ <u>with</u> $(\hat{X} \setminus X) \cap \mathbb{D}^2$ <u>non-empty</u>. <u>Is</u> $\hat{X} \setminus X$ <u>a limit of analytic subvarieties of</u> $\mathbb{D}^2$?

A rather particular case of this was considered by Alexander - Wermer [2].

REFERENCES

1. A l e x a n d e r H. Polynomial approximation and hulls in sets of finite linear measure in $\mathbb{C}^n$. - Amer.J.Math., 1971, 93, 65-74.
2. A l e x a n d e r H., W e r m e r J. On the approximation of singularity sets by analytic varieties. - Pacific J.Math., 1983, 104, 263-268.
3. Б е л о ш а п к а В.К. Об одном метрическом свойстве аналитических множеств. - Изв.АН СССР, сер.матем., 1976, 40, № 6, 1409-1415.

4. Г о л о в и н В.М. Полиномиальная выпуклость и множества конечной линейной меры в $\mathbb{C}^n$. - Сиб.матем.журн., 1979, 20, № 5, 990-996.

5. S t o l z e n b e r g G. A hull with no analytic structure. - J. of Math.and Mech., 1963, 12, 103-112.

6. В и т у ш к и н А.Г. Об одной задаче В.Рудина. - Докл.АН СССР, 1973, 213, 14-15.

7. W e r m e r J. Polynomially convex hulls and analyticity. - - Arkiv för mat., 1982, 20, 129-135.

H.ALEXANDER

Department of Mathematics
University of Illinois at Chicago
P.O. Box 4348
Chicago, Illinois 60680
USA

2.2. THE EXTREME RAYS OF THE POSITIVE PLURIHARMONIC FUNCTIONS
old

1. Let $n \geq 2$ and consider the class $N(\mathbb{B})$ of all holomorphic unctions g on $\mathbb{B}$ such that $\mathrm{Re}\, g > 0$ and $g(0) = 1$, where $\mathbb{B}$ is he open unit ball in $\mathbb{C}^n$. Thus $N(\mathbb{B})$ is convex (and compact in he compact open topology). We think that the structure of $N(\mathbb{B})$ s of interest and importance. Thus we ask:

<u>What are the extreme points of</u> $N(\mathbb{B})$?

'ery little is known. Of course if $n=1$, and if

$$g = (1+f)/(1-f), \tag{1}$$

then g is extreme if and only if $f(z) = cz$, where $c \in \mathbb{T}$. It is proved in [1] that if $f(z) = \sum_{1}^{n} z_j^2$ and if g is the Cayley transform (1) of f, then $g \in E(\mathbb{B})$, where $E(\mathbb{B})$ is the class of all extreme points of $N(\mathbb{B})$.

Let $k = (k_1, \dots, k_N)$ be a multi-index and consider monomials $f(z) = cz^k$ in $\mathbb{C}^n$ such that $|f(z)| \leq 1$ if $z \in \mathbb{B}$. Thus $|c| \leq \left(\frac{k}{|k|}\right)^{-\frac{k}{2}}$ where by $|k|$ we mean $k_1 + \dots + k_n$. Let $c_k = \left(\frac{k}{|k|}\right)^{-\frac{k}{2}}$ and let $g(z) = (1+c_k z^k)/(1-c_k z^k)$. It is proved in [2] that $g \in E(\mathbb{B})$ if and only if the components of k are relatively prime and positive.

2. We have $\mathbb{1} \notin E(\mathbb{B})$, however it is a corollary of the just mentioned theorem of [2] that $\mathbb{1} \in \mathrm{clos}\, E(\mathbb{B})$, where the closure is in the compact open topology. Thus $E(\mathbb{B}) \neq \mathrm{clos}\, E(\mathbb{B})$. (If $n=1$, then $E(\mathbb{B}) = \mathrm{clos}\, E(\mathbb{B})$).

It is also known that if g is an extreme point of $N(\mathbb{B})$ and if (1) holds(that is to say if $f = (g-1)/(g+1)$), then f is irreducible. This is a special case of Theorem 1.2 of [3]. The term "irreducible" is defined in [4]. If $n=1$, then g is extreme if and only if f is irreducible. However for $n \geq 2$, the fact that f is irreducible does not imply that g is extreme.

3. The extreme points g in section 1 and the extreme points that can be obtained from them by letting $\mathrm{Aut}(\mathbb{B})$ act on $N(\mathbb{B})$ have the property that the Cayley transform $f = (g-1)/(g+1)$ is holomorphic on $\mathbb{B} \cup \partial\mathbb{B}$.

<u>Is this the case for every g in</u> $E(\mathbb{B})$?

If the answer is yes, then it would follow (since $n \geq 2$) that the F. and M.Riesz theorem holds for those Radon measures on $\partial\mathbb{B}$ whose Poisson integrals are pluriharmonic. In particular there would be no singular Radon measures $\neq 0$ with this property, which in turn would imply that there are no nonconstant inner functions on $\mathbb{B}$.

REFERENCES

1. F o r e l l i F. Measures whose Poisson integrals are pluriharmonic II. - Illinois J.Math.,1975, 19, 584-592.
2. F o r e l l i F. Some extreme rays of the positive pluriharmonic functions. - Canad.Math.J., 1979, 31, 9-16.
3. F o r e l l i F. A necessary condition on the extreme points of a class of holomorphic functions. - Pacific J. Math.,1977, 73, 81-86.
4. A h e r n P., R u d i n W. Factorizations of bounded holomorphic functions. - Duke Math.J.,1972, 39, 767-777.

FRANK FORELLI

University of Wisconsin,
Dept. of Math.,
Madison, Wisconsin 53106,
USA

* * *

COMMENTARY

The second question has been answered in the negative. See Commentary in S.10.

2.3. PROPER MAPPINGS OF CLASSICAL DOMAINS

A holomorphic mapping $\varphi : \Omega \to \Omega$ of a bounded domain $\Omega \subset \mathbb{C}^n$ s called *proper* if $\operatorname{dist}(\varphi(z_\nu), \partial\Omega) \to 0$ for every se-uence $z_\nu \in \Omega$ with $\operatorname{dist}(z_\nu, \partial\Omega) \to 0$, $\nu \to \infty$.

A biholomorphism (automorphism) of Ω is called a *trivi-l proper mapping* of Ω. If Ω is the 1-dimen-ional disc $\mathbb{D}$ the non-trivial proper holomorphic mappings $\varphi : \mathbb{D} \to \mathbb{D}$ do exist. They are called finite Blaschke products.

The existence of nontrivial proper holomorphic mappings seems o be the characteristic property of the 1-dimensional disc in the lass of all irreducible symmetric domains.

CONJECTURE 1. <u>For an irreducible bounded symmetric domain Ω in $\mathbb{C}^n$, $\Omega \neq \mathbb{D}$, every proper holomorphic mapping $\Omega \to \Omega$ is an automorphism.</u>

According to the E.Cartan's classification there are six types of irreducible bounded symmetric domains. The domain $\Omega_{p,q}$ of the first type is the set of comlex $p \times q$ matrices Z, $p \geqslant q \geqslant 1$, such that the matrix $I - Z^* Z$ is positive. The following beautiful result of H.Alexander was the starting point for our conjecture.

THEOREM 1 (H.Alexander [1]). <u>Let Ω be the unit ball in $\mathbb{C}^p$, i.e. $\Omega = \Omega_{p,1}$ and let $p \geqslant 2$. Then every proper holomorphic mapping $\varphi : \Omega_{p,1} \to \Omega_{p,1}$ is an automorphism of the ball.</u>

Denote by S the distinguished boundary (Bergman's boundary) of the domain Ω. A proper holomorphic mapping $\varphi : \Omega \to \Omega$ is called *strictly proper* if $\operatorname{dist}(\varphi(z_\nu), S) \to 0$ for every sequence $z_\nu \in \Omega$ with the property $\operatorname{dist}(z_\nu, S) \to 0$. The next result generalizing Alexander's theorem follows from [2] and gives a convincing evidence in favour of CONJECTURE 1.

THEOREM 2 (G.M.Henkin, A.E.Tumanov [2]). <u>If Ω is an irreducible bounded symmetric domain in $\mathbb{C}^n$ and $\Omega \neq \mathbb{D}$, then any strictly proper holomorphic mapping $\varphi : \Omega \to \Omega$ is an automorphism.</u>

Only recently we managed to prove CONJECTURE 1 for some symmetric domains different from the ball, i.e. when $\partial\Omega \neq S$.

THEOREM 3 (G.M.Henkin, R.G.Novikov). <u>Let $\Omega \subset \mathbb{C}^n$, $n \geqslant 3$, be the classical domain of the 4-th type, i.e.</u>

$$\Omega=\{z: z\cdot\bar{z}'+((z\bar{z}')^2-|zz'|^2)^{1/2}<1\},$$

<u>where</u> $z=(z_1,\ldots,z_n)$ <u>and</u> z' <u>stands for the transposed matrix</u>. <u>Then every proper holomorphic mapping</u> $\Omega\to\Omega$ <u>is an automorphism</u>.

Note that the domain $\Omega_{2,2}$ of the first type is equivalent to a domain of the 4-th type. Hence Theorem 3 holds for $\Omega_{2,2}$.

We present now the scheme of the proof of Theorem 3 which gives rise to more general conjectures on the mappings of classical domains.

The classical domain of the 4-th type is known to have a realization as a tubular domain in $\mathbb{C}^n$, $n\geq 3$, over the round convex cone

$$\Omega=\{z=x+iy\in\mathbb{C}^n: y_0^2-y_1^2-\ldots-y_{n-1}^2>0,\ y_0>0\}.$$

The distinguished boundary of this domain coincides with the space $\mathbb{R}^n=\{z\in\mathbb{C}^n: y=0\}$. The boundary $\partial\Omega$ contains together with each point $z\in\partial\Omega\setminus S$ the 1-dimensional analytic component $\mathcal{D}_z=\{x+\lambda y:\lambda\in\mathbb{C},\ \operatorname{Im}\lambda>0\}$. The boundary $\partial\mathcal{D}_z$ of this component is the nil-line in the pseudoeuclidean metric $ds^2=dx_0^2-dx_1^2-\ldots-dx_{n-1}^2$ on the distinguished boundary S.

If φ is a map satisfying the hypotheses of the theorem, an appropriate generalization of H.Alexander's [1] arguments yields that outside of a set of zero measure on $\partial\Omega$ a boundary mapping (in the sense of nontangential limits) $\tilde{\varphi}:\partial\Omega\to\partial\Omega$ of finite multiplicity is well-defined. This mapping posseses the following property: for almost every analytic component $\mathcal{D}_z$ the restriction $\tilde{\varphi}|\mathcal{D}_z$ is a holomorphic mapping of finite multiplicity of $\mathcal{D}_z$ into some component $\mathcal{D}_W$. Furthermore, almost all points of $\partial\mathcal{D}_z$ are mapped (in the sense of nontangential limits) into points of $\partial\mathcal{D}_W$.

It follows then from the classical Frostman's theorem that $\tilde{\varphi}|\mathcal{D}_z$ is a proper mapping the half-plane $\mathcal{D}_z$ into the half-plane $\mathcal{D}_W$.

So it follows that the boundary map $\tilde{\varphi}$ defined a.e. on the distinguished boundary $S\subset\partial\Omega$ has the following properties:

a) $\tilde{\varphi}$ maps S into S outside a set of zero measure;

b) $\tilde{\varphi}$ restricted on almost any nil-line $\partial\mathcal{D}_z$ coincides (almost everywhere on $\partial\mathcal{D}_z$) with a piecewise continuous map of finite multiplicity of the nil-line $\partial\mathcal{D}_z$ into some nil-line $\partial\mathcal{D}_W$.

With the help of A.D.Alexandrov's paper [3] one can prove that the mapping $\tilde{\varphi}: S \to S$ satisfying a), b) is a conformal mapping with respect to the pseudoeuclidean metric on S. It follows that φ is an automorphism of the domain Ω.

To follow this sort of arguments, say, for the domains $\Omega_{p,2}$, where $p>2$, one should prove a natural generalization of H.Alexander's and O.Frostman's theorems.

Let us call a holomorphic mapping φ of the ball $\Omega_{p,1}$ almost proper if φ is of finite multiplicity and for almost all $z \in \partial\Omega_{p,1}$ we have $\tilde{\varphi}(z) \in \partial\Omega_{p,1}$, where $\tilde{\varphi}(z)$ is the nontangential limit of the mapping φ defined almost everywhere on $\partial\Omega_{p,1}$.

CONJECTURE 2. Let φ be an almost proper mapping of $\Omega_{p,1}$ and $p \geqslant 2$. Then φ is an automorphism.

If we remove the words "φ is of finite multiplicity" from the above definition, the conclusion of Conjecture 2 may fail, in virtue of a result of A.B.Aleksandrov [4].

Finally we propose a generalization of Conjecture 1.

CONJECTURE 3. Let Ω be a symmetric domain in $\mathbb{C}^n$ different from any product domain $\Omega_1 \times G$, $G \subset \mathbb{C}^{n-1}$ and let S be its distinguished boundary. Let $\mathcal{U}'$ and $\mathcal{U}''$ be two domains in $\mathbb{C}^n$ intersecting S and $\varphi: \Omega \cap \mathcal{U}' \to \Omega \cap \mathcal{U}''$ a proper mapping such that for some sequence $\{z_\nu\}$ with $z_\nu \in \Omega \cap \mathcal{U}'$, $z_\nu \to z' \in S \cap \mathcal{U}'$ we have: $\varphi(z_\nu) \to z'' \in S \cap \mathcal{U}''$. Then there exists an automorphism Φ of Ω such that $\Phi|\Omega \cap \mathcal{U}' = \varphi$.

The verification of Conjecture 3 would lead to a considerable strengthening of a result on local characterization of automorphisms of classical domains obtained in [2].

One can see from the proof of Theorem 3 that Conjecture 3 holds for classical domains of the fourth type. At the same time, it follows from results of [1] and [5] that Conjecture 3 holds also for the balls $\Omega_{p,1}$.

REMARK. After the paper had been submitted the authors became aware of S.Bell's paper [6] that enables, in combination with [2], to prove Conjecture 1.

REFERENCES

1. A l e x a n d e r H. Proper holomorphic mappings in $\mathbb{C}^n$. – Indiana Univ.Math.J., 1977, 26, 137-146.
2. Т у м а н о в А.Е., Х е н к и н Г.М. Локальная характеризация аналитических автоморфизмов классических областей. – Докл. АН СССР, I982, 267, № 4, 796–799.
3. А л е к с а н д р о в А.Д. К основам теории относительности. – Вестн.ЛГУ, I976, I9, 5–28.
4. А л е к с а н д р о в А.Б. Существование внутренних функций в шаре. – Матем.сб., I982, II8, I47–I63.
5. R u d i n W. Holomorphic maps that extend to automorphism of a ball. – Proc.Amer.Soc., 1981, 81, 429-432.
6. B e l l S.R. Proper holomorphic mapping between circular domains. – Comm.Math.Helv., 1982, 57, 532-538.

G.M.HENKIN
(Г.М.ХЕНКИН)

СССР, II74I8, Москва,
Центральный Экономико-математический институт АН СССР, ул.Красикова, 32

R.G.NOVIKOV
(Р.Г.НОВИКОВ)

СССР, II7234, Москва,
Ленинские Горы, МГУ,
мех.-мат.факультет

2.4. ON BIHOLOMORPHY OF HOLOMORPHIC MAPPINGS OF COMPLEX BANACH SPACES

Let f be a function holomorphic in a domain $\mathcal{D}, \mathcal{D} \subset \mathbb{C}$. It is well known that if f is univalent, then $f'(z) \neq 0$ in $\mathcal{D}$. Holomorphic mappings f of domains $\mathcal{D} \subset \mathbb{C}^n$ for $n > 1$ also possess a similar property: if $f: \mathcal{D} \longrightarrow \mathbb{C}^n$ is holomorphic and one-to-one then $\det f'(z) \neq 0$ at every point z of $\mathcal{D}$ ($f'(z) = \left(\frac{\partial f_j}{\partial z_k}\right)_{j,k=1}^{n}$ is the Jacobi matrix), or equivalently the differential $df(z)h = f'(z)h$ is an automorphism of $\mathbb{C}^n$, and then, by the implicit function theorem, f itself is biholomorphic.

Note that the continuity and the injective character of f immediately imply that f is a homeomorphism because $\dim \mathbb{C}^n = n < +\infty$. The proof of the diffeomorphic property depends essentially on the holomorphic properties of f. Such a result, as is known, for real spaces and mappings is wrong, which is clear from the example $(x_1, \dots, x_n) \longrightarrow (x_1^3, \dots, x_n^3): \mathbb{R}^n \longrightarrow \mathbb{R}^n$.

Let now X and Y be complex Banach spaces and f be a holomorphic mapping of the domain $\mathcal{D} \subset X$ into Y. Remind that the mapping f is called holomorphic in $\mathcal{D}$ if it is continuous and weakly Gâteaux differentiable, i.e. for any $x \in \mathcal{D}$ and $h \in X$ there exists

$$\lim_{t \to 0} \frac{f(x+th) - f(x)}{t} = f'(x)h \qquad (t \in \mathbb{C}). \qquad (*)$$

Then $df(x)h = f'(x)h$ is a linear operator $X \longrightarrow Y$. It is proved that in the complex case (*) implies the strong Fréchet differentiability of f:

$$\| f(x+h) - f(x) - df(x)h \|_Y \leq c(x) \| h \|_X^2$$

(for small $\|h\|_X$).

PROBLEM. <u>Let</u> f <u>be a holomorphic one-to-one mapping of the domain</u> $\mathcal{D} \subset X$ <u>onto the domain</u> $\mathcal{D}' \subset Y$. <u>Is the differential</u> $df(x)$ <u>an isomorphism (i.e. an injective and surjective mapping) of the space</u> X <u>onto</u> Y <u>at every point</u> $x \in \mathcal{D}$?

The positive answer and the implicit function theorem would imply that all one-to-one holomorphic mappings of domains of complex Banach spaces are diffeomorphisms.

Positive solution of the problem would allow to obtain, for instance, some important corollaries in the geometric theory of functions of a complex variable(in problems concerning univalency and quasiconformal extendability of holomorphic functions, characterization of boundary properties of functions starting from the interior properties; all that can be reduced to the consideration of some Banach spaces of holomorphic functions).

The author does not know any general result in this direction. It seems likely that the problem in the general statement must have a negative solution. The following conjecture can be formulated (at least as a stimulus to refute it).

CONJECTURE. Suppose conditions of the Problem are fulfilled. Then the mapping $df(x)$ $(x \in \mathcal{D})$ is injective but there exist spaces X, Y for which it is not surjective.

Then a QUESTION arises under what additional conditions of, maybe, geometric character, concerning the structure of the spaces X and Y, the mapping $df(x)$ is an isomorphism (for f satisfying assumptions of the Problem); will this be so at least for Hilbert spaces or spaces possessing some special convexity properties, etc?

S.L.KRUSHKAL
(С.Л.КРУШКАЛЬ)

СССР, 630072, Новосибирск
Институт математики СО АН СССР

C H A P T E R 13

MISCELLANEOUS PROBLEMS

13.1. BANACH ALGEBRAS OF FUNCTIONS GENERATED BY THE SET OF ALL ALMOST PERIODIC POLYNOMIALS WHOSE EXPONENTS BELONG TO A GIVEN INTERVAL

1. For any $A\in(0,\infty]$ let $\mathbb{P}_A$ denote the linear set of all almost periodic polynomials

$$\mathcal{P}(t)=\sum_K c_K e^{i\lambda_K t}$$

with exponents $\lambda_K\in(-A,A)$, endowed with the sup-norm $\|\mathcal{P}\|= =\sup\{|\mathcal{P}(t)|:-\infty<t<\infty\}$.

Evidently every linear functional Φ on $\mathbb{P}_A$ is completely defined by its χ-function(characteristic function): $f(\lambda)= =\Phi(e_\lambda)$ $(-A<\lambda<A)$, where $e_\lambda(t)=exp(i\lambda t)\,(-\infty<t<\infty)$ The χ-function may be an arbitrary function from $(-A,A)$ to $\mathbb{C}$.

Let us denote by $\mathcal{R}_A$ the set of all $f:(-A,A)\to\mathbb{C}$ generating linear continuous functionals Φ_f on $\mathbb{P}_A$ $(\Phi_f\in\mathbb{P}_A^*)$. For $f\in\mathcal{R}_A$ we put $\|f\|_A=\|\Phi_f\|$. It is easy to see that $\mathcal{R}_A$ is a Banach algebra of functions. By $\mathcal{R}_A^c$ $(\mathcal{R}_A^m)$ we denote the Banach subalgebras consisting of all continuous (measurable) $f\in\mathcal{R}_A$.

The following proposition was proved in [1] for $A=\infty$ and in [2] for $(0<)A<\infty$.

THEOREM 1. <u>Let</u> f <u>be a function from</u> $(-A,A)$ <u>to</u> $\mathbb{C}$. <u>Then</u> $f\in\mathcal{R}_A^c$ <u>iff it admits the representation</u>

$$f(\lambda)=\int_{-\infty}^{\infty}e^{i\lambda t}\,d\sigma(t)\quad\forall\lambda\in(-A,A),\tag{1}$$

<u>where</u> σ <u>is a complex measure on</u> $\mathbb{R}$ <u>of bounded variation</u> $Var\,\sigma<\infty$. <u>Moreover every</u> $f\in\mathcal{R}_A^c$ <u>admits representation (1) with</u> $Var\,\sigma= =\|f\|_A$.

Clearly it follows from this theorem, that every function $f\in\mathcal{R}_A^c$ $(0<A<\infty)$ admits an extension $\tilde{f}\in\mathcal{R}_\infty^c$ with $\|\tilde{f}\|_\infty= =\|f\|_A$. On the other hand every $f\in\mathcal{R}_A$ $(0<A<\infty)$ admits an extension $\tilde{f}\in\mathcal{R}_A$ with $\|\tilde{f}\|_\infty=\|f\|_A$ by the Hahn-Banach theorem.

QUESTION 1. <u>Does every</u> $f\in\mathcal{R}_A^m$ $(0<A<\infty)$ <u>admit an extension</u> $\tilde{f}\in\mathcal{R}_\infty^m$?

QUESTION II. <u>If every $f \in \mathcal{R}_A^m$ has an extension $\tilde{f} \in \mathcal{R}_\infty^m$, does there exist for this f such an extension $\tilde{f}_0$ that $\|\tilde{f}_0\|_\infty = \|f\|_A$?</u>

QUESTION III. <u>Does every $f \in \mathcal{R}_A^m$ $(0<A\leqslant\infty)$ admit a decomposition $f = f_c + f_m$, where $f_c \in \mathcal{R}_A^c$, $f_m \in \mathcal{R}_A^m$ and f_m equals zero a.e.?</u>

2. A functional $\Phi \in \mathbb{P}_A^*$ is said to be real if it takes real values on real $\mathcal{P}$'s, $\mathcal{P} \in \mathbb{P}_A$. A $\Phi \in \mathbb{P}_A^*$ is real iff its χ-function f $(\Phi = \Phi_f)$ is Hermitian: $f(-\lambda) = \overline{f(\lambda)}$ $\forall \lambda \in (-\infty,\infty)$. Every functional $\Phi \in \mathbb{P}_A^*$ admits a unique decomposition $\Phi = \Phi_R + i\Phi_J$, where $\Phi_R, \Phi_J (\in \mathbb{P}_A^*)$ are real. Therefore, it is easy to see, that in Questions I, II, III we may restrict ourselves to the case of Hermitian f only.

Denote by $\mathbb{K}_A^*$ $(0 < A \leqslant \infty)$ the cone of all non-negative $\Phi \in \mathbb{P}_A^*$. Naturally, a Φ $(\in \mathbb{P}_A^*)$ is said to be non-negative if $\Phi(\mathcal{P}) \geqslant 0$ as soon as $\mathcal{P}(t) \geqslant 0$ $\forall t \in (-\infty,\infty)$ $(\mathcal{P} \in \mathbb{P}_A)$.

Denote by $\mathcal{P}_A$ $(0<A\leqslant\infty)$ the cone of all χ-functions corresponding to the elements of $\mathbb{K}_A^*$. The subcone of all continuous (measurable) $f \in \mathcal{P}_A$ will be denoted by $\mathcal{P}_A^c$ $(\mathcal{K}_A^m)$.

It is easy to see that for every $f \in \mathcal{P}_A$ we have: $f(0) = \|f\|_A$. For any Hermitian $f \in \mathcal{R}_A$ there exists a decomposition $f = f_+ - f_-$, where $f_\pm \in \mathcal{P}_A$ and $\|f\|_A = f_+(0) + f_-(0)$.

To establish the last assertion it is sufficient to do this for $A = \infty$. In this case $\mathbb{P}_\infty$ forms a linear dense set in the Banach space $\mathbb{B}$ of all almost periodic Bohr functions with sup-norm. The cone $\mathbb{K}_\infty^*$ is dual to the cone $\mathbb{K}$ of all non-negative functions in $\mathbb{B}$. As $\mathbb{B}$ may be identified with a Banach space of all continuous functions on a compact space, the existence of the required decomposition for Hermitian $f \in \mathcal{P}_\infty$ follows. Moreover, this decomposition is unique and minimal in this case $(A=\infty)$.

If $f \in \mathcal{R}_A^c$ is Hermitian, then there exists a decomposition $f = f_+ - f_-$ with $f_\pm$ $\mathcal{P}_A^c$ and $\|f\|_A = f_+(0) + f_-(0)$. Indeed, for Hermitian $f \in \mathcal{R}_A^c$ we can obtain (1) with a real measure σ, $\mathrm{Var}\,\sigma = \|f\|_A$, admitting a unique decomposition $\sigma = \sigma_+ - \sigma_-$ with nonnegative measures $\sigma_\pm$ such that $\mathrm{Var}\,\sigma = \mathrm{Var}\,\sigma_+ + \mathrm{Var}\,\sigma_-$. This decomposition yields (via (1)) the required decomposition for f.

QUESTION IV. <u>Does every Hermitian $f \in \mathcal{R}_A^m$ admit a decomposition $f = f_+ - f_-$ with $f_\pm \in \mathcal{P}_A^m$</u> ?

It turns out, that the affirmative answer to this question implies the same for Question III. This connection is due to a theorem of [4] according to which every function $g \in \mathcal{P}_A^m$ admits a decomposition $g = g_c + g_m$, where $g_c \in \mathcal{P}_A^c$, $g_m \in \mathcal{P}_A^m$ and g_m equals zero a.e. This theorem has been generalized recently in [5].

It is plausible, that for any Hermitian $f \in \mathcal{R}_A^m$ in the decomposition $f = f_+ - f_-$ with $f_\pm \in \mathcal{P}_A$, $\|f\|_A = f_+(0) + f_-(0)$ (which always exists) automatically $f_\pm \in \mathcal{P}_A^m$.

3. For better orientation we will indicate that A.P.Artёmenko [6,7] has obtained a general proposition which contains, in particular, the following characterization of functions $f \in \mathcal{P}_A$.

THEOREM 2. Let $0 < A \leqslant \infty$ and let f be a function from $(-A, A)$ to $\mathbb{C}$. Then $f \in \mathcal{P}_A$ iff for any $\lambda_j \in [0, A]$ and any $\xi_j \in \mathbb{C}$ $(j = 1, 2, \ldots, n;\ n = 1, 2, \ldots)$:

$$\sum_{j,k=1}^{n} f(\lambda_j - \lambda_k)\xi_j \bar{\xi}_k \geqslant 0 .$$

The necessity of this condition is trivial. A transparent proof of it's sufficiency has been obtained by B.Ja.Levin [8].

For a continuous function f from $(-A, A)$ to $\mathbb{C}$ the assertion of theorem 2 (for $A = \infty$) is contained in the well-known theorem of Bochner and for $(0<)\ A < \infty$ in the author's corresponding theorem [2].

A series of unsolved problems concerning extensions of functions $f \in \mathcal{P}_A^m$ ($(0 < A < \infty)$ is formulated in [4]. In this connection we also mention [9].

REFERENCES

1. Bochner S. A Theorem on Fourier-Stieltjes Integrals. - Bull.Amer.Math.Soc., 1934, 40, N 4, 271-276.
2. Крейн М.Г. О проблеме продолжения эрмитово положительных непрерывных функций. - Докл.АН СССР, 1940, 26, № 1, 17-21.
3. Крейн М.Г. О представлении функций интегралами Фурье-Стилтьеса. - Ученые записки Куйбышевского ГПИ, 1943, № 7, 123-147.
4. Крейн М.Г. Об измеримых эрмитово-положительных функциях. - Матем.заметки, 1978, 23, № 1, 79-89.

5. L a n g e r H. On measurable Hermitian indefinite functions with a finite number of negative squares. – Acta Sci.Math.Szeged, 1983 (to appear).

6. А р т ё м е н к о А.П. О позитивных линейных функционалах в пространстве почти периодических функций H.Bohr'a . – Сообщ.Харьк. Матем.Об-ва, I940, (4), I6, III-II9.

7. А р т ё м е н к о А.П. Эрмитово положительные функции и позитивные функционалы I. – Теория функций, функц.анал. и их прил., I983 (в печати).

8. Л е в и н Б.Я., Об одном обобщении теоремы Фейера-Рисса. – Докл. АН СССР, I946, 52, 29I-294.

9. C r u m M.M. On positive definite functions, 1956. – Proc. London Math.Soc., 1956, (3) 6, 548-560.

M.G.KREIN
(М.Г.КРЕЙН)

СССР, 270057, Одесса,
ул.Артема I4, кв.6

13.2. SUPPORT POINTS OF UNIVALENT FUNCTIONS

Let $H(\mathbb{D})$ be the linear space of all functions analytic in the unit disk $\mathbb{D}$, endowed with the usual topology of uniform convergence on compact subsets. Set S be the class of functions $f \in H(\mathbb{D})$ which are univalent and normalized by the conditions $f(0)=0$ and $f'(0)=1$. Thus each $f \in S$ has an expansion of the form

$$f(z) = z + a_2 z^2 + a_3 z^3 + \dots, \quad |z| < 1.$$

Let L be a complex-valued continuous linear functional on $H(\mathbb{D})$ not constant on S. Because S is a compact subset of $H(\mathbb{D})$, the functional $\operatorname{Re}\{L\}$ attains a maximum value on S. The extremal functions are called *support points* of S. In view of the Krein-Milman theorem, the set of support points associated with each linear functional L must contain an extreme point of S. It is NOT KNOWN <u>whether every support point is an extreme point, or whether every extreme point is a support point</u>.

The support points of S have a number of interesting properties. It is known that each support point f maps $\mathbb{D}$ onto the complement of an analytic arc Γ which extends with increasing modulus from a point w_0 to ∞, satisfying

$$L\left(\frac{f^2}{f-w}\right) \frac{dw^2}{w^2} > 0. \tag{1}$$

The *radial angle* $\alpha(w) = \arg\left\{\frac{dw}{w}\right\}$ of Γ has the property $|\alpha(w)| \leqslant \frac{\pi}{4}$, $w \neq w_0$. The bound $\pi/4$ is best possible and in fact there are support points for which $|\alpha(w_0)| = \frac{\pi}{4}$. It is also known that $L(f^2) \neq 0$, from which it follows that Γ is asymptotic at infinity to the half-line

$$w = \frac{L(f^3)}{3L(f^2)} - L(f^2)t, \quad t \geqslant 0. \tag{2}$$

An exposition of these properties, with further references to the literature, may be found in [4].

Evidence obtained from the study of special functionals [1,2,,3] suggests the CONJECTURE that <u>the omitted arc</u> Γ <u>always has monotonic argument</u>. This is true for point-evaluation functionals $L(f) = f(\zeta)$, where $\zeta \in \mathbb{D}$; for derivative functionals $L(f)=e^{-i\sigma}f'(\zeta)$; for coefficient functionals $L(f) = a_3 + \lambda a_2$, where $\lambda \in \mathbb{C}$, and of course for coefficient functionals $L(f)=a_n$ with $2 \leqslant n \leqslant 6$, where the Bieberbach conjecture has been proved. A STRONGER CONJECTURE, supported by somewhat less evidence, is that <u>the radial angle</u> $\alpha(w)$ <u>tends monotonically to zero as</u> $w \longrightarrow \infty$ <u>along</u> Γ .

The Bieberbach conjecture asserts that $|a_n| \leqslant n$, with strict inequality for all n unless f is a rotation of the Koebe function $k(z)=z(1-z)^{-2}$. A geometric reformulation is that the arc Γ corresponding to each coefficient functional $L(f)=a_n$ is a radial half-line. It is essentially equivalent to say that the asymptotic half-line (2) is a trajectory of the quadratic differential (1). A weak form of the Bieberbach conjecture is that for each coefficient functional $L(f)=a_n$ the asymptotic half-line is radial. It is interesting to ask what relation this conjecture may bear to other weak forms of the Bieberbach conjecture, such as the asymptotic Bieberbach conjecture and Littlewood's conjecture on omitted values, now known [5] to be equivalent.

REFERENCES

1. Brown J.E. Geometric properties of a class of support points of univalent functions. - Trans.Amer.Math.Soc. 1979, 256, 371-382.
2. Brown J.E. Univalent functions maximizing $\mathrm{Re}\{a_3+\lambda a_2\}$. - Illinois J.Math. 1981, 25, 446-454.
3. Duren P.L. Arcs omitted by support points of univalent functions. - Comment.Math.Helv. 1981, 56, 352-365.
4. Duren P.L. Univalent Functions. Springer-Verlag, New York, 1983.
5. Hamilton D.H. On Littlewood's conjecture for univalent functions. - Proc. Amer. Math. Soc. 1982, 86, 32-36.
6. Pearce K. New support points of S and extreme points of HS. - Proc.Amer.Math.Soc. 1981, 81,425-428.

P.L.DUREN Department of Mathematics University of Michigan
Ann Arbor, Michigan 48109 USA

13.3. MORE PROBLEMS BY ALBERT BAERNSTEIN

Let Ω be a simply connected domain in $\mathbb{C}$ and F a conformal mapping from Ω onto $\mathbb{D}$, normalized by $|F'(a)|=1$ when $F(a)=0$. Hayman and Wu [1] proved that

$$\int_{\mathbb{R}\cap\Omega} |F'(x)|\,dx \leqslant A < \infty$$

for some constant A. A simpler proof has been given by Garnett, Gehring and Jones [2]. Is it true that

$$\int_{\mathbb{R}\cap\Omega} |F'(x)|^p\,dx \leqslant A_p < \infty \tag{1}$$

for some constant A_p when $1<p<2$? The example $f(z)=(1-z)^{1/2}$, $F(z)$ the inverse of f, shows that $\int_{\mathbb{R}\cap\Omega} |F'(x)|^p\,dx=\infty$ is possible when $2\leqslant p<\infty$.

Using the technique of [1] or [2] together with classical harmonic measure estimates, it can be shown that (1) is true for $\frac{2}{3}<p\leqslant 1$. The inverse function of $f(z)=(1-z)^{-2}$ shows that $\int_{\mathbb{R}\cap\Omega} |F'(x)|^p\,dx=\infty$ is possible when $0<p\leqslant\frac{2}{3}$.

Inequality (1) would follow if a symmetrization type inequality, (2) below, is true. Let $\{I_j\}$ be a Whitney type decomposition of $\mathbb{R}\cap\Omega$ as described in [2, §3]. Denote by x_j the center of I_j, ℓ_j the length of I_j, L_j the vertical half line starting from $x_j+i\ell_j$, and let Ω' be the domain obtained by deleting from $\mathbb{C}$ all the half lines L_j. Is it true that

$$g(x,a)\leqslant CG(x,a) \tag{2}$$

for every $x\in\mathbb{R}$ and $a\in\mathbb{C}$ with $\operatorname{Im} a\leqslant 0$, where g and G denote the Green's functions of Ω and Ω' respcetively?

REFERENCES

1. H a y m a n W.K., W u J.-M.G. Level sets of univalent functions. -Comm.Math.Helv. 1981, 56, 366-403.
2. G a r n e t t J.B., G e h r i n g F.W., J o n e s P.W. Conformally invariant length sums.-Indiana Univ.Math.J., to appear.

A.BAERNSTEIN Washington University
St.Louis, MO 63130, USA

3.4. SOME EXTENSION PROBLEMS

Let $K: G \to K(G)$ and $\tilde{K}: G \to \tilde{K}(G)$ associate with each open set $G \subset \mathbb{C} \equiv \mathbb{R}^2$ a class of complex-valued functions on G. A set $E \subset \mathbb{C}$ will be termed negligible $(K, \tilde{K})$ if, for each open set $G \subset \mathbb{C}$ and each $f \in K(G)$, the existence of an open set $\tilde{G} \subset G$ such that $f|\tilde{G} \in \tilde{K}(\tilde{G})$ and $f(G \setminus \tilde{G}) \subset E$ implies that $f \in \tilde{K}(G)$. For the case when $K = C$ (= sheaf of continuous functions) and $K = A$ (= sheaf of holomorphic functions), negligibility of finite sets was established by T.Radó in [4]. P.Lelong showed in [3] that also all polar sets are negligible (C, A).

PROBLEM 1. <u>What are necessary and sufficient conditions for $E \subset \mathbb{C}$ to be negligible</u> (C, A)?

For continuously differentiable functions some related results concerning harmonicity are known. If $\omega \geqslant 0$ is a continuous non-decreasing function on $\mathbb{R}_+$ with $\omega(t) > 0$ for $t > 0$, we denote by $C^\omega(G)$ the class of all functions f on G satisfying the condition

$$|f(u) - f(v)| = O(\omega(|u-v|)) \quad \text{as } |u-v| \downarrow 0$$

locally in G; $C^\omega_*(G)$ will stand for the subclass of all $f \in C^\omega(G)$ enjoying the property

$$|f(u) - f(v)| = o(\omega(|u-v|)) \quad \text{as } |u-v| \downarrow 0$$

locally in G. Further we denote by $C^{1,\omega}(G)$ and $C^{1,\omega}_*(G)$ the classes of all continuously differentiable real-valued functions whose first order partial derivatives belong to $C^\omega(G)$ and $C^\omega_*(G)$, respectively. If $H(G)$ denotes the class of all real-valued functions harmonic on G, then the following result holds (cf. [2]).

THEOREM. <u>A set $E \subset \mathbb{R}$ is negligible $(C^{1,\omega}, H)$ if (and also only if in case $\omega(0) = 0$) the Hausdorff measure corresponding to the measure function ω vanishes on all compact subsets of E. A necessary and sufficient condition for $E \subset \mathbb{R}$ to be negligible $(C^{1,\omega}_*, H)$ consists in σ-finiteness of the Hausdorff mea-</u>

sure corresponding to ω on all compact subsets of E.

For subharmonic functions similar question seems to be open. (Of course, necessity of the corresponding condition follows from the above theorem.) Let $S(G)$ denote the class of all subharmonic functions on G.

CONJECTURE. Any set $E \subset \mathbb{R}$ with vanishing Hausdorff measure corresponding to the measure function ω is negligible $(C^{1,\omega}, S)$. If $E \subset \mathbb{R}$ has σ-finite Hausdorff measure corresponding to ω, then E is negligible $(C_*^{1,\omega}, S)$.

PROBLEM 2. What are necessary and sufficient conditions for $E \subset \mathbb{C}$ to be negligible (C^{ω}, A) or (C_*^{ω}, A)?

Similar questions may be posed for various classes of functions in more general spaces (compare [1]).

REFERENCES

1. C e g r e l l U. Removable singularities for plurisubharmonic functions and related problem. - Proc.London Math.Soc.,1978,XXXVI, 310-336.
2. K r á l J. Some extension results concerning harmonic functions, to appear in J.London Math.Soc.,1983.
3. L e l o n g P. Ensembles singuliers impropres des fonctions plurisousharmoniques - J.Math.Pures Appl.,1957, 36, 263-303.
4. R a d ó T. Über eine nicht fortsetzbare Riemannsche Mannigfaltigkeit - Math.Z. 1924, 20, 1-6.

JOSEF KRÁL

Matematicky ústav ČSAV
Žitná 25
11567 Praha 1
Czechoslovakia

3.5. PARTITION OF SINGULARITIES OF ANALYTIC FUNCTIONS

Let S be a closed set in $\mathbb{R}^2$, and let $A^n(S)$ be the class [o]f functions f, holomorphic in $W=\mathbb{R}^2\setminus S$, such that $f,\dots,f^{(n)}$ [c]an be extended continuously to $W^-=W\cup\partial S$ $(n=0,1,2,\dots)$. [L]et $S=S_1\cup S_2$, each S_i being closed, and moreover $S=S_1^\circ\cup S_2^\circ$, [w]here S_i° is the interior of S_i relative to S.

In this situation IT IS NATURAL TO GUESS

$$A^n(S)=A^n(S_1)+A^n(S_2),\quad (n=0,1,2,3,\dots). \tag{1}$$

[T]o explain the difficulties involved in (1), we suppose that f is [c]ontinuous in $\mathbb{R}^2$ and use the operator $\bar\partial$ defined by $2\bar\partial\equiv\partial/\partial x+i\partial/\partial y$. [F]ollowing the classical method, we choose a function $\varphi\in C^\infty(\mathbb{R}^2)$ such that $0\leqslant\varphi\leqslant 1$ and $\varphi=0$ on $S\setminus S_1^\circ$, $=1$ on $S\setminus S_2^\circ$. [S]uppose that $\bar\partial h=f\bar\partial\varphi$ (in the sense of distributions, or the Cauchy-Green formula). Then $\bar\partial(\varphi f-h)=\varphi\bar\partial f$, and therefore $F=\varphi f-h$ belongs to $A^0(S_1)$; similarly $f-F=(1-\varphi)f+h$ belongs to $A^0(S_2)$. [T]herefore (1) is true for $n=0$ (a classical observation, to be sure) but the reasoning seems to fail when $n=1$ since f is generally not C^1 (or even Hölder-continuous) on $\mathbb{R}^2$.

If (1) were true (for some $n\geqslant 1$) it would imply that the triviality of $A^n(S)$ is a local property of S. (Triviality of $A^n(S)$ means of course that all of its elements are restrictions to W of entire functions.) Even this much is unknown.

R.KAUFMAN

Dept.Math.
Univ.of Illinois
Altegeld Hall
Urbana, Illinois 61801
USA

13.6. REARRANGEMENT-INVARIANT HULLS OF SETS

Let (S,Σ,μ) be a non-atomic finite measure space. Denote by $\Omega_\varepsilon\,(\varepsilon>0)$ a family consisting of all μ-preserving invertible transformations $\omega: S \to S$ such that $\mu\{x\in S : \omega(x)\neq x\}\leqslant \varepsilon$. Each $\omega\in\Omega_\varepsilon$ generates a linear operator $T_\omega : \mathfrak{M} \to \mathfrak{M}$ (where $\mathfrak{M}$ denotes the space of all measurable functions on S) by the formula $T_\omega f(x) = f(\omega(x)),\ x\in S,\ f\in\mathfrak{M}$. The elements of a set $R_\varepsilon \overset{\text{def}}{=} \{T_\omega : \omega\in\Omega_\varepsilon\}$ are called the ε-rearrangements. Each T_ω preserves the distribution of a function, hence the integrability properties of functions are also preserved.

Given a subset A of $\mathfrak{M}$ define the r e a r r a n g e - m e n t - i n v a r i a n t h u l l s of A as follows:

$$RH_0(A) \overset{\text{def}}{=} \bigcap_{\varepsilon>0} R_\varepsilon(A), \quad RH(A) = \bigcup_{\varepsilon>0} R_\varepsilon(A).$$

The general problem of characterization of such hulls for a given concrete set A has been posed by O.Cereteli. We refer to [5] for the contribution of O.Cereteli to the solution in some concrete cases.

The following results have been obtained in [2] and [3]. Consider for the simplicity the case when S is $[0,1]$ equipped with the usual Lebesgue measure.

a) Let $\Phi=\{\varphi_n\},\ n\geqslant 1$ be a family of bounded functions such that for any $f\in L^\infty$

$$\sum |c_n(f;\Phi)|^2 \leqslant \text{const}\ \|f\|_2^2$$

with a constant independent of f. Here $c_n(f,\Phi)=\int f\varphi_n dx,\ n\geqslant 1$. For a non-negative sequence $\{\rho_n\},\ n\geqslant 1$, such that $\rho_n\to 0$ when $n\to\infty$, define a class

$$A \overset{\text{def}}{=} \{f\in L^1 : \sum [c_n(f;\Phi)^*]^2\rho_n < \infty\},$$

where $c_n(f;\Phi)^*$ denotes the non-increasing rearrangement of $\{|c_n(f,\Phi)|\}$. Then $RH_0(A) = RH(A) = L^1$ [2].

b) Let $\Phi=\{\varphi_n\}$ be a complete orthonormal family of bounded functions in L^2. For given $p,\ 1\leqslant p<2$, define a class

$$A \overset{\text{def}}{=} \{ f \in L^1 : \| f - (S_n f) \|_p \to 0 \},$$

where $S_n f$ denotes the n-th partial sum of the Fourier series of f with respect to Φ. Then $RH_0(A) = RH(A) = L^1$ [3], i.e. any complete orthonormal family of bounded functions is in some sense a basis in L^p, $1 \leq p < 2$, modulo rearrangements.

A different effect occurs for the class $A = \{ f \in L^1 : \tilde{f} \in L^1 \}$, where L^1 is taken over the unit circle $\mathbb{T}$ and $\tilde{f}$ denotes the conjugate function of a function f. In that case [4, 5]

$$RH_0(A) = RH(A) = \{ f \in L^1 : \int_1^\infty \frac{dt}{t} \left| \int_{\mathbb{T}} {}^t f(x)\, dx \right| < \infty \}. \tag{1}$$

Here

$${}^t f(x) = \begin{cases} f(x), & |f(x)| > t \\ 0, & |f(x)| \leq t. \end{cases}$$

The class

$$M \ln^+ M \overset{\text{def}}{=} \{ f \in L^1 : \int_1^\infty \frac{dt}{t} \left| \int_{\mathbb{T}} {}^t f(x)\, dx \right| < \infty \}$$

arising in (1) (in [4, 5] this class is denoted by Z) coincides, on non-negative functions, with the class $L \ln^+ L$. Moreover, $L \ln^+ L \subset M \ln^+ M$.

In addition to (1) it has been proved in [3] that if $A \overset{\text{def}}{=} \{ f \in L^1 : \| \tilde{f} - \tilde{S}_n(f) \|_1 \to 0 \}$ then

$$RH_0(A) = RH(A) = M \ln^+ M,$$

i.e. any function from $M \ln^+ M$ can be rearranged on a set of small measure so that the obtained function has L^1-convergent conjugate trigonometrical series.

For any p, $p > 1$, define a class M^p over (S, Σ, μ) as follows

$$M^p \overset{\text{def}}{=} \{ f \in L^1 : \int_0^\infty t^{p-2}\, dt \left| \int_S {}^t f(x)\, dx \right| < \infty \}.$$

It is clear that $L^p \subset M^p$ and M^p coincides with L^p on non-negative functions. The class M^p in comparison with L^p takes into account not only the degree of integrability of function but also the degree of cancellation of the positive and negative values of the function. It is known [6] that $M\ln^+ M$ is linear. As for M^p, it has been proved in [3] that 1) M^p, $p>2$, is non-linear, moreover, there exists $f \in M^p$ such that $f+1 \notin M^p$;
2) $M^p + L^\infty \subset M^p$, $1<p\leq 2$.

PROBLEM 1. <u>Is the class</u> M^p <u>linear for</u> $1<p\leq 2$?

Consider a family $\Phi = \{\varphi_n\}$ such as in b). Denote by S^* the maximal operator for the sequence of partial sum operators $\{S_n\}$ with respect to Φ, i.e. $S^* f = \sup_n |S_n f|$, $f \in L^1$. The problem of finding the rearrangement invariant hulls of a set $A = \{f \in L^1 : S^* f \in L^1\}$ is not solved even for classical families Φ. Some partial results have been obtained in [3]. For the trigonometrical system $\{\exp inx\}$, $-\infty < n < +\infty$, on $\mathbb{T}$ the following inclusion holds:

$$M^p \subset RH_0(A), \quad p>1.$$

PROBLEM 2. <u>Find</u> $RH_0(A)$ <u>and</u> $RH(A)$ <u>for the trigonometrical system</u>.

Very interesting is the case when Φ is the family of Legendre polynomials $\{L_n\}$ on the interval $[-1,+1]$. It is true [3] that

$$M^p \subset RH_0(A), \quad p>\frac{8}{7}. \tag{2}$$

PROBLEM 3. <u>Find</u> $RH_0(A)$ <u>and</u> $RH(A)$ <u>for the family of Legendre polynomials</u>.

We pose also two easier problems related to Problem 3.

PROBLEM 3'. <u>Is the inclusion</u> $L^{8/7} \subset RH_0(A)$ <u>true</u>?

PROBLEM 3''. <u>Is the inclusion</u> $M^{8/7} \subset RH_0(A)$ <u>true</u>?

The number $\frac{8}{7}$ in (2) appears from the general theorem proved in [3]. The theorem states that if a sequence of integral operators $\{T_n\}$, $n\geq 1$, has a localization property in L^∞, and the maximal operator $T^* f = \sup |T_n f|$ has a weak type (p,p) with some $p>1$, then

$$M^{\frac{2p}{p+1}} \subset RH_0(A), \qquad (3)$$

where $A = \{ f \in L^1 : T^* f \in L^1 \}$. It is not known, whether the power $\frac{2p}{p+1}$ in (3) is sharp on the whole class of the operators T^* under consideration. The maximal operator with respect to the Legendre polynomial system has weak type (p, p), $4/3 < p < 4$ [1]. The number $\frac{8}{7}$ is the value of $\frac{2p}{p+1}$ at $p = \frac{4}{3}$.

The problems analogous to Problems 3, 3', 3'' can also be formulated for Jacobi polynomial systems.

REFERENCES

I. Б а д к о в В.М. Сходимость в среднем и почти всюду рядов Фурье по многочленам, ортогональным на отрезке.- Матем. сборн., I974, 95, № 2, 229-262.

2. Г у л и с а ш в и л и А.Б. Об особенностях суммируемых функций. - Зап. научн. семин. ЛОМИ, I98I, II3, 76-96.

3. Г у л и с а ш в и л и А.Б. Перестановки, расстановки знаков и сходимость последовательностей операторов.- Зап. научн.семин. ЛОМИ, I982, I07, 46-70.

4. Ц е р е т е л и О.Д. О сопряженных функциях. - Матем. заметки, I977, 22, № 5, 77I-783.

5. Ц е р е т е л и О.Д. О сопряженных функциях. - Докторская диссертация, Тбилиси, I976.

6. Ц е р е т е л и О.Д. Об одном случае суммируемости сопряженных функций.- Труды Тбилисского матем. ин-та АН Груз.ССР, I968, 34, I56-I59.

A.B.GULISASHVILI

(А.Б.ГУЛИСАШВИЛИ)

СССР, 380093, Тбилиси,

ул.Рухадзе

Математический институт

АН Грузинской ССР

13.7. NORMS AND EXTREMALS OF CONVOLUTION OPERATORS ON SPACES OF ENTIRE FUNCTIONS

Given a compact subset $K\subset\mathbb{R}^n$ let $B(K)$ be the Bernstein class of all bounded functions f on (the dual copy) $\mathbb{R}^n$ with Fourier transform $\hat{f}$ supported on K. In fact, every function $f\in B(K)$ can be extended to an entire function of exponential type on $\mathbb{C}^n$. The linear space $B(K)$ with the uniform norm on $\mathbb{R}^n$ is a Banach space (in fact, a dual Banach space).

EXAMPLE. Let K be the unit ball in $\mathbb{R}^n$, i.e.,

$$K=\left\{\xi\in\mathbb{R}^n:\ \sum_1^n|\xi_k|^2\leqslant 1\right\}.$$

Then $f\in B(K)$ if and only if the function f is a restriction to $\mathbb{R}^n\subset\mathbb{C}^n$ of an entire function on $\mathbb{C}^n$ satisfying

$$|f(z)|\leqslant C\exp|\operatorname{Im} z|,\quad z\in\mathbb{C}^n,$$

for some constant C.

We shall consider operators $T:\ B(K)\to B(K)$ of the form

$$(T_\mu f)(x)=\int_{\mathbb{R}^n} f(x-t)\,d\mu(t),$$

μ being a complex-valued regular Borel measure of bounded variation on $\mathbb{R}^n$. In other words, $T_\mu f=f*\mu$. The function $\tau=\hat{\mu}\,|\,K$ is said to be the s y m b o l of $T=T_\mu$ The representation $\mu\to T_\mu$ is not an isomorphism, but nevertheless the symbol τ is uniquely determined by T. The spectrum of T coincides with the range of τ and its norm with the norm of the functional $f\to(Tf)(0)$. If K is a set of spectral synthesis then the symbol τ uniquely determines the corresponding operator $T=\tau(D)$, $D=\frac{1}{i}\partial/\partial x$ in $B(K)$. Moreover, in this case

$$\|T\|=\inf\{\|\nu\|:\hat{\nu}\,|\,K=\tau\}.$$

DEFINITION. A normal extremal for T is any element $f\in B(K)$ such that $\|f\|=1$, $(Tf)(0)=\|T\|$.

It can be easily shown that the normal extremals always exist (and form a convex set). For example, in the case $n=1, K=[-\sigma,\sigma]$, $\sigma>0$, and $Tf=f'$ the classical result asserts that $\|T\|=\sigma$ and all normal extremals have the following form $a\exp i\sigma x+ b\exp(-i\sigma x)$.

A measure ν is called extremal for T if $Tf= =\nu * f, f\in B(K)$ and $\|T\|=\|\nu\|$.

The set of extremal measures may be empty even in case of finite K.

Such problems as calculation of norms and discription of extremals go back to the classical papers of S.N.Bernštein A survey of results obtained in the field up to the middle of 60-ies can be found in [1]. For additional aspects of the topic see [2], which is, unfortunately, flooded by misprints, so be careful.

A compact set K in $\mathbb{R}^n$ is said to be a star if with every ξ it contains $\rho\xi$ for each $\rho\in[0,1]$. Every star K is a set of spectral synthesis and $B(K)$ contains sufficiently many functions vanishing at infinity. If $\xi_0\in K$ and $\sup_{\xi\in K}|\tau(\xi)|=\tau(\xi_0)=1$ then $\|\tau(D)\|=1$ (i.e. the norm of $\tau(D)$ coincides with its spectral radius) if and only if the function $\xi\longrightarrow\tau(\xi+\xi_0)$ admits a positive definite extension to $\mathbb{R}^n$. The operator $\tau(D)$ has extremal measures. If $\|\tau(D)\|>1$ then every extremal measure is supported on a proper analytic subset of $\mathbb{R}^n$ and the extremal measure is unique provided $n=1$. For $n>1$ **the uniqueness does not hold (example: K is the unit ball and $\tau(D)$ is the Laplace operator).**

PROBLEMS IN THE ONE-DIMENSIONAL CASE.

Every polynomial (in one variable) is related to a wide stock of positive definite functions. Suppose that the zeros of a polynomial τ are placed in the half-plane $\operatorname{Re}\zeta\geqslant\frac{1}{2}\sigma>0$ and that $\tau(0)=1$. Then the restriction of τ to $[0,\sigma]$ **extends** to a positive definite function on $\mathbb{R}$. It follows that **for all linear** τ

$$\|\tau(D)\|_{B_\sigma}=max\{|\tau(\xi)|:\ \xi\in[-\sigma,\sigma]\},$$

where $B_\sigma\overset{def}{=}B([-\sigma,\sigma])$. In this case all normal extremals can be easily determined and there exists an extremal measure (at least one). At the same time <u>for polynomials</u> τ <u>of degree 2 these</u>

problems still do not have a full solution. The simplest operator is provided by $y \to -y'' + \lambda y$, $\lambda \in \mathbb{C}$. For $\lambda \in \mathbb{R}$ the problems are solved (see papers of Boas-Shaeffer, Ahiezer and Meiman). For some complex λ (in particular those for which the zeros of the symbol τ satisfy the above mentioned condition) τ admits (after a proper normalization) a positive definite extension, so that the norm of $\tau(D)$ coincides with its spectral radius.

Is it possible to calculate the norm for all $\lambda \in \mathbb{C}$? How do the "Euler equations" look in this case?

Note that according to the Krein theorem the extremal measure is unique provided $K=[-\sigma,\sigma]$ and the spectral radius is less than the norm. Of course, these problems remain open for polynomials of higher degree.

The Bernštein inequality for fractional derivatives leads to the following PROBLEM. Consider on $[-1,1]$ the function $\tau(\xi)=(1-|\xi|)^{\alpha}$, $\alpha>0$. The problem is to find

$$\inf\{\|\mu\|_{M(\mathbb{R})} : \mu \geqslant 0,\ \hat{\mu} \mid [-1,1] = \tau\}.$$

If $\alpha \geqslant 1$ then τ is even and is convex on $[0,1]$, and therefore coincides on $[-1,1]$ **with a restriction of a positive definite** function by the Polya theorem. For $\alpha<1$ τ becomes concave on $[0,1]$ and moreover τ cannot be extended to a positive definite function on $\mathbb{R}$. Indeed, if φ is positive definite then $-\varphi''$ is a positive definite distribution. At the same time, $-\tau''$ is non-negative, locally integrable on a neighbourhood of zero and non-integrable on a left neighbourhood of the point $\xi=1$. A positive definite function cannot satisfy this list of properties.

The best known estimate of the norm for $\alpha \in (0,1)$ is $2(1+\alpha)^{-1}$. It is evidently not exact but it is asymptotically exact for $\alpha \to 0$ and $\alpha \to 1$. It should be noticed that in the space of trigonometric polynomials of degree $\leqslant m$ the norm of the operator of fractional differentiation coincides with its spectral radius for $\alpha \geqslant \alpha_0$, where $\alpha_0 = \alpha_0(m) < 1$.

Another example is related to the family $\{\tau_\alpha\}$ of functions $\tau_\alpha(\xi) = 1-|\xi|^{\alpha}$, $\alpha>0$, $-1 \leqslant \xi \leqslant 1$. For $\alpha \leqslant 1$ they are positive definite. Consider the family for $\alpha>1$. Since every characteristic function φ satisfies the unequality $|\varphi(t)|^2 \leqslant \frac{1}{2}(1+|\varphi(2t)|)$, here are no positive definite extensions for $\alpha \geqslant 2$.

Consider now the case $1 < \alpha < 2$. The following idea has been suggested by A.V.Romanov Extend $\tau(\xi)$ to $(1,2)$ by the formula

$$\tau(1+\xi) = -\tau(1-\xi), \quad 0<\xi<1 .$$

Extend now the obtained function on $(0,2)$ to an even periodic function of period 4 keeping the same notation τ for this function. We have

$$\tau(\xi) = \sum_k a_k \cos\frac{\pi k}{2}\xi, \quad \xi \in \mathbb{R},$$

where the sum is taken over odd positive integers. It is easily verified (integration by parts) that a_k and

$$b_k = \int_0^{\frac{k\pi}{2}} \frac{\cos t}{t^{2-\alpha}}\, dt$$

are of the same sign. Clearly $b_1 > 0$ and $b_k \geqslant b_3$ for $k \geqslant 3$. Hence τ is positive definite if

$$\rho(\alpha) = \int_0^{\frac{3\pi}{2}} \frac{\cos t}{t^{2-\alpha}}\, dt \geqslant 0.$$

(cf. [4], Ch.V, Sec.2.29). The function $\rho(\alpha)$ decreases on $(1,2)$ and $\rho(1)=\infty$, $\rho(2)<0$. Therefore the equation $\rho(\alpha)=0$ has a unique solution $\alpha_0 \in (1,2)$ (Romanov's number). The function τ is positive definite on $[-1,1]$ if $\alpha \leqslant \alpha_0$. At the same time slightly modifying the arguments from [5], Th.4.5.2 one can easily show that for $0<\alpha_1<\alpha_2<2$ the function

$$\frac{1-|\xi|^{\alpha_1}}{1-|\xi|^{\alpha_2}}$$

is positive definite on $\mathbb{R}$. Hence τ_{α_1} is positive definite on $[-1,1]$ if so is τ_{α_2} . The "separation point" β_0 is clearly $\geqslant \alpha_0$

Is it true that $\beta_0 > \alpha_0$? For $\alpha > \beta_0$ the above problems remain open.

PROBLEMS IN THE MULTIDIMENSIONAL CASE.

For $n \geq 2$, except the case when the norm of $\tau(D)$ coincides with its spectral radius, very few cases of exact calculation of $\|\tau(D)\|$ and discription of extremals are known. The GENERAL PROBLEM here is to obtain proper generalizations of Boas-Schaeffer's and Ahiezer-Meiman's theorems, i.e. to obtain "Euler's equations" at least for real functionals. Our problems concern concrete particular cases; however it seems that the solution of these problems may throw a light on the problem as a whole.

If $\psi : \mathbb{R}^n \to \mathbb{R}$ is a linear form then the operator with the symbol $\psi \mid K$ is hermitian on $B(K)$ and hence its norm coincides with the spectral radius. This again will be the case for some operators with the symbol of the form $(p \circ \psi) \mid K$ where p is a polynomial. The following simple converse statement is true . If τ is a polynomial and

$$\|\tau(D)\|_{B(K)} = \max\{|\tau(\xi)| : \xi \in K\}$$

for every symmetric convex star K then $\tau = p \circ \psi$ where ψ is a linear form and p is a polynomial.

Does the similar converse statement hold when K ranges over the balls?

Let

$$K_p = \{\xi \in \mathbb{R}^n : (\sum |\xi_k|^p)^{1/p} \leq 1\},$$

where $p \geq 1$ and $\tau(\xi) = -\sum_1^n \xi_k^2$. The operator $\tau(D)$ is obviously the Laplacian Δ . The norm of Δ coincides with the spectral radius in the following cases: $n=1$; $n=2, p=1$; $n \geq 2, p=\infty$. The proof is based on the following well known fact: if μ is a probability measure then $\{\xi : |\hat{\mu}(\xi)| = 1\}$ is a subgroup. The case $p=2$ turns out to be the most pathological and perhaps the most interesting. We have $\|\Delta\|_{B(K_2)} = n$. In this case extremal measure is not unique and it would be interesting to describe all extremal measures (notice that they form a compact convex set). The problem of calculation of the norm can be reduced to the one-dimensional case for operators of the form $p(\Delta)$ in $B(K_2)$. It is possible to calculate the norms by operators with linear symbol in the space $B(K_2)$ explicitly. For example, the norm of Cauchy-Riemann operator equals 2 and its normal extremal is unique. Namely,

$$(x_1 - ix_2)\frac{\sin(x_1^2+x_2^2)^{1/2}}{(x_1^2+x_2^2)^{1/2}}$$

However, for the operators of the second order the things are more complicated. If the symbol $\tau(\xi) = (A\xi, \xi)$ is a positive real quadratic form then $\|\tau(D)\|_{B(K_2)} = \operatorname{tr} A$, the spectral radius of $\tau(D)$ coincides with that of A and normal extremals are of the form

$$\cos|x| + i(a, x)\frac{\sin|x|}{|x|} ,$$

where $|x|$ is the norm, $a \in \mathbb{R}^n$, $|a| \leq 1$.

At the same time <u>nothing is known about extremals and norm of the operator</u> $\frac{\partial^2}{\partial x_1 \partial x_2}$ <u>in</u> $B(K_2)$ <u>for</u> $n=2$.

REFERENCES

1. А х и е з е р Н.И. Лекции по теории аппроксимации. Москва, Наука, 1965.
2. Г о р и н Е.А. Неравенства Бернштейна с точки зрения теории операторов. - Вестн.Харьк.ун-та, № 205. Прикладная математика и механика, вып.45. - Харьков, Вища школа, Изд-во Харьк.ун-та, 1980, 77-105.
3. Г о р и н Е.А., Н о р в и д а с С.Т. Экстремали некоторых дифференциальных операторов. - Школа по теории операторов в функциональных пространствах, Минск, 4-11 июля 1982. Тезисы докл., 48-49.
4. Z y g m u n d A., Trigonometric Series, vol.I. Cambr Univ.Press, 1959.
5. L u k a c s E . Characteristic functions, 2 nd ed., Griffin, London, 1970.

E.A. GORIN
(Е.А.ГОРИН)

СССР, 117233, Москва,
Ленинские горы, МГУ
Механико-математический факультет

13.8. old ALGEBRAIC EQUATIONS WITH COEFFICIENTS IN COMMUTATIVE BANACH ALGEBRAS AND SOME RELATED PROBLEMS

The proposed questions have arisen on the seminar of V.Ya.Lin and the author on Banach Algebras and Complex Analysis at the Moscow State University.

In what follows A is a commutative Banach algebra (over $\mathbb{C}$) with unity and connected maximal ideal space M_A. For $a \in A$, $\hat{a}$ denotes the Gelfand transform of a.

A polynomial $p(\lambda) = \lambda^n + a_1\lambda^{n-1} + \ldots + a_n$, $a_i \in A$ is said to be *separable* if its discriminant d is invertible(i.e. for every ξ in M_A the roots of $\lambda^n + \hat{a}_1(\xi)\lambda^{n-1} + \ldots + \hat{a}_n(\xi)$ are simple); p is said to be *completely reducible* if it can be expanded into a product of polynomials of degree one. The algebra is called *weakly algebraically closed* if all separable polynomials of degree greater than one are reducible over it.

In many cases there exist simple (necessary and sufficient) criteria for all separable polynomials of a fixed degree n to be completely reducible. A criterion for $A = C(X)$, with a finite cell complex X, consists in triviality of all homomorphisms $\pi_1(X) \longrightarrow B(n)$, $B(n)$ being the Artin braid group with n threads [1]. If (and only if) $n \leqslant 4$ this is equivalent to $H^1(X, \mathbb{Z}) = 0$ (which is formally weaker). The criterion fits as a *sufficient* one for arbitrary arcwise connected locally arcwise connected spaces X.

It can be deduced from the implicit function theorem for commutative Banach algebras that if the polynomial with coefficients $\hat{a}_i$ is reducible over $C(M_A)$ then the same holds for the original polynomial p over A. On the other hand (cf.[2],[3]) for arbitrary integers k, n, $4 < k \leqslant n < \infty$ there exists a pair of uniform algebras $A \subset B$, with the same maximal ideal space, such that $\dim B/A = 1$, all separable polynomials of degree $\leqslant n$ are reducible over A, but there exists an irreducible (over B) separable polynomial of degree k.

WE INDICATE A CONSTRUCTION OF SUCH A PAIR. Let G_k be the collection of all separable polynomials $p(\lambda) = \lambda^k + z_1\lambda^{k-1} + \ldots + z_k$ with complex coefficients $z_1, \ldots, z_k$, endowed with the complex structure induced by the natural embedding into $\mathbb{C}^k$, $p \mapsto (z_1, \ldots, z_k)$. Define X as the intersection of G_k, the submanifold $\{z_1 = 0$,

$d(z)=1\}$ and a ball $\{z : \|z-z^0\| \leq C(n) < \infty\}$. Note that X is a finite complex. The algebra B is the uniform closure on X of polynomials in $z_1,\dots,z_k$ and A consists of all functions n B with an appropriate directional derivative at an appropriate oint equal to zero. With the parameters properly chosen, (A,B) is pair we are looking for (the proof uses the fact that the set of olomorphic functions on an algebraic manifold which do not take alues 0 and 1 is finite, as well as some elementary facts of Morse heory and Montel theory of normal families that enable to control he Galois group).

Do there exist examples of the same nature with A weakly algebraically closed? We do not even know any example in which A is weakly algebraically closed and $C(M_A)$ is not. A refinement of the construction described in [4] and [5] may turn out to be sufficient.

If X is an arbitrary compact space such that the division by 6 is possible in $H^1(X,\mathbb{Z})$ then all separable polynomials of degree 3 are reducible over $C(X)$. The situation is more complicated for polynomials of degree 4: there exists a metrizable compact space X of dimension two such that $H^1(X,\mathbb{Z})=0$ but some separable polynomial of degree 4 is irreducible over $C(X)$ [6]. On the other hand, the condition that all elements of $H^1(X,\mathbb{Z})$ are divisible by $n!$ is necessary and sufficient for all separable polynomials of degree $\leq n$ to be completely reducible, provided X is a homogeneous space of a connected compact group (and in some other cases). These type's results are of interest, e.g., for the investigation of polynomials with almost periodic coefficients.

Is it possible to describe "all" spaces X (not necessarily compact) for which the problem of complete reducibility over $C(X)$ of the separable polynomials can be solved in terms of one-dimensional cohomologies? In particular, is the condition $H^1(X,\mathbb{Z})=0$ sufficient in the case of a (compact) homogeneous space of a connected Lie group? (Note that the answer is affirmative for the homogeneous spaces of c o m p l e x Lie groups and for the polynomials with h o l o m o r p h i c coefficients [9]).

Though the question of complete reducibility of separable polynomials in its full generality seems to be transcendental, there is an encouraging classical model, i.e. the polynomials with holomorphic coefficients on Stein (in particular algebraic) manifolds. Note that the known sufficient conditions [9] for holomorphic polynomials are essentally weaker than in general case.

The peculiarity of holomorphic function algebras is revealed in a very simple situation. Consider the union of m copies of the annualus $\{z: R^{-1} < |z| < R\}$ identified at the point $z=1$. It can be shown that a separable polynomial of prime degree n with coefficients holomorphic on these space, and with discriminant $d=1$ is reducible if $R \geqslant R_0(m,n)$, primarity of n being essential for $m \geqslant 2$ [10]. If $m=1$, n can be arbitrary [2], and we denote by $R_0(n)$ the corresponding least possible constant. Now if n is even then $R_0(n)=1$, and so the holomorphity assumption is superfluous. However $R_0(k\ell) \geqslant C(k)^{k\ell}$ if k and ℓ are odd, with $C(k)>1$ for $k \geqslant 3$. At the same time $R_0(n) \leqslant C^n$ for all n . These results, as well as the fact that $R_0(p)^{1/p} \to 1$ as p tends to infinity along the set of prime numbers, have been proved in [10]. <u>Nevertheless the exact asymptotic of</u> $R_0(p)$ <u>remains unkown, it is unknown even whether</u> $R_0(p) \to \infty$ <u>as</u> $p \to \infty$.

If X is a finite cell complex with $H^1(X,\mathbb{Z})=0$ then each completely reducible separable polynomial over $C(X)$ is homotopic in the class of all such polynomials to one with constant coeffinients (the reason is that $\pi_q(G)=0$ for $q>1$). Let $X=M_A$ and consider a polynomial completely reducible over A .

<u>Is it possible to realize the homotopy within the class of polynomials over</u> A ?

Such a possibility is equivalent, as a matter of fact, to the holomorphic contractibility of the universal covering space $\widetilde{G}_n$ for G_n . It is known [11] that $\widetilde{G}_n = \mathbb{C}^2 \times V^{n-2}$, V^{n-2} being a bounded domain of holomorphy in $\mathbb{C}^n$ homeomorphic to a cell [12]. In $\mathbb{C}^n$ there are contractible but non-holomorphically contractible domains [12], though examples of bounded domains of such a sort seem to be unknown (that might be an additional reason to study the above question). Evidently $\widetilde{G}_3 = \mathbb{C}^2 \times \mathbb{D}$ is holomorphically contractible.

<u>Is the same true for</u> $\widetilde{G}_n$ <u>with</u> $n \geqslant 4$?

There are some reasons to consider also transcendental equations $f(w)=0$, where $f: A \to A$ is a Lorch holomorphic mapping (i.e. f is Fréchet differentiable and its derivative is an opera-

or of multiplication by an element of A). In [13] the cases when equations of this form reduce to albebraic ones have been treated (in this sence the standard implicit function theorem is nothing but a reduction to a linear equation). A systematic investigation of such trancendental equations is likely to be important. This might require to invent various classes of Artin braids with an infinite set of threads.

REFERENCES

1. Горин Е.А., Лин В.Я. Алгебраические уравнения с непрерывными коэффициентами и некоторые вопросы алгебраической теории кос. - Матем.сб., 1969, 78, 4, 579-610.
2. Горин Е.А., Лин В.Я. О сепарабельных полиномах над коммутативными банаховыми алгебрами. - Докл.АН СССР, 1974, 218, 3, 505-508.
3. Горин Е.А. Голоморфные функции на алгебраическом многообразии и приводимость сепарабельных полиномов над некоторыми коммутативными банаховыми алгебрами. - В кн.: Тезисы докл.7-й Всесоюзной топ.конф., Минск, 1977, 55.
4. Горин Е.А., Караханян М.И. Несколько замечаний об алгебрах непрерывных функций на локально связном компакте. - В кн.: Тезисы докл. 7-й Всесоюзной топ.конф., Минск, 1977, 56.
5. Караханян М.И. Об алгебрах непрерывных функций на локально связном компакте. - Функц.анал. и его прил., 1978, 12, 2, 93-94.
6. Лин В.Я. О полиномах четвертой степени над алгеброй непрерывных функций. - Функц.анал. и его прил., 1974, 8, 4, 89-90.
7. Зюзин Ю.В. Алгебраические уравнения с непрерывными коэффициентами на однородных пространствах.- Вестник МГУ, сер.мат.мех., 1972, № 1, 51-53.
8. Зюзин Ю.В., Лин В.Я. Неразветвленные алгебраические расширения коммутативных банаховых алгебр. - Матем.сб., 1973, 91, 3, 402-420.
9. Лин В.Я. Алгеброидные функции и голоморфные элементы гомотопических групп комплексного многообразия. - Докл.АН СССР, 1971, 201, 1, 28-31.
10. Зюзин Ю.В. Неприводимые сепарабельные полиномы с голоморфными коэффициентами на некотором классе комплексных пространств. - Матем.сб., 1977, 102, 4, 159-591.

II. К а л и м а н Ш.И. Голоморфная универсальная накрывающая пространства полиномов без кратных корней. – Функц. анал. и его прил., 1975, 9, I, 7I.

I2. H i r c h o w i t z A. À propos de principe d'Oka.– C.R.Acad. sci. Paris, 1971, 272, A792-A794.

I3. Г о р и н Е.А., С а н ч е с К а р л о с Ф е р н а н д е с. О трансцендентных уравнениях в коммутативных банаховых алгебрах. – Функц.анал. и его прил., 1977, II, I, 63–64.

E.A.GORIN
(Е.А.ГОРИН)

СССР, II7234, Москва
Ленинские Горы
Московский Государственный
Университет
Механико-Математический факультет

* * *

COMMENTARY BY THE AUTHOR

Bounded contractible but non-holomorphically contractible domain of holomorphy in $\mathbb{C}^2$ have been constructed in [14]. All other questions, including that of contractibility of the Teichmüller space $\widetilde{G}_n$, seem to rest open.

A detailed exposition of a part of [13] can be found in [15].

REFERENCES

14. З а й д е н б е р г М.Г., Л и н В.Я. О голоморфно не стягиваемых ограниченных областях голоморфности.– Докл.АН СССР, 1979, 249, № 2, 28I–285.

15. F e r n á n d e z C. S á n c h e z , G o r i n E.A. Variante del teorema de la función implícita en álgebras de Banach conmutativas. - Revista Ciencias Matemáticas (Univ. de La Habana, Cuba), 1983, 3, N 1, 77-89.

13.9. old

HOLOMORPHIC MAPPINGS OF SOME SPACES CONNECTED WITH ALGEBRAIC FUNCTIONS

1. For any integer n, $n \geq 3$ and any $z = (z_1, \ldots, z_n) \in \mathbb{C}^n$ consider the polynomial $P(\lambda) = \lambda^n + z_1 \lambda^{n-1} + \ldots + z_n$, and let $d_n(z)$ be the discriminant of P. Then d_n is a polynomial in $z_1, \ldots, z_n$ and the sets $G_n = \{z : d_n(z) \neq 0\}$, $G_n^\circ =$

$$= G_n \cap \{z : z_1 = 0\}, \quad SG_n = \{z : z_1 = 0, \ d(z) = 1\}$$

are non-singular irreducible affine algebraic manifolds, G_n being isomorphic to $G_n^\circ \times \mathbb{C}$. The restriction $d_n^\circ = d_n \mid G_n^\circ$: $G_n^\circ \to \mathbb{C}^* = \mathbb{C} \setminus \{0\}$ is a locally trivial holomorphic fibering with the fiber SG_n. These three manifolds play an important role in the theories of algebraic functions and of algebraic equations over function algebras. Each of the manifolds is $K(\pi_1, 1)$ for its fundamental group π_1, $\pi_1(G_n)$ and $\pi_1(G_n^\circ)$ being both isomorphic to the Artin braid group $B(n)$ with n threads and $\pi_1(SG_n)$ being isomorphic to the commutator subgroup of $B(n)$, denoted $B'(n)$ ([1],[2]). $\mathbb{Z}$- and $\mathbb{Z}_p$-cohomologies of G_n are known [1],[3],[4]. However, our knowledge of analytic properties of G_n, G_n°, SG_n essential for some problems of the theory of algebraic functions is less than satisfactory ([5]-[10]). We propose several conjectures concerning holomorphic mappings of G° and SG_n Some of them have arisen (and all have been discussed) on the Seminar of E.A.Gorin and the author on Banach Algebras and Analytic Functions at the Moscow State University.

2. A group homomorphism $H_1 \to H_2$ is called *abelian* (resp. *integer*) if its image is an abelian subgroup of H_2 (resp. a subgroup isomorphic to $\mathbb{Z}$ or $\{0\}$). For complex spaces X and Y, $C(X,Y)$, $Hol(X,Y)$ and $Hol^*(X,Y)$ stand for the sets of, respectively, continuous, holomorphic and *non-constant* holomorphic mappings from X to Y. A mapping $f \in C(G_k^\circ, G_n^\circ)$ is said to be *splittable* if there is $h \in C(\mathbb{C}^*, G_n^\circ)$ such that f is homotopic to $h \circ d_k^\circ$, d_k° : $G_k^\circ \to \mathbb{C}^*$ being the standard mapping defined above; f is splittable if and only if the induced homomorphism $f_* : B(k) \approx \pi_1(G_k^\circ) \to \pi_1(G_n^\circ) \approx B(n)$ is integer. There exists a simple explicit description of splittable elements of $Hol(G_k^\circ, G_n^\circ)$ [6].

CONJECTURE 1. Let $k>4$ and $n \neq k$. Then (a) every $f \in Hol(G_k^\circ, G_n^\circ)$ is splittable; (b) $Hol^*(SG_k, SG_n) = \emptyset$.

It is easy to see that (b) implies (a).

Let $\Pi(k)$ be the union of four increasing arithmetic progressions with the same difference $k(k-1)$ and whose first members are k, $(k-1)^2$, $k(k-1)$, $k(k-1)+1$. According to [6], if $k>4$ and $n \notin \Pi(k)$ then all f in $Hol(G_k^\circ, G_n^\circ)$ are splittable. A complete description of all non-splittable f in $Hol(G_k^\circ, G_n^\circ)$ has been also given in [6]. If $k>4$ and $n<k$, there are only trivial homomorphisms from $B'(k)$ to $B(n)$ [11]. Thus for such n and k all elements of $C(G_k^\circ, G_n^\circ)$ are splittable and all elements of $C(SG_k, SG_n)$ are contractible. The last assertion implies rather easily that $Hol^*(SG_k, SG_n) \neq \emptyset$. It is proved in [10] that for $k \neq 4$ each $f \in Hol^*(SG_k, SG_k)$ is biholomorphic and has the form $f(z_2, \dots, z_k) = (\varepsilon^2 z_2, \varepsilon^3 z_3, \dots, \varepsilon^k z_k)$ with $\varepsilon^{k(k-1)} = 1$.

3. Let $\mathbb{C}^{**} = \dot{\mathbb{C}} \setminus \{0, 1\}$. A useful technical device in the topic we are discussing is provided by explicit descriptions of all functions $f \in Hol^*(X, \mathbb{C}^{**})$ for some algebraic manifolds X associated with G_n, G_n° and SG_n ([6],[8],[9],[10]). This has led to the questions and results discussed in this section.

Let $\mathcal{A}$ be the class of all connected non-singular affine algebraic manifolds. For every $X \in \mathcal{A}$ the cardinality $q(X)$ of $Hol^*(X, \mathbb{C}^{**})$ is finite (E.A.Gorin). Besides, if $m > max\{\tau(X), 1\}$ ($\tau(X)$ is the rank of the cohomology group $H^1(X, \mathbb{Z})$) then $Hol^*(X, \mathbb{C} \setminus \{\xi_1, \dots, \xi_n\}) \neq \emptyset$ for any distinct points $\xi_1, \dots, \xi_n \in \mathbb{C}$. Using these two assertions, it is not difficult to prove that, given $X \in \mathcal{A}$ and $n \geq 3$, the set $Hol^*(X, SG_n)$ is finite. In particular, for every $k, n \geq 3$ the set $Hol^*(SG_k, SG_n)$ is finite. Let $Top(X)$ be the class of all Y in $\mathcal{A}$ homeomorphic to X; it is plausible that for any $x \in \mathcal{A}$ the function $q: Top(X) \to \mathbb{Z}_+$ is bounded. I even do not know any example disproving the following stronger

CONJECTURE: there exists a function $\nu: \mathbb{Z}_+ \to \mathbb{Z}_+$ such that $q(X) \leq \nu(\tau(X))$ for all $X \in \mathcal{A}$.

A function $\nu_1: \mathbb{Z}_+ \to \mathbb{Z}_+$ with $q(\Gamma) \leq \nu_1(\tau(\Gamma))$ for all urves $\Gamma \in \mathcal{A}$ does exist. It has been proved in [14] that here exists a function $V_2 : \mathbb{Z}_+ \times \mathbb{Z}_+ \to \mathbb{Z}_+$ such that $q(X) \leqslant V_2(r_1(X), r_2(X))$ for all manifolds $X \in \mathcal{A}$ of dimension two (here $r_i(X) = \text{rank } H^i(X, \mathbb{Z})$, $i = 1, 2$). For $X \in \mathcal{A}$ it is known that (i) if $r(X) \leqslant 1$ then $q(X) = 0$; (ii) if $r(X) = 2$ then $q(X)$ is 0 or 6; (iii) if $r(X) = 3$ then $q(X)$ is 0, 6, 24 or 36 (all cases do occur).

4. SG_k contains a curve $\Gamma_k' = SG_k \cap \{z : z_1 = \ldots = z_{k-2} = 0\}$ isomorphic to $\Gamma_k = \{(x, y) \in \mathbb{C}^2 : x^k + y^{k-1} = 1\}$. It can be proved that if f_1, f_2 in $Hol(SG_k, SG_n)$ agree on Γ_k' then $f_1 \equiv f_2$. Since $Hol^*(SG_k, SG_n) = \emptyset$ provided $k > 4$ and $n < k$, the following assertion admittedly implies conjecture 1.

CONJECTURE 2. <u>If</u> $n > k > 4$ <u>then</u> $Hol^*(\Gamma_k, SG_n) = \emptyset$.

The curve Γ_k can be obtained from a non-singular projective curve of genus $(k-1)(k-2)/2$ by removing a single point. It seems <u>plausible that</u> $Hol^*(\Gamma^{(g)}, SG_n) = \emptyset$ <u>for all</u> $n > 4$ <u>and all curves</u> $\Gamma^{(g)} \in \mathcal{A}$ <u>of genus</u> $g < (n-1)(n-2)/2$. In any case the following weaker conjecture is likely to be true (this is really the case if $m \leqslant 1$ or $n \leqslant 4$, E.A.Gorin).

CONJECTURE 3. <u>Let</u> $n \geqslant 3$, $m \geqslant 0$ <u>and</u> $\zeta_1, \ldots, \zeta_m \in \mathbb{C}$. <u>Then</u> $Hol^*(\mathbb{C} \setminus \{\zeta_1, \ldots, \zeta_m\}, S(G_n)) = \emptyset$.

5. Even the following weakened variant of CONJECTURE 1 would be useful for applications. Let $X^\ell \in \mathcal{A}$, $\dim X^\ell = \ell$; $k > 4$, $n \geqslant 3$.

CONJECTURE 4. (a) <u>Let</u> $f_1 \in Hol(G_k^\circ, X^\ell)$, $f_2 \in Hol(X^\ell, G_n^\circ)$. <u>If</u> $\ell \leqslant k-2$ <u>then</u> $f_2 \circ f_1$ <u>is splittable</u>.

(b) <u>Let</u> $f_1 \in Hol(SG_k, X^\ell)$, $f_2 \in Hol(X^\ell, SG_n)$; <u>if</u> $\ell \leqslant k-3$ <u>then</u> $f_2 \circ f_1$ <u>is a constant mapping</u>.

It follows from results of [6], [7], [10], [11] that the assertions 4(a) and 4(b) hold if either $k > 4$, $n \leqslant k$ or $k > 4$, $\ell = 1$ (of course, 4(a) is true for $k > 4$ and $n \notin \Pi(k)$). Maybe even the following sharpenings of 4(a) and 4(b) hold, though they look less probable.

CONJECTURE 5. (a). <u>If</u> $k > 4$, $\ell \leqslant k-2$ <u>and</u> $f \in Hol(G_k^\circ, X^\ell)$ <u>then the induced homomorphism</u> $f_* : \pi_1(G_k^\circ) \to \pi_1(X^\ell)$ <u>is abelian</u>.

(b) <u>If</u> $k>4$, $l \leqslant k-3$ <u>and</u> $f \in \mathrm{Hol}(SG_k, X^l)$ <u>then the induced homomorphism</u> $f_* : \pi_1(SG_k) \longrightarrow \pi_1(X^l)$ <u>is trivial</u>.

If $k>4$ and $l=1$, f_* really has these properties. It can be proved also that if $k \geqslant 4$ and $l \leqslant k-3$ then for any r a t i o n a l $f \in \mathrm{Hol}(SG_k, X^l)$ the kernel of f_* is non-trivial. Conjecture 5 looks a little more realistic in case when X^l is the complement to an algebraic hypersurface in $\mathbb{C}^l$ and f is holomorphic and rational.

6. We formulate here an assertion concerning algebraic functions. To prove this assertion it suffice to verify CONJECTURE 1 for p o l y n o m i a l mappings from G_k^o to G_n^o . Let $\lambda_n = \lambda_n(z)$ be an algebraic function in z $(\in \mathbb{C}^n)$ defined by the equation $\lambda^n + z_1 \lambda^{n-1} + \dots + z_n = 0$ and let Σ_n be the discriminant set of this function, i.e. $\Sigma_n = \{z : d_n(z) = 0\}$.

CONJECTURE 6. <u>For</u> $n>4$ <u>there exists no entire algebraic function</u> $F = F(z)$ <u>with the following properties: (1)</u> F <u>is a composition of polynomials, and entire algebraic functions in less then</u> $n-1$ <u>variables; (2) the discriminant set of</u> F <u>coincides with</u> Σ_n ; (3) <u>in some domain</u> $U \subset \mathbb{C}^n$ <u>the functions</u> λ_n <u>and</u> F <u>have at least one joint irreducible branch</u>.

Condition (2) means that F is forbidden to have "extra" branching points (compared with λ_n). It is known that CONJECTURE 6 becomes true if this condition is replaced by that of absence of "extra branchs" (which is much stronger) [7,13]

REFERENCES

I. А р н о л ь д В.И. О некоторых топологических инвариантах алгебраических функций. - Труды Моск.матем.об-ва, 1970, 2I, 27-46.

2. Г о р и н Е.А., Л и н В.Я. Алгебраические уравнения с непрерывными коэффициентами и некоторые вопросы алгебраической теории кос. - Матем.сб., 1969, 78 (I20), № 4, 579-6I0.

3. Ф у к с Д.Б. Когомологии группы кос по модулю 2. - Функц. анал. и его прил., 1970, 4, № 2, 62-73.

4. В а й н ш т е й н Ф.В. Когомологии групп кос. - Функц.анал. и его прил., 1978, I2, № 2.

5. Л и н В.Я. Алгеброидные функции и голоморфные элементы гомотопических групп комплексного многообразия. - Докл.АН СССР, 1971, 201, № 1, 28-31.

6. Л и н В.Я. Алгебраические функции с универсальным дискриминантным многообразием. - Функц.анал. и его прил., 1972, 6, № 1, 81-82.

7. Л и н В.Я. О суперпозициях алгебраических функций. - Функц. анал. и его прил., 1972, 6, № 3, 77-78.

8. Калиман Ш.И. Голоморфная универсальная накрывающая пространства полиномов без кратных корней. - Функц.анал. и его прил., 1975, 9, № 1, 71.

9. К а л и м а н Ш.И. Голоморфная универсальная накрывающая пространства полиномов без кратных корней. - Теор.функций, функц. анал. и их прилож., вып.28, Харьков, 1977, 25-35.

10. К а л и м а н Ш.И. Голоморфные эндоморфизмы многообразия комплексных полиномов с дискриминантом 1. - Успехи матем.наук, 1976, 31, № 1, 251-252.

11. Л и н В.Я. О представлениях группы кос перестановками. - Успехи матем.наук, 1972, 27, № 3, 192.

12. Л и н В.Я. Представления кос перестановками. - Успехи матем. наук, 1974, 29, № 1, 173-174.

13. Л и н В.Я. Суперпозиции алгебраических функций. - Функц.анал. и его прил., 1976, 10, № 1, 37-45.

14. Б а н д м а н Т.М. Голоморфные функции без двух значений на аффинной поверхности. - Вестник Моск.унив., сер.1, матем.,механ., 1980, № 4, 43-45.

15. Л и н В.Я. Косы Артина и связанные с ними группы и пространства. - В кн.: Итоги науки и техники, сер."Алгебра. Топология. Геометрия", Москва, 1979, т.17, 159-227.

V.Ya.LIN
(В.Я.ЛИН)

СССР, 117418, Москва
ул. Красикова 32,
Центр.Эконом.-Матем.Институт
АН СССР

13.10. ON THE NUMBER OF SINGULAR POINTS OF A PLANE AFFINE ALGEBRAIC CURVE

Let $p(x,y)$ be an irreducible polynomial on $\mathbb{C}^2$. It has been proved in [1] that if the algebraic curve $\Gamma_0=\{(x,y)\in\mathbb{C}^2 : p(x,y)=0\}$ is simply connected then there exist a polynomial automorphism α of the space $\mathbb{C}^2$ and positive integers k, ℓ with $(k,\ell)=1$ such that $p(\alpha(x,y))=x^k-y^\ell$. It follows from this theorem that an irreducible simply connected algebraic curve in $\mathbb{C}^2$ cannot have more than one singular point. (Note that such a curve in $\mathbb{C}^3$ may have as many singularities as you like.)

In view of this result the following QUESTION arises: does there exist a connection between the topology of an irreducible plane affine algebraic curve and the number of its irreducible singularities? Is it true, for example, that the number of irreducible singularities of such a curve Γ does not exceed $r+1$, where $r=\operatorname{rank} H^1(\Gamma,\mathbb{Z})$?

The above assertion on the singularities of the irreducible simply connected curve may be reformulated as follows: let u and v be polynomials in one variable $z\in\mathbb{C}$, such that for any distinct points $z_1,z_2\in\mathbb{C}$ either $u(z_1)\neq u(z_2)$ or $v(z_1)\neq v(z_2)$; then the system of equations $u'(z)=0$, $v'(z)=0$ has at most one solution. It would be very interesting to find a proof of this statement not depending on the above theorem about the normal form of a simply connected curve Γ_0. Maybe such a proof will shed some light onto the following question (which is a slightly weaker form of the question about the irreducible singularities of a plane affine algebraic curve). Let X be an open Riemann surface of finite type (g,n) (g is its genus and $n\geq 1$ is the number of punctures), and let u, v be regular functions on X (i.e. rational functions on X with poles at the punctures only). Suppose that the mapping $f: X\to\mathbb{C}^2$, $f(x)=(u(x),v(x))$, $x\in X$, is injective.

How many solutions (in X) may have the system of equations $\partial u=0, \partial v=0$? (Here $\partial=\frac{\partial}{\partial z}\cdot dz$, where z is a holomorphic local coordinate on X.)

REFERENCE

. З а й д е н б е р г М.Г., Л и н В.Я. Неприводимая односвязная алгебраическая кривая в $\mathbb{C}^2$ эквивалентна квазиоднородной. - Доклады АН СССР, I983, 27I, №5, I048-I052.

V.Ya.LIN
(В.Я.ЛИН)

СССР, II74I8, Москва,
ул.Красикова 32,
Центр.Эконом.-Матем.Институт
АН СССР

M.G.ZAIDENBERG
(М.Г.ЗАЙДЕНБЕРГ)

СССР, 302015, Орел,
Комсомольская ул., I9
Педагогический институт

S O L U T I O N S

Under this title those "old" problems are collected which have been completely solved (the "new" problem S.11 is an exception). All are accompanied with commentary - except for S.9 where commentary by the author is incorporated into the text. Problems S.1-S.10 follow exactly the same order as in the first edition.

5.1.
old

ABSOLUTELY SUMMING OPERATORS FROM THE DISC ALGEBRA

Let A denote the Disc Algebra i.e. the subspace of the Banach space $C(\mathbb{T})$ consisting of all functions which are boundary values of uniformly continuous analytic functions in the open unit disc $\mathbb{D}$. Let $H_0^1=\{g\in L^1(\mathbb{T}):\int_{\mathbb{T}} fg\,dm=0$ for every f, $f\in A\}$.

Recall that a bounded linear operator $u:X\to Y$ (X, Y - Banach spaces) is p - a b s o l u t e l y s u m m i n g ($0<p<\infty$) if there is a constant $K=K(u)$ such that for every finite sequence (x_j),

$$\sum_j \|u(x_j)\|^p \leq K^p \sup \sum_j |x^*(x_j)|^p ,$$

where the supremum is extended over all x^* in the unit ball of the dual of X. Finally by ℓ^p we denote the Banach space of p-absolutely summable complex sequences ($1\leq p<\infty$).

We would like to understand what differences and what similarities there are between the properties of bounded linear operators from the Disc Algebra to Banach spaces and the operators from $C(S)$-spaces. The results of Delbaen [1] and Kisliakov [2] characterizing weakly compact operators and the results by Pełczyński-Mitjagin [3] that for $1<p<\infty$, p-absolutely summing operators from A into a Banach space are p-integral (i.e. these operators extend to p-absolutely summing operators from $C(\mathbb{T})$) are examples of similar properties while the existence of an absolutely summing surjection from A onto ℓ^2 (cf [3]) indicates differences between A and spaces of continuous functions.

The problems discussed below if they would have positive answers will indicate furher similarities. Roughly speaking the positive answers would mean that properties of 2-absolutely summing operators from A are the same as the properties of 2-absolutely summing operators from $C(S)$-spaces. The situation is clear for translation invariant operators (cf. [4]).

Let us consider the following statements:

(α) For every sequence (g_i) in $L^1(\mathbb{T})$ such that

$$\sum_j \left|\int_{\mathbb{T}} g_j f\,dm\right|^2<\infty \qquad \text{for every } f,\ f\in A, \tag{1}$$

there exists a sequence (h_j) in H_0^1 such that

$$\sum_j \left| \int_{\mathbb{T}} (g_j + h_j) f \, dm \right|^2 < \infty \quad \text{for every } f, f \in C(\mathbb{T}) ;$$

(β) for every bounded linear operator $u: A \to \ell^2$ there exists a finite non-negative Borel measure μ on $\mathbb{T}$ such that

$$\|uf\|^2 \leq \int |f|^2 d\mu \quad \text{for every } f, f \in A ;$$

(γ) for every sequences (g_j) in $L^1(\mathbb{T})$ satisfying (1) and every sequence (f_K) in A with $\sup_{z \in \mathbb{T}} \sum_K |f_K(z)|^2 < \infty$,

$$\sum_j \sum_K \left| \int_{\mathbb{T}} g_j f_K \, dm \right|^2 < \infty ;$$

(δ) every bounded linear operator $u: A \to \ell^2$ extends to a bounded linear operator from $C(\mathbb{T})$ into ℓ^2;

(ε) every bounded linear operator from A into ℓ^2 is 2-absolutely summing;

(η) for every bounded linear operators $v: \ell^2 \to A$ and $u: A \to \ell^2$ the composition $uv: \ell^2 \to \ell^2$ belongs to the Hilbert-Schmidt class.

(a) For every sequence (g_j) in $L^1(\mathbb{T})$ such that

$$\sum_j \left| \int_{\mathbb{T}} g_j f \, dm \right| < \infty \quad \text{for every } f, f \in A, \tag{2}$$

there exists a sequence (h_j) in H_0^1 such that

$$\sum_j \left| \int_{\mathbb{T}} (g_j + h_j) f \, dm \right| < \infty \quad \text{for every } f, f \in C(\mathbb{T}) ;$$

(b) for every bounded linear operator $u: A \to \ell^1$ there exists a non-negative finite Borel measure μ on $\mathbb{T}$ such that

$$\|uf\|^2 \leq \int_{\mathbb{T}} |f|^2 d\mu \quad \text{for } f, f \in A ;$$

(c) for every sequence (g_j) in $L^1(\mathbb{T})$ satisfying (2) and for every sequence (f_k) in A with $\sup_{z\in\mathbb{T}} \sum_k |f_k(z)|^2 < \infty$,

$$\sum_j \sqrt{\sum_k \left| \int_{\mathbb{T}} f_k g_j \, dm \right|^2} < \infty ;$$

(d) every bounded linear operator $u: A \to \ell^1$ extends to a bounded linear operator from $C(\mathbb{T})$ into ℓ^1;

(e) every bounded linear operator from A into ℓ^1 is 2-abslolutely summing,

(f) every bounded linear operator from ℓ^2 into A has absolutely summing adjoint.

(A) Every bounded linear operator $u: A \to \ell^1$ is Hilbertian [*],

(B) For every sequence (g_j) in $L^1(\mathbb{T})$ satisfying (2) and for every sequence (f_k) in A with $\sum_k \sup_{z\in\mathbb{T}} |f_k(z)|^2 < +\infty$,

$$\sum_j \sqrt{\sum_k \left| \int_{\mathbb{T}} f_k g_j \, dm \right|^2} < +\infty ;$$

(*) the space L^1 / H^1_0 is of cotype 2 i.e. there is a K, $K>0$, such that for every positive integer n and every $g_1, g_2, \dots, g_n$ in $L^1(\mathbb{T})$

$$2^n \sum_{j=1}^{n} \sup_{\substack{f\in A \\ \|f\|\le 1}} \left| \int_{\mathbb{T}} g_j f \, dm \right|^2 \le c \sum_{\varepsilon} \sup_{\substack{f\in A \\ \|f\|=1}} \left| \int_{\mathbb{T}} \Big(\sum_{j=1} \varepsilon_j g_j \Big) f \, dm \right| ,$$

where the sum $\sum_{\varepsilon}$ extends for all sequences $\varepsilon = (\varepsilon_j)_{j=1}^{n}$ with $\varepsilon_j = \pm 1$ for $j = 1, 2, \dots, n$.

Using the standard technique of absolutely summing operators one can prove

---

*) i.e. can be factored through a Hilbert space. - Ed.

PROPOSITION 1. <u>The following implications hold</u>

$$(*) \Longleftarrow (a) \Longleftrightarrow (b) \Longleftrightarrow (c) \Longleftrightarrow (d) \Longleftrightarrow (e) \Longleftrightarrow (f)$$

$$(a) \searrow (\alpha) \Longleftrightarrow (\beta) \Longleftrightarrow (\gamma) \Longleftrightarrow (\delta) \Longleftrightarrow (\varepsilon) \Longleftrightarrow (\eta)$$

$$(a) \Downarrow (A) \Longleftrightarrow (B).$$

PROBLEM 1. <u>Is</u> (a) <u>true</u>?

PROBLEM 2. <u>Is</u> (d) <u>true</u>?

PROBLEM 3. <u>Is</u> (A) <u>true</u>?

PROBLEM 4. <u>Is</u> (*) <u>true</u>?

REFERENCES

1. D e l b a e n F. Weakly compact operators on the disc algebra. - Journ. of Algebra, 1977, 45, N 2, 284-294.
2. К и с л я к о в С.В. Об условиях Данфорда-Петтиса, Пелчинского и Гротендика. - Докл.АН СССР, 1975, 225, 6, 1252-1255.
3. P e ł c z y ń s k i A. Banach spaces of analytic functions and absolutely summing operators. CBMS, Regional Confer. Ser. in Math. N 30, AMS, Providence, Rhode Island 1977.
4. K w a p i e ń S., P e ł c z y ń s k i A. Remarks on absolutely summing translation invariant operators from the disc algebra and its dual into a Hilbert space. - Mich.Math.J. 1978, 25, N 2, 173-181.

A. PEŁCZYŃSKI

Institute of Mathematics
Polish Academy of Sciences
Śniadeckich 8,
00-950 Warsaw, Poland

* * *

COMMENTARY

J.Bourgain has answered ALL QUESTIONS IN THE AFFIRMATIVE. A summary of his main results on the subject with brief indications of the proofs can be found in [5]. The proofs are to appear in " Acta

Mathematica".

Quite recently Bourgain obtained further improvements of his results. So those who are interested in the question have to follow his forthcoming publications. We review here some "Hard Analysis" aspects of this new work.

First of all Bourgain has proved that given a positive $w \in L^1(\mathbb{T})$ there exist $W \geq w$, with $\int_{\mathbb{T}} W \leq c \int_{\mathbb{T}} w$, and a projection $P: L^2(W) \longrightarrow H^2(W)$ satisfying the weak type estimate

$$W\left(\left\{\zeta \in \mathbb{T} : |Pf(\zeta)| > t\right\}\right) \leq \frac{c}{t} \cdot \|f\|_{L^1(W)}$$

and which is bounded simultaneously in $L^2(W)$. This leads to a conceptual simplification of the methods used in [5].

Further, Bourgain has proved that any operator mapping a reflexive subspace of L^1/H^1 to H^∞ admits an extension to an operator from L^1/H^1 to H^∞. This result has an interesting application to the Fourier Analysis. Namely, for every sequence $\{g_n\}_{n \geq 1}$ of H^∞- functions satisfying $\sup_{t \in \mathbb{T}} \sum_n |g_n(t)|^2 < +\infty$ there exists $F \in H^\infty(\mathbb{T}^2)$ with

$$\int_{\mathbb{T}} F(t,\zeta)\bar{\zeta}^{2n}\, dm(\zeta) = g_n(t), \quad t \in \mathbb{T}, \quad n = 1, 2, \ldots .$$

REFERENCE

5. B o u r g a i n J. Opérateurs sommants sur l'algèbre du disque. - C.R.Acad.Sc.Paris, 1981, 293, Sér I, 677-680.

S.2. old GOLUBEV SERIES AND THE ANALYTICITY ON A CONTINUUM

The collection of all open neighbourhoods of a compact set $K(\subset \mathbb{C})$ will be denoted by $\mathcal{V}(K)$. A function analytic on a set belonging to $\mathcal{V}(K)$ will be called analytic on K. It will be called τ - a n a l y t i c o n K $(\tau>0)$ if

$$\overline{\lim}\left(\frac{|f^{(k)}(t)|}{k!}\right)^{\frac{1}{k}} \leqslant \tau \text{ for every } t, t\in K.$$

DEFINITION. A compact set $K(\subset \mathbb{C})$ is r e g u l a r if there exists a mapping $R_K : \mathbb{R}_+ \longrightarrow \mathcal{V}(K)$ enjoying the following property: for every $\tau>0$ and for every f τ-analytic on K there exist a function g analytic in $R_K(\tau)$ and a set W, $W\in \mathcal{V}(K)$ such that

$$W\subset R_K(\tau), \quad f|W = g|W.$$

The set

$$S=\{j^{-1}: j=1,2,\dots\}\cup\{0\} \tag{1}$$

is not regular. Indeed, putting

$$f_j(t)=\begin{cases} 0, & \operatorname{Re} t<\frac{1}{2}(j^{-1}+(j+1)^{-1}) \\ 1, & \operatorname{Re} t>\frac{1}{2}(j^{-1}+(j+1)^{-1}) \end{cases} \qquad j=1,2,\dots$$

we see that f_j is τ-analytic on S for all values of τ and j but $\mathcal{V}(S)$ contains no set where a l l f_j are analytic.

QUESTION. <u>Is every plane continuum (i.e. a compact connected set) regular?</u>

This question related to the theory of analytic continuation robably can be reformulated as a problem of the plane topology. Its ppearance in the chapter devoted to spaces of analytic functions the first edition of the collection is meant – Ed.] seems natural be-ause of the following theorem, a by-product of a description of the ual of the space $\mathcal{H}(K)$ of all functions analytic on K.

THEOREM. Let K be a regular compact set and μ a positive orel measure on K such that $clos(K \setminus e) = K$ for every e, $e \subset K$, with $\mu(e) = 0$. Then every function u analytic in $\hat{\mathbb{C}} \setminus K$ is representable by the following formula

$$u(t) = u(\infty) + \sum_{n \geq 0} \int_K \frac{y_n(\zeta)\, d\mu(\zeta)}{(\zeta - t)^{n+1}}, \quad t \in \hat{\mathbb{C}} \setminus K, \qquad (2)$$

$(y_n)_{n \geq 0}$ being a sequence of $L^2(\mu)$-functions and $\lim_{n \to \infty} \|y_n\|_{L^2(\mu)}^{1/n} = 0$.

This theorem was proved in [1]. The regularity of K leads to a definition of the topology of $\mathcal{H}(K)$ explicitly involving convergence radii of germs of functions analytic on K.

Unfortunately, the regularity assumption was omitted in the statement of the Theorem as given in [1] (though this assumption was essentially used in the proof - see [1], the beginning of p.125). The compact K was supposed to be nothing but a continuum. A psychological ground (but not an excuse) of this omission is the problem the author was really interested in (and has solved in [1]), namely, the question put by V.V.Golubev ([2], p.111): is the formula (2) valid for every function u analytic in $\hat{\mathbb{C}} \setminus K$ provided K is a rectifiable simple arc and μ is Lebesgue measure (the arc-length) on K ? The regularity of a simple arc (and of every l o c a l l y - c o n n e c t e d plane compact set) can be proved very easily, see e.g., [3], p.146. The Theorem reappeared in [4] and [5] and was generalized to a multidimensional situation in [3]. It was used in [6] as an illustration of a principle in the theory of Hilbert scales.

We have not much to add to our QUESTION and to the Theorem. The local-connectedness is not necessary for the regularity: the closure of the graph of the function $t \mapsto \sin 1/t$, $t \in (0,1]$ is regular. The definition of the regularity admits a natural multidimensional generalization. A non-regular continuum in $\mathbb{C}^2$ was constructed in [7]. The regularity is essential for the possibility to represent functions by Golubev series (2): a function analytic in $\hat{\mathbb{C}} \setminus S$ (see (1)) and with a simple pole of residue one at every point j^{-1} $(j = 1, 2, \ldots)$ is not representable by a series (2). Non-trivial examples of functions analytic off an everywhere discontinuous plane compactum and not representable by a Golubev series (2) were given in [8].

REFERENCES

1. Хавин В.П. Один аналог ряда Лорана. – В кн.: "Исследования по современным проблемам теории функций комплексного переменного". М., Физматгиз, 1961, 121-131.
2. Голубев В.В. Однозначные аналитические функции. Автоморфные функции. М., Физматгиз, 1961.
3. Трутнев В.М. Об одном аналоге ряда Лорана для функций многих комплексных переменных, голоморфных на сильно линейно выпуклых множествах. – В сб."Голоморфные функции многих комплексных переменных". Красноярск, ИФ СО АН СССР, 1972, 139-152.
4. Baernstein A. II. Representation of holomorphic functions by boundary integrals.-Trans.Amer.Math.Soc.,1971,169,27-37.
5. Baernstein A. II. A representation theorem for functions holomorphic off the real axis. - ibid. 1972,165, 159-165.
6. Митягин Б.С., Хенкин Г.М. Линейные задачи комплексного анализа. – Успехи матем.наук, 1971, 26, 4, 93-152.
7. Zame R. Extendibility, boundedness and sequential convergence in spaces of holomorphic functions. - Pacif.J.Math., 1975, 57, N 2, 619-628.
8. Витушкин А.Г. Об одной задаче Данжуа. – Изв.АН СССР, сер.матем., 1964, 28, № 4, 745-756.

V.P.HAVIN
(В.П.ХАВИН)

СССР, 198904, Ленинград
Петергоф, Библиотечная площадь, 2
Ленинградский государственный
университет, Математико-механический
факультет

* * *

COMMENTARY BY THE AUTHOR

The answer to the above QUESTION is YES. It was given in [9] and [10]. Thus the word "regular" in the statement of the Theorem can be replaced by "connected" (as was asserted in [1]).

REFERENCES

9. В а р ф о л о м е е в А.Л. Аналитическое продолжение с континуума на его окрестность. – Записки научн.семин.ЛОМИ, 1981, 113, 27-40.
10. R o g e r s J.T., Z a m e W.R. Extension of analytic functions and the topology in spaces of analytic functions. – Indiana Univ. Math.J., 1982, 31, N 6, 809-818.

S.3. old

THE VANISHING INTERIOR OF THE SPECTRUM

Let A and B be complex unital Banach algebras and let $x \in A \subset B$, then it is well known that

$$\sigma_A(x) \supset \sigma_B(x) \qquad \text{and } \partial\sigma_A(x) \subset \partial\sigma_B(x),$$

where $\sigma_A(x)$ is the spectrum of x relative to A and $\partial\sigma_A(x)$ is its boundary. Taking A to be the unital Banach algebra generated by x, in this context we say that x i s n o n-t r i v i a l if $\operatorname{int}\sigma_A(x) \neq \emptyset$. Šilov [1] has proved that if $\lambda \in \mathbb{C}$, $\lambda - x$ is permanently singular in A (i.e. $\lambda - x$ is not invertible in any superalgebra B of A) if, and only if, $\lambda - x$ is an approximate zero divisor (AZD) of A i.e. if $\exists y_n \in A, \|y_n\| = 1$, such that $y_n(\lambda - x) \to 0$ $(n \to \infty)$.

Let $\{\alpha_n\}_{n \geqslant 0}$ be a sequence in $\mathbb{R}$ with $\alpha_0 = 1$ and $\alpha_{n+m} \leqslant \alpha_n \alpha_m$ $n, m \geqslant 0$. Then the power series algebra

$$A = \Big\{ \sum_{n \geqslant 0} a_n z^n : \sum_{n \geqslant 0} |a_n| \alpha_n < \infty \Big\}$$

($\{a_n\}$) denotes a sequence of complex numbers) is a Banach algebra under the norm $\|\sum_{n \geqslant 0} a_n z^n\| = \sum_{n \geqslant 0} |a_n| \alpha_n$ which is generated by z and $\sigma_A(z)$ is a disk of a radius r. Šilov [1] shows that for an appropriate choice of the sequence $\{\alpha_n\}_{n \geqslant 0}$ $\exists r_0$ such that $0 < r_0 < r$ and if $\mu \in \{\lambda : r_0 \leqslant \lambda \leqslant r\}$, $\mu - z$ is an ADZ in A. Thus in every superalgebra B this annulus is contained in $\sigma_B(x)$ and we say that $\sigma_A(x)$ has a n o n - v a n i s h i n g i n t e r i o r.

If A is a uniform algebra then it is easy to show [5] that for each non-trivial element x we can construct a superalgebra B such that $\operatorname{int}\sigma_B(x) = \emptyset$. If T is a subnormal operator on a Hilbert space (i.e. T has a normal extension in a larger Hilbert space) then the algebra which it generates is a uniform algebra [2] hence the same is true of T.

Šilov's theorem has been extended by Arens [3] to commutative Banach algebras which are not necessarily singly generated and Bollobas [4] has shown that it is not, in general, possible to construct a superalgebra B of a Banach algebra A in which all the elements which are not AZD's in A become simultaneously invertible.

QUESTIONS. Let A be generated by the non-trivial element x such that $\lambda - x$ is an ADZ in A if, and only if, $\lambda \in \partial \sigma_A(x)$. Can one construct a superalgebra B such that $\sigma_B(x) = \partial \sigma_A(x)$ i.e. superalgebra B in which $\operatorname{int} \sigma_A(x)$ vanishes simultaneously? If x is a non-trivial element of a C^*-algebra A does there exist a superalgebra B of A_x such that $\sigma_B(x) = \partial \sigma_{A_x}(x)$, where A_x is the unital Banach algebra generated by x in A ?

REFERENCES

1. Шилов Г. Е. О нормированных кольцах с одной образующей. - Матем.сб., 1947, 21 (63), 25-47.
2. Bram J. Subnormal operators. - Duke Math.J.,1955, 22, 75-94.
3. Arens R. Inverse producing extensions of normed algebras. - Trans.Amer.Math.Soc.,1958, 88, 536-548.
4. Bollobas B. Adjoining inverses to commutative Banach algebras. - Trans.Amer.Math.Soc.,1973, 181, 165-174.
5. Murphy G.J., West T.T. Removing the interior of the spectrum. - Comment.Math.Univ.Carolin., 1980, 21, N 3, 421-431.

G.J.MURPHY
T.T.WEST

39 Trinity college
Dublin 2
Ireland

* * *

COMMENTARY

The first problem has been completely solved by C.J.Read [5]. Moreover he has proved that for any commutative Banach algebra A and for any $a \in A$ there exists a superalgebra B such that for $\lambda \in \mathbb{C}$, $a - \lambda$ is not invertible in B if and only if $a - \lambda$ is an ADZ in A.

This result solves the problem posed earlier by B.Bollobás in [6].

The second question has a negative answer Indeed, let T be a weighted shift operator defined on ℓ^2 by $Te_k = \lambda_k e_k$, where $\{e_k\}_{k \geq 1}$ is the standard orthogonal basis of ℓ^2 and $\lambda_k = 1 + 1/n$ if $2^n < k < 2^{n+1}$, $\lambda_{2^n} = \left(1 + \frac{1}{n-1}\right)^{-(2^{n-1}-1)}$ It is easy to check that the spectral radius of T equals 1 and $\lim \|T^{2^n-1}\| \cdot \|T^{2^n}\|^{-1} = 0$ It follows that in the algebra A_T generated by T the spectrum of T coincides with $\{\zeta : |\zeta| \leq 1\}$ and T is an ADZ since for $T_n \overset{def}{=} T^{2^n} / \|T^{2^n}\|$ we have $T_n \in A$, $\|T_n\| = 1$ but $\lim_n \|T_n T\| = 0$.

REFERENCES

5. R e a d C.J. Inverse producing extension of a Banach algebra eliminates the residual spectrum on one element. - Trans.Amer.Math. Soc. (to appear).
6. B o l l o b á s B. Adjoining inverses to commutative Banach algebras, Algebras in Analysis, Acad.Press 1975, edited by J.H.Williamson, 256-257.

.4.
1d

ON THE UNIQUENESS THEOREM FOR MEAN PERIODIC FUNCTIONS

A function f continuous on $\mathbb{R}$ is called [1] ω-mean periodic if it satisfies the integral equation

$$\int_0^{\omega} f(x+t)\,d\sigma(x) = 0 \qquad (t \in \mathbb{R}), \tag{1}$$

σ being a function of bounded variation with 0 and ω as its growth points. In the particular case of $\sigma(x) \equiv x$ (1) becomes

$$\int_0^{\omega} f(x+t)\,dx = 0,$$

i.e.

$$\int_t^{t+\omega} f(x)\,dx = 0,$$

which implies $f(t+\omega) = f(t)$, the usual periodicity.

An ω-periodic function vanishing on the "principal" period $\Omega = [0,\omega]$ is identically zero. It is not hard to prove, using Titchmarsh convolution theorem [2], that any ω-mean periodic function is also completely determined by its restriction onto Ω. Put $\{x\} \overset{def}{=\!=} x - \left[\frac{x}{\omega}\right]\omega$ ($[a]$ is the largest integer $\leqslant a$). Suppose the set M, $M \subset \mathbb{R}$, satisfies $\{M\} = \Omega$. Then an arbitrary ω-periodic function vanishing on M is identically zero.

<u>Is the same true for ω-mean periodic functions?</u>

REFERENCES

1. D e l s a r t e J. Les fonctions "moyenne-périodiques". - J. Math.Pures Appl., 1935, Sér.14, N 9, 409-453.

2. Л ю б и ч Ю.И. Об одном классе интегральных уравнений. -

Матем.сб. I956, 38, I83-202.

Yu.I.LYUBICH
(Ю.И.ЛЮБИЧ)

СССР, 3I0077, Харьков
пл.Дзержинского 4
Харьковский государственный
университет

* * *

COMMENTARY

The answer is NO. P.P.Kargaev [3] has constructed a non-zero mean $(\pi+\varepsilon)$-periodic function (for every $\varepsilon > 0$) vanishing on $[2\pi, 3\pi] \cup [-3\pi, -2\pi]$.

REFERENCE

3. К а р г а е в П.П. О нулях функций, периодических в среднем. – Вестник ЛГУ (to appear)

.5. L^2-BOUNDEDNESS OF THE CAUCHY INTEGRAL ON LIPSCHITZ GRAPHS
ld

Let φ be a real C_0^∞-function defined on $\mathbb{R}$, γ the path n the complex plane defined by the equation $\gamma(t)=t+i\varphi(t)$ $(t\in\mathbb{R})$, nd

$$(K_\gamma\psi)(s)\overset{def}{=}\text{v.p.}\int_{\mathbb{R}}\frac{\psi(t)\cdot(1+i\varphi'(t))\,dt}{(t-s)+i(\varphi(t)-\varphi(s))}\qquad(s\in\mathbb{R})$$

the Cauchy integral of ψ taken along the graph of φ). I have roved [1] that

$$\|K_\gamma\psi\|_{L^2(\mathbb{R})}\leq C(\sup_{\mathbb{R}}|\varphi'|)\|\psi\|_{L^2(\mathbb{R})}\qquad(\psi\in C_0^\infty(\mathbb{R})),$$

where the finite positive function C is defined on an interval $[0,a)$, a being an absolute positive constant, $\lim_{t\to a-0}C(t)=+\infty$ (provided $\sup|\varphi'|<a$).

THE PROBLEM is to know <u>whether</u> C <u>can be replaced by a function defined o n t h e w h o l e half-line</u> $[0,+\infty)$, i.e. whether the L^2-boundedness of K_γ can be proved for a l l Lipschitz graphs γ (not only for those with the slope not exceeding a).

REFERENCE

1. C a l d e r ó n A.P. Cauchy integrals on Lipschitz curves and related operators. - Proc.Nat.Acad.Sci. USA, 1977, 74, N 4, 1324-1327.

A.P.CALDERÓN

The University of Chicago
Department of Mathematics
5734 University avenue
Chicago, Illinois 60637
USA

* * *

COMMENTARY

The PROBLEM (coinciding with problem 1 of 6.1) has been solved in [3]: <u>the Cauchy integral defines a bounded linear operator in L^2 o n e v e r y Lipschitz graph (for any value of its slope)</u>. The proof is based on an estimate of the operator A_n (see a) in Problem III of 6.1) with $A' \in L^\infty(\mathbb{R})$. It is proved in [3] that $\|A_n\| \leqslant C(1+n^4)\|A'\|_\infty^n$. Using results of [2] Guy David found a lucid geometrical description of the class R_p, $1<p<\infty$ (we use the notation from 6.2). Associate with every simple curve Γ a maximal function $M_\Gamma : \Gamma \to (0,+\infty]$,

$$M_\Gamma(\zeta) = \sup_{r>0} \frac{|\{z\in\Gamma : |z-\zeta| \leqslant r\}|}{r},$$

where $|\cdot|$ stands for the one dimensional Hausdorff measure. G.David proved ([4], [6]) that

$$\Gamma \in R_2 \iff M_\Gamma \in L^\infty(\Gamma).$$

After this result it seems very probable that the weight ω_Γ discussed in 6.1, II is connected with M_Γ.

It is proved in [5] that the singular operator with the kernel $F\left(\frac{A(x)-A(y)}{x-y}\right)\frac{1}{x-y}$ is continuous in $L^2(\mathbb{R})$ whenever $F\in C^\infty(\mathbb{R}^n)$ and $A: \mathbb{R}\to\mathbb{R}^n$ satisfies the Lipschitz condition on $\mathbb{R}$. The proof is based on the quoted estimate of the operator A_n. The work [5] contains also a proof of the continuity of a singular Calderón-Zygmund operator with odd kernel on $L^2(\Gamma)$, Γ being the graph of a Lipschitz function $\varphi : \mathbb{R}^n \longrightarrow \mathbb{R}$.

Articles [2] and [6] show that the more the spaces $H^2_+(\Gamma)$ and $H^2_-(\Gamma)$ are close to be orthogonal the more Γ is close to a straight line (and vice versa).

And now one more interesting

QUESTION. <u>Which closed Jordan rectifiable curves Γ have the following property: all Cauchy integrals of measures on Γ belong to the Nevanlinna class N_Γ in $\mathrm{Int}\,\Gamma$ (i.e. are quotients of bounded functions analytic in $\mathrm{Int}\,\Gamma$</u>)?

A.B.Aleksandrov has pointed out that no non-Smirnov curve enjoys this property. Moreover if Γ is not a Smirnov curve then there exist $f\in L^1(\Gamma)$ and a discrete measure μ on Γ whose Cauchy integrals

.o not belong to N_Γ (f is found by a simple closed-graph argument, he existence of μ uses some results from [7]).

REFERENCES

2. C o i f m a n R.R., M e y e r Y. Une généralisation du théorème de Calderon sur l'intégrale de Cauchy. Fourier Analysis, Proc. Sem. El Escorial, Spain, June 1979. (Asociación Matemática Española, Madrid, 1980).

3. C o i f m a n R.R., M c I n t o s h A., M e y e r Y. L'intégrale de Cauchy définit un opérateur borné sur L^2 pour les courbes Lipschitziennes. - Ann.Math., 1982, 116, N 2, 361-388.

4. D a v i d G. L'intégrale de Cauchy sur les courbes rectifiables. Prépublication Orsay, 1982, 05, N 527.

5. C o i f m a n R.R., D a v i d G., M e y e r Y. La solution des conjectures de Calderón. Prépublication Orsay, 1982, 04, N 526.

6. D a v i d G. Courbes corde-arc et espaces de Hardy généralisés.- Ann.Inst.Fourier, 1982, 32, N 3, 227-239.

7. А л е к с а н д р о в А.Б. Два аналога теоремы М.Рисса о сопряженных функциях для пространств В.И.Смирнова E^p, $0<p<1$. - В сб. "Теория операторов и теория функций", 1983, № I, Изд-во ЛГУ, 9-20.

S.6. old

SETS OF UNIQUENESS FOR QC

By QC is meant the space of functions on $\mathbb{T}$ that belong together with their complex conjugates to $H^\infty + C$. Here, H^∞ is the space of boundary functions on $\mathbb{T}$ for bounded holomorphic functions in $\mathbb{D}$, and C denotes $C(\mathbb{T})$. It is well known [1] that $H^\infty + C$ is a closed subalgebra of L^∞ (of Lebesgue measure on $\mathbb{T}$). Thus, QC is a C^*-subalgebra of L^∞. The functions in QC are precisely those that are in L^∞ and have vanishing mean oscillation [2]; see [3] for further properties.

A measurable subset E of $\mathbb{T}$ is called a *set of uniqueness* for QC if only the zero function in QC vanishes identically on E. The PROBLEM I propose is that of

<u>characterizing the sets of uniqueness for</u> QC.

There are two extreme possibilities, neither of which can be eliminated on elementary grounds:

1. The only sets of uniqueness are the sets of full measure;
2. A set meeting each arc of $\mathbb{T}$ in a set of positive measure is a set of uniqueness.

If possibility 1 were the case then, in regard to sets of uniqueness, QC would resemble L^∞, while if possibility 2 were the case it would resemble C. One can, of course, inquire about sets of uniqueness for H^∞ and for $H^\infty + C$. For H^∞ the answer is classical: any set of positive measure is a set of uniqueness. In view of this, it is quite surprising that, for $H^\infty + C$, the first of the two extreme possibilities listed above is the case. In fact, S.Axler [4] has shown that any nonnegative function in L^∞ - in particular, any characteristic function - is the modulus of a function in $H^\infty + C$.

Concerning QC, I have been able to rule out only the second of the two extreme possibilities: I can show that there are nonzero functions in QC that are supported by closed nowhere dense subsets of $\mathbb{T}$. The construction is too involved to be described here. It suggests to me that the actual state of affairs lies somewhere between the two extreme possibilities. However, I have not yet been able to formulate a plausible conjecture.

REFERENCES

1. S a r a s o n D. Algebras of functions on the unit circle. - Bull.Amer.Math.Soc.,1973, 79, 286-299.

. S a r a s o n D. Functions of vanishing mean oscillation. - Trans.Amer.Math.Soc.,1975, 207, 391-405.

. S a r a s o n D. Toeplitz operators with piecewise quasi-continuous symbols. - Indiana Univ.Math.J.,1977, 26, 817-838.

. A x l e r S. Factorization of L^∞ functions. - Ann. of Math., 1977, 106, 567-572.

DONALD SARASON

University of California,
Dept.Math., Berkeley,
California, 94720, USA

* * *

COMMENTARY BY THE AUTHOR

T.H.Wolff [5] has shown that the only sets of uniqueness for QC are the sets of full measure. He did this by establishing the remarkable result that every function in L^∞ can be multiplied into QC by an outer function in QC. The result says, roughly speaking, that the discontinuities of an arbitrary L^∞ function form a very small set. Wolff makes the preceding interpretation precise and gives other interesting applications of his result in his paper.

REFERENCE

5. W o l f f T.H. Two algebras of bounded functions. - Duke Math.J., 1982, 49, N 2, 321-328.

EDITORS' NOTE: S.V.Kisliakov has shown that for every set $E\subset\mathbb{T}$ of positive length there exists a non-zero function f in VMO supported on E and such that the Taylor series $\sum_{n\geq 0}\hat{f}(n)z^n$ converges uniformly in the closed disc. Let $\varphi(x)=\min\{|x|,1\}$. Then $\varphi\circ f\in VMO\cap L^\infty=QC$. However it is not clear if it is possible to find a function f in QC satisfying the mentioned conditions (see С.В.Кисляков. Еще раз о свободной интерполяции функциями регулярными вне предписанного множества. - Зап.научн.семин.ЛОМИ,1982, 107, 71-88).

S.7.
old

ANOTHER PROBLEM BY R.KAUFMAN

Let f be meromorphic in the disk $\mathbb{D}$, and $|f'| \geqslant |f|$. What can be concluded about the growth of f ?

R.KAUFMAN

University of Illinois
at Urbana-Champaign
Department of Mathematics
Urbana, Illinois 61801
USA

* * *

COMMENTARY

The PROBLEM has been solved by A.A.Goldberg [1] with a later improvement of the proposer. The answer is "NOTHING". Namely, A.A.Goldberg had shown [1] that for every function $\Phi(r)$ tending to $+\infty$ as $r \to 1$ there exists a function f meromorphic in $\mathbb{D}$ satisfying $|f'| \geqslant |f|$ in $\mathbb{D}$ and such that

$$\overline{\lim_{r\to 1}} \frac{T(r,f)}{\Phi(r)} \geqslant 1 , \qquad (1)$$

where $T(r,f)$ denotes the Nevanlinna characteristic of f. R.Kaufman has strengthened this. He has constructed an f [2] with

$$\underline{\lim_{r\to 1}} \; N(r,0,f)/\Phi(r) \geqslant 1$$

instead of (1). It is shown in [1] that for an ANALYTIC f satisfying $|f'| \geqslant |f|$ in $\mathbb{D}$ the following (precise) estimate holds:

$$\log\log M(r,f) = O((1-r)^{-1}) , \quad r \to 1 . \qquad (2)$$

This estimate is implied by a weaker assumption $T(r,\frac{f'}{f}) = O(1)$, $r \to 1$.

In [2] R.Kaufman has proposed A NEW PROBLEM for functions meromorphic in $\mathbb{C}$. Suppose $h(r)>1$ $(r>0)$, $r^{-n}h(r) \underset{r\to+\infty}{\longrightarrow} \infty$ $(n \in \mathbb{N})$.

hat can be said about $N(r,f)$ and $T(r,f)$ if

$$\left|\frac{f'(z)}{f(z)}\right| > \frac{1}{h(|z|)} \quad ?$$

REFERENCES

I. Г о л ь д б е р г А.А. О росте мероморфных в круге функций с ограничениями на логарифмическую производную. - Укр.мат.ж., 1980, 32, № 4, 456-462.

2. К а у ф м а н Р. Некоторые замечания об интерполяции аналитических функций и логарифмических производных. - Укр.матем.ж., 1982, 34, № 5, 616-617.

S.8. old

RATIONAL FUNCTIONS WITH GIVEN RAMIFICATIONS

Let n , q be positive integers, and $\nu=\{\nu_j : 1\leqslant j\leqslant q\}$, $\lambda=\{\lambda_{kj} : 1\leqslant k\leqslant \nu_j ;\ 1\leqslant j\leqslant q\}$ be two numerical systems satisfying

$$\sum_{j=1}^{q}\sum_{k=1}^{\nu_j}\lambda_{kj}=2(n-1); \tag{1}$$

$$\sum_{k=1}^{\nu_j}\lambda_{kj}\leqslant n-\nu_j ,\quad j=1,\dots,q. \tag{2}$$

We say that the problem $\mathcal{P}[n,\nu,\lambda]$ is solvable, if there exists a rational function $\mathcal{R}$ of degree n and complex numbers z_{kj} , $1\leqslant k\leqslant \nu_j$, $1\leqslant j\leqslant q$, so that $\mathcal{R}(z_{1j})=\mathcal{R}(z_{2j})=\dots=\mathcal{R}(z_{\nu_j j})$, $1\leqslant j\leqslant q$, and the derivative $\mathcal{R}'$ has a zero of order λ_{kj} at the point z_{kj} . Conditions (1) and (2) are necessary for the solvability of the problem $\mathcal{P}[n,\nu,\lambda]$ ((1) is the well-known formula of Riemann-Hurwitz).

PROBLEM. Find efficient criteria of non-solvability of the problem $\mathcal{P}[n,\nu,\lambda]$.

It is known that the problem $\mathcal{P}[n,\nu,\lambda]$ is not always solvable. For example, the problem with parameters $n=4$, $q=3$, $\nu_1=\nu_2=2$, $\nu_3=1$, $\lambda_{11}=\lambda_{21}=\lambda_{12}=\lambda_{22}=1$, $\lambda_{13}=2$ has no solutions ([1], p.468). On the other hand, if all $\nu_j=1$, $1\leqslant j\leqslant q$, then the problem $\mathcal{P}[n,\{1\},\lambda]$ is always solvable ([1], p.469, th.4.1). A series of sufficient conditions for solvability of the problem $\mathcal{P}[n,\nu,\lambda]$ has been obtained by A.Hurwitz [2], [3]. The solution of the posed problem should follow from one general result of H.Weyl [4], but that result is formulated in a very inefficient form so that - according to the author - it remains unclear how can one derive concrete corollaries from it. B.L.Van der Waerden wrote on that result: "Leider kann man mit der Schlussformel noch nicht viel anfangen".

REFERENCES

1. Гольдберг А.А., Островский И.В. Распределение значений мероморфных функций. М., Наука, 1970.
2. Hurwitz A. Ueber Riemann'sche Flächen mit gegebenen Verzweigungspunkten. - Math.Ann., 1891, 39, 1-61.
3. Hurwitz A. Ueber die Anzahl der Riemann'schen Flächen mit gegebenen Verzweigungspunkten. - Math.Ann., 1902, 55, 53-66.
4. Weyl H. Ueber das Hurwitzsche Problem der Bestimmung der Anzahl Riemannscher Flächen von gegebener Verzweigungsart. - Comment.math.helv., 1931, 3, 103-113.

A.A.GOL'DBERG (А.А.ГОЛЬДБЕРГ)

СССР, 290602, Львов
Львовский Государственный
Университет

* * *

COMMENTARY BY THE AUTHOR

The problem has been solved by S.D Bronza and V G.Tairova ([5] - [7]). They proposed an effective algorithm which permits to decide whether the problem $\mathcal{P}[n, \nu, \lambda]$ is solvable and in case it is, the algorithm permits to describe all solutions.

REFERENCES

5. Бронза С.Д., Таирова В.Г. Профили римановых поверхностей. - Теория функций, функц. анал. и их прил., Харьков, 1980, вып. 33, 12-17.
6. Бронза С.Д., Таирова В.Г. Конструирование римановых поверхностей класса $\mathfrak{F}_q^*$. - ibid., 1983, вып. 40 (to appear).
7. Бронза С.Д., Таирова В.Г. Конструирование римановых поверхностей класса $\mathfrak{F}_q^*$. II. - ibid., 1984, вып. 41 (to appear).

S.9. TWO PROBLEMS ON ASYMPTOTIC BEHAVIOUR OF ENTIRE FUNCTIONS old

1. Let $SH(\rho)$ be the class of subharmonic functions in $\mathbb{C}$ of order ρ and of normal type. Let V_t, $t \in (0,\infty)$ be the one-parameter group of rotations of $\mathbb{C}$ defined by

$$V_t z = e^{i\alpha \ln t} z, \quad \alpha \in \mathbb{R}^1, \quad P_t = tV_t .$$

Given $u \in SH(\rho)$ put

$$u_t(z) = u(P_t z)\, t^{-\rho}.$$

Let $\mathcal{D}'$ be the space of Schwartz distributions.

It is known [1] that the family $\{u_t\}$ is compact in $\mathcal{D}'$ as $t \to \infty$ i.e. for each sequence $t_j \to \infty$ there is a subsequence $t_j' \to \infty$ and a function v subharmonic in $\mathbb{C}$ such that $u_{t_j'} \to v$ in $\mathcal{D}'$. The set of all limits v is called the cluster set and is denoted by $Fr[u, V_t]$ or $Fr[u]$. It describes the asymptotic behaviour of u along the spirals $\ell_\varphi = \{z = P_t e^{i\varphi} : t \in (0,\infty)\}$ and, in particular, (when $\alpha = 0$) along the rays starting from the origin.

Let $U(\rho, \sigma)$ be the class of subharmonic functions v satisfying $v(0) = 0$; $v(z) \le \sigma |z|^\rho$, $z \in \mathbb{C}$. The set $Fr[u]$ is closed in $\mathcal{D}'$, invariant with respect to the transformations $(\cdot)_t$; further, $Fr[u] \subset U[\rho, \sigma]$ and $Fr[u]$ is connected in $\mathcal{D}'$.

Let $v \in U[\rho, \sigma]$. The simplest set with the mentioned properties, which contains v, is

$$\Lambda(v) = \operatorname{clos}_{\mathcal{D}'} \{ v_t : t \in (0,\infty) \}. \tag{1}$$

Let $A(\rho)$ be the class of entire functions f of order ρ and of normal type. Set $Fr[f] \overset{\text{def}}{=} Fr[\ln|f|]$.

PROBLEM 1. <u>Does there exist an entire function $f \in A(\rho)$ such that $Fr[f] = \Lambda(v)$?</u>

We denote by $Fr_0[v]$, $Fr_\infty[v]$ the sets of all limits in $\mathcal{D}'$ of the families $\{v_t\}$ as $t \to 0$ and $t \to \infty$ respectively. The following theorem solves Problem 1.

THEOREM 1 [2]. <u>A necessary and sufficient condition for the existence of a function f for Problem 1 is</u>

$$Fr_0[v] \cap Fr_\infty[v] \neq \emptyset. \qquad (2)$$

The paper [2] contains examples which show that condition (2) may fail for some v.

2. Let $f \in A(\rho)$ and $h_f(\varphi)$, $\underline{h}_f(\varphi)$ be the indicator and the lower indicator respectively. One of the possible (equivalent) definitions of the indicators is

$$\left.\begin{array}{l} h_f(\varphi) = \sup \\ \underline{h}_f(\varphi) = \inf \end{array}\right\{ v(e^{i\varphi}) : v \in Fr[f] \right\} . \qquad (3)$$

The equality

$$h_f(\varphi) = \underline{h}_f(\varphi) \qquad (4)$$

shows that f is a function of completely regular growth (cf. [3], p. 139) on the ray $\{\arg z = \varphi\}$ ($f \in A_{reg,\varphi}$). It is known [3], [1] that $f \in A_{reg}$ implies the equalities

$$h_{fg}(\varphi) = h_f(\varphi) + h_g(\varphi) \ \forall g \in A(\rho), \qquad (5)$$

$$\underline{h}_{fg}(\varphi) = \underline{h}_f(\varphi) + \underline{h}_g(\varphi) \ \forall g \in A(\rho). \qquad (6)$$

It is also known [4] that (5) $\Longrightarrow$ (4).

PROBLEM 2. <u>Prove that</u>

$$(6) \Longrightarrow (4). \qquad (7)$$

ANSWER. A necessary and sufficient condition for (6) to hold, expressed in terms of $Fr[f]$ (cf. [5]), shows that (7) is not true. But if we suppose that (6) holds for $\varphi \in E \subset [0, 2\pi]$ and the set $e^{iE} \overset{\text{def}}{=} \{e^{i\varphi} : \varphi \in E\}$ is thick at the point $e^{i\varphi_0}$ then (4) is true for $\varphi = \varphi_0$.

REFERENCES

I. А з а р и н В.С. Теория роста субгармонических функций, II, конс-

пект лекций, Харьков, ХГУ, 1982.
2. Азарин В.С., Гинер В.Б. О строении предельных множеств целых и субгармонических функций.- Теор.функций, функциональн. анал. и их прил., вып. 38, Харьков, 1982, 3-12.
3. Левин Б.Я. Распределение корней целых функций. М., 1956
4. Азарин В.С. Об одном характеристическом свойстве функций вполне регулярного роста.-Теор. функций, функциональн. анал. и их прил., вып. 2, Харьков 1966, 55-66.
5. Гинер В.Б., Подошев Л.Р., Содин М.Л., О сложении нижних индикаторов целых функций.-Теор.функций, функциональн.анал. и их прил., вып. 43, Харьков, 1984 (В печати).

V.S.AZARIN
(В.С.АЗАРИН)

СССР, 310050, Харьков, Харьковский институт инженеров железнодорожного транспорта

3.10. old THE INNER FUNCTION PROBLEM IN BALLS

The open (euclidean) unit ball in $\mathbb{C}^n$ (with n at least 2) is denoted by B . A n o n - c o n s t a n t bounded holomorphic function f with domain B is called i n n e r if its radial limits $f^*(\omega) = \lim_{r\to 1} f(r\omega)$ satisfy $|f^*(\omega)| = 1$ a.e. on $S = \partial B$, where "a.e." refers to the rotation-invariant probability measure σ on S .

CONJECTURE 1. <u>There are no inner functions in</u> B .

Here is some evidence in support of the conjecture:

(i) <u>If</u> f <u>is inner in</u> B , <u>and if</u> V <u>is an open subset of</u> $\mathbb{C}^n$ <u>that interesects</u> S , <u>then</u> $f(B \cup V)$ <u>is dense in the unit disc</u> $\mathbb{D}$.

PROOF. If not, then V contains one-dimensional discs $\mathcal{D}$ with $\partial\mathcal{D} \subset S$, such that $f|\mathcal{D}$ is a one-variable inner function whose range in not dense in $\mathbb{D}$, an impossibility. ●

In other words, at every boundary point of B , the cluster set of f is the whole closed unit disc. No inner function behaves well at any boundary point.

(ii) <u>If</u> f <u>is inner in</u> B <u>and if</u> E <u>is the set of all</u> w , $w \in S$, <u>at which</u> $\lim_{r\to 1} |f(rw)| = 1$, <u>then</u> E <u>has no interior (relative to</u> S).

PROOF. If not, an application of Baire's theorem leads to a contradiction with (i). ●

CONJECTURE 1 could be proved by proving it under some additional hypotheses, for if there where an inner function f in B , then there would exist

(a) a zero-free inner function, namely $exp\left(\frac{f+1}{f-1}\right)$;

(b) an inner function g with $\lim_{r\to 1} \int_S \log|g_r|\, d\sigma = 0$,

via Frostman's theorem; for almost all one-dimensional discs $\mathcal{D}$ through the origin, $g|\mathcal{D}$ would be a Blaschke product;

(c) an inner function h that satisfies (b) and is not a product of two inner functions i.e., h is irreducible, in the terminology of [1].

(d) a non-constant bounded pluriharmonic function u with $u^* = 1$ or 0 a.e. on S, namely $u = \mathrm{Re}(\varphi \circ f)$, where φ is a conformal map of $\mathbb{D}$ onto the strip $0 < x < 1$ (i.e., there would be a set E, $E \subset S$, $\sigma(E) = \frac{1}{2}$, whose characteristic function has a pluriharmonic Poisson integral);

(e) a function $F = \dfrac{1+f}{1-f}$ with $\mathrm{Re}\, F > 0$ in B but $\mathrm{Re}\, F^* = 0$ a.e. on S.

This $\mathrm{Re}\, f$ would be the Poisson integral of a singular measure. Hence CONJECTURE 1 is equivalent to

CONJECTURE 1'. If μ is a positive measure on S whose Poisson integral is pluriharmonic, then μ cannot be singular with respect to σ.

Forelli [3], [4] has partial results that support the following conjecture (which obviously implies 1'):

CONJECTURE 2. If μ is a real measure on S, with pluriharmonic Poisson integral, then $\mu \ll \sigma$.

CONJECTURE 2 leads to some related H^1-problems:

CONJECTURE 3. If f is holomorphic in B and $\mathrm{Re}\, f > 0$, then $f \in H^1(B)$.

CONJECTURE 3'. There is a constant c, $c < \infty$ (depending only on the dimension n) such that $\int_S |f|\, d\sigma \leq c \int_S |\mathrm{Re}\, f|\, d\sigma$ for all f, $f \in A(B)$ (the ball algebra).

CONJECTURE 3''. If f is holomorphic in B, $f = u + iv$, and $|v| \leq u$, then $f \in H^2(B)$.

Clearly, 3' implies 3, and 3'' is a reformulation of 3 that might be easier to attack. Let $N(B)$ be the Nevanlinna class in B ($\int_S \log^+ |f_r|\, d\sigma$ is bounded, as $r \to 1$), and let $N_*(B)$ consist of all f, $f \in N(B)$, for which $\{\log^+ |f_r|\}$ is uniformly integrable.

CONJECTURE 4. $N(B) = N_*(B)$.

This would imply 1', hence 1.

CONJECTURE 1 leads to the problem of finding the extreme points of the unit ball of $H^1(B)$. (When $n = 1$, these are exactly the uter functions of norm 1.) Let $\Omega = \{ f \in H^1(B) : \|f\|_1 \leq 1 \}$.

CONJECTURE 5. <u>Every</u> f, $f \in H^1(B)$, <u>with</u> $\|f\|_1 = 1$ <u>is an xtreme point of</u> Ω.

It is very easy to see that 5 implies 1. If $f \in A(B)$ it is nown (and easy to prove) that $f(S) = f(\overline{B})$. It is tempting to try o extend this to $H^\infty(B)$: <u>Is it true for every</u> f, $f \in H^\infty(B)$, <u>hat the essential range of</u> f^* <u>on</u> S <u>is equal to the closure of</u> $f(B)$?

An affirmative answer would of course be a much stronger result than CONJECTURE 1. To prove it, one would presumably need a more quantitative version of $f(S) = f(\overline{B})$. For example: <u>Does there exist</u> c, $c > 0$ (<u>depending only on the dimension</u> n) <u>such that</u> $\sigma\{|f| < 1/2\} > c$ <u>for every</u> f, $f \in A(B)$, <u>with</u> $f(0) = 0$, $|f| < 1$? Finally, call a holomorphic mapping $\Phi : B \to B$ i n n e r if $\lim_{r \to 1} \Phi(r\omega) \in S$ for almost all ω, $\omega \in S$.

CONJECTURE 6. <u>If</u> Φ <u>is inner, then</u> Φ <u>is one-to-one and onto.</u>

This implies 1, as well as the conjecture that every isometry of $H^p(B)$ into $H^p(B)$ is actually onto, when $p \neq 2$. See [5]. If "inner" is replaced by "proper", then CONJECTURE 6 is true, as was proved by Alexander [2].

REFERENCES

1. A h e r n P.R., R u d i n W. Factorizations of bounded holomorphic functions. - Duke Math.J., 1972, 39, 767-777.
2. A l e x a n d e r H. Proper holomorphic mappings in $\mathbb{C}^n$. - Indiana Univ.Math.J., 1977, 26, 137-146.
3. F o r e l l i F. Measures whose Poisson integrals are pluriharmonic. - Ill.J.Math., 1974, 18, 373-388.
4. F o r e l l i F. Measures whose Poisson integrals are pluriharmonic II. - Ill.J.Math., 1975, 19, 584-592.
5. R u d i n W. L^p-isometries and equimeasurability. - Indiana Univ.Math.J., 1976, 25, 215-228.

WALTER RUDIN

Department of Mathematics University of Wisconsin. 110 Marinette Trail
Madison, WI 53705, USA

* * *

COMMENTARY

The existence of non-constant inner functions in the ball of $\mathbb{C}^n$ was proved independently (and by different methods) by A.B.Aleksandrov [7] and E.Løw [9] (see also [18]). Both articles are refinements of preceding papers by A.B.Aleksandrov [6] and M.Hakim - N.Sibony [8] respectively, where the problem has been solved "up to ε " (but in different senses). Here are principal results of [7], [8], [9].

THEOREM 1 ([7]). Let φ be a positive lower semicontinuous function on S , $\varphi \in L^1(S)$. There exists a singular positive measure ν on S such that $\nu(S) = \int_S \varphi\, d\sigma$, and the Poisson integral of $\varphi\sigma - \nu$ is pluriharmonic.

THEOREM 2 ([8]). For every continuous positive function φ on S and for every positive number ε there exist a compact set K, $K \subset S$, with $\sigma(K) = 0$ and a function f , $f \in C(S \setminus K) \cap H^\infty(B)$ such that $\varphi - \varepsilon \leqslant |f| \leqslant \varphi$ everywhere on $S \setminus K$ and $f(0) = 0$.

THEOREM 3 ([7], [9]). Let φ be as in Th.1, $\varphi \in L^p(S)$. Then there is a function $f \in H^p(B)$ such that $|f| = \varphi$ a.e. on S , and $f(0) = 0$.

Th.1 implies Th.3 and Th.3 implies the existence of non-trivial inner functions in B . The last fact DISPROVES ALL CONJECTURES of the PROBLEM and yields a negative answer to the second question of 12.2. Some of CONJECTURES of S.10 have been disproved in [6].

THEOREM 4 ([10]). Let φ be a positive lower semicontinuous function on S , $\varepsilon > 0$. Then there exists a function $f \in A(B)$ (= the ball algebra) such that $|f| \leqslant \varphi$ everywhere on S and $\sigma\{|f| \neq \varphi\} < \varepsilon$.

The following two theorems can be viewed as a multidimensional analogue of the Schur theorem (on the approximation by inner function) and of the Nevanlinna - Pick interpolation theorem.

THEOREM 5 ([12]). Let φ be as in Th.4, $\varphi \in L^p(S)$, $1 \leqslant p \leqslant +\infty$, $\frac{1}{p} + \frac{1}{p'} = 1$. Then the weak $(L^p, L^{p'})$ -closure of the set $\{f \in H^p(B) : |f| = \varphi$ a.e.$\}$ coincides with $\{f \in H^p(B) : |f| \leqslant \varphi\}$.

This theorem was proved in [7] for $\varphi \equiv 1$, and then, independently, proved in [18].

Denote by $H^\infty_{p.c.}(B)$ (see [19]) the set of all f, $f \in H^\infty(B)$

such that $\lim_{z\in B, z\to\zeta} f(z)$ exists for almost all $\zeta\in S$. In other words the boundary values of a $H^\infty_{p.c.}(B)$ -function agree a.e. on S with a Riemann integrable function.

THEOREM 6 ([12]). Let φ be as in Th.5 and suppose $g, f\in H^\infty_{p.c.}(B)$, $g\neq 0$. Suppose there exists a function $h\in H^\infty_{p.c.}(B)$ such that $h\neq 0$, $|f|+|h|\leqslant\varphi$ a.e. on S. Then there exists an $F\in H^p(B)$ such that $|F|=\varphi$ a.e. and $(F-f)g^{-1}\in N_*(B)$.

This theorem has been proved in [7] for $\varphi\equiv 1$. Some particular cases (also for $\varphi\equiv 1$) have been independently rediscovered in [18] and [20]. Th.6 has been rediscovered (for $\varphi\equiv 1$) in [19].

Natural analogues of Theorems 1-6 hold for strictly pseudoconvex bounded domains with a C^2-boundary and for pseudoconvex bounded domains with a C^∞-boundary. In these situations theorems 2,3,4 are due to Løw [11]. These results combined with those of [12] imply analogues of Theorems 1,5,6 as well. In [12] analogues of Theorems 1-6 for Siegel domains of the first and the second kind are established (and in particular for bounded symmetric domains).

Th.4 implies the existence of $f\in A(B)$, $\|f\|_\infty\leqslant 1$, $f(0)=0$, $\sigma\{|f|=1\}>0$. G.M.Henkin has remarked (see also [13]) that methods of N.Sibony [14] can be used to show that such a function cannot satisfy the Lipschitz condition of order $>\frac{1}{2}$. Whether it can be lipschitzian of order $\leqslant\frac{1}{2}$ remains unclear.

The following result by Tamm yields a precise sufficient condition for the equality $\operatorname{Clos} f(B)=f(S)$ with $f\in H^\infty(B)$. Here $f(S)$ denotes the set of all essential values of $f(\in L^\infty(S))$.

THEOREM 7 [15]. If $f\in H^\infty(B)$ and $\|f-f_r\|_{H^2(B)}=O((1-r)^{1/4})$ then $\operatorname{Clos} f(B)=f(S)$.

On the other hand it is stated in [15] that "a majority" of elements of the set $\{f\in H^\infty(B): \|f-f_r\|_{H^2}=O((1-r)^\alpha)\}$ do not satisfy this equality whenever $\alpha<1/4$.

A.B.Aleksandrov [16] has proved analogues of Theorems 2,3,5 for gradients of harmonic functions. Note also a very simple proof [12] of Theorem 4 based on the following nice result of J.Ryll and P.Wojtaszczyk: there exists a homogeneous polynomial $P_N(z_1,\dots,z_n)$ of degree N such that $|P_N|\leqslant 1$ in B and $\int_S |P_N|^2 d\sigma\geqslant \pi 4^{-n}$ $(0\leqslant N<+\infty,\ n\geqslant 1)$.

W.Rudin has proved in [18] that the linear span of inner functions is not norm-dense in $H^\infty(B)$. This gives a negative answer to the question posed in [7] and shows there is no analogue of the Marshall theorem for $n \geq 2$.

The following QUESTION has been put in [18]: does the H^∞-norm-closure of the linear span of inner functions contain $A(B)$ (or-what is the same ([18]) - at least one non-constant element of $A(B)$)?

The results of H.Alexander [21] and of M.Hakim - N.Sibony [22] show that $A(B)$ and $H^\infty(B)$-functions may have zero sets "as large as functions of the Nevanlinna class $N(B)$". Recall that zeros of $N(B)$-functions are completely described by a well-known G.M.Henkin-H.Skoda theorem. The work [21] uses Ryll - Wojtaszczyk polynomials and [22] uses techniques of [8].

REFERENCES

6. Александров А.Б. Классы Харди H^p и полувнутренние функции в шаре. - Докл.АН СССР, 1982, 262, № 5, 1033-1036.

7. Александров А.Б. Существование внутренних функций в шаре. - Матем.сборник.,1982, 118, № 2, 147-163.

8. Hakim M., Sibony N. Fonctions holomorphes bornées sur la boule unité de $\mathbb{C}^n$. - Inv.math., 1982, 67, N 2, 213-222.

9. Løw E. A construction of inner functions on the unit ball in $\mathbb{C}^n$. - Inv.math., 1982, 67, N 2, 223-229.

10. Александров А.Б. О граничных значениях голоморфных в шаре функций. - Докл.АН СССР, 1983, 271, № 4.

11. Løw E. Inner functions and boundary values in $H^\infty(\Omega)$ and $A(\Omega)$ in smoothly bounded pseudoconvex domains. Dissertation. Princeton University. June 1983.

12. Александров А.Б. Внутренние функции на компактных пространствах.-Функц.анализ и его прил. (to appear).

13. Rudin W. Function theory in the unit ball of $\mathbb{C}^n$. N.Y. - Heidelberg - Berlin: Springer-Verlag, 1980.

14. Sibony N. Valeurs au bord de fonctions holomorphes et ensembles polynomialement convexes. Lect.Notes Math., 1977, 578, 300-313.

15. Tamm M. Sur l'image par une fonction holomorphe bornée du bord d'un domaine pseudoconvex. - C.R.Ac.Sci., 1982, 294, Sér.I, 537-540.

16. Александров А.Б. Внутренние функции на пространствах однородного типа. - Зап.научн.семин.ЛОМИ, 1983, 126, 7-14.

17. Ryll J., Wojtaszczyk P. On homogeneous polyno-

mials on a complex ball.– Trans Amer Math Soc., 1983, 276, N 1, 107–116

8. R u d i n W. Inner functions in the unit ball of $\mathbb{C}^n$. – J.Funct. Anal., 1983, 50, N 1, 100–126.

9. H a k i m M., S i b o n y N. Valeurs au bord des modules de fonctions holomorphes. Prépublication Orsay. 1983, 06.

0. T o m a s z e w s k i B. The Schwarz lemma for inner functions in the unit ball in $\mathbb{C}^n$. Preprint (Medison, WI.) 1982.

1. A l e x a n d e r H. On zero sets for the ball algebra. – Proc. Amer.Math.Soc., 1982, 86, N 1, 71–74.

2. H a k i m M., S i b o n y N. Ensemble des zéros d'une fonction holomorphe bornée dans la boule unité. – Math.Ann.,1982,260, 469–474.

S.11. HOMOGENEOUS MEASURES ON SUBSETS OF $\mathbb{R}^n$.

A locally finite positive measure μ supported by a closed subset E of $\mathbb{R}^n$ and satisfying "the doubling - condition"

$$\mu(B(x,2r)) \leq K\mu(B(x,r)), \quad x\in E \ , \ r>0 \qquad (\mathcal{D})$$

is called h o m o g e n e o u s. Here $B(x,r)$ denotes the ball centered at x with radius r. Evidently, supp $\mu = E$. A set E supporting a homogeneous measure becomes a space of homogeneous type in the sense of Coifman and Weiss [1-4]. The theory of Hardy spaces H^p, $0<p\leq 1$, can be extended to such sets. On the other hand, the existence of a homogeneous measure is important for the description of traces of smooth functions on E and for free interpolation problems [4,5].

CONJECTURE. Each closed subset of $\mathbb{R}^n$ supports a homogeneous measure.

Except for some evident examples of sets with constant dimension (Lipschitz manifolds, Cantor sets), the existence of a homogeneous measure has been proved (up to the present) only for subsets E of $\mathbb{R}^1$ satisfying the following condition:

$$\sup_{x\in I} \operatorname{dist}(x,E) \geq \mathrm{const}\,|I|$$

for any interval $I\subset \mathbb{R}^1$ [5]. This condition means that the dimension of E is in some sense less than 1. For general sets on the line and for sets in $\mathbb{R}^n$ the problem is open.

Our conjecture has an interesting dual reformulation. Let $\{B_j\}_{j=1}^N = \{B(x_j,r_j)\}_{j=1}^N$ be a finite set of balls in $\mathbb{R}^n$. Define $\widetilde{B}_j = B(x_j, 2r_j)$ and write

$$P_i = \sum \chi_{B_j}(x_i), \quad i=1,\dots,N,$$

$$Q_i = \sum \chi_{\widetilde{B}_j}(x_i), \quad i=1,\dots,N,$$

i.e. P_i is the multiplicity of the covering $\{B_j\}$ at x_i, and Q_i is the analoguous multiplicity of $\{\tilde{B}_j\}$.

CONJECTURE. $\inf_i Q_i/P_i \leq K$, where the constant K depends only on n.

The equivalence of these two conjectures follows by Hanh-Banach theorem.

Independently of the general conjecture, it is interesting to connect the properties of a homogeneous measure μ (if it exists) with geometric characteristics of E. In particular, it is interesting to estimate the growth of $\mu(B(x,r))$ in r in term of the Lebesgue measure of ε-neighbourhoods of E.

REFERENCES

1. C o i f m a n R.R., W e i s s G. Extensions of Hardy spaces and their use in analysis. - Bull.Amer.Math.Soc., 1977, 83, 569-645.
2. M a c i a s R.A., S e g o v i a C. A decomposition into atoms of distributions on spaces of homogeneous type. - Adv.Math., 1979, 33, 271-309.
3. F o l l a n d G.B., S t e i n E.M. Hardy spaces on homogeneous groups. Princeton, 1972.
4. J o n s s o n A., S j ö g r e n P., W a l l i n H. Hardy and Lipschitz spaces on subsets of $\mathbb{R}^n$. - Univ.Umeå Dept.Math. Publ. , 1983, N 8.
5. Д ы н ь к и н Е.М. Свободная интерполяция функциями с производной из H^1. - Записки научн.семин.ЛОМИ, 1983, 126, 77-87.

E.M.DYN'KIN
(Е.М.ДЫНЬКИН)

СССР, Ленинград, 197022
Ленинградский электротехнический институт

* * *

COMMENTARY

Recently S.V.Konyagin and A.L.Vol'berg have proved that any closed $E\subset\mathbb{R}^n$ carries a probabilistic measure μ satisfying $\mu(B(x,kr))\leq C(n)k^n\mu(B(x,r))$ $(x\in E, r>0, k>1)$. They proved also a more precise assertion for E's of a lower $(\leq n)$ dimension and a generalization to metric spaces.

SUBJECT INDEX

a.b means Problem a.b, a.o means Preface to Chapter a, o.o means Preface

AUTHOR INDEX

a.b means Problem a.b, a.0 means Preface to Chapter a, Ack. means Acknowlegement.

STANDARD NOTATION

Symbols $\mathbb{N}, \mathbb{Z}, \mathbb{R}, \mathbb{C}$ denote respectively the set of positive integers, the set of all integers, the real line, and the complex plane.

$$\mathbb{R}_+ = \{t \in \mathbb{R} : t \geqslant 0\}, \quad \mathbb{Z}_+ = \mathbb{Z} \cap \mathbb{R}_+$$

$$\mathbb{T} = \{z \in \mathbb{C} : |z| = 1\}$$

$$\mathbb{D} = \{z \in \mathbb{C} : |z| < 1\}$$

$\hat{\mathbb{C}}$ stands for the one-point compactification of $\mathbb{C}$, m denotes the normed Lebesgue measure on $\mathbb{T}$ ($m(\mathbb{T}) = 1$), $f|X$ is the restriction of a mapping (function) f to X. $\operatorname{clos}(\cdot)$ is the closure of the set $(\cdot)$. $\bigvee(\cdot)$ is the closed span of the set $(\cdot)$ in a linear topological space.

$\|T\|$ denotes the norm of the operator T.

$\hat{f}(\cdot)$ denotes the sequence of Fourier coefficients of f.

$\mathcal{F}f$ denotes the Fourier transform of f.

$\mathfrak{S}_p$ is a class of operators A on a Hilbert space satisfying trace $(A^*A)^{p/2} < +\infty$.

H^p is the Hardy class in $\mathbb{D}$, i.e. the space of all holomorphic functions on $\mathbb{D}$ with

$$\|f\|_p \overset{\text{def}}{=\!=} \sup_{0<r<1} \left(\int_{\mathbb{T}} |f(r\zeta)|^p \, dm(\zeta) \right)^{1/p} < +\infty, \quad p > 0.$$